The Potatoes of South America: Bolivia

Frontispiece: Plate I. – *Solanum acaule* Bitter.

The Potatoes of South America: Bolivia

CARLOS M. OCHOA

Translated by
DONALD UGENT

Color Illustrations by
Franz Frey

Published in collaboration with the International Potato Center

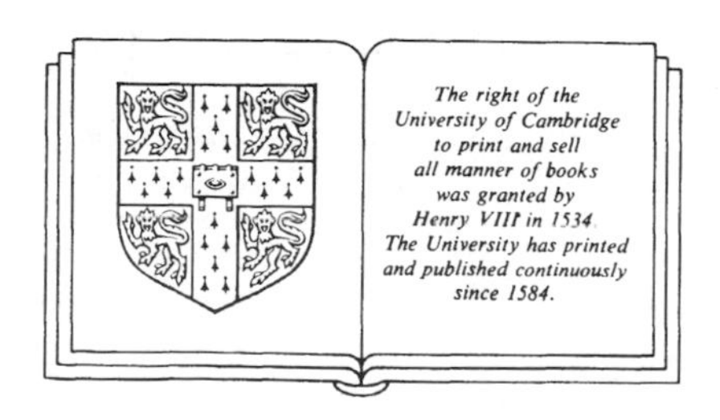

CAMBRIDGE UNIVERSITY PRESS

Cambridge

New York Port Chester Melbourne Sydney

CAMBRIDGE UNIVERSITY PRESS
Cambridge, New York, Melbourne, Madrid, Cape Town, Singapore,
São Paulo, Delhi, Dubai, Tokyo, Mexico City

Cambridge University Press
The Edinburgh Building, Cambridge CB2 8RU, UK

Published in the United States of America by Cambridge University Press, New York

www.cambridge.org
Information on this title: www.cambridge.org/9780521380249

First published 1990

A catalogue record for this publication is available from the British Library

Library of Congress Cataloguing in Publication data

Ochoa, Carlos M.
The potatoes of South America: Bolivia/Carlos M. Ochoa: translated by Donald Ugent: color
illustrations by Franz Frey.
p. cm.
"Published in collaboration with the International Potato Center."
Includes bibliographical references.
1. Potatoes–Bolivia–Classification. 2. Solanum–Bolivia–Classification.
3. Potatoes–Bolivia–Germplasm resources. 4. Solanum–Bolivia–Germplasm resources.
I. International Potato Center. II. Title.
SB211.P8024 1990
635'.21'0984–dc20 90-33077 CIP

ISBN 978-0-521-38024-9 Hardback

Additional resources for this publication at www.cambridge.org/9780521380249

Dedication

Donovan Stewart Correll

Contents

List of Plates

These plates are available for download in colour from
www.cambridge.org/9780521380249

List of Maps

List of Figures

Foreword

Potatoes, a wonderful food, are found on dinner plates around the world, but where do they come from? What were they like when they grew wild? And what are their wild relatives like? This volume addresses interesting and important questions about the potato by examining the potato and its wild relatives in Bolivia.

Even though the potato was introduced to Europe in the eighteenth century and to North America somewhat later, controversy still surrounds the exact geographic source of the potatoes that first graced European gardens and larders. Surely it was somewhere in western parts of South America, but just where? Part of the problem is that potatoes are not all alike, and they were not always just as they are now. Modern scientific works list about 200 different species of potatoes, most, but not all of which are from western South America. Some writers have shown that even today wild potatoes are undergoing change as they are accidentally or deliberately hybridized with stocks cultivated by farmers nearby. This means that in searching for wild ancestors of the potato, one must look for wild potatoes that may now be slightly altered from those that were first brought into European cultivation four centuries ago.

Scientists differ when it comes to pointing out which wild potatoes most closely resemble the present cultivated potatoes. Although scientific papers have become more precise in describing characters, mechanisms, chemical responses, and breeding behavior of the potato, this has not yet clarified the origin of the cultivated potato. Surely a clear cataloging of the wild species and their distinctions is a first step to assessing the ancestral group and the closest relatives to the potatoes we eat. Such cataloging is part of the science of systematics, or taxonomy.

As the systematic study of the potato has evolved over time, potatoes have been studied by different kinds of taxonomic botanists. At first, general botanists described potatoes in the course of describing a wide range of plants. Carl Linnaeus described the cultivated potato, *Solanum tuberosum*, in 1753. During the past century, general taxonomists such as D. F. L. von Schlechtendal and Asa Gray each described potato species from the north of Colombia, and Alphonse de Candolle described a species from South America. Perhaps the first specialist on the potato family Solanaceae was Michel-Felix Dunal, who described potato species between 1813 and 1852. But the notion that *Solanum tuberosum* was not the only species in South America did not come into being until the time of Berthault in 1911, and the next great specialist in the family, Georg Bitter, who described many species between 1911 and 1914. These workers all worked with the collections of others, but they were never able to study the plants under native field conditions.

Beginning in the 1920s, expeditions were sent from Europe and North America to study potatoes in the lands where they occur. Among the most noteworthy of these potato-expedition leaders were those that came from Russia, America, and England. Also in this century, botanists, native or resident in areas where wild potatoes occur, some of them still living, for example, Cesar Vargas and Heinz Brücher, studied potatoes and described new species. The results of these South American workers have been largely accepted and incorporated into the taxonomic revisions of the potato group, which were prepared by workers elsewhere. This volume is the first 'magnum opus' to be written by an established South-American expert on potato taxonomy. The author has been publishing descriptions of South American potatoes for over 30 years and actively studying them in the field for much longer.

Much of the germplasm used in this century in breeding better potatoes and in studying the biochemical, inheritance, pathological, and other relationships of the potato has come from material collected by the potato expeditions from Europe and North America. Our descriptions of the group as a whole and of how the species occur in their native lands have also come largely from workers living elsewhere who either made relatively short forays into the source areas of the potato or never came to South America at all. That our knowledge of such an important crop plant should stand on such a limited base is quite undesirable. For the kind of inventory that will really document and analyze the variability, habitats, and geography of the wild potatoes, more continuous surveys are clearly needed. In recent years, the International Potato Center, headquartered in Lima, Peru, has provided facilities and encouragement for such efforts, and the Center's long-standing interest in the taxonomy of native potatoes has made this present work possible.

This volume is the first in a series on the taxonomy of the potato in South America; future volumes are promised that will cover Peru, Ecuador, and other countries as well. The three-country area hosts perhaps one hundred different potato species. To have this important series about the taxonomy of the potato written by a native and resident of the area is a welcome development. The author has for many years been ideally placed to study the different species and varieties in the wild, under cultivation, and in native markets.

This first volume catalogues the potato species occurring in Bolivia. The wild species are placed in a traditional taxonomic framework, and the taxonomy of the cultivated species is expanded greatly from what has been done in the past. This achievement is based on the author's long-standing and intimate knowledge of the group. The notations on chromosome numbers, taxonomic affinities, distributions, and habitat preferences that appear throughout the book will be most helpful to future workers, as will be the full specimen citations.

Most texts tend to lose something in translation from one language to another; however, for this book it is more a case of two professional potato taxonomists collaborating to create a text based on the work of one of them. Ugent brings seasoned and expert skills to the translation task.

The abundant illustrations add a special utility to this book that is lacking in almost all other taxonomic texts. Clear and diagnostic line drawings, specially commissioned watercolor prints, and photographs of living material of all the species provide a useful illustrated guide to the information given in the dichotomous keys. The photographs of type specimens, and the fact that the drawings and photographs were made by the author himself, engender confidence in the scientific decisions underlying the taxonomy of the group. Colored illustrations tying together the color and shape of tubers of naturally occurring species with the taxonomic descriptions and nomenclature is a novel feature of the book. This should be especially valuable to potato scientists elsewhere who try to assess and diagnose the variability in unusual and important cultivated strains for breeding and other purposes. A series of maps show the distribution of Bolivia's potatoes and climates.

This volume is a milestone for texts published on the wild relatives of any cultivated plant. One looks forward to seeing the following volumes.

W. G. D'Arcy

Preface

Much has been accomplished in the interrelated fields of potato systematics and crop breeding since the appearance of Vavilov's monumental 1935 work on the *Origin, Variation, Immunity and Breeding of Cultivated Plants.* Following the publication of this classic work, scientists from many countries joined the Russians in their efforts to collect, classify, and maintain samples of South American potato germplasm. Especially valued by the breeders of that era were the samples of wild and primitively cultivated potato species that had been collected in the central Andes. This latter region was recognized by Vavilov as an important center of domestication for the cultivated potato.

Although the great majority of germplasm collecting expeditions of the 1930s and early 1940s were organized and carried out by European and North American scientists, an increasing number of the later expeditions were made by South American researchers, most notably from Colombia, Chile, Bolivia, and Peru. One result of these later expeditions was a rather dramatic upsurge in the number of papers published in Spanish. Today, approximately one half of the papers that appear in the fields of potato systematics and breeding carry the names of scientists from Latin American countries.

Despite a spectacular rise in the number of potato introductions that are currently maintained on a worldwide basis by germplasm stations, there remains a critical need for additional collecting. Exploration of the Andean potato zones, and particularly of the remote, isolated valleys of the arid coastal ranges and the moist western slopes of the central cordilleras, will probably continue to rank as an important priority for future investigators.

It must be pointed out, however, that the Andes constitute a formidable barrier for those wishing to engage in plant exploration. Distances between the various potato collecting localities can run into hundreds and even thousands of kilometers, and roads and communication lines are often poorly or not at all developed. Many mountain valleys and crests are still awaiting exploration by the hardy adventurer.

My interest in the study of wild and cultivated species of potato springs from my experiences as a plant breeder with the Peruvian Ministry of Agriculture. Although I was originally assigned in 1946 to a wheat improvement project, it soon became evident that a more critical breeding situation existed with respect to the leading native root crop of Peru, the potato. This was brought home dramatically to me shortly after my employment began with the Ministry of Agriculture, when the potato plants in the area where I was working, the Mantaro Valley of Central Peru, were destroyed by an unusually early and severe frost. With the loss of the entire Mantaro Valley potato crop that year – and the resultant misery that fell upon its human population – I began the long search for new potato varieties that could improve the economy of the area. Unfortunately, there were no other Peruvian stocks suitable for the Mantaro Valley, nor were there other Peruvian potato breeders who could help me with the problem. Thus, I eventually began to experiment with the breeding of new hybrid potato varieties that had some resistance to frost, this derived from the transference of germplasm from such usually hardy wild species as *Solanum acaule* and *S. bukasovii*. But, the identification of wild forms suitable for these breeding studies was, in the beginning, a difficult task. Few taxonomic studies on the wild and cultivated potatoes of South America were available to me in the mid-1940s, and those that I had were found to have many omissions and errors. I therefore soon realized the necessity for studying both the horticultural and taxonomic characteristics of the plants on hand. In the first instance, this led eventually to the development of six potato varieties that still remain popular in Peru today (Renacimiento, Mantaro, Tomasa Condemayta, Yungay, Cusco, and Micaela Bastidas); while in the second, it led ultimately to the publication of this book.

In my more than thirty years of active botanical exploration of the central Andes, I have attempted to visit the regions that had been little collected by previous workers. My travels have taken me to the high mountains of Peru, Bolivia, Ecuador, Colombia, and Venezuela, as well as to the seaside habitats of wild potato species in south-central Chile and the Atlantic Basin, mainly the territories of Paraguay, Uruguay, and southern Brazil. These South American collections have formed the basis of numerous systematic, cytological, and horticultural papers.

Although many of my dried plant collections are found in some of the world's major herbaria, the living collections of seeds and tubers that I assembled for breeding purposes and for use in biosystematic studies on the Andean potatoes will be stored on a permanent basis by the germplasm bank of the International Potato Center (CIP), with its main headquarters located in Lima, Peru. The identification of the germplasm collections at CIP has been a job that has occupied my attention since the opening of the Center in 1972. Today, about 90% of the collections at CIP have been classified.

By having access to CIP's excellent facilities and the important world germplasm collections, I have been able to study the phenotypes of plants grown under uniform climatic conditions. As pointed out frequently by other authorities of *Solanum*, Sect. *Potatoe*, it is extremely important to take environmental plasticity into account when revising the species of this group. Often, plants that appear very different in nature are seen to be nearly identical when grown under standard field conditions. Moreover, the growth of plants under uniform field conditions facilitates the study of hybridization, introgression, and ecotypic differentiation. Lastly, I would like to point out that without the above facilities at my disposal, my work on the crossability, fertility, and chromosomal relationships of the various species of this section would probably never have come to fruition.

The present book constitutes the first volume of a series of works that will deal ultimately with all the wild and cultivated potato species of the central and northern Andes. Bolivia was chosen as the first country for study, as its potatoes are not only among the more interesting for the region, but are also the least known.

C. M. Ochoa
International Potato Center, Lima, Peru
Research Associate, Department of Botany
Smithsonian Institution, Washington, D.C.

Acknowledgments

The author wishes to thank Richard L. Sawyer, Director General of the International Potato Center, Lima, Peru, for his kind encouragement and support of this study. Without his able administrative assistance, my work would have been impossible.

Also, the author is grateful to Dieter Wasshausen, Curator, Department of Botany, Smithsonian Institution, Washington, D.C., for making the facilities, outside loans, and collections of his museum available for my studies.

José Cuatrecasas, Research Associate, Department of Botany, Smithsonian Institution, contributed many valuable suggestions as to the improvement of this work in its early stages. The author is especially indebted to him for his critical review of the final manuscript.

Special thanks go to Donald Ugent, Ethnobotanist and Potato Systematist, Southern Illinois University, Carbondale, Illinois, for translating the manuscript into English and for his major contribution in editing and reviewing this work. I am also grateful to him for computer coding of the manuscript required for the publication of this book.

The curators of all herbaria and institutions referred to in this work were most helpful to me and kindly put at my disposal the material that I have reviewed.

My assistants at the International Potato Center are Alberto Salas, who helped both in the organization and computer processing of data and in the collection of plants in the field; Matilde Orrillo de Jara, who contributed her skills in the counting of chromosomes; and Jesús Amaya Castillo, who made artificial crosses and drew some of the illustrations found on these pages.

Franz Frey, Plant Pathologist, deserves special recognition for contributing the very fine series of watercolors that adorn this work.

Finally, a word of thanks goes to Linda W. Peterson, Senior Science Editor, International Potato Center, Lima, Peru, for her assistance in the development and organization of this manuscript and its final editing; to Rosa Ng Ying de Salazar, my secretary at the International Potato Center, for her aid in the word processing of the manuscript; and to countless other people who have aided me in small ways during the years that this book has been in the making.

C. M. Ochoa

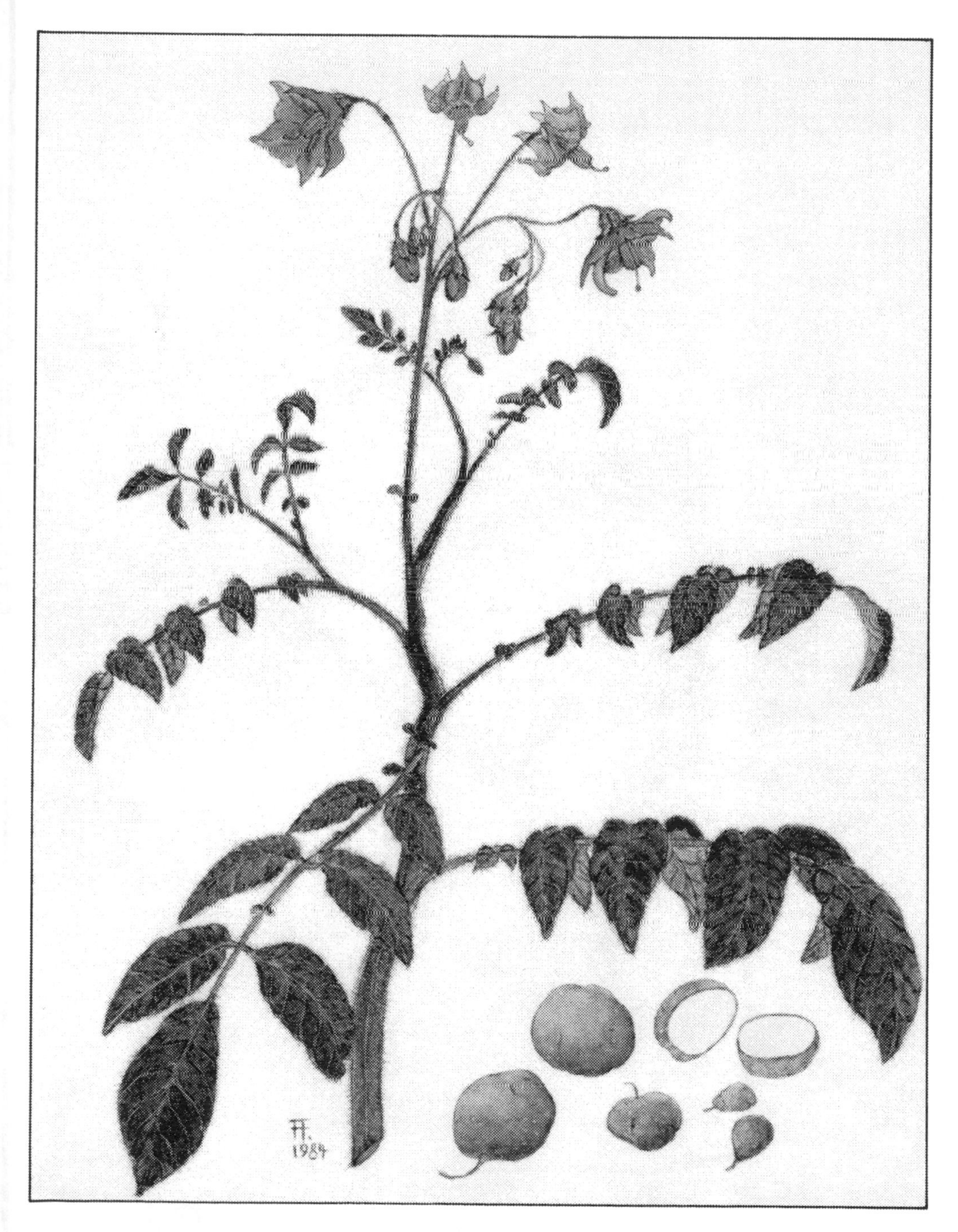

Plate II. – *Solanum berthaultii* Hawkes. (See frontispiece for plate I.)

Plate III. – *Solanum chacoense* Bitter.

Plate IV. – *Solanum litusinum* Ochoa.

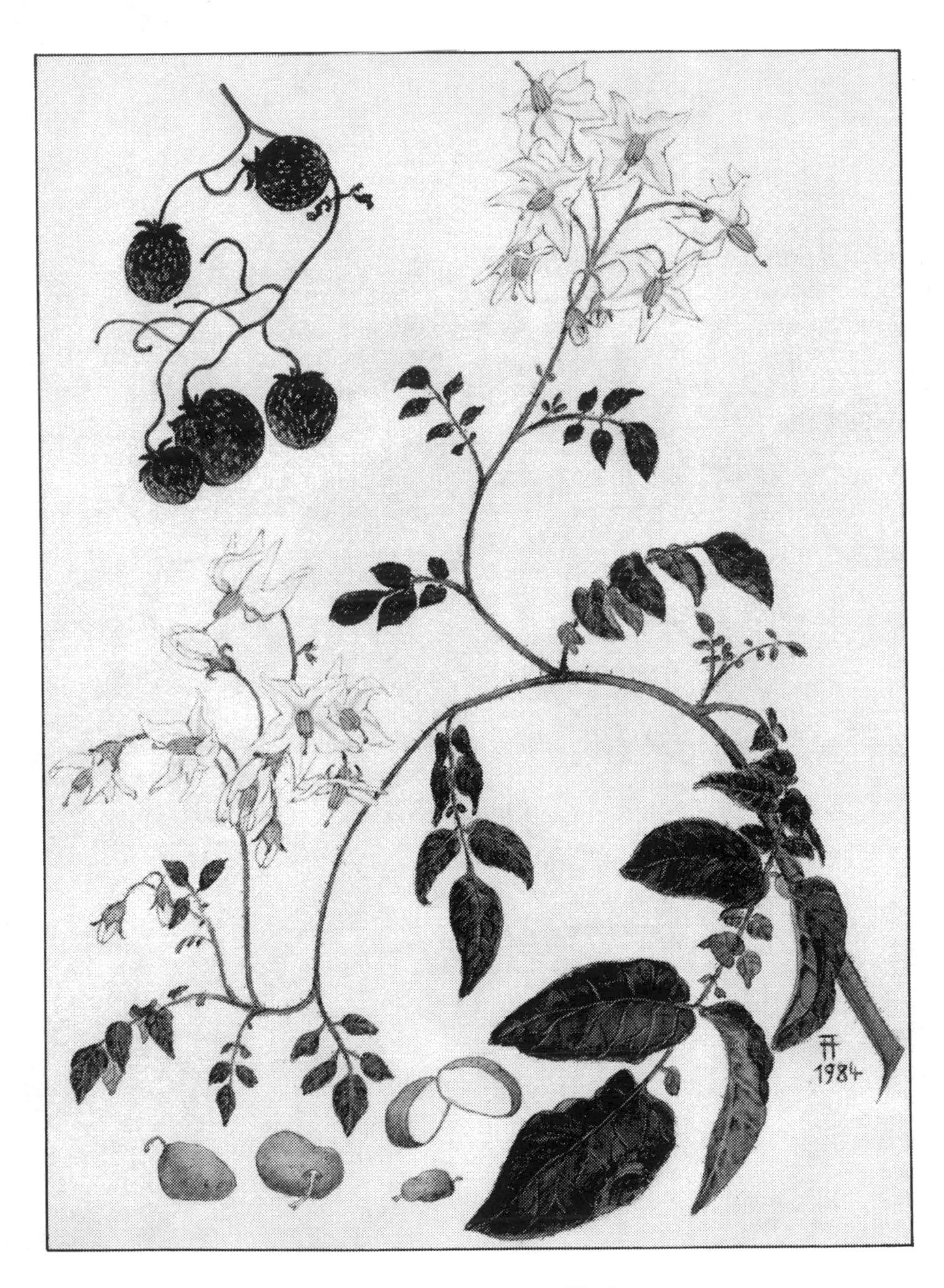

Plate V. – *Solanum tarijense* Hawkes.

Plate VI. – *Solanum yungasense* Hawkes.

Plate VII. – *Solanum infundibuliforme* Philippi.

Plate VIII. – *Solanum circaeifolium* Bitter.

Plate IX. – *Solanum bombycinum* Ochoa.

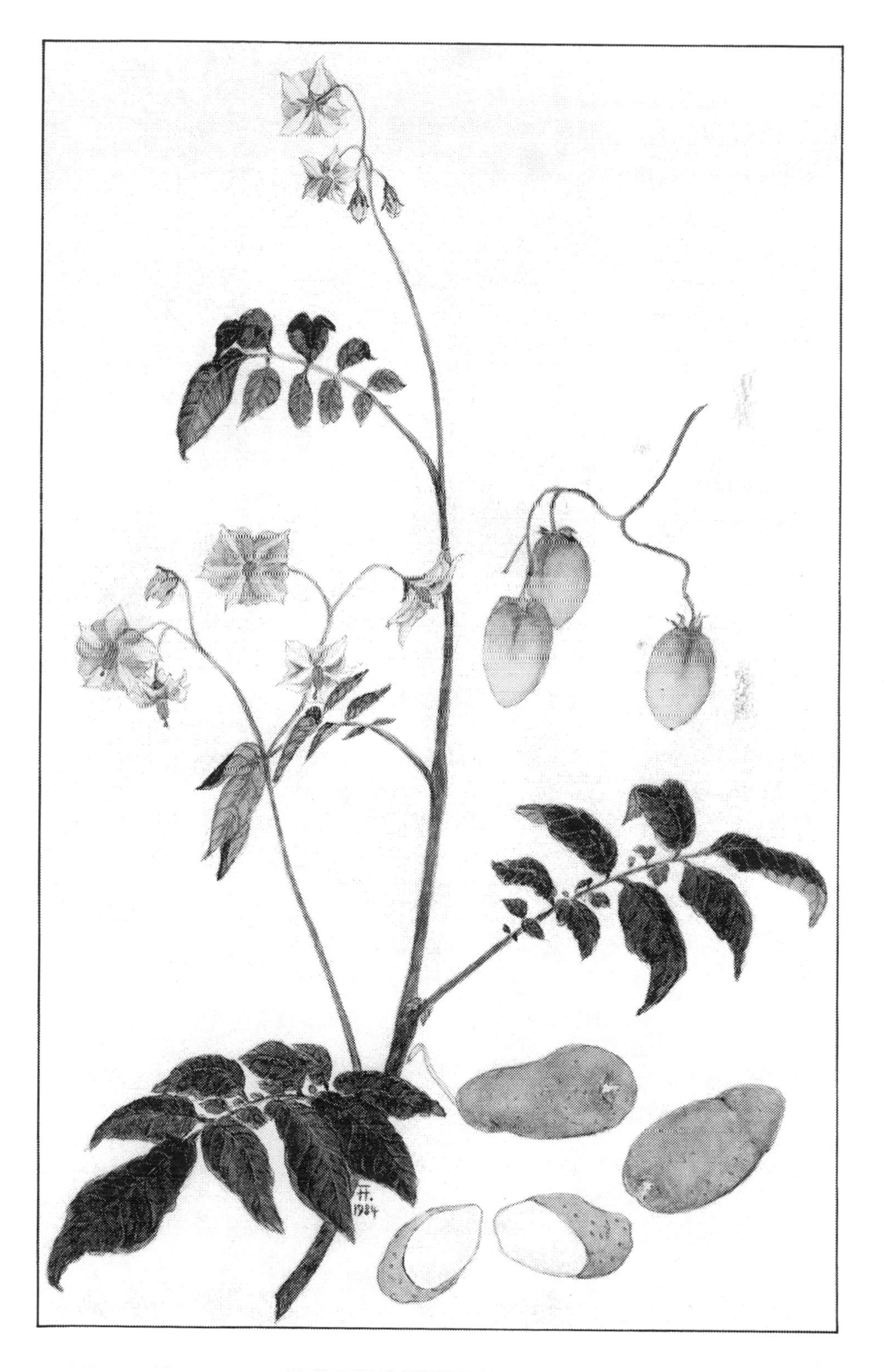

Plate X. – *Solanum neovavilovii* Ochoa.

Plate XI. – *Solanum boliviense* Dunal.

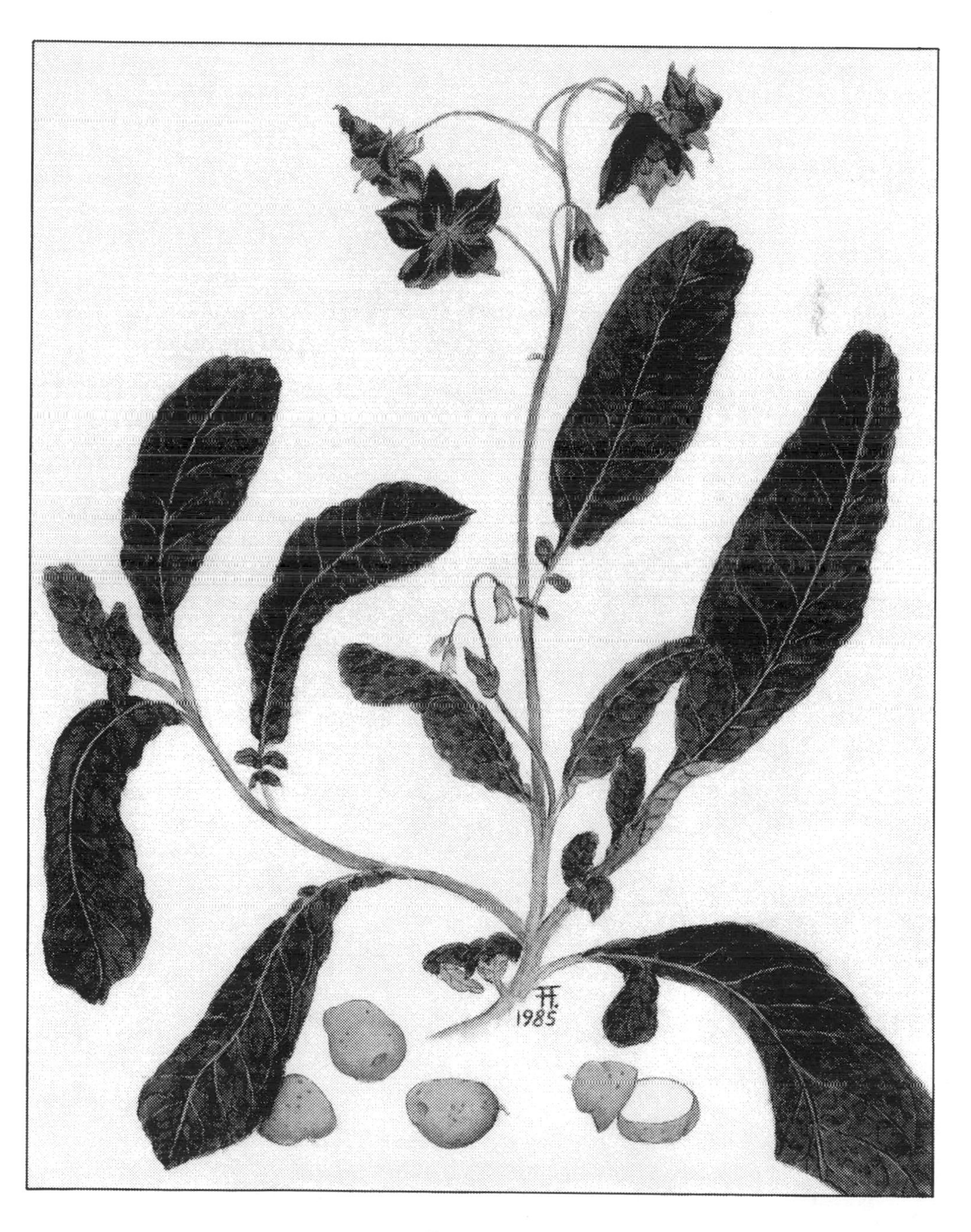

Plate XII. – *Solanum megistacrolobum* Bitter var. *toralapanum* (Cárdenas and Hawkes) Ochoa.

Plate XIII. – *Solanum alandiae* Cárdenas.

Plate XIV. – *Solanum gandarillasii* Cárdenas.

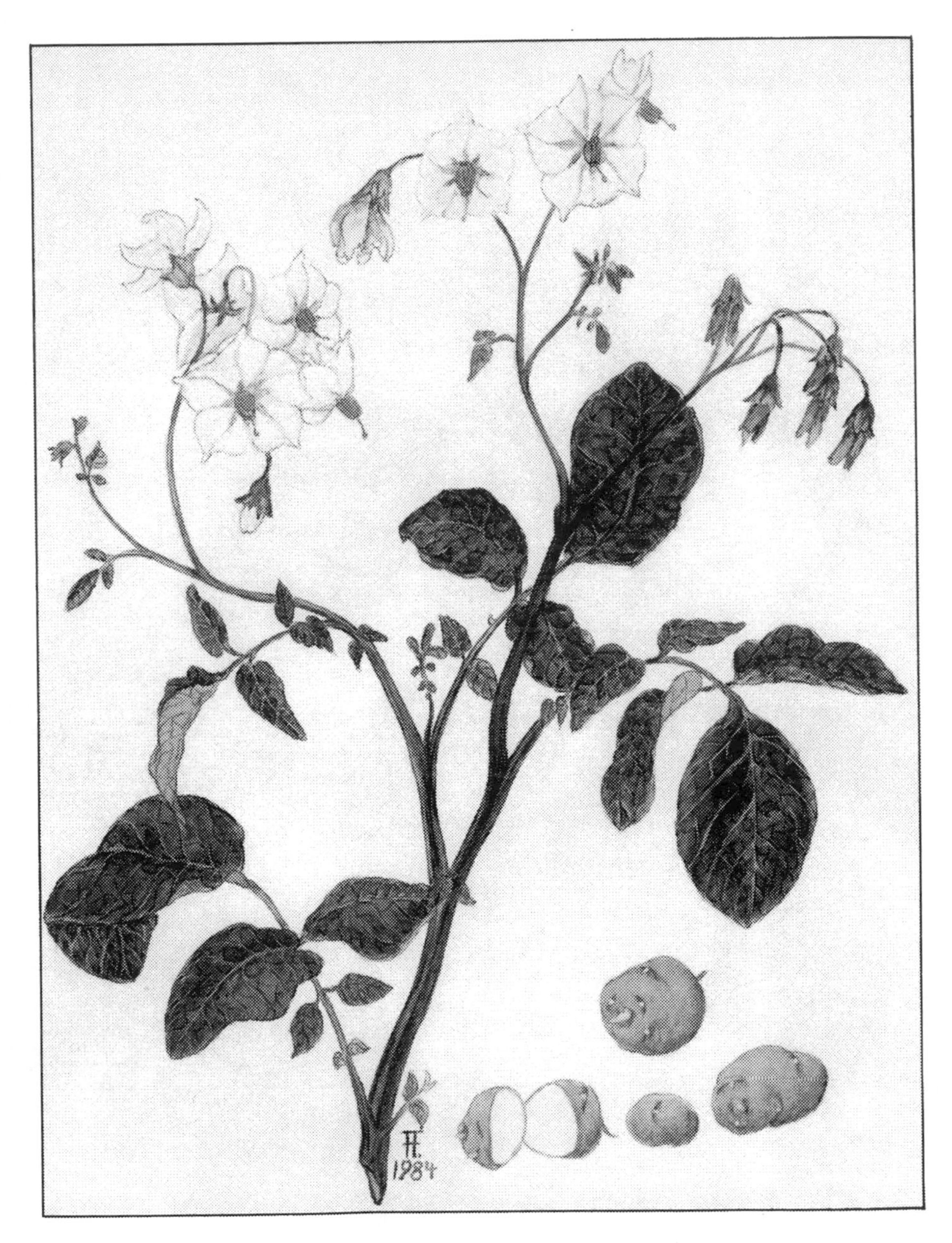

Plate XV. – *Solanum okadae* Hawkes and Hjerting.

Plate XVI. – *Solanum oplocense* Hawkes.

Plate XVII. – *Solanum sparsipilum* (Bitter) Juzepczuk and Bukasov.

Plate XVIII. – *Solanum vidaurrei* Cárdenas.

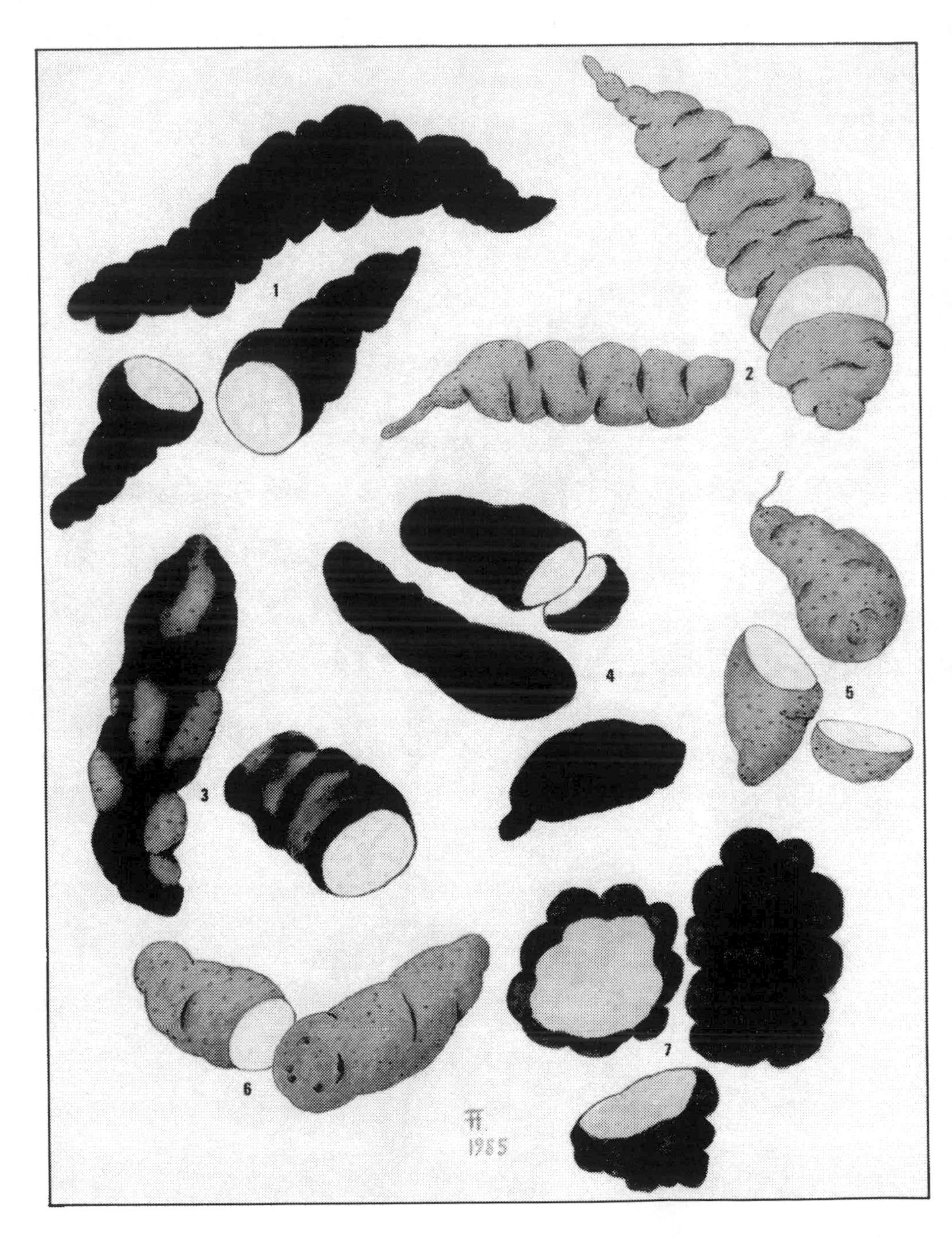

Plate XIX. – Tubers of Bolivian potatoes. 1. *Solanum* × *ajanhuiri*. 2. *S.* × *ajanhuiri* f. *janck'o-ajanhuiri*. 3 & 4. *S.* × *ajanhuiri* var. *yari*. 5. *S.* × *juzepczukii* var. *lucki* f. *lucki-pinkula*. 6. *S.* × *juzepczukii*. 7. *S.* × *chaucha* var. *piña* f. *chulluco*.

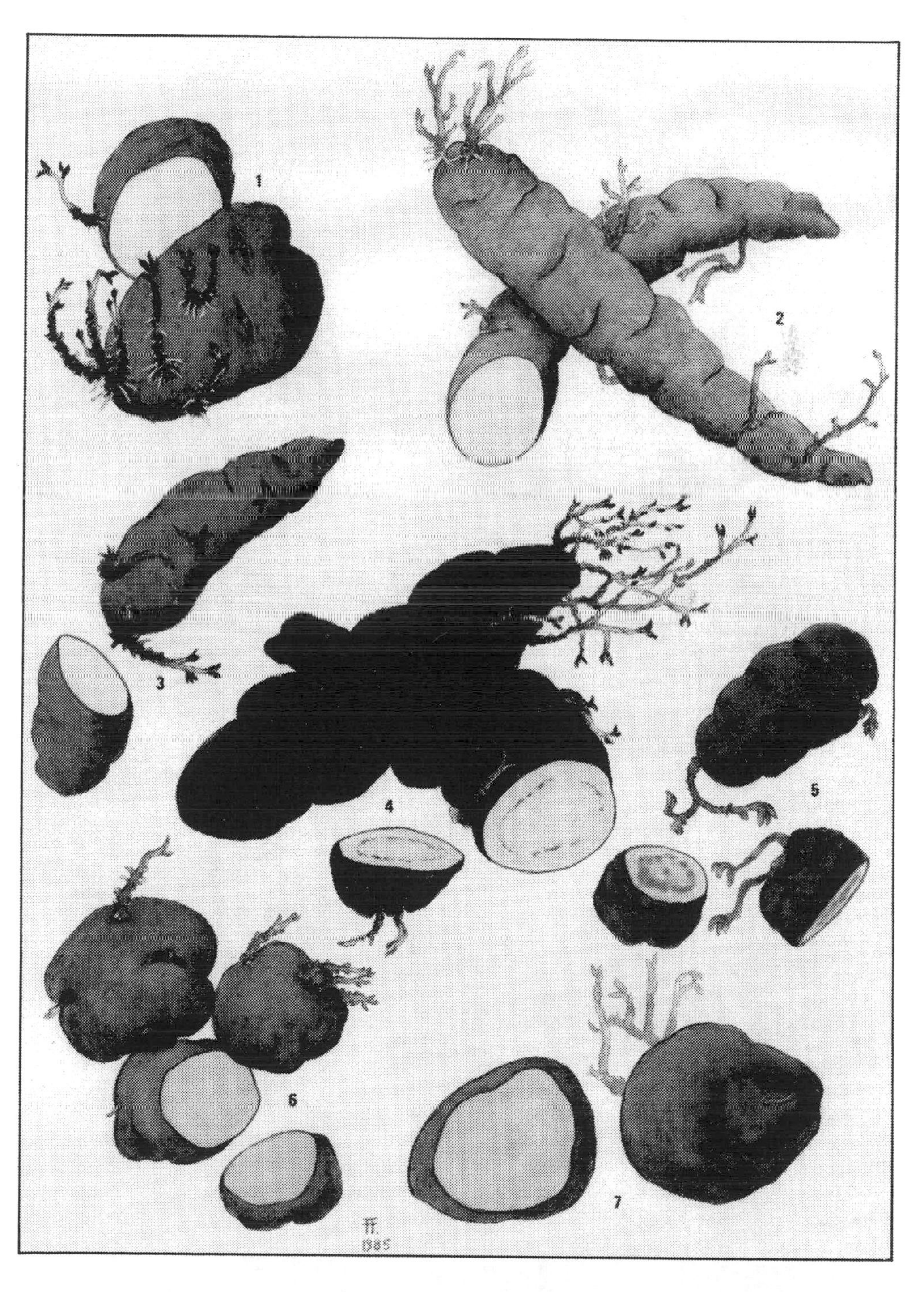

Plate XX. – Tubers of *Solanum phureja:* 1. var. *sanguineus* f. *puina;* 2. var. *quearanum;* 3. var. *rubro-rosea;* 4. var. *caeruleus;* 5. f. *puca chaucha* (endemic to Peru); and 6 & 7. var. *flavum*.

Plate XXI. – Tubers of *Solanum phureja* and *Solanum stenotomum*. 1. *S. phureja* var. *phureja*; 2. var. *janck'o-phureja* f. *timusi*; 3. var. *sanguineus*; and 4. var. *phureja* f. *viuda*. 5. *S. stenotomum* var. *pitiquiña* f. *phiti-kalla*; 6. var. *pitiquiña* f. *puca-pitiquiña*; and 7. var. *chojllu* f. *janck'o-chojllu*.

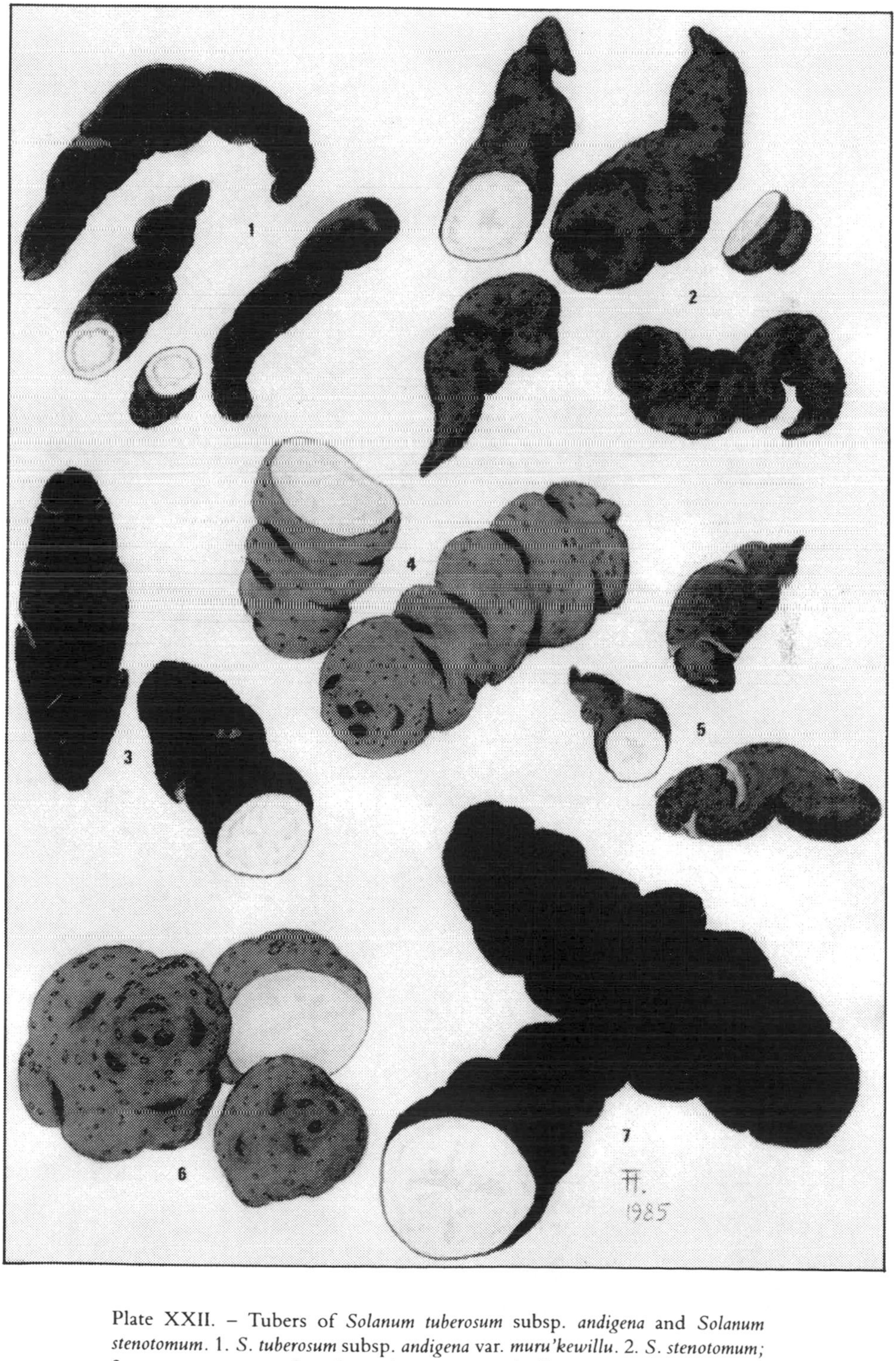

Plate XXII. – Tubers of *Solanum tuberosum* subsp. *andigena* and *Solanum stenotomum*. 1. *S. tuberosum* subsp. *andigena* var. *muru'kewillu*. 2. *S. stenotomum;* 3. var. *stenotomum* f. *pulu-wayk'u;* 4. var. *chojllu;* 5. var. *stenotomum* f. *alkka-phiñu;* 6. var. *zapallo;* and 7. var. *kkamara*.

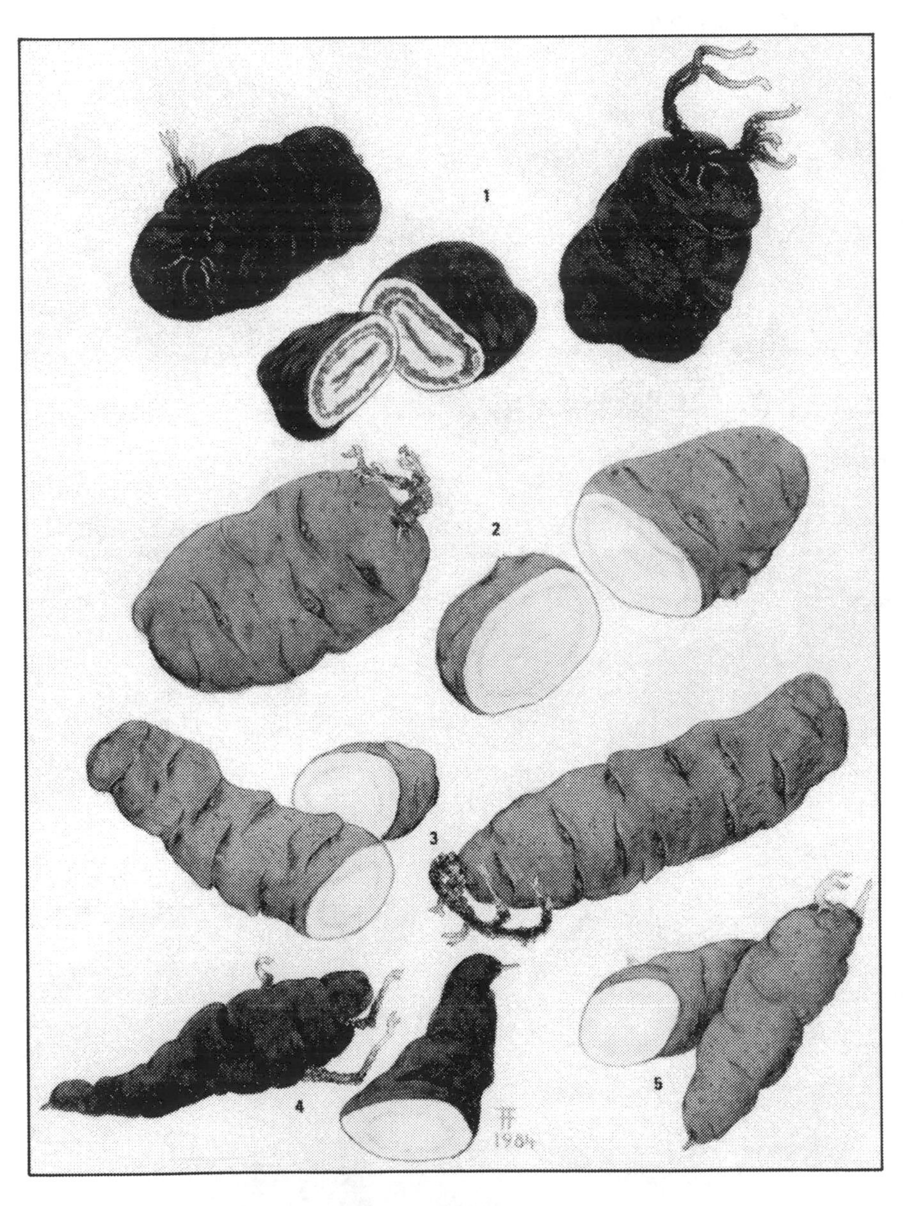

Plate XXIII. – Tubers of Bolivian potatoes. 1. *Solanum* × *curtilobum*. 2&3. *S.* × *curtilobum* var. *curtilobum* f. *china-malko*. 4. *S.* × *ajanhuiri*. 5. *S.* × *juzepczukii* var. *sisu* f. *janck'o-sisu*.

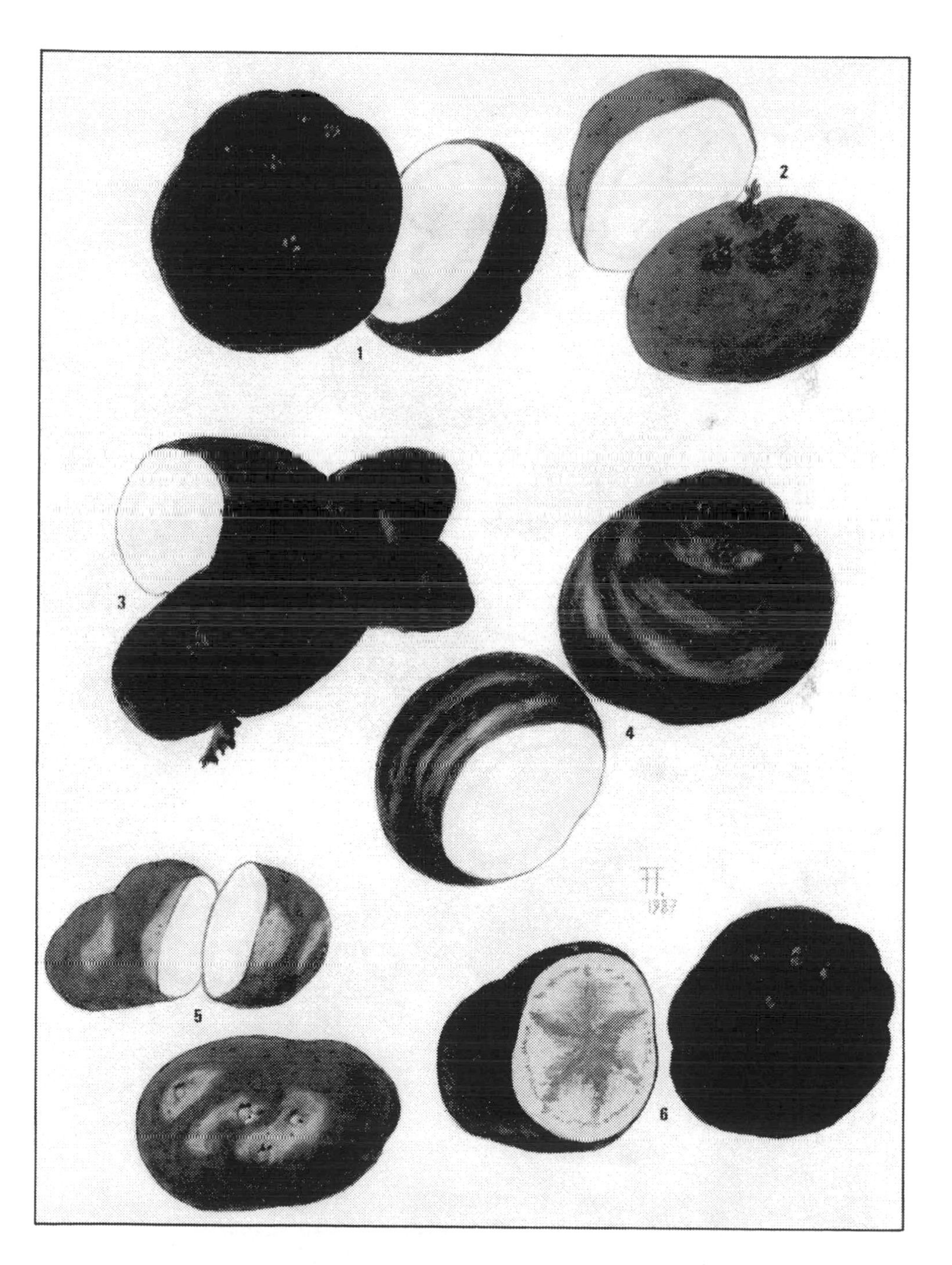

Plate XXIV. – Tubers of Bolivian potatoes. 1. *Solanum tuberosum* subsp. *andigena* var. *chiar-imilla* f. *nigrum;* and 2. f. *wilamonda*. 3. *S. stenotomum* var. *stenotomum* f. *chiar-ckati*. 4. *S. tuberosum* subsp. *andigena* var. *chiar-imilla* f. *sani-imilla;* 5. f. *alkka-imilla;* and 6. var. *sicha*.

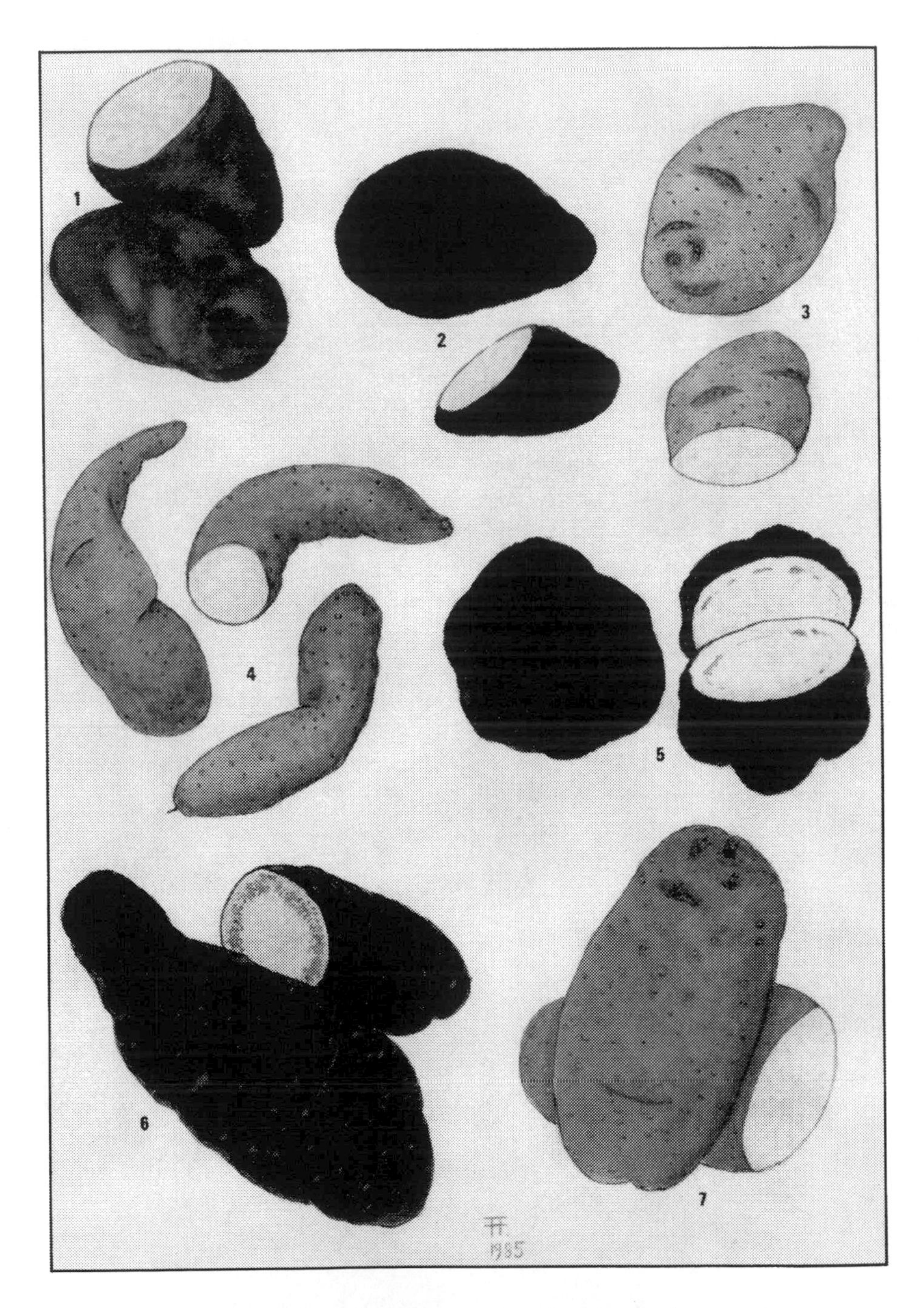

Plate XXV. – Tubers of Bolivian potatoes. 1. *Solanum tuberosum* subsp. *andigena* var. *bolivianum* f. *wila-pala;* 2. f. *chiar-pala;* 3. f. *janck'o-pala;* and 4. var. *longibaccatum* f. *cevallossi*. 5. *S.* × *chaucha* var. *kkoyllu*. 6. *S.* × *tuberosum* subsp. *andigena* var. *aymaranum* f. *huaca-zapato;* and 7. var. *runa*.

1

Historical Review

Although many explorers and naturalists have visited Bolivia in the past, only the most important will be treated in the following account. Those included have each contributed in some significant way to our scientific knowledge of Bolivian potatoes, either directly through the collection and study of various wild and cultivated species, or indirectly through the general sampling of the flora.

Many new species of tuber-bearing *Solanum* have been discovered in Bolivia since the beginning of the last century. With some exceptions, as will be pointed out again later under the individual treatments of species, neither the holotypes nor the isotypes of the Bolivian plants have remained within the country of their origin, all having been deposited instead in the larger herbaria of Europe and the United States.

Our story of potato collecting in Bolivia begins with the celebrated French explorer, Alcides D'Orbigny. From his travelogues to South America, covering the years 1830–1833, we have learned a great deal about the geography and natural history of this country during an era when it was first being organized as a republic. His monumental travelogue, *Voyage dan's la Amerique Meridionale*, remains as an important general reference today not only for Bolivia, but also for Argentina, Brazil, Chile, Uruguay, and Peru (D'Orbigny, 1945).

Although not trained originally as a botanist, D'Orbigny collected an enormous quantity of material of all classes during his three years of explorations in Bolivia. Among his collections were more than 3000 herbarium specimens that have been deposited today in the Natural History Museum of Paris. The great majority of these plants turned out to be new species. This explorer made the first collection ever of a wild potato species for Bolivia, *S. boliviense*, which was described by Michel-Felix Dunal in 1852. Although neither the exact place nor the date of this collection are known, it is probable that this plant was collected in the vicinity of Chuquisaca, or Sucre as it is known today, during

D'Orbigny's two-month visit to this historic locality from December 1832 to March 1833 (D'Orbigny, 1945, Vol. IV, pp. 1477-1486).

Gilbert Mandon, a French traveler and collector of plants in Bolivia, as well as in Madeira and the Canary Islands, lived during the second half of the nineteenth century in Sorata, Bolivia, where he was the business director of a gold mine from 1855 to 1861. During this period, he collected numerous botanical novelties in the vicinities of Sorata and Mapiri. According to Hugh Weddell (1867), Mandon collected more than 1800 herbarium specimens, which are now deposited in the Museum of National History of Paris. Among the many new species discovered for the flora of Bolivia by this dedicated collector were *S. candolleanum* Berthault (1911), *S. circaeifolium* Bitter (1912), *S. alticolum* Bitter (1913), and *S. boliviense* subsp. *virgultorum* Bitter (1913).

Toward the end of the nineteenth century, Miguel Bang, a Danish gardener and botanical collector, who had been trained at Kew Gardens, London, was sent to Bolivia by Kew to collect living orchids. Later, when he was already a resident of Bolivia, he came in contact with one of the greatest botanical explorers of this country, the American physician, Henry H. Rusby. Rusby's first expedition to Bolivia was in 1885-1886, by which time Bang had already made a substantial start on the classification of the flora of Bolivia. In 1889, Rusby and Bang agreed to collaborate in a joint effort to continue their floristic collections of Bolivia. Bang's collections were eventually sent to Rusby, and the largest of these sets is now housed permanently in herbaria of the New York Botanical Garden and the Smithsonian Institution in Washington, D.C. The results were published in four parts, mainly in the *Memories of the Torrey Botanical Club* (Rusby, 1893).

Of the 50 new Bolivian species of plants discovered by Bang, perhaps the most notable is *S. brevicaule*. This wild potato species, known from the vicinity of Cochabamba, was collected by him in 1891 under the No. *1100*. It was described as new by the German solanologist, Georg Bitter, in 1912.

Among the German collectors of the early twentieth century, Karl Fiebrig and Otto Buchtien both hold a prominent place. In 1903, Fiebrig undertook an expedition to the Andean region of southern Bolivia and northwestern Argentina for the purpose of making phytogeographic studies. The results of his expedition, which was sponsored by the Botanical Museum of Berlin, were later published in Leipzig (Fiebrig, 1910). To this explorer and phytogeographer we owe the discovery of two other important tuber-bearing species, *S. acaule* and *S. megistacrolobum*, both of which were collected by him in the Punas of Patanca, Department of Tarija, Bolivia. He also collected two species, *S. bijugum* and *S. microdontum*, at Toldos, in the vicinity of Bermejo, Bolivia (located today in the Province of Jujuy in the northwest territory of Argentina). These four species were described by Bitter in 1912.

Buchtien collected actively in both Bolivia and Chile. During his long residence in Bolivia, covering almost three decades, he made extensive field trips throughout the country and completed his botanical studies in the mid-1930s. Between 1911 and 1915, he issued a series identified as *Herbarium bolivianum*. Buchtien's contributions to the flora of Bolivia list 45 new species of various genera and families (Buchtien, 1910). These plants were originally described by specialists from the University of Breslau, Germany, in the botanical journal *Fedde Repertorium*, between 1908 and 1909. Buchtien had probably collected more abundant and diverse material from Bolivia than any other botanist. His valuable herbarium, consisting of more than 45,600 specimens, was acquired in 1922 by the Smithsonian Institution. Duplicates also exist in the University of Breslau Herbarium and in other German herbaria.

Among the type collections of Buchtien are *S. violaceimarmoratum*, a species from Unduavi, Province of Nor Yungas, and *S. acaule* var. *subexinterruptum* from the Bolivian altiplano, both of which were described by Bitter in 1912. Buchtien also discovered *S. leptophyes* and *S. tuberosum* subsp. *sparsipilum* in the vicinity of La Paz; these latter two were described by Bitter in 1913. Buchtien also collected the type specimen of *S. yungasense*, near Milluhuaya, Province of Nor Yungas, in 1917. This species was described by Hawkes (1954) almost 40 years later. Also to Bitter (1913), we owe the description of *S. acaule* var. *caulescens*. The latter plant was grown in Grenoble, France, from tubers that were collected originally at Viacha, near La Paz, by Claude Verne in 1913.

In 1914, the Bolivian agronomist, Walter Cevallos Tovar, published a work on the classification of Bolivian cultivated potatoes, which was the first work of this type for South America. These investigations are based largely on tuber shape, in which he presents a total of 184 perfectly distinct clones grouped in round, long, flat, and irregular shapes, and he also indicates for each tuber the flesh and skin color, eye characteristics, and other attributes. Moreover, he gives the native name for each sample, mostly in the Aymara language. Although Cevallos Tovar does not indicate the exact locality from whence the material came, the majority was probably collected from the Bolivian altiplano (Cevallos Tovar, 1914).

Among other travelers to Bolivia between 1913 and 1920 were the American plant and seed collector, William F. Wight (1916), and the Swedish botanist, Eric Asplund (1926). These two explorers made general collections of the Bolivian flora, and a few of their herbarium specimens of potato can be found today in European and American museums. In addition, Wight, in 1913, collected nearly 50 samples of cultivated potatoes in La Paz and Oruro for the United States Department of Agriculture (Washington, D.C.).

The first expedition to collect living material in South America for inclusion in potato improvement programs was probably made by the Soviet Union in 1925. This work, which evolved under the leadership of Nicolai I. Vavilov, included Sergei M. Bukasov, who made collections in Mexico, Guatemala, and Colombia between 1925 and 1926, and Sergei W. Juzepczuk, who did the same for Peru, Bolivia, and Chile between 1927 and 1928. Vavilov, in addition to making extensive trips in 1932 to North and Central America, collected in Ecuador with the assistance of E. Kesselbrenner (Bukasov and Lechnovitch, 1935). Many new wild and cultivated potato species resulted from these explorations (Juzepczuk and Bukasov, 1929; Bukasov, 1933, 1934, 1971b).

After nearly eight months of work in Peru, Juzepczuk arrived in La Paz, Bolivia, by mid-August of 1927, via the Puno-Guaqui route. In Bolivia, he worked mainly in the vicinities of Lake Titicaca, Viacha, La Paz, and Sorata. Given the season of his visit and the briefness of his stay in Bolivia, Juzepczuk dedicated himself exclusively to the collection of native cultivated potatoes.

Juzepczuk, in collaboration with Bukasov (1929), described for Bolivia the following new species: *S. ajanhuiri*, *S. andigenum*, *S. curtilobum*, *S. chaucha*, *S. juzepczukii*, *S. phureja*, and *S. tenuifilamentum*. In 1937, they also elevated *S. tuberosum* subsp. *sparsipilum* to the rank of *S. sparsipilum*, the name by which it is known today (Bukasov, 1937). In conection with their cultivated collections, later cytological studies by the Russians revealed the presence of a polyploid series in which the chromosome number varied from $2n=2x=24$ to $2n=3x=36$, $2n=4x=48$, and $2n=5x=60$ (Rybin, 1929, 1933).

A more complete classification of the South American cultivated potatoes, mostly based on the Juzepczuk and Bukasov collections, was carried out many years later by V. S. Lechnovitch (1971), also from Vavilov's Institute.

Shortly after the Soviet efforts in 1929, expeditions were undertaken by other countries with the purpose of collecting potatoes in the South American Andes. In 1930, the German Institute of Erwin Bauer sent Rudolph Schick to Bolivia and Peru (Schick, 1931). In 1932, an American expedition was made by H. G. MacMillan and C. O. Erlanson (Anonymous, 1934). In 1933, the Swedish Svalof Institute sent Karl Hammarlund to the same countries (Hammarlund, 1943). These expeditions collected primarily in the region of the Bolivian altiplano and southern Peru.

Hammarlund worked in Bolivia during 1933 and part of 1934. While in the altiplano, he worked principally in Lucurmata, Tiahuanacu, and Viacha. Later, he traveled to Oruro via Eucaliptus and Huancaroma. Afterward, he traveled from Rio Mulatos to Potosi, visiting along the way, Laja, Tambo, and Otavi in the Province of Linares. Later, he went from San Lucas to Irukasa, Tambillo, Sivingamayo, Tacaquira, and Camargo (Province of Nor Cinti, Department of Chuquisaca), to San Pedro and Culpina. Little is known about the results of this Swedish expedition; however, ten years later, Hammarlund reported a total of 800 general collections of the flora, which included many wild and cultivated potato species. He did not indicate the native names of the cultivated varieties or the precise localities where these were collected, nor did he undertake the taxonomic classification of this material.

In 1939, an English expedition was undertaken by the horticulturalist, Edward K. Balls, the physician William Balfour Gourlay, and the botanist Jack G. Hawkes. Their goal was to collect cultivated potatoes and their wild relatives, as well as other native cultivated species. The results of these explorations, which covered essentially the area from Colombia to northwest Argentina, are given in detail by Hawkes (1944). The expedition entered Bolivia in late January 1939 by the Guaqui-La Paz route and then traveled to the Argentine Provinces of Salta and Jujuy. From there they returned to Bolivia by the Villazon-Tarija route, and from Tarija they later traveled to Potosi and Sucre. Afterward, they visited Oruro and Cochabamba, where they visited with the Bolivian botanist, Martín Cárdenas. Later, they traveled from Colomi to the eastern side of the Cordillera of Tunari. Of the original group, only Balls returned to the altiplano from Cochabamba. After making short trips to the Lake Titicaca region, he completed his travels in Bolivia by late April 1939 and then proceeded to Peru, Ecuador, and Colombia.

The material collected by this two-month British expedition to Bolivia enabled Hawkes (1944) to describe the following wild species: *S. anomalocalyx*, *S. berthaultii*, *S. brevimucronatum*, *S. lapazense*, *S. oplocense*, *S. pachytrichum*, *S. platypterum*, *S. subandigena*, *S. sucrense* and its variety *brevifoliolum*, *S. tarijense*, and *S. violaceimarmoratum* var. *papillosum*. For the cultivated group, he described *S. cardenasii* and two new varieties, *S. phureja* var. *pujeri* and *S. chaucha* var. *roseum*. Also, Hawkes (1954) proposed the new Bolivian series Circaeifolia, under which were placed *S. circaeifolium* and *S. capsicibaccatum*.

Martín Cárdenas, the well-known Bolivian botanist (Argandoña, 1971), began floristic studies of his country in July of 1921, when he was assigned by the Bolivian government to assist Rusby on his record year-long 'Mulford Biological Exploration of the Amazon Basin.' It was during this expedition that Cárdenas learned, in particular, how to identify the complex flora of the Amazon Basin. At a later date, he also accompanied Asplund on plant collecting trips in Bolivia. Cárdenas was also in contact with many other foreign botanists, including Buchtien, the German botanist who resided in Bolivia for more than 30 years. The contributions that Cárdenas made to the flora of Bolivia were principally in the families Amaryllidaceae and Cactaceae, plus he also did important studies on the wild and cultivated potato species of this country.

In 1944, Cárdenas described his first two new Bolivian species of wild potatoes, *S. capsicibaccatum* and *S. pinnatifidum* (Cárdenas, 1944), both names based on type collections made by Humberto Gandarillas. In 1945, Cárdenas also published a work in collaboration with Hawkes concerning new or little known potato species from Bolivia and Peru (Cárdenas and Hawkes, 1945). In this work, they described for the first time, *S. decurrentilobum*, *S. toralapanum* and its variety *subintegrifolium*, and *S. ellipsifolium*, all of the series Megistacroloba. In addition, they proposed the following Bolivian species for series Tuberosa: *S. mollepujroense*, *S. liriunianum*, and three subtaxa of *S. anomalocalyx*, vars. *llallaguanianum*, *brachystila*, and *muralis*. Also proposed was the species *S. virgultorum*, which was based on *S. boliviense* subsp. *virgultorum* Bitter. Moreover, there also appeared under the sole authorship of Hawkes, the species *S. xerophyllum*, which he placed in the series Cuneoalata Hawkes.

Ten years after the appearance of this joint paper by Cárdenas and Hawkes, the most extensive work published by Cárdenas on the Bolivian wild species of potato was published in the 1956 *Bulletin of the Peruvian Society of Botany*. In this publication, the author (Cárdenas, 1956), based on his own type collections, described the following new species: *S. achacachense*, *S. alandiae*, *S. arnezii*, *S. caipipendense*, *S. candelarianum*, *S. cevallos-tovari*, *S. colominense*, *S. cuevoanum*, *S. gandarillasii*, *S. higueranum*, *S. subandigena* var. *camarguense*, *S. torrecillasense*, *S. trigalense*, *S. ureyi*, *S. uyunense*, *S. vallegrandense* and its var. *pojoense*, *S. vidaurrei*, and *S. zudanense*. With the description of *S. ruiz-zeballosii* in 1968 (Cárdenas, 1968), he ended his contributions to the taxonomy of the Bolivian potatoes.

Also worthy of mention here are two Bolivian agronomists, Humberto Gandarillas and Segundo Alandia. These two scientists assisted, for many years, in the field collections of Cárdenas, as well as in those of other botanists, and their names have been cited frequently in works by Donovan Correll and Hans Ross. Individually, both have contributed to the production and improvement of Bolivian potato varieties (Gandarillas, 1961). Moises Zavaleta, another Bolivian agronomist, is known to have made a collection of cultivated potatoes which he maintained for a long time at Caquiaviri, a locality southeast of Lake Titicaca (Zavaleta, 1968). Unfortunately, these collections have been lost and are no longer available.

It is also important to mention here that a large cultivated germplasm collection was gathered by the Bolivian Ministry of Agriculture at the Experimental Station of Toralapa, near Cochabamba (Cárdenas, 1963).

In 1955, under the sponsorship of the Plant Improvement Institute of Wageningen, Holland, the author of the present work made a two-month trip to Bolivia to collect native cultivated potatoes. A year later, this work on the potatoes of the border region of Lake Titicaca, Department of Puno, Peru, was completed (Ochoa, 1958), in which were given the descriptions and chromosomal determinations of more than 500 collections of potato. The living materials collected in connection with this study were accessioned by the Dutch Potato Improvement Program. In 1956, the author published a single illustration and complete description of the little-known wild diploid species, *S. candolleanum*, which was collected in 1955 during a visit to Tacacoma, north of Sorata (Ochoa, 1956).

In February of 1959, a German expedition, headed by Hans Ross and including the scientists R. Rimpau and Ludwig Diers, entered southern Bolivia (Ross and Rimpau, 1959; Ross, 1960a, b). They were sent to South America by the Max-Planck-Institute for the sole purpose of collecting germplasm of Bolivian potatoes for inclusion in a plant improvement program. They entered Bolivia from La Quiaca, Argentina, and collected initially in the Bolivian localities of Tarija, Potosi, and Sucre. Later, they continued on to Cochabamba via Aiquile-Totora. After making collections in the above localities, including the tributaries of the Tunari River, they headed to La Paz via the Oruro-Belen route, visiting Sorata and Los Yungas along the way. Finally, they crossed from Tiahuanacu to Desaguadero and arrived in Puno, Peru, on the first of April 1959, where they worked for several months collecting material. The living collections made during these travels were deposited in the Max-Planck-Institute, near Cologne, West Germany.

In 1960, the American botanist, Donovan S. Correll, undertook an expedition to collect the potatoes of Argentina, Bolivia, and Peru under the auspices of the Texas Research Foundation, the National Science Foundation, and the U.S. Department of Agriculture (Correll, 1962). Besides Correll, the expedition included Kenneth S. Dodds, Graham J. Paxman, and Heinz H. Brücher. Even though this expedition was in Bolivia for barely two weeks in February, they covered an enormous stretch of mountainous territory, traveling from Lake Titicaca and Los Yungas to La Paz, Oruro, Cochabamba, Sucre, Potosi, Tarija, and Villazon, and entering Argentina through La Quiaca. One result of this expedition was a new wild potato species for Bolivia that was described by Correll in 1961 under the name of *S. doddsii* (Correll, 1961).

In addition to the Andean trips already mentioned, Dodds and Paxman also spent several months collecting samples of the cultivated potatoes, 'chaucha' and 'phureja.' Dodds' expedition led ultimately to a revision of the cultivated potatoes based upon their inferred evolution. His work is explained and illustrated in a chapter in Correll (Dodds, 1962) entitled, 'Classification of Cultivated Potatoes.'

In 1962, the American botanist, Donald Ugent, organized a year-long expedition to collect the potatoes of Mexico, Ecuador, Peru, and Bolivia. This expedition was sponsored by the University of Wisconsin and the National Science Foundation. After five months in Peru, Ugent traveled to Bolivia via the Puno-Desaguadero route, arriving at La Paz in February of 1963. In Bolivia, Ugent collected principally in the mountainous zone from the vicinity of La Paz to Santa Cruz in the southeast. Some collections were made in collaboration with the Bolivian botanist, Martín Cárdenas, and the biology teacher, Arturo Vidaurre.

Although new species of potato were not collected during Ugent's expedition to Bolivia, the results of his work contributed valuable information on the distribution and ecology of many species. Tuber and seed collections from this expedition are maintained by the U.S.D.A. Potato Introduction Station at Sturgeon Bay, Wisconsin.

In 1971, a potato collecting expedition to Peru and Bolivia was organized by J. G. Hawkes and Philip J. Cribb of the University of Birmingham, United Kingdom. This expedition, which was supported in part by the International Potato Center (CIP) in Lima, Peru, also included Jean P. Hjerting of the Copenhagen Botanical Garden and Zósimo Huamán of CIP. The expedition began in La Paz, Bolivia, in January of 1971, and followed a route that took them as far south as Entre Rios, Argentina, concluding several months later in April. Summarized results of the potato collecting expeditions in Bolivia, Chile, and Peru were also published by the author (Ochoa, 1975b).

Hawkes also organized other trips to Bolivia between the years 1978 and 1981. Participating at various times in these expeditions were Dave Astley (England), W. Hondelman and Jean P. Hjerting (Denmark), A. M. van Harten and J. M. Soest (Holland), Zósimo Huamán and Juan Landeo (Peru), Israel Aviles, Carlos Alarcón, Arturo Moreira, and Gerardo Caero (Bolivia), and Armando Okada (Argentina). Complete information on these expeditions has been given in recent publications by Soest et al. (1980, 1983).

Hundreds of samples of Bolivian wild and cultivated potato species resulted from the scientific expeditions of Hawkes and his various collaborators. Many of these collections were accessioned by the Germplasm Station of the University of Birmingham, and by the more recently organized Potato Bank of Germany and Holland, located in Braunsweig, West Germany. Some of the above collections were also received by the Potato Introduction Station at Wisconsin in the United States and by CIP in Peru. From the above-mentioned explorations came the following new species, all of which were described by Hawkes and Hjerting (1983, 1985a, b): *S. astleyi*, *S. avilesii*, *S. circaeifolium* subsp. *quimense*, *S. hondelmannii*, *S. neocardenasii*, *S. okadae*, and *S. soestii*.

In conclusion, it should be noted that the number of potato collecting expeditions to Peru and Bolivia increased substantially between 1971 and 1981. In this respect, credit must be given here to the valuable role played by the International Potato Center (CIP) (1973, 1976, 1979a, b). Thus, a major potato collecting expedition to Bolivia was organized by the author of this work in 1978 under the sponsorship of the International Potato Center. The expedition route included every major potato collecting locality from the border of northern Peru to Tarija in southeastern Bolivia. A detailed account was later published in the journal *Biota* (Ochoa, 1979d).

In 1983-1984, with the assistance of Alberto Salas of CIP, the explorations of the author were expanded to include areas never before collected. Moreover, the goals of these explorations were broadened to include the collection of topotypes of rare or little-known species. Tuber and seed collections made in connection with these studies were incorporated into the CIP germplasm bank in Peru, and duplicate collections were sent to the Potato Introduction Station in Wisconsin. Included among the many species collected in Bolivia by the author are the following new taxa: *S. bombycinum*, *S. flavoviridens*, *S. litusinum*, *S. neovavilovii*, *S. venatoris*, *S. capsicibaccatum* var. *latifoliolatum*, *S. infundibuliforme* var. *albiflorum*, and *S. microdontum* var. *montepuncoense* (Ochoa, 1980a, b, 1981, 1982, 1983b, c, 1984a, b).

From D'Orbigny to the present time, more than 60 wild species of tuber-bearing *Solanum* have been described for Bolivia. In the following revision, this number has been reduced to approximately one half of the original proposals.

2

Geography and Climate

GEOGRAPHY

Bolivia, a land of diversified habitats and climates, is situated between latitude 9°38′ and 22°53′ S and longitude 57°26′ and 69°38′ W. It borders Brazil on the north and east, Peru and Chile on the west, and Argentina and Paraguay on the south (Map 1), with a total land area of 1,098,521 km^2 (Munoz Reyes, 1977).

According to the last census made in 1983, the population of Bolivia was 5,966,000, with the majority of the country's population living in the highland areas. Even though more than half of Bolivia is either uninhabited or isolated from the main centers of production and economic importance and the country lacks a seaport and access to the coast, its railroads, highways, rivers, and airlines facilitate communication with neighboring countries. Traditionally, there has been free and open boat traffic on Lake Titicaca between Bolivia and southeastern Peru.

According to the Bolivian geographer, Jorge Muñoz Reyes (1977), this country is divided into several physiographic regions (Map 2), in which he refers to the high, cold tableland, or altiplano of Bolivia, as the 'Bloque Andino.' This great plateau, which is situated at an elevation between 3200 and 4400 m, dominates approximately one-third of the country. To the west of the altiplano is the Cordillera Occidental de los Andes, which is also known locally as either the 'Cordillera Occidental' or the 'Cordillera Volcanica,' while to the east of this vast plain is the Cordillera Real de los Andes (Map 2). These two major mountain chains converge in the north toward the Nudo de Vilcanota in La Raya, south of Cusco, Peru (14°31′ S), and in the south toward the Nudo de Licancabur in Chile (23° S), where the two ranges then trend south to the Punas de Atacama.

At still higher elevations in Bolivia, according to Cabrera and Willink

Map 1. – Political divisions of Bolivia (after Muñoz Reyes, 1977).

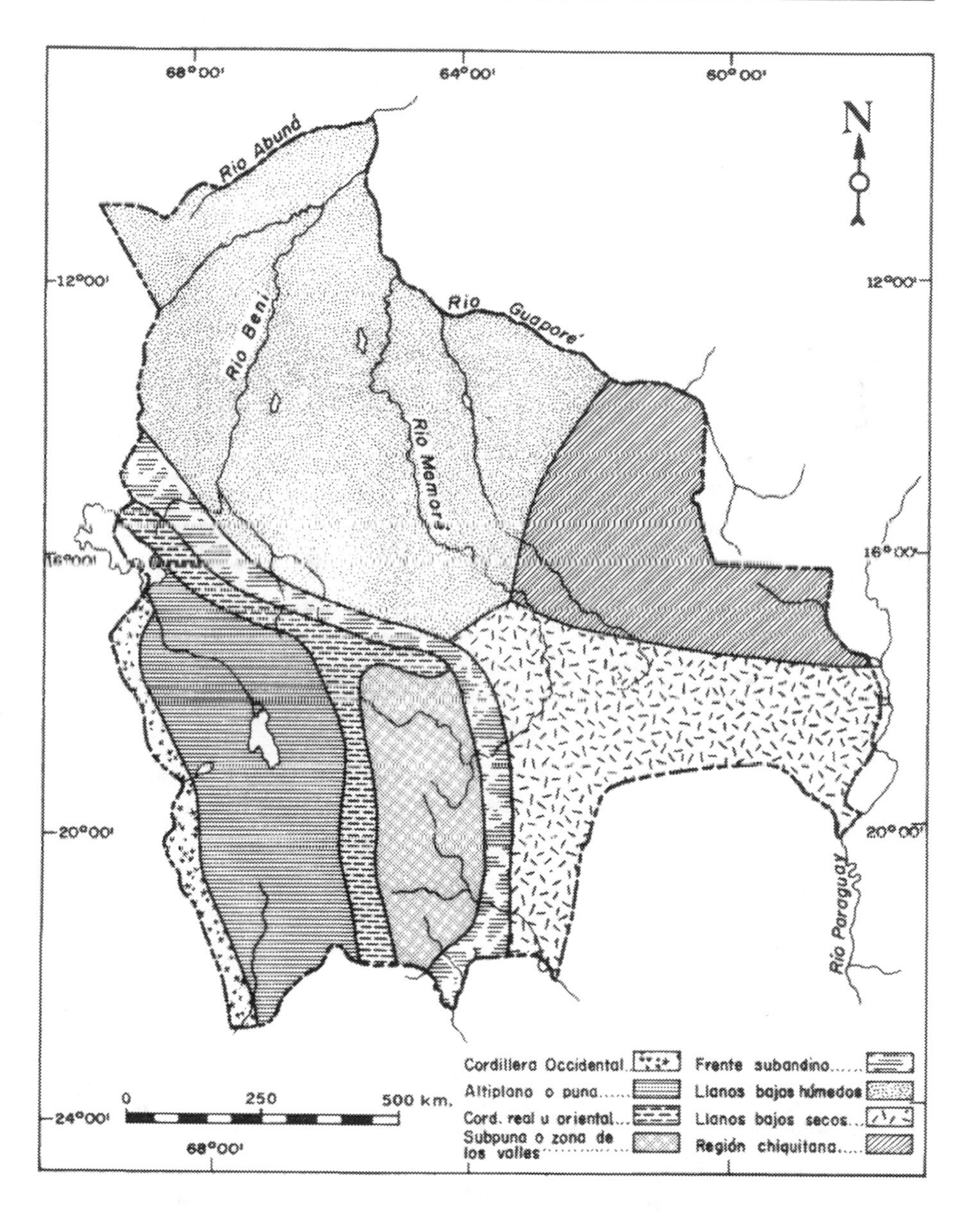

Map 2. – Physiographic regions of Bolivia (after Muñoz Reyes, 1977).

(1973), is the Provincia Altoandina. This zone, which includes some very high mountain ranges that jut out from the flat lands of the altiplano, extends north along the Andean cordillera to Venezuela and south to Tierra del Fuego in Chile. These same two authors, incidentally, also refer to the altiplano as the 'Provincia Puneña.'

Cordillera Occidental de los Andes

This great range, which is unique for its volcanic activity, includes the Sajama, the highest peak of the country, 6532 m according to Muñoz Reyes (1977). On the eastern side of this range are grown an assortment of Andean crop plants, including potatoes, oca, and quinoa. The total length of the mountain chain is 620 km, which ranges from southern Peru to Volcan Copiapo in northern Chile (27°30′ S). The Chilean side of this cordillera, which borders the Atacama Desert, is too dry for agriculture.

In contrast to the above mountain chain, which has a very dry climate, the Cordillera Real de los Andes is very moist and constitutes the lowland *divortium aquarium* of Bolivia.

Cordillera Real de los Andes

This broad mountain chain is composed of two major ranges, the Cordillera Real and the Cordillera Central (or Meridional), which, in turn, are each divided into a number of lesser chains. The first major range, which is the highest and the most important of the two, extends diagonally across southwestern Bolivia from the vicinity of Lake Titicaca to the Argentine border. The second major range, which extends in a north-south direction, divides the altiplano from the Subpuna, or Zona de los Valles. These greater and smaller mountain ranges are discussed in the following sections.

I. Cordillera Real

Apolobamba. – This minor and little known cordillera is situated largely in Peru, but its crests extend southeast to the vicinity of Lake Titicaca in northwest Bolivia. The most important peaks of this chain are Chauppi Orcco (6040 m), Palomani Cunca (5808 m), Palomani Grande (5920 m), and Palomani Tranca (5651 m).

Muñecas. – This small mountain chain is situated near the northeastern shores of Lake Titicaca. It is separated from the Cordillera Apolobamba by a deep gorge cut by the Camata River.

Real de La Paz. – This cordillera is probably the most spectacular in the Bolivian Andes. It runs 200 km from the steep Sorata River Valley, which separates this range from the previous cordillera, to the La Paz River Valley.

Included in this range are the peaks of Illampu (6424 m), Illimani (6322 m), and Huayna Potosi (6095 m).

Tres Cruces, or Quimsa Cruz. – This range, which extends from the southeastern part of the La Paz River Valley to the Abra de Quime, includes the snow-capped peaks of Jachakunokollo (5900 m), Gigante Grande (5807 m), Atoroma (5700 m), and San Juan (5700 m).

Nudo de Santa Vera Cruz. – This fairly small range, located in the northeast corner of the Department of La Paz, should probably be called a 'macizo' rather than a cordillera, according to Muñoz Reyes (1977). The principal peak of this chain is Santa Vera Cruz (5520 m). Other mountain chains of moderate height occur toward the southeast.

Cochabamba. – This cordillera includes a number of lengthy subranges of fairly low elevation. The most important peak is the Nevado Tunari (5200 m), which is situated northeast of the city of Cochabamba. Its final spur ends with the Cerro Volcan (4000 m), north of the city of Pojo.

II. Cordillera Central

This cordillera is the point of origin of several isolated mountain chains that follow a more or less parallel north-south course. The western chain of this cordillera divides the northern sector of the altiplano from the Subpuna, or Zona de los Valles. It includes the following three subdivisions.

Azanaques. – This cordillera originates near Oruro (18° S) and terminates south of Challapata (19° S). With the exception of Negro Pabellon (5383 m), this fairly low cordillera lacks peaks of high elevation. Bordering this region is a high plateau and the beforementioned *divortium aquarium*, a low-lying zone between the Amazon and Plata Basins.

Los Frailes. – Extending from the south of Challapata (19°30′ S) to a point just north of Huanchaca (20°30′ S), this cordillera skirts the western border of the great salt lake, Uyuni, and the fresh water lake, Poopó. The highest peak of the chain is Michaga (5300 m).

Meridional. – This cordillera includes the ranges Chichas and Lipez. Chichas is dominated by the peaks Tazma (5805 m) and Chorolque (5630 m), while the mountain range, Lipez, is topped by the peaks Nuevo Mundo (6020 m), Lipez (5903), and Zapaleri (5650 m). The peak Zapaleri, which terminates this chain, is located in the northern border region of Argentina.

Southern Cordillera Ranges

The small mountain ranges that occur south of the Cordillera Real form the eastern border of the Subpuna, or Zona de los Valles. These mountain ranges extend in the direction of Colcha, Colquechaca, and the Cordillera de Potosi. Two of the main peaks of this area are the Cerro Hermoso (5100 m) and the Cerro Malmisa (5150 m). Mochara (4600 m), one of the minor mountain chains south of the Cordillera de Potosi, is situated between the Province of Sud Cinti, Department of Chuquisaca, and the Province of Chichas, Department of Potosi. Toward the eastern part of this cordillera, in the Province of Aviles, Department of Tarija, are found the Punas de Patanca and the Serranias de Tajsara (5000 m). To the south of these mountains are located the Serranias de Santa Victoria in the Argentine Province of Jujuy.

The subandean sierras that slope toward the Amazon Basin are formed by parallel north-south chains. These extend from the Argentine border (22° S) to Santa Cruz de la Sierra (17°45′ S). From the latter point, they trend northwest through Bolivia to 13° S. In the southern part of this range occur the smaller ranges of Serranias Huacareta, Charagua, Incahuasi, Florida, Espejos, and Amboro, while in the northern part are found the Serranias of Mosetenes, Sejeruna, Jatunari, and Eslabon, as well as others.

The Sierras Chiquitanas, which are located in the extreme eastern portion of the country, are situated between longitude 58° and 61° W and latitude 17°30′ S. The highest peak, Chochis (1490 m), is located between the mountains of San Jose (550 m) and Santiago (930 m).

CLIMATE

According to Muñoz Reyes (1977), the climate of Bolivia can be treated under the following three categories: 1) *Cold*, which includes the altiplano and the high peaks of the cordilleras; 2) *Temperate*, which includes the Zona de los Valles and the portion of the subandean region that borders the low, eastern grasslands of the Chaco; and 3) *Warm*, which includes the humid savannas of Beni and Pando, as well as the dry savannas of Santa Cruz and Chaco.

For purposes of this monograph, however, the author will consider the more detailed contributions of Köppen (1931), and Cabrera and Willink (1973). In Cabrera and Willink's work on the biogeography of Latin America, Bolivia is divided into regions called 'dominios' and 'provincias,' which is based on the use of dominant plant and animal species as indicators.

Climatic Zones of Bolivia

Cordillera Region, or High Mountain Climate

This climatic region is represented by the key EB on Map 3 (Köppen, 1931). The upper zones of the high mountain regions have a permanent cover of snow and ice. The lower levels, which are cooler and more humid, have a climate that corresponds to the northern tundra. Examples of regions that have this type of climate are the Cordilleras Occidental and Real. In some respects, the climate of this region is similar to that of the southwestern part of the Altiplano Region, which is discussed later in this chapter.

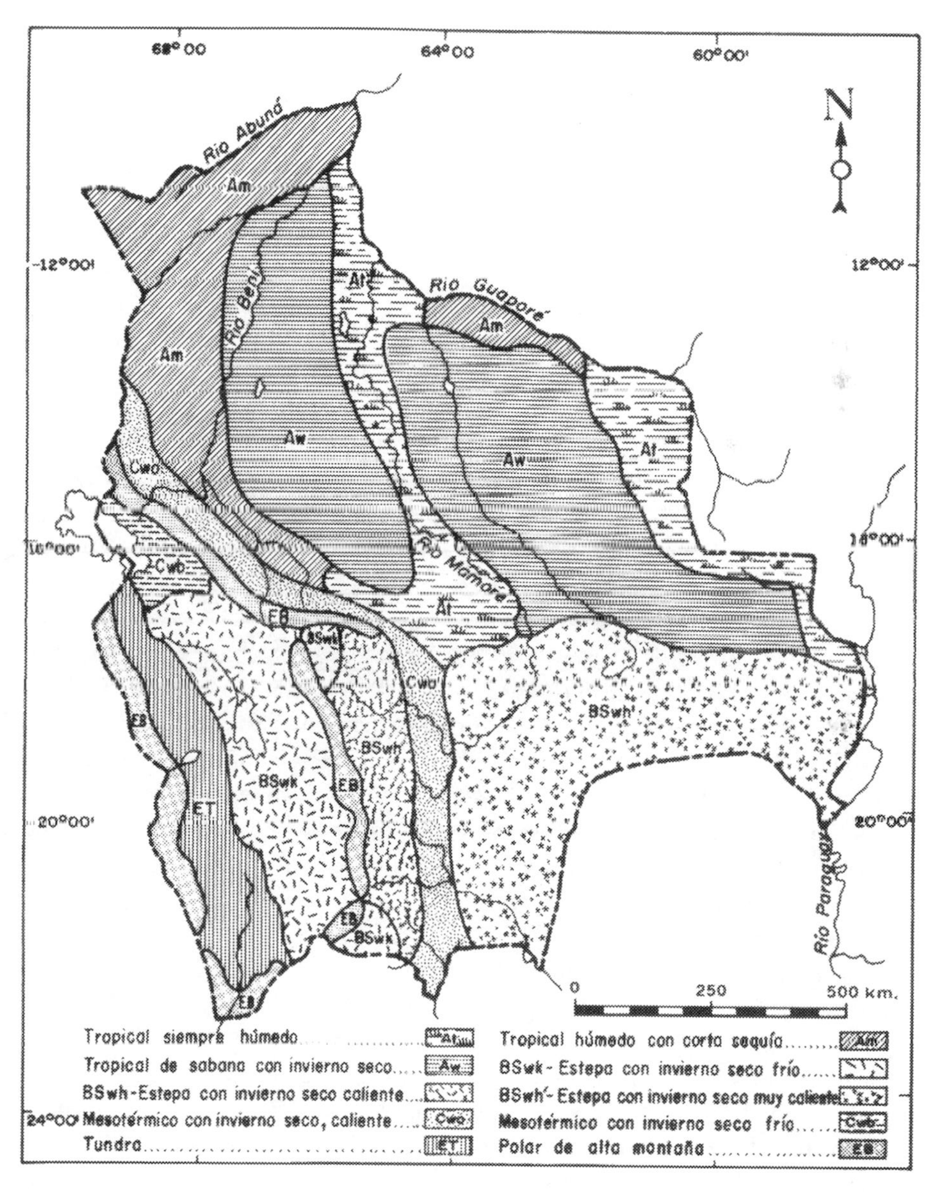

Map 3. – Climatic regions of Bolivia (after Köppen, 1931).

In the classification by Cabrera and Willink (1973), the Cordillera Region is treated as the Provincia Altoandina. This region begins at 4200 m elevation in Bolivia, which is a zone of perpetual ice and snow, and descends toward the south to 500 m altitude in Tierra del Fuego. Temperatures within this zone are predominantly cold throughout the year, and precipitation is largely snow or hail; thus, few plants are capable of growing in this area. However, the wild potato species, *S. acaule*, which is tolerant to cold, is occasionally found here. Other common plants of the Provincia Altoandina include the grass, *Aciachne* sp., which forms cushion-like colonies; two genera of the pink family, *Pycnophyllum* and *Pentacaena*; the mallow genus, *Nototriche*; and the composite, *Werneria*.

Altiplano Region, or Intermountain Climate

This climatic zone includes the high plain that is situated between the Cordillera Occidental and the Cordilleras Azanaques and Los Frailes (or Cordillera Central). In the vicinity of Lake Titicaca and the lower levels of the Cordillera Real, the climate of the altiplano is mesothermic, or temperate (Map 3, key Cwb), with dry winters and temperatures ranging from 10° to 22° C.

On El Alto (4071 m), situated in the altiplano above the city of La Paz, the average summer precipitation (October-March) between the years 1950-1982 was 593.5 mm, while the average summer temperature for the period 1963-1982 was 8° C. By contrast, the average summer precipitation for La Paz (3632 m), between the years 1960-1983, was 505.5 mm, with an average temperature of 11.3° C. In addition, valuable information on geomorphology, geology, climate, and plant communities in the Bolivian altiplano, particularly Huaraco, Antipampa, Machacamarca, and Angostura, can be found in Fisel and Hanagarth (1983).

According to Lorini and Liberman (1983), it rarely rains in the altiplano between Eucaliptos and Patacamaya (Province Aroma, Department La Paz). In that area, the average annual precipitation ranges from a high of 400 mm for Patacamaya to a low of 270 mm for Sica Sica. The cool, rainy season in this region, termed 'libre de heladas' by the above authors, lasts for only four months from December to March. January is the wettest and warmest month of the year, with temperatures ranging from 22° to -0.7° C. The agricultural season is therefore very limited. In Oruro, which is to the south of Patacamaya and goes up to an altitude of 3701 m, the average annual precipitation from October to March was reported to be 573.5 mm (1943-1983). The average annual temperature for these same six months, on the other hand, was 10.6° C (1946-1982).

The central part of the altiplano has a warm, semi-desert, steppe climate, with cold, dry winters (Map 3, key BSwk). However, both the southern region and the lower elevational levels along the western side of this high plateau have a tundra-like climate. There are enormous diurnal differences in temperatures in these regions. Muñoz Reyes (1977), for example, points out that in Uyuni (Department Potosi), a temperature of 30° C at noon and -25° C at midnight was registered for one given day in July, with a thermal difference of 55° C.

According to the phytogeographic system of Cabrera and Willink (1973), the altiplano falls under the classification, 'Provincia Puneña.' However, in the map treatment of latitudinal zones by Unzueta et al. (1975), the altiplano is included in a Subtropical Region (ST) having high grasslands and scrub forests. The latter subdivision, which falls under the altitudinal category of Montano Bajo, is also treated under the ecological subdivision of 'Bosque Humedo.'

Due to the cold, dry climate, the vegetation of the Provincia Puneña is only poorly developed. The average annual precipitation of this region oscillates between 700 mm and 50 mm or less, and the temperature varies between 8° and 9° C. Throughout most of the year the soil lacks plant cover; however, during the brief rainy season, some annual species appear that have fleshy underground parts. These plants, however, are always distantly spaced across the surface of the ground. Typical plants of this region include the grasses, *Stipa ichu*, *S. inconspicua*, *Festuca dolichophylla*, *Poa subspicata*; the composites, *Senecio pampae*, *Tagetes multiflora*, *Hypochaeris meyeniana*, *Bidens andicola*; the legumes, *Astragalus garbancillo*, *Trifolium amabile*, *Lupinus microphyllum*; the vervain, *Junellia minima*; and the Iris genus, *Sisyrinchium*.

As previously mentioned, the humid plains found in the northern reaches of the altiplano and in the vicinity of Lake Titicaca serve partly as pasture lands. Large areas, however, are also devoted to the cultivation of native Andean crop plants, such as potatoes (*Solanum tuberosum* subsp. *andigena*, *S. stenotomum*, *S.* × *ajanhuiri*, *S.* × *juzepczukii*, and *S.* × *curtilobum*); and many other varieties and species such as oca (*Oxalis tuberosa*), olluco (*Ullucus tuberosus*), *isaño* or *mashua* (*Tropaeolum tuberosum*). Also grown in these plains are pseudocereals (*Chenopodium quinoa* and *C. pallidicaule*), as well as the legume, tarwi (*Lupinus mutabilis*).

On the slopes and walls of quebradas in the vicinity of the altiplano between 3500 and 4400 m altitude, the author noted relic patches of forest trees, such as *ckehuiña* (*Polylepis incana*), *ccolli* (*Margyricarpus setosus*), and *quisuar* (*Buddleja* sp.), and several species of lupine (*Lupinus*) and shrubs (*Cassia*, *Berberis*, and *Baccharis*).

Of the wild potato species of the Altiplano Region, or Provincia Puneña, the most important is *S. acaule*. This species forms extensive colonies in the more humid zones of the altiplano and becomes progressively less abundant toward the drier, southern and southwestern parts of the altiplano, where it eventually disappears. Other tuberous species of *Solanum* found in the altiplano are

S. megistacrolobum, which, like *S. acaule*, is also frost-tolerant and the wild potatoes, *S. sparsipilum*, *S. leptophyes*, *S. achacachense*, and *S. brevicaule*. These tender, herbaceous potato species, which provide fine forage for grazing animals, grow commonly beneath nettles and various spiny plants, such as *Cajophora*, *Urtica*, *Lobivia*, *Opuntia*, and *Tephrocactus*, or between the stiff bundles of the grass, *Stipa ichu*, or between stones and rocks.

Subandean Zone

This narrow belt divides the high altiplano into western plateaus (dry) and northern plateaus (humid), which form a barrier to the winds that blow from the Amazon toward the south. On the northern side of this zone are the humid valleys called 'Yungas.' The Yungas, according to Muñoz Reyes (1977), are subdivided into the following three divisions based on altitude: Ceja de Yungas (2500-3000 m), Yungas (1500-2000 m), and Vegas (700-1500 m).

The northern part of the Bolivian subandean zone is very humid. This is due, partly, to warm, moisture-laden winds from the Amazon that contact the cool air masses of the cordillera. Following the deposition of their moisture upon the northern Bolivian Subandean Zone, these winds desiccate the high lands of the altiplano and the southern Subpuna.

Another name for the Subandean Zone of Bolivia, according to the phytogeographic classification of Cabrera and Willink (1973), is the 'Provincia de Las Yungas'; *yunga*, it should be noted, is a Quechuan word for 'hot land.' According to Unzueta et al. (1975), this region should also be called 'Piso Montano Bajo,' the latter being a subdivision of Bosque Húmedo, or Bosque Húmedo Alterado as indicated on Unzueta's map of vegetation. This narrow zone extends from Venezuela along the eastern side of the Andes to northwest Argentina and forms the western border of the Provincia Amazónica (Cabrera and Willink, 1973).

Because of abundant rain and persistent fog, the Yunga region is perpetually humid. Its altitudinal limits lie between 500 and 3500 m. Within the upper region of this zone (1800-2500 m) are cloud forests rich in plants of the myrtle and laurel families. Above this, there are grasslands and forests of *Podocarpus*, *Polylepis* sp., and *Alnus jorullensis*.

Typical plants of the Bolivian Subandean Zone, or Yungas, are coca (*Erythroxylum coca*), cacao (*Theobroma cacao*), gualusa (*Dioscorea* sp.), and yacon (*Polymnia sonchifolia*). Among the fine wood-producing trees of the region are walnut (*Juglans neotropica*), white cedro (*Cedrela fissilis*), and red cedro (*Cedrela brasiliensis*).

In addition to the above, the general flora of this region includes many wild potato species. The author has seen in the Provinces of Nor Yungas and Sud Yungas near La Paz, *S. circaeifolium*, *S. yungasense*, and *S. violaceimarmoratum*. Other species and their localities include *S. bombycinum* (Yungas of Puina), *S.*

flavoviridens and *S. virgultorum* (Yungas of Larecaja and Camata), *S. soestii* and *S. okadae* (Yungas of Inquisivi). In the humid forests that stretch between Cochabamba and Santa Cruz are also found *S. circaeifolium* and *S. microdontum*.

Subpuna, or Zona de Los Valles

This zone, which occurs on the eastern side of the Cordillera Central and occupies a major portion of the altiplano, has an abundant and varied flora. It is called Subpuna, or Zona de Los Valles, because it is situated at a lower altitude than the main portion of the altiplano, and also because in this region are found many inhabited valleys. The major cities and their ecological zones, according to Unzueta et al. (1975), are as follows: 1) Cochabamba (2548 m), Piso Montano Bajo, Estepa Espinosa Montano, and Montano Bajo Subtropical; 2) for both Sucre (2750 m) and Tarija (1957 m), Piso Montano Bajo and Bosque Seco Templado; and 3) Potosi (4040 m), Piso Montano and Estepa Montano Templada. Incidentally, it should be noted that these valleys are drained by major tributaries of the Amazon and La Plata Rivers.

The valleys where the above cities are located have a warmer climate than that of the Subpuna. In Cochabamba, for example, the average annual precipitation during the summer season, October-March, has been 482 mm, with an average annual temperature of 18.1° C (Metereological Data, 1943-1983). In Sucre, on the other hand, the average annual precipitation for the same season has been 646 mm, with an average temperature of 15.3° C. The comparative figures for two other localities in the same region are as follows: Tarija (649.8 mm, 18.6° C) and Santa Cruz (414 m, 1265.7 mm, 24.5° C). The latter city is situated in an ecological zone called Bosque Húmedo.

The Subpuna, with its dry, warm winters, is noted for its steppe-like climate (Map 3, key BSwh). The temperature varies from 25° C in the summer to 5° C in the winter. However, the deep valleys of the Subpuna are warmer and more humid than the upper levels in the summer, and drier in the winter. The climate, therefore, may be termed 'mesothermic' (Map 3, key Cwo). Throughout the greater part of the year, the temperatures of these valleys average 20° C.

The Subpuna Region, or Zona de los Valles (Map 3, key Cwo), is the most important zone in Bolivia from the standpoint of the number and distribution of wild potato species. Herbaceous, shrubby and tuber-bearing plants are more abundant in the valleys than in the upper levels of that zone. Plants common in the higher and dryer sectors of the Subpuna include the bunch grass *ichu* (*Stipa ichu*); the *t'tola* (*Lepidophyllum quadrangulare*); and trees such as *churqui* (*Prosopis alba*) and the common algaroba (*Prosopis juliflora*). Certain wild potato species of the altiplano, such as *S. acaule* and *S. megistacrolobum*, are also found in this region.

Other wild potato species found in the watershed and lower slopes of the

Subpuna are *S. leptophyes* and *S. brevicaule*. These potato species often grow under the thorny branches of algaroba species (*Prosopis* subsp.); or under the protection of such forest trees as *aliso* (*Alnus jorullensis*); or under the branches of the shrubs *Cantua buxifolia* and *C. pendens*; or under the tall, leafy stems of species belonging to such herbaceous plant genera as *Rumex*, *Plantago*, and *Tagetes*.

The wild potato species, *S. oplocense*, which frequently grows under the protective branches of chañar trees (*Gourliea geoffrey* and *G. decorticans*), is common in the vicinity of Cinti (Department Chuquisaca). The latter potato species, which is distributed widely throughout Bolivia, reaches a maximum altitude of 3700 m on Oro Ingenio, near Tupiza in southern Bolivia.

Extensive formations of cacti occur between 2400 and 2700 m of altitude in the semiarid or arid valleys of the Zona de los Valles, which is a transition region between the watersheds of the Subpuna and the Subandean Regions. Plants of the cactus family that occur in the valleys of Arque, Capinota, Ayopaya, and Cochabamba include *ulala* (*Cereus haenkeanum*), *pasacana* (*Trichocereus lamploclorus*), *tuna blanca* (*Opuntia arcei*), *tunilla* (*Opuntia pampeana*), *airampu del valle* (*Opuntia cochabambensis*), and *Cleistocactus buchtienii*.

Among the more important native trees of the warm valleys of central and southern Bolivia are the Peruvian pepper tree (*Schinus molle*), *churqui* (*Prosopis ferox*), *palqui* (*Acacia feddeana*), and *tara* (*Caesalpinia tinctoria*). The ornamental tree *chilijchi* (*Erythrina falcata*) is also found here. This tree, along with a huge variety of other woody plants, herbs and shrubs, is endemic to Cochabamba. Common wild potato species of this zone include *S. sparsipilum*, *S. brevicaule*, *S. berthaultii*, and *S. infundibuliforme*. Other less widely spread species of this region are *S. doddsii*, *S. alandiae*, *S. gandarillasii*, and *S. sucrense*. These wild potatoes all occur between Totora, Aiquile, and Sucre at elevations between 2800 and 2300 m. On the other hand, two other species of limited distribution in this region, *S. vidaurrei* and *S. tarijense*, occur between Sucre and Tarija.

Humid Plains Region

This tropical-savanna zone, which is also called the 'Provincia Amazónica' (Cabrera and Willink, 1973), has dry winters and a rainy season of approximately eight months. The region includes the Beni Plain, the eastern half of Chiquitos, the upper course of the Guapore River (west of the Beni River, Department Pando), and a large part of the Chapare Region (Department Cochabamba). This rainy and very humid forested region has an average annual temperature of 26° C (which varies little yearly), and a short dry season. The minimal temperatures of this region are always above freezing.

Some regions of the Chapare, as in the vicinities of the Mamoré River Valley and eastern Chiquitos, have a tropical humid climate. In these areas, the annual rainfall is more than 3000 mm during the lengthy growing season, which averages about 200 days.

The complex rain forests of the Humid Plains Region is composed of woody species arranged in several strata. These range from trees of 50 m to low palms, shrubs, and herbaceous species. Wild tuber-bearing species of *Solanum* are not found in this very humid and warm valley region.

Dry Plains Region

According to Cabrera and Willink (1973), this area may be termed the 'Provincia Chaqueña.' The plains of this region, which also include some low mountain ranges, extend from southern Bolivia and western Paraguay to Cordoba, San Luis, and Santa Fe in northern Argentina. In Bolivia, this region includes the plateaus of El Chaco and the plains of the Provincia Cordillera in the southern part of the Department of Santa Cruz.

The above steppe region is warm and has dry winters (Map 3, key BSwh). Its climate, which may be termed 'continental,' is hotter than that of the valleys of the Subpuna. Rainfall ranges between 500 and 1200 mm each year, while the average annual temperature varies between 20° and 23° C. It rains almost continuously throughout the year in the eastern part of this zone, while in the western sector it rains only during the summer, from November to March.

The vegetation of the Dry Plains Region is predominantly xerophytic, deciduous forest with an understory of grasses, many terrestrial bromeliads, and cacti. Among the typical woody species are *quebracho blanco* (*Aspidosperma quebracho-blanco*), *quebracho colorado* (*Schinopsis* sp.), *toboroche* (*Chorisia insignis*), various species of algaroba (*Prosopis*), and *luchan* (*Ceiba pentandra*). One of the more important wild potato species of this region is *S. chacoense*.

Summers in the high mountains surrounding the Dry Plains Region of southern Bolivia are more humid than in the Chaco. The winter season, however, is dry, and the region as a whole may be termed 'temperate.' The hottest month of the year averages 22° C, with occasional daily extremes of 38°, or 45° C. The temperature of the coldest month varies between 3° and 18° C. However, temperatures can fall to 2° C when the cold winds (*surazo*) blow in abruptly from the Antarctic.

According to Muñoz Reyes (1977), the mountainous region surrounding the Chaco constitutes a transition zone between the Subpuna of southwest Bolivia and the lower, dry plains of the Chaco and the Santa Cruz area. In the middle and lowermost foothills of the mountains of the Province of Valle Grande are found the wild tuber-bearing species *S. gandarillasii*, *S. neocardenasii*, *S. litusinum*, *S. chacoense*, and *S. tarijense*. In the upper and more humid zones of the cordillera (2000–2800 m), however, one finds *S. microdontum*, *S. candelarianum*, and *S. circaeifolium* var. *capsicibaccatum*.

Additional or more complete general information on the phytogeography, climate, and ecology of Bolivia can be found in works by Buchtien (1910), Fiebrig (1910), Herzog (1923), Asplund (1926), Knoch (1930), Unzueta et al. (1975), and Hueck (1978).

In the following pages, the systematics of the wild and cultivated potato species of Bolivia will be treated.

3

Systematic Treatment: Wild Species

More than 300 tuber-bearing species of the genus *Solanum* have been currently proposed. However, this genus, which was considered by Linnaeus (1753) to be highly polymorphic and very complex, contains presently more than 2400 species distributed throughout the world, with the largest representation in tropical and subtropical regions. The greater part of this group is comprised of herbaceous and shrubby plants, often protected with spines. Some of these are represented by species that produce edible tubers (*S. tuberosum*), while others such as eggplant (*S. melongena*) produce edible fruits, and still others bear poisonous fruits.

Dunal (1852) divided the genus *Solanum* into the following two sections: Pachystemonum, which includes all species lacking spines and having short, thick anthers; and Leptostemonum, which includes all the spiny-leaved and stemmed species with long, narrow anthers. Moreover, the second section was divided by Dunal into the following five subsections: Tuberarium, Morella, Dulcamara, Micranthes, and Lycianthes.

Toward the beginning of the present century, Bitter (1912) elevated Dunal's sections Leptostemonum and Pachystemonum to the rank of subgenera, and his five subsections to the rank of section. Bitter also divided section Tuberarium into two great subsections: Basarthrum and Hyperbasarthrum. Subsection Basarthrum included species having the pedicel articulated near the base, bayonet-type hairs, and pubescent stems. These species lack stolons and tubers. Subsection Hyperbasarthrum, on the other hand, included species that had the pedicel articulated at least some distance from the base. Plants that were included in this group also lacked bayonet hairs, and although most were provided with stolons and tubers, some were observed to be non-tuber-bearing. The subsection was divided further by Bitter into smaller taxonomic groups called series. Although probably of slight taxonomic value, the rank of series has been employed in classifications by Rydberg (1924), Juzepczuk and Bukasov (1929), Bukasov (1937, 1971a, b), and many others to the present day.

Toward the beginning of the last decade, D'Arcy (1972) published what is probably the most extensive revision to date of the genus *Solanum*. His objectives were 1) to bring some order to the large number of previously described subdivisions of the genus, 2) to confirm the validity and typification of the various taxa, and 3) to clarify the relationships of the species and sections.

D'Arcy's work, which follows the latest rules of the *International Code of Botanical Nomenclature*, has been widely consulted by modern-day authorities of the genus. The present work, which follows this treatment, places the tuber-bearing species of *Solanum* in the section Petota (Dumortier, 1827), of the subsection Potatoe (Don, 1838).

Several different taxonomic treatments of the various series of tuber-bearing species of *Solanum* presently exist. Correll (1962), for example, groups 159 species into 26 series, while Hawkes (1963) places approximately the same number in 17 series. Bukasov (1971b), on the other hand, treats 94 species in 32 series.

Classified in this chapter are 31 wild species of the Bolivian tuber-bearing *Solanum*. These plants, following the previous treatment of this group by the author (Ochoa, 1984b, 1985), are grouped into 7 series. In addition, the 7 known cultivated potatoes of Bolivia are treated in a separate chapter.

The author, who has studied herbarium material as well as living collections of the various species maintained in the germplasm banks at the International Potato Center (CIP) in Lima, Peru, and the U.S.D.A. IR-1 Project, at Sturgeon Bay, Wisconsin, has arrived at the following conclusions with regard to the classification of the Bolivian potato series and species:

Series Acaulia: *Solanum acaule*.
Series Circaeifolia: *S. circaeifolium* and *S. soestii*.
Series Commersoniana: *S. berthaultii*, *S. chacoense*, *S. flavoviridens*, *S. litusinum*, *S. tarijense*, and *S. yungasense*.
Series Conicibaccata: *S. bombycinum*, *S. neovavilovii*, and *S. violaceimarmoratum*.
Series Cuneoalata: *S. infundibuliforme*.
Series Megistacroloba: *S. boliviense* and *S. megistacrolobum*.
Series Tuberosa:
Wild Species: *S. achacachense*, *S. alandiae*, *S. brevicaule*, *S. candelarianum*, *S. candolleanum*, *S. doddsii*, *S. gandarillasii*, *S. leptophyes*, *S. microdontum*, *S. neocardenasii*, *S. okadae*, *S. oplocense*, *S. sparsipilum*, *S. sucrense*, *S. vidaurrei*, and *S. virgultorum*.

Cultivated Species: *S. tuberosum subsp. andigena*, *S.* × *ajanhuiri*, *S.* × *curtilobum*, *S.* × *chaucha*, *S.* × *juzepczukii*, *S. phureja*, and *S. stenotomum*.

Key to Series

1. Leaves with strongly winged rachis, or with decurrent leaflets.
    2. Leaves simple or compound, but with the terminal leaflet approximately two to three times the length of the lateral leaflets VI. Series *Megistacroloba* (p. 147).
    2. Leaves compound, with the terminal leaflets equal or slightly larger in size than the laterals .... III. Series *Cuneoalata* (p. 96).
1. Leaves slightly or not at all winged, or occasionally only with decurrent upper leaflets.
    3. Plants acaulescent, not more than 20 cm in height; the leaves forming a prostrate rosette, the floriferous branches axillary and few-flowered; pedicels articulated at or near the base of calyx, lacking a conspicuous abscission ring .... I. Series *Acaulia* (p. 28).
    3. Plants with leafy stems, erect or ascending, usually more than 20 cm in height, pedicels conspicuously articulated.
        4. Fruit ellipsoid-conical to long-conical, the apex obtuse and mucronate.
            5. Plants narrow-stemmed, delicate, up to 1 m high; fruit ellipsoid-elongate or ellipsoid-conical .... IV. Series *Circaeifolia* (p. 105).
            5. Plants thick stemmed, robust, up to 3 m high; fruit ellipsoid .... V. Series *Conicibaccata* (p. 128).
        4. Fruit globose to ovoid, the apex variously tipped.
            6. Plants wild, the corolla stellate or substellate; calyx provided mostly with short apiculate lobes .... II. Series *Commersoniana* (p. 46).
            6. Plants wild or cultivated, the corolla rotate or pentagonal; calyx mostly with long acuminate lobes .... VII. Series *Tuberosa* (p. 182).

I. SERIES ACAULIA

ACAULIA Juzepczuk, Bull. Acad. Sci. U.S.S.R., ser. Biol. 2:316. 1937, *nom. nud.*; ex Bukasov and Kameraz, Bases of Potato Breeding 21. 1959.

Plants small, rosette-forming, stemless or with an extremely abbreviated stem, or occasionally with elongated internodes. Tubers very small, white or milky white, borne at the ends of long stolons. Leaves odd-pinnate, bearing obtuse leaflets. Peduncle very short or absent, few-flowered, strongly recurving, i.e., showing geotrophic tendencies. Pedicel articulation usually absent or marked only by a ring of pigment, the latter occurring near base of calyx. Corolla small, rotate, light purple, blue, white or whitish, with very small lobes. Fruit globose to ovoid, dark green, persistent at maturity, occasionally embedded partially in the soil.

Bukasov and Kameraz (1959), and Lechnovitch (1971) include *S.* × *juzepczukii* and *S.* × *curtilobum* in series Acaulia. However, as these two plants are natural hybrids of *S. acaule* with cultivated species of series Tuberosa, they are perhaps better treated under series Tuberosa rather than Acaulia. The latter classification has already been in use for some time by Hawkes and Hjerting (1969).

The series Acaulia consists of two species: *S. albicans* (Ochoa, 1983a), a hexaploid (2n=72), and *S. acaule*, a tetraploid (2n=48). *Solanum albicans* occurs from northern Peru to Ecuador. This species, discovered and collected for the first time in Peru by the author, will be treated in another volume of this work.

Solanum acaule is widely distributed in the Andes of central and southern Peru, as well as in the highlands of Bolivia and northern Argentina (Brücher 1959a; Hawkes 1956a, 1963; Ochoa, 1962). As interpreted here, this species consists of the following two varieties: *S. acaule* var. *acaule*, and *S. acaule* var. *aemulans*, the latter only from Argentina.

1. Solanum acaule Bitter, Repert. Sp. Nov. 11:391-393. 1912; 12:453. 1913.
FIGS. 1-3; MAP 4; PLATE I.

S. acaule var. *subexinterruptum* Bitt., Rep. Sp. Nov. 11:393-394. 1912. TYPE: *Buchtien 5858*, Bolivia (US).

S. acaule var. *caulescens* Bitt., Rep. Sp. Nov. 12:453-454. 1913. TYPE: *Verne s.n.*, Bolivia. Viacha near La Paz (GOET).

S. punae Juz., Bull. Acad. Sci. U.S.S.R., ser. Biol. 2:316. 1937. TYPE: *Juzepczuk 10414, 10435*, Peru. Cerro de Pasco (LE, OCH, WIR).

S. depexum Juz., Izv. Akad. Nauk U.S.S.R. 2:317-318. 1937. TYPE: *Rosenstiel s.n.*, Argentina. La Peña (LE, plants grown near Leningrad from tubers originally collected by K. von Rosenstiel sent to R. Schick).

S. schreiteri Buk., Sov. Pl. Ind. Rec. 4:13. 1940, *nom. nud.* Based on *Schick s.n.*, Argentina. San Jose (WIR = K-3, plants grown near Leningrad from tubers originally collected by R. Schick in 1932).

S. acaule var. *checcae* Hawkes, Bull. Imp. Bur. Pl. Breed. & Genet., Cambridge 2:22, 115, Fig. 5. 1944. TYPE: *Vargas 2005*, Peru. Checca (K, plants grown in Cambridge, U.K., from seeds originally collected by C. Vargas).

S. depexum var. *chorruense* Hawkes, Bull. Imp. Bur. Pl. Breed. & Genet., Cambridge 2:23, 115. 1944. TYPE: *Balls 6026*, Argentina. Near Tilcara, Province Jujuy (CPC, not seen).

S. acaule var. *punae* (Juz.) Hawkes, Bull. Imp. Bur. Pl. Breed. & Genet., Cambridge 2:23, 115. 1944. TYPE: *Juzepczuk 10414, 10435*, Peru. Cerro de Pasco (LE, OCH, WIR).

S. uyunense Cárd., Bol. Soc. Peruana Bot. 5:33-35, Pl. III, C, Figs. 1-3, Pl. V, A. 1956. TYPE: *Cárdenas 5474*, Bolivia. Between Uyunito and Chocaya (CA, LL).

S. acaule subsp. *punae* (Juz.) Hawkes et Hjerting, Scottish Pl. Breed. Stn. Rec. 117. 1963. TYPE: *Juzepczuk 10414, 10435*, Peru. Cerro de Pasco (LE, OCH, WIR).

Plants small, sparsely and roughly glandular-pubescent, forming rosettes 15-20(-40) cm in diameter, usually semiprostrate or prostrate; normally stemless, or, in open fields, with extremely abbreviated stems, or when shaded by vegetation or walls of cattle corrals with stems 8-20 cm or more tall. Stolons long, thick, occasionally forming secondary plantlets. Tubers small, white, oval, round, or flattened, 1.5-2(-4) cm long. Leaves imparipinnate, (2.5-)6-15(-25) cm long and (1.8-)4-7(-9) cm wide with petioles 3-6(-8) cm long, with a broadened rachis and 0-4 interjected leaflets. Leaflets 5-6 paired, rugose, sessile, and more or less auriculate at base. Terminal leaflets (2-)3.5-4.5(-7.5) cm long and (1.5-)2.5-3.5(-5.5) cm wide, slightly larger to much larger than the laterals, broadly ovate-rhomboid to nearly orbicular, with an obtuse apex and abruptly cuneate base. Lateral leaflets (1.5-)3-3.5(-4) cm long and (0.8-)1.5-2.5(-3) cm wide, ovate-elliptic to broadly ovate-elliptic, the apex rounded to obtuse, the base lobed on the acroscopic side and lobed or variously decurrent on the basiscopic side. Pseudostipular leaves very small and inconspicuous, or absent. Inflorescence with 1-4(-7) flowers; the peduncle very short, 3-7 mm long, originating usually in the crown of the plant near the center of the rosette, or, in the case of taller plants, in the axils of leaves. Pedicels glabrous or sparsely pilose, short and thick, 1-2.5(-4) mm long, the articulation unmarked or only barely visible, or at times marked by a ring of pigment, the articulation when present, situated normally 2-3(-5) mm below base of calyx. Calyx sparsely pilose, 3-4(-6) mm long, the lobes ovate-triangular to lanceolate-subobtuse, acuminate, or subacuminate. Corolla

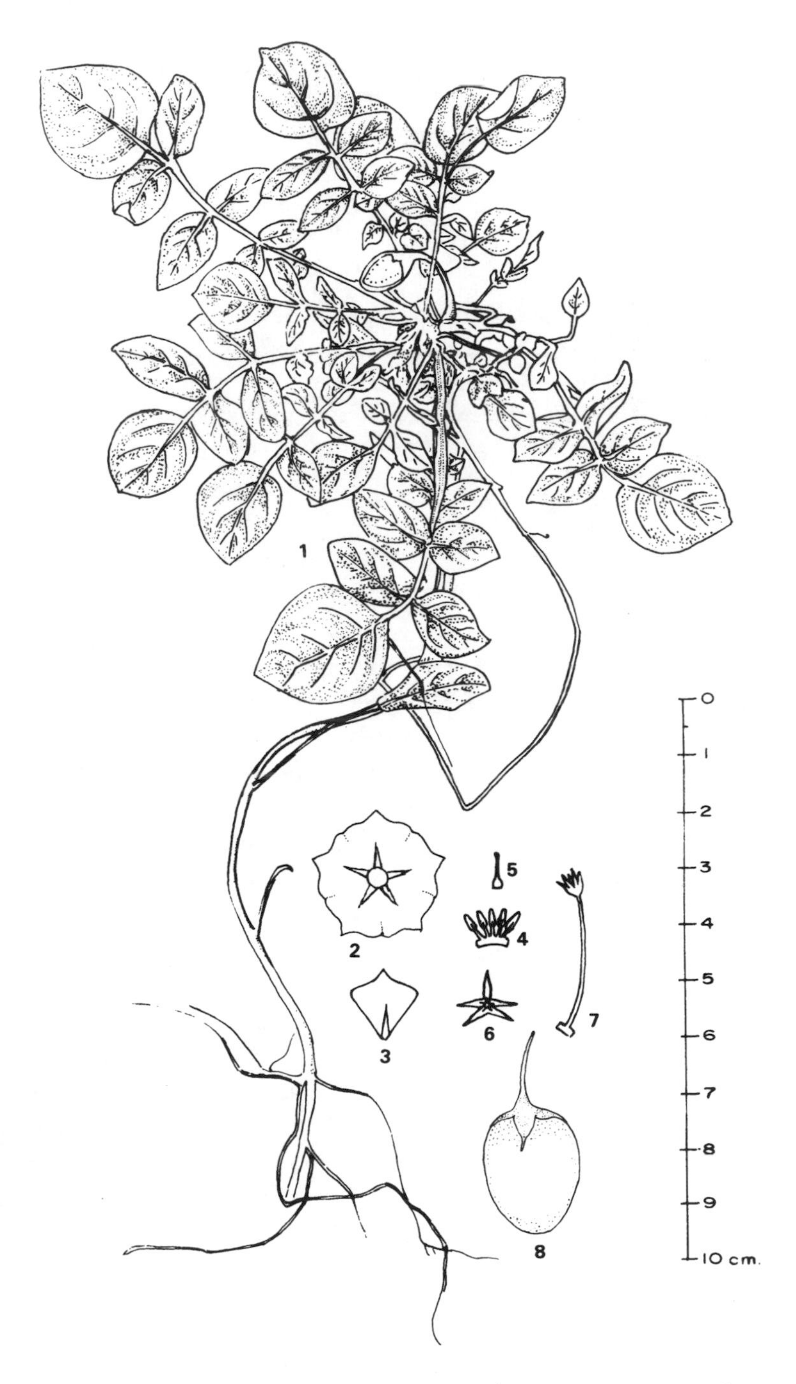

Figure 1. – *Solanum acaule* (*Ochoa 11961*). 1. Plant. 2. Corolla. 3. Petal. 4. Stamens, dorsal view. 5. Pistil. 6. Calyx. 7. Pedicel and calyx. 8. Fruit. All ×1.

rotate, (1-)1.5-2(-2.5) cm in diameter, blue-violet, dark purple, violet to lilac with whitish acumens, or more rarely white-flowered, the lobes extremely short, broadly triangular, 1.4-2.2 mm long. Anthers short, ellipsoid to oblong in dorsal view, 2.5-4 mm long, borne on narrow, (1-)1.5-2.5(-3) mm-long, glabrous filaments. Style (3.5-)4.5-5.5(-7) mm long, thick, papillose on lower half; stigma about 1 mm in diameter, globose to ovoid. Fruit to 2 cm long, usually ovoid or subglobose or occasionally long-conical with pointed apex, green or dark green, often tinged with light purple. Fruits and pedicels with geotropic tendencies, especially in biotypes that grow at high elevations in open fields.

Chromosome number: 2n=4x=48.

Type: BOLIVIA. Department Tarija, Province Aviles, Puna Patanca, 3700 m alt., January 8, 1904; *K. Fiebrig 3429* (lectotype of *S. acaule* subsp. *acaule* S, designated by Hawkes and Hjerting, S, isotypes, F, G, HBG, M, NY, O, SI, W).

Environmental modifications of this species are not unknown. In the high Andes, for example, this species appears as a small, stemless, rosette-forming plant that produces fruits at, near, or slightly below ground level. Frequently, the down-turning peduncles of this species embed the immature fruits in the soil, and thus insure seed production in areas of early frost. At lower elevations in the Andes, the plants of this species tend to have longer stems, as well as leaves or leaflets of different shape or dissection. Similar appearing forms of this species are known from cultivation.

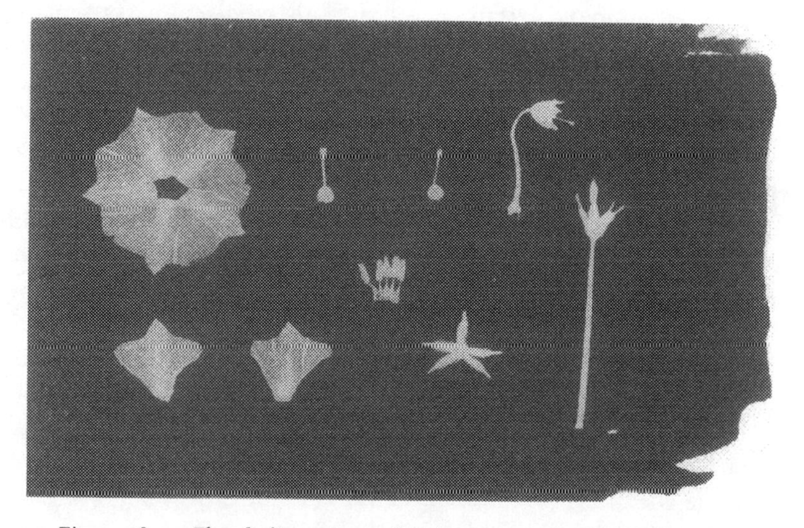

Figure 2. – Floral dissection of *Solanum acaule* (*Ochoa and Salas 11818*), ca. ×1.

One reported case where the growth form of *S. acaule* was modified by culture involved plants grown from tubers in a field at Grenoble, France. The original tubers, which were collected near Viacha, Bolivia (3900 m alt.), by Claude Verne, gave rise to a growth form that strongly resembled *S. acaule* var. *caulescens* Bitter (Bitter, 1913). However, caulescent plants of *S. acaule* are often formed at the lower and warmer altitudinal levels in the Andes. They are also seen at high elevations in the warm, manured soil that occurs along the stone or adobe walls of corrals where animals are penned-up for the night.

The author has observed an enormous range of variation in *S. acaule* in both dried and pressed herbarium specimens of this species and in living collections grown under uniform environmental conditions. These variations involve the color of the flowers, pubescence and pigmentation of the stem, nature of the pedicel articulation, and presence or absence of the interjected leaflets. Despite these major variations, this species may be generally recognized by its basal rosette of leaves, abbreviated peduncles, lateral leaflets that are eared acroscopically, and small rotate corollas. Moreover, all living collections of *S. acaule* that have been examined cytologically to date by the author have proved to be tetraploid (2n=48).

Affinities

Solanum acaule is related to plants in the series Megistacroloba and Tuberosa. Though there are no lack of theories regarding the origin of this plant, its ancestry remains unknown. Rybin (1929, 1933), who was the first to determine that it was tetraploid (2n=48), suggested it may be an amphidiploid, but he was unable to determine its ancestry. Bukasov (1960) also believed that this species was of amphidiploid origin. He suggests that the members of this series evolved by polyploidy from South American diploids. Swaminathan (1954), who made an analysis of the meiotic chromosomes of *S. acaule*, discovered 24 pairs of bivalents, and thus confirmed its amphidiploid origin.

In still an earlier report, Swaminathan (1951) suggested that *S. acaule* may be of hybrid origin. His hypothesis was based on an appraisal of the growth form of this plant and its fertility. He noted that crosses were easily made between artificially doubled forms of *S. acaule* (used as a female parent) and *S. tuberosum* and yielded abundant seed; whereas normal crosses between the two species resulted in no seed. Similar conclusions were obtained by Wangenheim (1954). According to Camadro and Peloquin (1981), these results may be explained by the presence of dominant genes (*CI*) that control an incompatibility reaction between the pistil and pollen grain.

Although *S. acaule* is the South American phytogeographic equivalent (or vicariad) of the Mexican *S. demissum* (Correll, 1962; Hawkes, 1963; Hawkes and Hjerting, 1969), these two heteroploid species appear to have had an independent evolutionary origin.

According to Bukasov (1960) and Ugent (1970, 1981), the series Acaulia and Megistacroloba are closely related. The type species of both series, *S. acaule* and *S. megistacrolobum*, are rosette-like plants having similar geographic distributions and ecological preferences. However, according to Ugent (1970), the maximum correlation of *acaule*-like characters is found in plants derived from natural crosses of *S. megistacrolobum* and *S. canasense*, or of *S. megistacrolobum* and *S. brevicaule*. These natural crosses have resulted in some segregate types bearing a marked resemblance to the tetraploid species *S. acaule*, and others resembling the Peruvian diploid species *S. raphanifolium*. Thus, Ugent (1981) believes that *S. acaule* arose from a complex of natural crosses between these species in a process that included the segregation of *acaule*-like plants, the doubling of the chromosome number, the restoration of fertility, and natural selection.

The diploid species, *S. raphanifolium* (2n=24), is very abundant in the Department of Cusco in southern Peru. Small, *acaule*-like forms of this species are frequently found in nature, with few but often strongly decurrent leaflets. Some of these plants strongly resemble *S. acaule*, while others are similar to *S. megistacrolobum*. Thus, it is not surprising to learn that these segregate forms

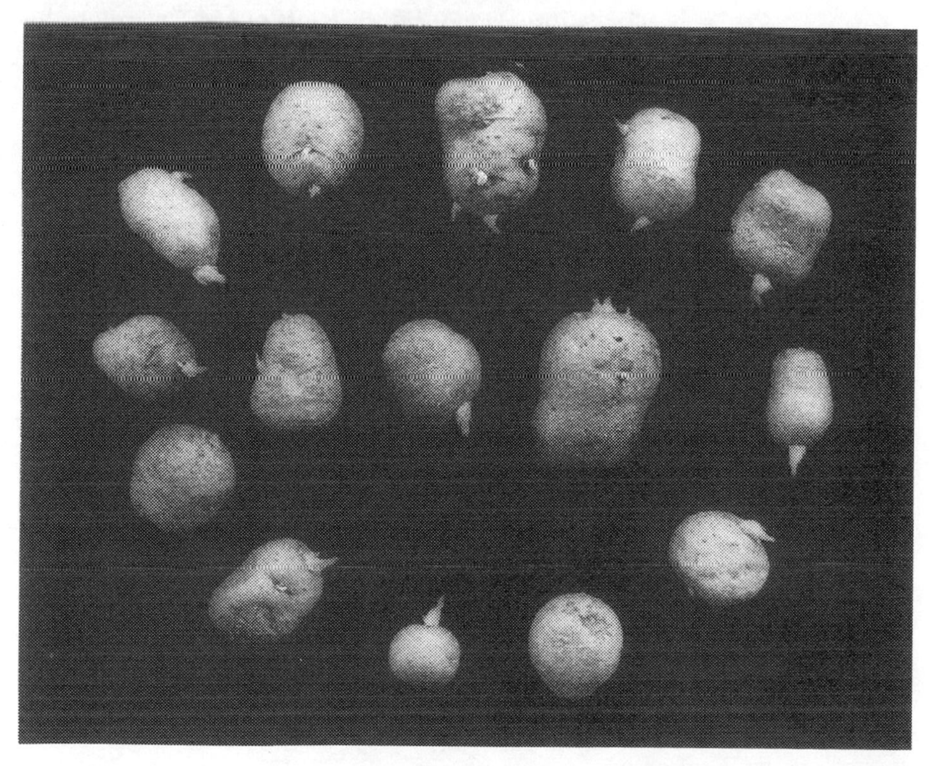

Figure 3. – Tubers of *Solanum acaule* (*Ochoa 12033*), ca. ×1.

and species have been occasionally confused with one another in the herbarium.

Another species that appears to have played an important role in the origin of *S. acaule* is *S. bukasovii* (=*S. canasense*). Both species occur frequently in the same habitats in central and southern Peru. Here, they occur at elevations ranging from 3600 to 3900 m. Occasionally, immature, high altitudinal forms of *S. bukasovii* are confused with *S. acaule*. However, at flowering time, the elongated central stalks of *S. bukasovii* clearly differentiate this plant from the former species.

The diploid, *S. leptophyes*, may also have played a role in the evolution of *S. acaule*. This species, which is of wide distribution in southern Peru, extends from Bolivia into northern Argentina. In Peru, it appears to have the same habitat preferences as *S. bukasovii*.

Solanum acaule, particularly through its var. *acaule*, appears to be related very closely to *S. albicans*. Both have similar leaves, a rosette-like habit, short peduncles, and small, rotate corollas. The pedicel articulation of these species is either present or absent.

Through its var. *aemulans*, *S. acaule* also exhibits many features in common with diploid species of the series Megistacroloba. For this reason, it may be that *S. acaule* var. *aemulans* has differentiated from plants of Megistacroloba by a process that has included hybridization and spontaneous polyploidy. Possible ancestors of *S. acaule* var. *aemulans* include *S. megistacrolobum*, *S. sancta-rosae*, and *S. acaule* var. *acaule*.

A strong relationship also exists between *S. acaule* var. *aemulans* and *S. brucheri*; however, according to Hawkes (1963), Hawkes and Hjerting (1969), and Ugent (1981), this last plant is considered to be merely a triploid hybrid (2n=36) of *S. acaule* and *S. megistacrolobum*. But, here it is necessary to add some clarification regarding *S. brucheri*. According to a personal communication from Heinz Brücher, the epithet *brucheri* was used simultaneously by two different authors. The true *S. brucheri* (or *S. bruecheri* [note orthographic error] in Index Kewensis, Suppl. XIV, pg. 126) corresponds to the species as described by Correll (1961). This was based on a diploid plant collection made by Brücher (No. 557) at Puesto I, Tilcara, Argentina. On the other hand, the name *S. bruecheri* as published by Bukasov (in Bukasov and Kameraz, 1959) represents a *nomen nudum*. It was based on a triploid hybrid of *S. acaule* and *S. megistacrolobum* that was collected by Brücher at Mina Aguilar, Tres Cruces, Argentina.

As mentioned earlier, *S. acaule* also bears a superficial resemblance to *S. demissum*, a wild potato species of the highlands of Mexico and Guatemala. Both plants are small and rosette-like, with diminutive, rotate corollas.

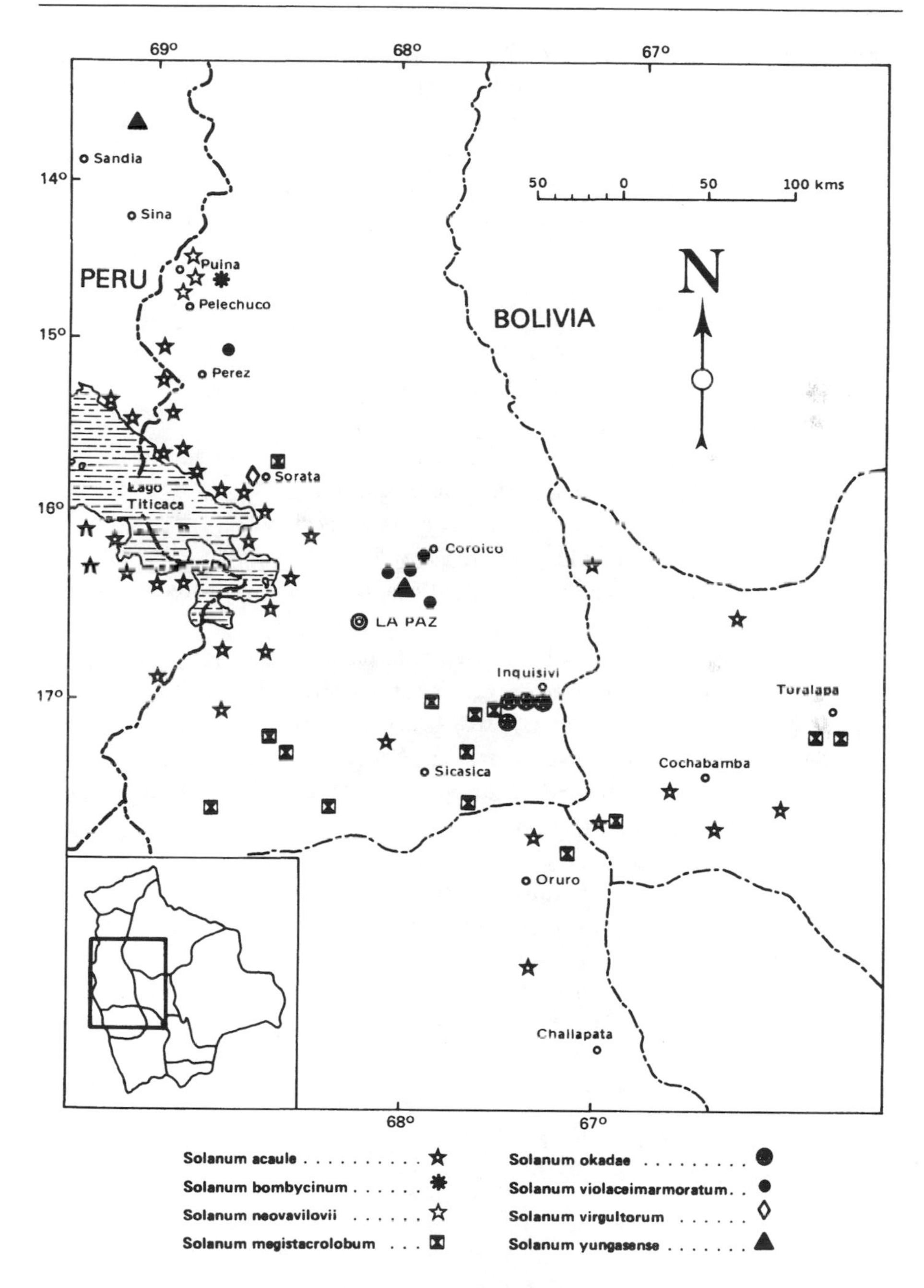

Map 4. – Distribution of wild potato species in the Lake Titicaca region of Bolivia and Peru, showing *Solanum acaule* (★).

Habitat and Distribution

The habitat of *S. acaule* lies in the Puna, a high, cold grassland. Frequently, this species grows at higher elevations than that known for any other wild potato species, reaching, in some instances, an altitude of 4600 m. *Solanum acaule* var. *aemulans*, however, grows only to an altitude of 4000 m. The lower limits of distribution for this species lie, as a whole, between 2600 and 2800 m.

Geographically, the distribution of this species is greater than that known for any other wild potato. This species occurs from the Province of Recuay in the Department of Ancash in northern Peru to the sierras of Famatina in the Province of Rioja in northwest Argentina.

Solanum acaule grows in open fields or between bunches of the grass species *Stipa ichu*, or between rock crevices, or along stone walls, trails, and the borders of cattle corrals. It is also found frequently beneath such nettles as *Cajophora* and *Urtica*, or beneath the spiny stems of cacti, such as *Lobivia*, *Opuntia*, or *Ferocactus*. Many specimens of this species have been collected from cultivated fields located on the altiplano in the vicinity of Lake Titicaca, or in the vicinities of Lake Junin in central Peru, or Lake Tajsara near the Punas of Patanca in Tarija, Bolivia.

This species is highly adapted for growth in pasture lands. Thus, it appears to better withstand the effects of grazing than any other known wild or cultivated potato species. In fact, grazing of the mature fruits by sheep and goats serves merely to disperse the seeds. According to Ugent (1981), seed dispersal by animals has played an important role in the evolution of many diverse forms and species of *Solanum*.

Throughout its range, *S. acaule* also grows in close association with several other plants. In central and southern Peru, for example, it occurs with *S. bukasovii* at altitudes ranging from 3600 to 3900 m. In Bolivia, this species grows with *S. megistacrolobum*, or, occasionally in the more arid areas of the country, with *S. infundibuliforme*. And lastly, on the altiplano of Peru and Bolivia near Puno and Copacabana, this plant grows in close association with *S. sparsipilum* and *S. leptophyes*.

Specimens Examined

Department Cochabamba

Province Arani: Koari, 3550 m alt., 87 km from Cochabamba on the road to Santa Cruz, by wall in enclosed field of Hacienda amongst other tall plants. Long stems and leaves. Berries. January 26, 1971; *Hawkes, Hjerting, Cribb and Huamán 4243* (CIP).

Province Ayopaya: Puente San Miguel, 3800 m alt., 17°10′ S and 66°25′ W, above Liriuni on the Cochabamba-Vizcachas road, about 25 km NNW of Cochabamba (air dist.). Rocky slope with *Solanum brevicaule*, *S. capsici-*

baccatum, S. acaule, Lupinus, Chenopodium, Salvia, Oxalis, Calceolaria, and nontuberous *Solanum* sp. April 9, 1963; *Ugent 4772, 4773, 4774* (WIS).

Province Capinota: Near Parotani, 2450 m alt., 1-2 km after turning to Capinota off the Cochabamba-Oruro road, km 30, under trees in ploughed field. Large narrow leaflets. February 15, 1971; *Hjerting, Cribb and Huamán 4416* (CIP).

Province Chapare: Colomi, about 3650 m alt. [12,000 feet according to the collector], 48 km ENE of Cochabamba, on open slopes. January 9, 1949; *Brooke 3004* (BM, F). Colomi. January, 1949; *Cárdenas s.n.* (HNF, LL).

Province Quillacollo: 28.3 km on road from Quillacollo to Morachata, 4000 m alt., between Janco Cala and Puente San Miguel. Growing in stony ground amongst rocks by roadside. Flat rosettes, with well-dissected leaves. Flowers deep rich purple. January 27, 1971; *Hawkes, Hjerting, Cribb, and Huamán 4271* (CIP).

Province Tapacari: Challa, 3800 m alt., just past Cami on the route Confital-Oruro. Associated with *Stipa ichu* in moist soil. March 2, 1984; *Ochoa and Salas 15569* (CIP, OCH). Between Confital and Japu, 4100 m alt., on the route Confital and Caracollo. March 2, 1984; *Ochoa and Salas 15570* (CIP OCH).

Department Chuquisaca

Province Nor Cinti: Near Padcaya (just north), 3500 m alt., 126 km on road from Potosi to Camargo, on edges of field only. Chromosome number 2n=48. February 1, 1971; *Hawkes, Hjerting, Cribb, and Huamán 4324* (CIP).

Department La Paz

Province Aroma: Between Eucaliptus and Kelkata, 3750 m alt., in dry, rocky soil; common name: *Apharuma.* April 2, 1939; *Balls 6404* (NA, US). Baños de Viscachani, 3800 m alt., 17°10′ S; 67°58′ W, among stones of low wall in bathing enclosure. January 23, 1971; *Hawkes, Hjerting, Cribb, and Huamán 4236* (CIP). Tablachaca, 4100 m alt., between Quime and Eucaliptos, via Colcha along the highway route Troncal-Oruro. Chromosome number 2n=4x=48. March, 1978; *Ochoa 11917-A* (CIP, OCH). Patacamaya, Luribay, 3980 m alt. February 5-April 13, 1980; *Soest, Okada, and Alarcon (S.O.A.)-53* (CIP). Patacamaya, Luribay, 4020 m alt. February 5-April 13, 1980; *Soest, Okada, and Alarcon (S.O.A.)-54* (CIP).

Province Camacho: Ascending from Italaque toward Hualpacayo, 3850 m alt. Chromosome number 2n=4x=48. February 1978; *Ochoa 11899-A* (CIP, OCH). Carabuco, 3820 m alt. Chromosome number 2n=4x=48. March 7, 1978; *Ochoa and Salas 11912* (CIP, OCH).

Province Ingavi: Viacha, 3900 m alt. January 29, 1921; *Asplund 2107* (UPS). Guaqui, 3900 m alt. 1921; *Asplund 5832* (US). Near Tiahuanacu, 3800 m alt. [12,500 feet, according to the collector], in open, sandy soil, between cultivated potatoes. March 29, 1939; *Balls 6339* (US). Viacha, 3800 m alt. From tubers collected originally in Viacha (without date and collector number) and grown in Grenoble, France, by Claude Verne. Corresponds to type collection of *Solanum acaule* var. *caulescens* Bitter (GOET).

Province Inquisivi: Probably in the district of Quime, 1949; *Brooke 3063* p.p. (BM, F). Huañacta or Huañaccota, 4000 m alt., mountain downslope, between Quime and Tres Cruces. February 24, 1984; *Ochoa and Salas 15519* (CIP, OCH). Collected on the mountainous ascent from Quime to Tres Cruces, near Huañacta, 4000 m alt., on a smooth slope. February 24, 1984; *Ochoa and Salas 15520* (CIP, OCH).

Province Larecaja: Near Sorata, April 1946; *Cárdenas 3685* (LL). Cruzpata, 3800 m alt., about 4 km SE of Ancoma, in moist, black soils associated with *Stipa ichu*. February 17, 1984; *Ochoa and Salas 15465* (CIP, OCH). Huancuiri, 4000 m alt., between Ancoma and Paso de Chucho, in an open field associated with *Stipa ichu*. February 17, 1984; *Ochoa and Salas 15466* (CIP, OCH). Near Huancuiri, 4000 m alt., along the route Ancoma to Paso de Chucho, in steppe lands of *Stipa ichu*. February 17, 1984; *Ochoa and Salas 15467, 15468* (CIP, OCH).

Province Loayza: Viloco, 4560 m alt., large tin mine, approximately 160 km from Oruro via Eucaliptos-Cacsata, near Araca, slopes of hills in a cloudy region. March 18, 1949; *Brooke 3031* (BM). Viloco, 4560 m alt., large tin mine, approximately 160 km from Oruro via Eucaliptos-Cacsata, near Araca, slopes of hills in a cloudy region. March 19, 1949; *Brooke 3032* (BM, F). Viloco, 4450 m alt., large tin mine, approximately 160 km from Oruro via Eucaliptos-Cacsata, Araca. March 27, 1950 [possibly 1949]; *Brooke 3035* (BM).

Province Los Andes: On rocky slopes of altiplano along the route La Paz to Huarina, approximately 4000 m alt. February 10, 1960; *Correll, Dodds, Brücher, and Paxman B601* (LL, S, US).

Province Manco Kapac: Cusarata, 3800 m alt., about 5 km from Copacabana on the road to Los Baños del Inca. Chromosome number 2n=2x=48. December 25, 1977; *Ochoa and Salas 11823* (CIP, OCH). Near Rio Lacko, 3800 m alt., between Copacabana and Casani. Chromosome number 2n=2x=48. December 25, 1977; *Ochoa and Salas 11824* (CIP, OCH). Near the banks of the Rio Lacko, 3800 m alt., between Copacabana and Casani. Chromosome number 2n=2x=48. December 25, 1977; *Ochoa and Salas 11825* (CIP, OCH). Vicinity of Copacabana, 3820 m alt. Chromosome number 2n=2x=48. December 25, 1977; *Ochoa and Salas 11826* (CIP, OCH). Near Tiquina, 3825 m alt. Chromosome number 2n=2x=48. February 27, 1978; *Ochoa and Salas 11895* (CIP, OCH, US).

Province Murillo: La Paz, approximately 3700 m alt. January 26, 1921; *Asplund 2048, 2049* (LL, UPS). La Paz, 3900 m alt. January 26, 1921; *Asplund 5835* (A.2148) (US). Palca, 4000 m alt. January, 1909; *Pflanz 95* (F). [This was distributed erroneously as a type of *S. acaule* by the Museum of Natural History, Chicago.]

Province Omasuyos: Huatajata, 3900 m alt., on the SE shore of Lake Titicaca, rocky shore of the lake. May 19, 1950; *Brooke 6380* (BM). Near Pucarani, flowers white. February 1946; *Cárdenas 3684* (LL). Puna, 3900–4000 m alt. December-April, 1857; *Mandon 401* (GOET; P, drawing). Near Achacachi, 3800 m alt. Chromosome number 2n=4x=48. December 23, 1977; *Ochoa and Salas 11818* (CIP, OCH). Coromata, 3950 m alt. Chromosome number 2n=4x=48. December 25, 1977; *Ochoa and Salas 11829, 11830, 11831* (CIP, OCH). Coromata, 3950 m alt. Chromosome number 2n=4x=48. December 26,1977; *Ochoa and Salas 11832* (CIP, OCH). Chachacomani, 3950 m alt., between Coromata and Huarina. Chromosome number 2n=4x=48. December 26, 1977; *Ochoa and Salas 11833* (CIP, OCH). Between Achacachi and Sorata, 4000 m alt. Chromosome number 2n=4x=48. March 7, 1978; *Ochoa and Salas 11910* (CIP, F, OCH, US). Near the summit of the pass between Achacachi and Sorata, 4000 m alt. Chromosome number 2n=4x=48. March 7, 1978; *Ochoa and Salas 11910-A* (CIP, OCH). Parahua, 3700 m alt., vicinity of Huarisata, in moist soil in an open field. February 20, 1984; *Ochoa and Salas 15476* (CIP, OCH).

Province Pacajes: Corocoro, 4000 m alt. February 15, 1921; *Asplund 2436, 2437* (UPS). Iracollo, 3870 m alt., common name *Apharu*, on open plateau, between pastures. May, 1955; *Ochoa CO-2414* (OCH). Rosario, 3800 m alt., on sandy soil, flowers purple, 1920–1921; *Shepard 231* (GH, US).

Province Sud Yungas: Chulumani. November, 1948; *Bridarolli 4322* (LP). In the vicinity of a small lagoon E of La Cumbre, along the highway from La Paz to Unduavi, approximately 4500 m alt. Flowers purple. February 10, 1960; *Correll, Dodds, Brücher, and Paxman B600* (LL, NY).

Province Franz Tamayo: Huaylapuquio, 4200 m alt., descending from Catantika toward Pelechuco. Chromosome number 2n=4x=48. February, 1983; *Ochoa and Salas 14945, 14946, 14947, 14948, 14949* (CIP, OCH). Viscachani, 4000 m alt., about 3 km from Huaylapuquio on road to Pelechuco. Chromosome number 2n=4x=48. February, 1983; *Ochoa and Salas 14950, 14951, 14952, 14953, 14954* (CIP, OCH). Chuncho Apacheta, 4000 m alt., along the road from Pelechuco to Amajara and Queara. February, 1983; *Ochoa and Salas 14960* (CIP, OCH).

Province unknown: On the altiplano, 4100 m alt. [without precise locality]. March, 1910; *Buchtien 5858* (US, type collection of *S. acaule* var. *subexinterruptum*).

Department Oruro

Province Cercado: Dry, rocky slope, Oruro suburb, 4030 m alt. February 1, 1949; *Brooke 3014 p.p.* (BM). Aguas de Castilla, 4000 m alt. [possibly Aguas de

Castilla, of the Province Quijarro, Potosi, C.O.], among rocks near stream. February 1, 1949; *Brooke 3016, 3019* (BM). Hacienda Huancaroma, 3800 m alt., near Eucaliptos. February 19-24, 1934; *Hammarlund 128* (S). Near Machacamarca, 3700 m alt., between Oruro and Patacamaya. Chromosome number 2n=4x=48. March, 1978; *Ochoa 12033* (CIP, F, OCH, US).

Department Potosi

Province Frias: Potosi, 4400 m alt., in railway yards, in loose stony soil or clinkers. March 10, 1939; *Balls 6201* (K, US). Potosi, 4000 m alt., on sand. May, 1932; *Cárdenas 185* (GH). Potosi, 4000 m alt., in wet soil, without tubers. February, 1933; *Cárdenas 606* (US). Near Potosi. February, 1949; *Cárdenas s.n.* (LL). Laguna de Potosi, Cerro de Potosi. March [without exact date]; *D'Orbigny 1494* (P, W). Estación de Agua Dulce, 4100 m alt. February 27, 1953; *Hjerting, Petersen, and Reche 1035* (OCH). Potosi, 3945 m alt., 19°35′ S, 65°45′ W, just N of the city. Rocky banks of a cool stream, with *Solanum acaule, Stipa, Erodium, Nicotiana,* and *Senecio.* April 30, 1963; *Ugent and Vidaurrei 5153* (WIS), *5154* (WIS, OCH), *5156* (WIS), *5163-64* (OCH, WIS), *5165* (WIS), *5166* (WIS, US), *5167-68* (F, WIS), and *5170* (WIS).

Province Linares: Puna. March 11, 1892; Herb. *Kuntze s.n.* (NY, US).

Province Quijarro: On the road between Uyunito and Chocaya, 3800 m alt. On sandy, humid crevices. March, 1952; *Cárdenas 5474* (HNF, LL, type collection of *S. uyunense.* However, in CA the specimen representing the '*Typus*' of *S. uyunense* is identified as *Cárdenas 5074*). Among piles of rocks on the altiplano about 25 km from Potosi on road to Camargo, 4000 m alt. Flowers purple, rotate-pentagonal. February 20, 1960; *Correll, Dodds, Brücher, and Paxman B636* (K, LL, MO, UC).

Province Sud Chichas: Paso de Machu Cruz, 4250 m alt., Cordillera de Mochara, between Impora and Tupiza, associated with *Solanum megistacrolobum* and near *S. infundibuliforme,* in poor, sandy soil. Chromosome number 2n=4x=48. March 17, 1978; *Ochoa 11961* (CIP, OCH). Descending from Paso de Machu Cruz toward Tupiza, 3800 m alt., along the road from Impora to Tupiza. Chromosome number 2n=4x=48. March 17, 1978; *Ochoa 11965* (CIP, OCH).

Department Tarija

Province Aviles: Puna Patanca. January 8, 1904; *Fiebrig 3429* (G, M, NY, O, S, SI, U, W, Z, type collection of *S. acaule*). San Miguel, 3840 m alt., a few km from Yunchara, along the road from Tojo to Iscayachi. March 19, 1978; *Ochoa 11979* (CIP, F, OCH, US). San Miguel, 3840 m alt., near Yunchara along the road from Tojo to Iscayachi, growing along walls of sheep corrals, in loose, fertilized soil. Plants and leaves well developed, without flowers. March 19, 1978; *Ochoa 11980* (CIP, OCH, US). Between Cerro Tajsara and Patanca,

3700 m alt., associated with *Solanum megistacrolobum*, on grassy plains. Chromosome number 2n=4x=48. March 19, 1978; *Ochoa 11982* (CIP, OCH). Punas de Patanca, 3750 m alt., 14 km SE of Copacabana, associated with *S. megistacrolobum*, among large bunches of *Stipa ichu*. Chromosome number 2n=4x=48. March 20, 1978; *Ochoa 11983* (CIP, OCH, topotypes). Punas de Patanca, 3780 m alt., 14 km SE of Copacabana, associated with *S. megistacrolobum* and *Stipa ichu*. Chromosome number 2n=4x=48. March 21, 1978; *Ochoa 11984* (CIP, OCH, topotypes). Punas de Patanca, 3780 m alt., 10 km NE of Copacabana, associated with *S. megistacrolobum* and *Stipa ichu*. Chromosome number 2n=4x=48. March 21, 1978; *Ochoa 11986* (CIP, OCH, topotypes). Five km W of Punas de Patanca, 3720 m alt., along the road to Lagunas de Tajsara. Chromosome number 2n=4x=48. March 21, 1978; *Ochoa 11989* (CIP, OCH). Six km W of Punas de Patanca, 3700 m alt., along the road to the Lagunas de Tajsara. Chromosome number 2n=4x=48. March 21, 1978; *Ochoa 11990* (CIP, OCH).

Province Mendez: Iscayachi, about walls and on rock piles. Flowers purplish. February 22, 1960; *Correll, Dodds, Brücher, and Paxman B653* (LL).

Crossability

Previous reports.– *Solanum acaule* represents a valuable resource for genetic programs aimed at the improvement of the cultivated potato. This species, native of the high Andes (3400 and 4000 m), is well able to withstand the sudden changes of temperature that occur frequently in the mountains during flowering season. Along with the hexaploid *S. albicans*, a related species from northern Peru and Ecuador that bears a rosette habit similar to that of *S. acaule*, this plant can be used for the improvement of frost resistance in the cultivated potato. It should be noted that the leaves of this species are little damaged by a temporary blanket of snow or hail.

Both *S. acaule* and *S. albicans* are self-compatible. Moreover, both are frequently self-pollinated in the bud.

The unusually strong resistance of *S. acaule* to frost is a physiological trait of this plant that has been much discussed by Bukasov (1932, 1933, 1939), Dremliug (1937), Stelzner (1943), Ochoa (1951), Mastenbroek (1956), Firbas and Ross (1961), Blomquist and Lauer (1962), Ross (1966), Dearborn (1969), Estrada (1980), and Ugent (1981). Recently, Estrada (1984) has obtained a fertile, tetraploid hybrid line, *Acaphu*, from a cross between *S. acaule* and the cultivated species *S. phureja*, that retains the best characteristics of each. This selection is not only self-fertile and cross-compatible with tetraploid cultivars of the Andean potato, but is also resistant to frost and such bacterial diseases as *Pseudomonas solanacearum* or fungal as *Phytophthora infestans*, as well as diseases of viral origin and resistance to the nematode *Globodera pallida*.

In the investigations by Hermsen and Ramana (1969), hybrids of *S. acaule* ×

S. bulbocastanum were compared with triploid F_1 populations of *S. acaule* × haploids of *S. tuberosum*. Some favorable results were obtained among these various chromosomal segregates (3x, 4x, and 6x) for resistance to *P. infestans*, the potato late blight disease.

Solanum acaule is reported to be resistant to potato virus X (PVX) (Ross, 1954). Recently, still other studies performed at CIP have uncovered clones of *S. acaule* that are also resistant to potato leafroll virus (PLRV) (Brown et al., 1984a); whereas still other investigations performed at this institute have shown that some other clones of *S. acaule* are resistant to potato viruses PVY and PVX-cp. as well as to potato spindle tuber viroid (PSTV), the latter being responsible for producing fusiform tubers (Intl. Pot. Ctr., 1982, 1983; Brown et al., 1984b).

Various other valuable traits have been discovered in *S. acaule*, including resistance to *G. pallida*, *G. rostochiensis* (Schmiediche, 1977b), *P. infestans* (Intl. Pot. Ctr., 1983), and *P. solanacearum* (Schmiediche, personal communication, 1983). This species also shows resistance to *Premnotrypes suturicallus* (Intl. Pot. Ctr., 1980), *Myzus persicae*, and *Empoasca fabae* (Intl. Pot. Ctr., 1978).

Solanum acaule has also been artificially crossed with numerous wild and cultivated species. In the horticultural or plant improvement literature of Europe, America, and the Soviet Union, there are many reports of hybrids of *S. acaule* with other Andean wild potato species, such as *S. bukasovii*, *S. brevicaule*, *S. multidissectum*, *S. megistacrolobum*, *S. microdontum*, *S. sparsipilum*, *S. goniocalyx*, *S. phureja*, *S. stenotomum*, and *S. tuberosum*. However, crosses have also been made between *S. acaule* and Mexican species such as *S. bulbocastanum*, *S. cardiophyllum*, *S. demissum*, and *S. pinnatissectum*.

According to Hawkes and Hjerting (1969), some collections that were determined by Correll (1962) as *S. brucheri* are really natural hybrids of *S. acaule* and *S. infundibuliforme*. The same authors also report that *S. acaule* hybridizes naturally with *S. megistacrolobum* and *S. spegazzinii* in northern Argentina.

Hawkes and Hjerting (1969) also report that they have obtained triploid hybrids (2n=36) from experimental crosses of *S. acaule* with two sympatric species, *S. sancta-rosae* and *S. megistacrolobum*. Triploids were also obtained from crosses of *S. acaule* var. *aemulans* and *S. kurtzianum*. Crosses between the latter two taxa yielded fertile and vigorous F_1 and F_2 populations.

Okada (1973), who had obtained experimental crosses of *S. acaule* with *S. megistacrolobum*, reports in still another publication (1979) that *S. acaule* subsp. *acaule* forms natural hybrids with *S. megistacrolobum* and *S. infundibuliforme* in northwestern Argentina. This same author also believes that plants of *S. acaule* subsp. *aemulans* from Tilcara, Argentina, have arisen from natural crosses of *S. acaule* with *S. megistacrolobum* via the union of a normally reduced gamete from *S. acaule* with a nonreduced one from *S. megistacrolobum*. Okada (1979), on the other hand, reports that plants of *S. acaule* subsp. *aemulans* from the Sierra de Famatina have probably had a different origin. With respect to this variety, the present author believes that its origin may have involved *S. sancta-rosae*.

Experimental work at CIP.– Numerous experimental crosses of tetraploid *S. acaule* (2n=48) were made in connection with the present work. These involved the species *S. alandiae* (2x), *S. bukasovii* (2x), *S. brevicaule* (2x), *S. soukupii* (2x), *S. megistacrolobum* (2x), *S. raphanifolium* (2x), *S. sparsipilum* (2x), *S. tarijense* (2x), and *S. litusium* (2x). With the exception of *S. acaule* (11986) × *S. litusinum* (12027), which developed parthenocarpic fruits, the remaining crosses bore abundant fruits, containing viable seeds. Moreover, crosses between *S. acaule* and certain Mexican species (*S. demissum* (6x), *S. stoloniferum* (4x), and *S. polytrichon* (4x)) yielded viable and fairly abundant seed. The results of these investigations are summarized in the following sections.

Solanum acaule × *S. tarijense*.– When used as a mother plant in crosses with *S. tarijense*, *S. acaule* bore abundant fruits that contained an average of 100 seeds per berry. The hybrids of this cross developed very vigorous, sparsely pilose plants up to 50 cm high, with dark purple stems up to 12 mm in diameter at the rosette-like base, and broadly decurrent leaflets. In general, these hybrids produced long, narrow leaves, with 3–4 pairs of very narrowly decurrent leaflets and 2–6 interjected leaflets. Additional characteristics of these plants included inflorescences with peduncles up to 7 cm long and 6–8 flowers, pedicels clearly articulated near the base of the calyx, and small, light blue, purple, or lilac corollas, similar in form to those of *S. acaule*. In summary, the dominant traits of *S. acaule* that were manifest in these hybrids included a rosette-like habit, and similar leaves, leaflets, and corolla. The influence of *S. tarijense* shows up primarily in the vigorous habit of the plants, in the uniform presence of interjected leaflets, and in the development of a longer floral peduncle with clearly visible articulation. In some combinations involving these two species, very different results were obtained. Thus, plants derived from crosses between *S. acaule* (14947) and *S. tarijense* (11994) were similar to *S. tarijense* in having white flowers, as well as light green and densely pubescent leaves. However, these plants varied in habit from rosette-like or sub-rosette-like to simple-stemmed and erect. The small, rotate corolla varied in color from light blue, to sky blue, to white. These were borne upon 45 mm long pedicels that were articulated 5 mm below the base of the calyx. The long, narrow leaves of the hybrids had 5–7 interjected leaflets and up to 6 pairs of elliptic-lanceolate, sessile to subsessile lateral leaflets, which were nondecurrent upon the rachis and had rounded bases and pointed apices. It should be noted that the reciprocal cross, *S. tarijense* × *S. acaule*, yielded only parthenocarpic berries.

Solanum acaule × *S. brevicaule*.– Crosses between *S. acaule* (13865) and *S. brevicaule* (11934) yielded abundant, fertile seed. Plants that developed from these crosses were 40 cm tall, very compact, with branched and sparsely pilose, pigmented stems 6–8 cm in diameter. While resembling *S. brevicaule* in leaf and leaflet shape, these hybrids had 4–5 pairs of sessile, narrowly decurrent leaflets, and were provided with only few interjected leaflets. Moreover, while the inflorescences and small, light blue to dark purple, rotate flowers of these hybrids were similar to *S. acaule*, the articulation was clearly visible and occurred in the upper one-third of the stalk.

Solanum acaule × *S. soukupii*.– Hybrids of *S. acaule* (11602) and *S. soukupii* (13543) also yielded abundant, viable seed. Although the leaves of these hybrids were very similar to those of *S. soukupii*, the plants were more branched and rosette-like. The hybrids had elliptic-lanceolate leaflets with rounded, sessile or subsessile bases, and pointed apices. The inflorescence and floral details of these plants, on the other hand, were typical of those of *S. acaule*. In all cases, the articulation occurred 4–5 mm below the calyx, and was clearly visible.

Solanum acaule × *S. megistacrolobum*.– Segregates of the cross *S. acaule* (8611) × *S. megistacrolobum* (14272) yielded uniformly small and rosette-like or sub-rosette-like populations of plants with long, narrow leaves, these appearing more similar to *S. acaule* than to *S. megistacrolobum*. Some hybrid individuals of this population were provided with leaves having 3–4 pairs of lateral leaflets, having the upper one or two pairs broadly decurrent about the rachis. In these plants, the terminal leaflet was much larger than the laterals, but not as exaggerated as in the case of *S. megistacrolobum*. All lacked interjected leaflets. Other plants of this same F_1 hybrid population had small, poorly dissected leaves and small leaflets, these strongly decurrent upon the cuneate leaf base, as in the case of *S. infundibuliforme*. However, some dominant characteristics of these hybrids included their rotate corollas, which were of the same size or slightly larger than those of *S. acaule*, and their short, 3 mm-long anthers borne on narrow, 6 mm-long filaments. In all cases, the pedicel articulation of these hybrids was clearly visible and situated at 4–5 mm below the calyx. All chromosome counts yielded $2n=3x=36$. The above characteristics of these plants may be of value in detecting natural hybrids of *S. acaule* × *S. megistacrolobum*.

Solanum acaule × *S. raphanifolium*.– Crosses between *S. acaule* (13865) and *S. raphanifolium* (12040) were not difficult to obtain. These retained the small rotate corolla of *S. acaule*, and the size and shape of the terminal leaflets of *S. raphanifolium*. These hybrids were 25–40 cm tall, and had branched, pigmented stems. Their dark-green leaves lacked interjected leaflets and were provided with 3–4 pairs of lateral leaflets. The broadly elliptic, rhomboid-lanceolate terminal leaflets of these plants were much larger than the laterals. These had rounded leaf bases that were broadly decurrent on the rachis. All had a 6–8 flowered, paniculate inflorescence. The clearly visible pedicel articulation of these plants occurred 4-5 mm below calyx.

Solanum acaule × *S. sparsipilum*.– Hybrids of the cross *S. acaule* (13865) × *S. sparsipilum* (13564) developed 25-50 cm-tall stems. These, for the most part, had well developed, cymose-paniculate inflorescences that bore abundant small, blue or purple flowers. The pedicels of these plants were clearly articulated 20-30 mm above their base, or 4-5 mm below the point of attachment to calyx. Dominant in these hybrids was the purplish pigmentation of the stems, peduncles, calyces, and leaf rachises. All had light-green pedicels and leaves similar in form and dissection to *S. acaule*. However, they differed in having the upper lateral leaflets conspicuously decurrent on the rachis. In summary, the leaves and flowers of these hybrids were similar to those of *S. acaule*, while the inflorescences and general habit of these plants were like *S. sparsipilum*.

Solanum acaule × *S. bukasovii*.– Hybrids between *S. acaule* (14391, 14402) and the Peruvian *S. bukasovii* (13862, 13859) are sterile. Chromosome counts of these hybrids suggest they are triploid (2n=36). In appearance, the hybrids of this cross are mostly intermediate between their parental species. They range between 20 and 40 cm tall, and have short internodes along the lower third of their pigmented, shortly pilose, simple or few-branched stems and are subrosette-like in habit. The leaves of these plants are similar in appearance to those of *S. acaule*, as are the flowers. The latter, while of different shade of blue or light purple, are like those of *S. acaule* in being small and rotate. The 30 mm-long flowering pedicels of these plants are clearly articulated 2-3 mm below calyx. In crosses of *S. acaule* (14402), with still another introduction of *S. bukasovii* (13166), the resultant plants are more similar to *S. bukasovii* than in the previously described combination. Thus, these hybrids, which are provided with highly dissected leaves, have up to 7 pairs of lateral leaflets and numerous interjected leaflets. Moreover, these hybrids have long peduncled, cymose-paniculate inflorescences that bear many flowers. However, their small rotate corollas are similar to those of *S. acaule*. All have pedicels clearly articulated near the base of calyx.

In conclusion, the vast majority of the above hybrids of *S. acaule* var. *acaule*, if not all, have the small, rotate, corolla, which is so characteristic of that species. In all of these crosses, the pedicels of the hybrids are articulated very clearly near the base of calyx.

Finally, the relative ease of crossing of *S. acaule*, as observed in these and other hybrid combinations, can be explained in part by a recent hypothesis that deals with the numerical chromosomic balance of the endosperm, or EBN (Johnston and Hanneman, 1982). For *S. acaule* this value is EBN=2. This means that while this species is a tetraploid it can probably be easily crossed with diploid species having the same endosperm numerical balance.

II. SERIES COMMERSONIANA

COMMERSONIANA Bukasov, Bull. Acad. Sci. U.S.S.R., ser. Biol. 2:714. 1938, *nom. nud.*; ex Bukasov and Kameraz, Bases of Potato Breeding 19. 1959.

Glabrescentia Buk., Problemý Bot. 2. 1955, *nom. nud.*; ex Buk. and Kameraz, Bases of Potato Breeding 19. 1959.
Tarijensa Corr., Tex. Res. Found. Contrib. 4:233. 1962.
Yungasensa Corr., Tex. Res. Found. Contrib. 4:220–422. 1962.
Berthaultiana Buk., Bull. Appl. Bot. Genet. & Pl. Breed. 46(1):24–26. 1971, *nom. nud.*.

Plants erect to erect-spreading up to 1.2 m high. Leaves imparipinnate, usually with interjected leaflets. Leaflets typically ovate to elliptic, obtuse or pointed, rarely lanceolate and acuminate. Pedicels articulated near midpoint. Calyx usually with short apiculate lobes. Corolla stellate, pure white or tinged with light blue, or more rarely, lilac to purple-violet, the broad, or rarely narrowly triangular to ovate triangular lobes being twice as long as their width. Fruit globose to ovate-ellipsoid, light green, frequently mottled or speckled with white.

Chromosome number: $2n=2x=24$ and $2n=3x=36$.

A number of fairly recent treatments of this series are available (Hawkes, 1956a; Correll, 1962; Brücher, 1964; Hawkes and Hjerting, 1969). The one that is used here, however, is based largely on the older, classical work of Bukasov (1938) and Bukasov and Kameraz (1959).

Habitat and Distribution

The species of this series extend from Peru, Bolivia, and Argentina to Brazil, Paraguay, and Uruguay. They grow from nearly sea level (50 m) to 2300 m in exceptional circumstances. Plants of this series prefer the low plains of eastern South America.

Key to Species

1. Corolla purple or lilac, rarely whitish, as in the case of *S. berthaultii*.
    2. Corolla pentagonal or substellate with broad, triangular lobes 8–12 mm long, 7–15 mm wide; leaves with 4–5(–6) pairs of leaflets and 26 or more interjected leaflets .... 2. *S. berthaultii*.
    2. Corolla stellate, the narrow lanceolate lobes 15–17 mm long, 7–9 mm wide; leaves with 2–3 pairs of leaflets, and few or no interjected leaflets .... 5. *S. litusinum*.
1. Corolla white, creamy white, or yellowish as in the case of *S. yungasense*.

3. Corolla pentagonal or rarely substellate; leaves densely glandular pubescent, with 3-4(-5) pairs of leaflets .... 4. *S. flavoviridens.*
3. Corolla stellate, the lobes narrow or broadly triangular.
   4. Leaves with 4-5(-6-7) pairs of leaflets, glabrous or subglabrous; calyx small, 4.5-5.5 mm long.
      5. Corolla white, the lobes 8-15 mm long, 5-12 mm wide; the calyx generally glabrous, with apiculate lobes .... 3. *S. chacoense.*
      5. Corolla creamy white, the lobes 7-9(-11) mm long and 4-5 mm wide; the calyx rarely glandular pubescent and with short acuminate lobes 7. *S. yungasense.*
   4. Leaves with (3-)4-5 pairs of leaflets, glandular pubescent; the calyx 8-12 mm long .... 6. *S. tarijense.*

2. Solanum berthaultii Hawkes, Bull. Imp. Bur. Pl. Breed. & Genet., Cambridge (June):45, 122. 1944. FIGS. 4-8; PLATE II.

S. vallegrandense Córd., Bol. Soc. Peruana Bot. 5(1-3):23-24, Pl. II (B), Figs. 1-2. 1956. TYPE: *Cárdenas 5070*, Bolivia. Between Trigal and Mataral, Dept. Santa Cruz, Prov. Valle Grande, March, 1955 (CA, LL).

S. vallegrandense Cárd. var. *pojoense* Cárd., Bol. Soc. Peruana Bot. 5(1-3):24, Pl. II (C), Figs. 1-3. 1956. TYPE: *Cárdenas 5071*, Bolivia. Above Pojo, Dept. Cochabama, Prov. Carrasco, March, 1955 (CA, LL).

S. tarijense Hawkes var. *pojoense* (Cárd.) Corr., Wrightia 2:173. 1961. TYPE: *Cárdenas 5071*, Bolivia. Above Pojo, Dept. Cochabamba, Prov. Carrasco, March, 1955 (CA, LL).

Plants robust, up to 1 m or more tall, densely pubescent with simple, glandular hairs; stem winged, erect, somewhat flexible, simple or branched, 10-20 mm thick at base. Stolons 1.5 m long; tubers ovoid to ovoid-compressed, large, 4-5(-8) cm long, light brown or brownish yellow. Leaves light green, imparipinnate, smooth, and densely glandular pubescent, (10-12.5-)18-22(-30.5) cm long by (3.5-8.5-)11-15(-18) cm wide, the shortly pilose rachis provided with 26 or more interjected leaflets of various sizes and borne on petioles, (1.5-)2.5-3.5(-4.5) cm long. Lateral leaflets with undulate or crenulate margins, 4-5(-6) paired, narrowly elliptic-lanceolate at the pointed tip and strongly asymmetric at base, the petiolules sometimes with secondary interjected leaflets. Terminal leaflet of slightly larger size than the upper laterals, broadly elliptic-lanceolate, (4-)6-7.5(-10) cm long by (2-)3-4(-5) cm wide, the apex pointed to subacuminate and the base subcordate. Pseudostipular leaves eared, broadly falcate to subfalcate, 25 mm or more long by 10-12 mm wide. Inflorescence 5-15-flowered, cymose or cymose-paniculate, borne on peduncles, (4-5-)7-12(-16) cm long. Pedicels shortly pilose, articulate about

Figure 4. – *Solanum berthaultii*, topotype (*Ochoa 668*). 1. Upper part of flowering plant. 2. Corolla. 3. Petal. 4. Stamen, dorsal view. 5. Pistil. 6. Calyx. 7. Pedicel and anthers. 8. Fruit. All ×½.

Figure 5. – Living collection of *Solanum berthaultii* (*Ochoa and Salas 15599*).

the center or upper one-third, the lower part to 50 mm or more long and the upper 7-9(-13) mm long. Calyx shortly pilose, 7-10 mm long, the lobes elliptic-lanceolate and abruptly narrowed, with acumens 1.5 mm long. Corolla pentagonal or substellate, 2.5-3.5 cm in diameter, purple or lilac, or occasionally whitish, with prominent, broadly triangular lobes, 8-12 mm long and 7-15 mm wide, with grayish-yellow nectar guides. Anthers lanceolate, 6.5-7 mm long, cordate at base, borne on glabrous, 1-1.8 mm-long filaments. Style curved, 12-15 mm long, exerted 6-7 mm, glabrous or rarely papillose along the lower one-third of its length; stigma ovoid, recurved at tip. Fruit green, globose to ovoid, 2-2.5 cm in diameter, white-speckled.

Chromosome number: 2n=2x=24.

Type: BOLIVIA. Department Cochabamba, Province Cercado, Cerro San Pedro, 2625 m alt., near Cochabamba. March 18, 1939; *E. K. Balls, W. B. Gourlay, and J. G. Hawkes 6297* (lectotype, here designated, US; isotypes, BM, CPC, UC).

The plants described as *S. vallegrandense* var. *vallegrandense* and *S. vallegrandense* var. *pojoense* by Cárdenas (1956) are well within the realm of variation of *S. berthaultii*. This is confirmed by the author's own collection of specimens from Pojo (Department Cochabamba). It should be noted, however, that in the original description of *S. vallegrande* var. *pojoense*, Cárdenas indicated that the corolla was rotate. His illustration of the plant, however – which is correct

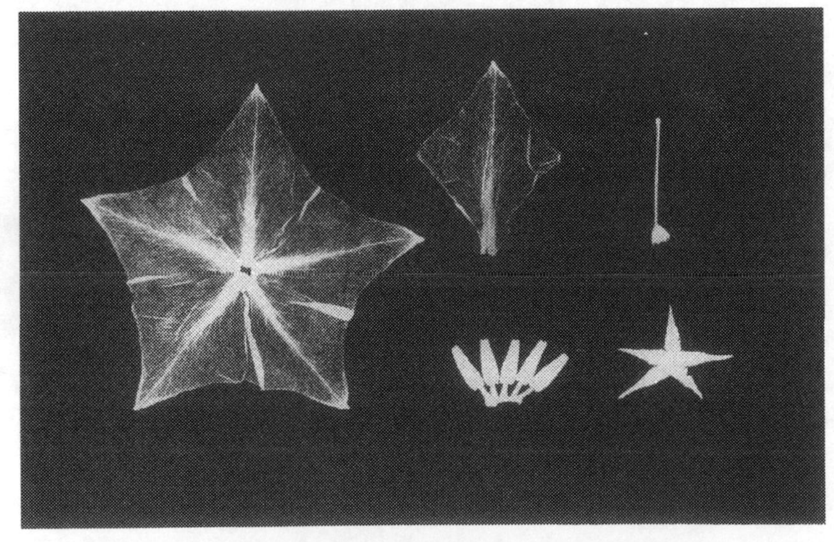

Figure 6. – Floral dissection of *Solanum berthaultii* (*Ochoa 12006*), ×1.

– has it stellate. This error was also noted by Correll (1962), who incorrectly classified the Pojo specimens under the white-flowered species *S. tarijense* var. *pojoense* (Cárd.) Correll.

The potential economic value of *S. berthaultii* lies in its multicellular, glandular hairs. These, according to Gibson (1971, 1974, 1976), are of the following two types: 1) a large hair with an ovate or spherical vessicle at its apex (TYPE A); and 2) a short hair with a four-lobed vessicle at its apex (TYPE B). These vessicles or glands produce a sticky, gummy exudate that immobilizes and eventually kills many species of insects, such as the aphids that are known to transmit certain viral diseases of potato. The value of this particular character of *S. berthaultii* lies in its use in potato improvement programs that are aimed at the elimination of this viral transmitting organism (Tingey et al., 1981, 1982; Mehlenbacher et al., 1983; Mehlenbacher and Plaisted, 1983). Moreover, these glandular hairs may be of value in protecting the potato plant from such other destructive or disease-carrying insect pests such as *Tetranichus urticae*, *Thrips tabaci*, *Polyphagotarsonemus latus*, *Liriomyza huidobrensis* (Intl. Pot. Ctr., 1977), and *Myzus persicae* and *Empoasca fabae* (Intl. Pot. Ctr., 1978).

Affinities

As suggested by its habit as well as by its dense, glandular pubescense and pentagonal or substellate corolla, *S. berthaultii* is related to *S. tarijense* and *S. flavoviridens*, both of series Commersoniana. Its indumentum, leaflet shape, and form and color of the fruit also suggest that it is related to *S. litusinum*. On

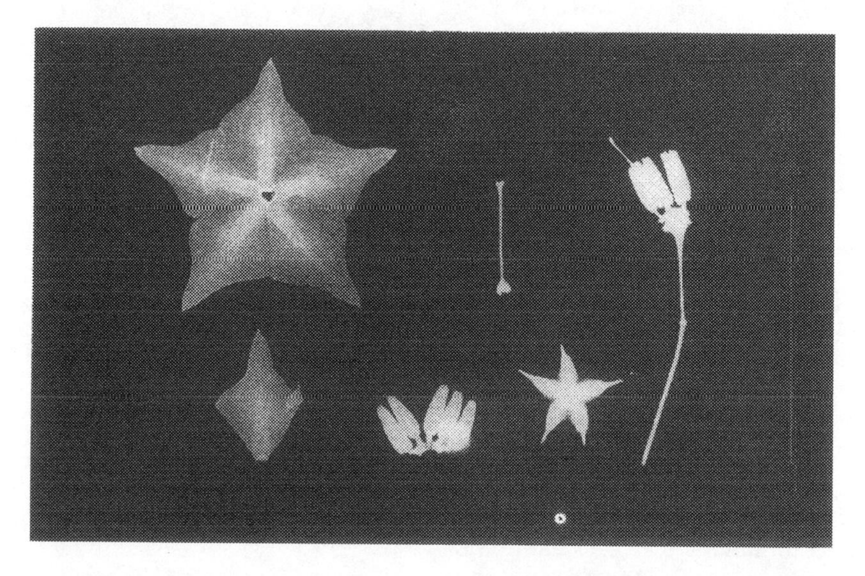

Figure 7. – Floral dissection of *Solanum berthaultii* (*Ochoa 12029*), ×1.

the other hand, this species differs from *S. chacoense* and other species of this series by its conspicuous glandular hairs and distinctive calyx shape.

This species shows some introgression from *S. doddsii*. This is apparent in specimens of *S. berthaultii* that were collected by Ochoa and Salas (15528) in the vicinity of Pojo, along the road from Cochabamba to Santa Cruz. These characteristics of *S. doddsii* show up particularly in the narrow leaflets of these plants, which are also finely denticulate. However, the dense glandular pubescence and lilac-pentagonal corolla are characteristics more typical of *S. berthaultii*. Still one other specimen that shows some introgression from wild plants of this series is a collection by Ochoa and Salas (15578), which has sharp, finely denticulate leaflet margins.

In the present work, both *S. berthaultii* and *S. tarijense* are treated under the series Commersoniana. Although it is typical for the plants of this series to have a stellate corolla, one classical species of this group, *S. chacoense*, differs in having both rotate and pentagonal flowers.

One suggested origin of *S. berthaultii* lies in a natural cross between *S. tarijense* and a blue-flowered species of the series Tuberosa. Hawkes (1963) and Hawkes and Hjerting (1969) have suggested *S. sparsipilum* as the second parental species. Up to the present time, however, it has not been possible to synthesize artificial hybrids between *S. tarijense* and *S. sparsipilum*, which compare favorably with *S. berthaultii*. *Solanum tarijense*, however, has been used in other crosses involving Bolivian Andean species.

Habitat and Distribution

Solanum berthaultii prefers dry places such as open fields where it forms isolated colonies or areas dominated by xerophytic or subxerophytic vegetation, including plants of cacti and algaroba. It occurs, moreover, in natural forests of *Schinus molle* and *Acacia*. It is also frequent in poor, dry lands, on stony hillsides, or in clay or clay-sandy soils.

Frequently, *S. berthaultii* is found in temperate valleys between 2000 and 2800 m or in Prepuna between 2400 and 2600 m. Its upper altitudinal limits reach 3400 m. Geographically, its distribution extends from Cochabamba in central Bolivia to Yotala in the south, the latter is located in the Province of Oropeza, Department of Chuquisaca.

Specimens Examined

Department Cochabamba

Province Cercado: Cerro de San Pedro, near Cochabamba, 2592 m alt. [8500 feet according to the collectors], under bushes in sandy loam or among loose stones and herbaceous plants (plants small). March 18, 1939; *Balls, Gourlay, and Hawkes 6297*, type collection of *S. berthaultii* (lectotype, US; BM, CPC, K,

UC). Near La Recoleta, Cochabamba. Flowers white. January, 1949; *Cárdenas s.n.* (HNF No. 0080). On brushy roadside banks, 4 km from Cochabamba on road to Santa Cruz. Plants glandular, smooth; flowers light lilac; fruit ovoid, white-mottled, 2 cm long; tubers approximately 3 cm long, white. February 13, 1960; *Correll, Dodds, Brücher and Paxman B606* (K, LL, US). Cerro San Pedro, 2700 m alt., above botanic garden and in upper part of garden itself (natural hillside area). Dry stony hillside amongst bushes and below by road, in waste places and in far end of garden. Flowers ranging from almost white to purple, and from rotate to stellate. January 28, 1971; *Hawkes, Hjerting, Cribb, and Huamán 4279* (CIP). Lower levels of Cerro San Pedro, 2750 m alt., near Cochabamba, in poor, sandy, gravelly soil. Flowers pale lilac, almost white. Plants glandular up to 60 cm tall. January 27, 1949; *Ochoa 666* (OCH, US, topotypes). On the lower talus slopes of Cerro de San Pedro, 2700 m alt., near Cochabamba. Plants glandular, 0.5-0.7 m tall; flowers violet, abundant. Growing on sandy-clay or stony soil. Fruit globose, 1.5 cm in diameter. January, 1949; *Ochoa 668* (GH, OCH, topotypes). Cochabamba, at base of Cerro de San Pedro, 2625 m alt. With *Acacia, Baccharis, Chenopodium, Tagetes, Salvia,* nontuberous *Solanum* sp. and *Lycopersicon esculentum* (apparently escaped from cultivation). March 31, 1963; *Ugent and Cárdenas 4560 (US); 4561, 4561-1, 4561-2, 4562-1, 4562-2* (WIS); *4562* (US, type locality of

Figure 8. – Berries of *Solanum berthaultii* (*Ochoa and Salas 15528*), ×1.

S. berthaultii Hawkes). Two km from Tupuraya, 2700 m alt., on the road from Cochabamba to Sacaba, between woody shrubs. Plants glandular, 0.8 m tall. Infrequent. Flowers white, tubers large, oval-compressed, whitish. January 27, 1949; *Ochoa 671* (GH, OCH, US). Vicinity Cerveceria Taquina, near Cochabamba, 2820 m alt. March 30, 1978; *Ochoa 12029* (CIP, OCH). Cerveceria Taquina, near Cochabamba, 2700 m alt. Roadside. With *Solanum berthaultii, Lycopersicon esculetum* (apparently escaped from cult.), *Chenopodium, Salvia, Tagetes,* and nontuberous *Solanum* spp. April 6, 1963; *Ugent 4614, 4615-1, 4615-2, 4616-1, 4616-2* (WIS). Near Cochabamba. Without date; *Cárdenas 3513* (CPC).

Province Arque: Oyuni, 2700 m alt., between Parotani and Caroa, on road from Cochabamba to Challa, before Punco. Very rare, plants without flowers. March 1, 1984; *Ochoa and Salas 15568* (CIP, OCH). Oyuni, 2800 m alt., between Parotani and Confital. In dry places, in clay and lateritic soils, between large rocks in shade of *Schinus molle* and associated with shrubs of *Nicotiana glauca* and herbs such as *Verbena* sp. and various composites. Chromosome count $2n=2x=24$. November, 1984; *Ochoa and Salas 15599* (CIP, F, OCH, US).

Province Capinota: On road from Paratoni to Capinota, 2400 m alt., 2 km past Itapaya, on edge of maize field and road amongst bushes. February 15, 1971; *Hjerting, Cribb, and Huamán 4422* (CIP).

Province Carrasco: Vicinity of Pojo, 2300 m alt., in sandy soil among bushes. March, 1955; *Cárdenas 5071* (HNF = ex CA, LL type collection of *S. vallegrandense* var. *pojoense*). Near Pojo, 2000 m alt., between km 190–191 on Cochabamba-Santa Cruz road. On hillside slopes covered with weeds and especially composites and chenopods. Margins of leaflets minutely denticulate. February 26, 1984; *Ochoa and Salas 15528* (CIP, OCH, US).

Province Chapare: Abra de Sacaba, 2580 m alt., sandy slope. Plant up to 30 cm tall; flowers blueish-white. March 1941; *Cárdenas 2208* (HNF, US). In thicket along road to Sacaba from Cochabamba, 2500 m alt. [8200 feet, according to the collector]. January 11, 1949; *Brooke 3005* (BM, F). Cochabamba-Aguirre-Colomi road, 3100–3400 m alt., 23 km E of Cochabamba, near Tutimayo, roadside thicket with *Solanum berthaultii, Tagetes* and nontuberous *Solanum* sp. April 7, 1963; *Ugent 4644-1, 4644-2, 4645, 4645-1, 4645-2, 4645-3* (WIS).

Province Jordan: Vicinity of Angostura Dam, near Cochabamba, 2750 m alt., in crop lands. Flowers light violet or lilac; plants glandular, 60 cm tall. January, 1949; *Ochoa 657* (OCH, US).

Province Mizque: Three km W of Palcca and about 14 km SSW of Aiquile, 2400 m alt. Plants robust, glandular, up to 1 m or more tall; flowers white, corolla pentagonal to substellate, tubers large, 7-8 cm long and about 50–60 g of weight, round to oval to oval-compressed, light brown. Common name:

Apharu (='wild'). Chromosome number: 2n=2x=24. March 25, 1978; *Ochoa 12008* (CIP, OCH).

Province Quillacollo: At km 46 from Cochabamba on the road to Oruro, 2950 m alt. On rocky, bushy slopes; associated with *S. tarijense*. February 4, 1971; *Hawkes, Hjerting, Cribb, and Huamán 4349* (CIP). Between Suticollo and Parotani, 2450 m alt., where oil pipe crosses road, on rocky slopes under bushes and by roadside. Pale blue stellate flowers. February 15, 1971; *Hjerting, Cribb, and Huamán 4411* (CIP). Vicinity of Liriuni, 3000 m alt., along road from Liriuni to Tunari. January, 1949; *Ochoa 674* (OCH).

Department Chuquisaca

Province Oropeza: On brushy slope just south of Sucre. Plants tall, about 1 m or more high; flowers white, stellate. February 17, 1960; *Correll, Dodds, Brücher, and Paxman B621* (LL, S). On brushy slope just south of Sucre. Plants large, flowers light purple, stellate. February 17, 1960; *Correll, Dodds, Brücher, and Paxman B622* (LL, S). At km 28 on the road Sucre-Aiquile, 1950 m alt., about 15 km ENE from Sucre (air distance), 19°1′ S and 65°7′ W, on rocky slope with *Solanum zudanense, S. gandarillasii, Solanum* sp. (non-tuber-bearing), *Cleome aculeata* spp. *cordobensis, Polygonum*, and *Cerastium*. April 11, 1963; *Ugent and Cárdenas 4890* (WIS). On road from Sucre to Potosi, 2500 m alt., 14.5 km from Sucre, in valley bottom among thorn bushes, growing with *S. tarijense*. Leaves like *S. tarijense*, but some flowers quite blue. Few berries. March 6, 1971; *Hjerting, Cribb, and Huamán 4563* (CIP). Vicinity of Cabezas, 2380 m alt., on road Sucre-Yotala. On poor, rocky slopes, under shade of *Schinus molle*. March, 1984; *Ochoa and Cagigao 15578* (CIP, F, GH, NY, OCH, US). Between Cabezas and Yotala, 2370 m alt., on road Sucre-Yotala, between thicket bushes and in very dry places, associated with *Opuntia tuna* and *Opuntia exaltata*. Flowers light lilac to whitish. March, 1984; *Ochoa and Cagigao 15579* (CIP, NY, OCH). Near Yotala, 2360 m alt., on talus slopes along road Sucre-Yotala, associated with a white-flowered, non-tuber-bearing *Solanum* shrub. Common name: *Ita papa*. March, 1984; *Ochoa and Cagigao 15580* (OCH).

Province Yamparaez: Near Yamparaez, 16 km SE of Sucre on the road Sucre-Tarabuco. Growing on very rocky slopes along the margins of the road. April 12, 1963; *Ugent 4994* (WIS).

Province Zudañez: Between Zudañez and Tomina, 2200 m alt. February, 1949; *Cárdenas 4503* (CPC). Between Sunchu Tambo and La Quebrada (near Jatum Ckasa), 2600 m alt. Dry region with xerophytic vegetation. Plants robust, very pubescent, glandular. Flowers purple, corolla pentagonal to substellate. Tubers large, tan, 8 cm long, oval-compressed. Common name: *Lluttu papa*. Chromosome number 2n=2x=24. March 24, 1978; *Ochoa 12006* (CIP, OCH).

Department Potosi

Province Saavedra: On brushy slope about 32 km from Sucre on road to Potosi. Plants tall and leafy; flowers white. February 17, 1960; *Correll, Dodds, Brücher, and Paxman B623* (LL, MO). On brushy slope about 32 km from Sucre on road to Potosi. Plants tall, up to 1 m high; flowers dark lilac, rotate-stellate. February 17, 1960; *Correll, Dodds, Brücher, and Paxman B624* (LL, US). Among acacias about stone wall, about 100 km from Sucre on road to Potosi. Plants tall; flowers pink-lilac, rotate-stellate. February 18, 1960; *Correll, Dodds, Brücher, and Paxman B630* (LL, US).

Department Santa Cruz

Province Valle Grande: Under bushes forming dense thickets on the way from Trigal to Mataral, 2000 m alt. March, 1955; *Cárdenas 5070* (CA, type of *S. vallegrandensis*).

Province Manuel Maria Caballero: Aguadilla, 1800 m alt., near the merge of Cochabamba-Santa Cruz road to Pulquina, associated with xerophitic vegetation. February 29, 1984; *Ochoa and Salas 15565* (CIP, OCH, US).

Crossability

At present, the agronomic value of *S. berthaultii* lies in its glandular hairs, which secrete a gummy substance that traps harmful aphids and other insects. Although sucseptible to potato late blight, *P. infestans*, it is possible that this species has some resistance to drought as it occurs in dry valleys.

Solanum berthaultii is a self-incompatible species. In crosses involving introductions of this species (No. 12029) and *S. alandiae* (No. 12014 and 12018) and *S. tarijense* (No. 12001), fruits containing abundant seeds (average of 150) were easily obtained. Similar results were had in crosses of this species with *S. chacoense* (No. 12026). On the other hand, crosses of *S. berthaultii* (No. 12029) with the hexaploid *S. oplocense* (No. 11972) were negative. In the latter, about 50% of the pollinations yielded apparently normal-looking fruits up to 20 mm long, but lacking seed. This can be explained by Johnston and Hanneman's (1982) previously cited hypothesis in regard to the numerical balance of the endosperm. Thus, *S. berthaultii*, a diploid, is probably EBN=2, whereas *S. oplocense*, a hexaploid, is EBN=4.

Solanum berthaultii × *S. tarijense*. – Artificial F_1 hybrids of *S. berthaultii* and *S. tarijense* are similar to *S. tarijense* in the form and dissection of their leaves. The segregates of this cross, however, vary greatly with respect to pubescence – ranging from sparsely pubescent, and provided with both simple and short glandular hairs, to very pilose and invested with simple, long, and short glandular hairs. All the various plants of this hybrid population had a light violet to whitish-blue, substellate to stellate corolla, 3 cm or more in diameter. None were observed to be white-flowered.

Solanum berthaultii × *S. chacoense*. – Artificial crosses between these two species also show much hybrid vigor. The plants are typically tall and have many dominant characteristics of *S. chacoense*, including leaf shape and dissection, and a small calyx with glabrous or glabrescent lobes, the latter somewhat apiculate. The pubescence of these hybrids is extremely variable, ranging from densely pilose in certain individuals, to glabrescent in others. The corolla varies from strongly pentagonal to strongly stellate, and from pure white to bluish-white or indigo.

Solanum berthaultii × *S. alandiae*. – Plants derived from this cross recombine some characteristics of each parental species. However, while these segregates are more like *S. berthaultii* in leaf characteristics, they have an indumentum consisting mostly of simple hairs, but including some tetra-lobed glandular hairs as well. Articulation of the pedicel occurs in the upper one-third of the stalk. All have a pentagonal corolla that varies in color from purple to light violet.

Solanum berthaultii × *S. sparsipilum*. – Hybrids of this cross yielded few fruits with an average of 8 seeds per berry.

3. Solanum chacoense Bitter, Repert. Sp. Nov. 11:18. 1912.

FIGS. 9, 10; PLATE III.

S. tuberosum L. var. *glabriusculum* Dun., in DC Prodr. 13(1):32. 1852. TYPE: *Bacle 117*, Argentina. Buenos Aires (G).

S. guaraniticum Hassl., Repert. Sp. Nov. 9:115. 1911 (not *S. guaraniticum* A. Saint Hilaire, Mem. Mus. 12:321–322, 1825). TYPE COLLECTION: *Hassler 470*, Paraguay. Villeta (G); and *Hassler 2805*, Paraguay. Gran Chaco, Santa Elisa (G, lectotype of *S. chacoense*, proposed by Correll).

S. guaraniticum Hassl. var. *angustisectum* Hassl., Repert. Sp. Nov. 9:115. 1911. TYPE COLLECTION: *Hassler 470*, Paraguay. Villeta (G); and *Hassler 2805*, Paraguay. Gran Chaco, Santa Elisa (G, lectotype of *S. chacoense*, proposed by Correll).

S. guaraniticum Hassl. var. *latisectum* Hassl. f. *glabrescens* Hassl., Repert. Sp. Nov. 9:115. 1911. TYPE: *Hassler 326*, Paraguay. Cordillera de Altos (G).

S. bitteri Hassl., Repert. Sp. Nov. 11:190. 1912. TYPE COLLECTION: *Hassler 470*, Paraguay. Villeta (G); and *Hassler 2805*, Paraguay. Gran Chaco, Santa Elisa (G, lectotype of *S. chacoense*, proposed by Correll).

S. subtilius Bitt., Repert. Sp. Nov. 12:6–7. 1913. TYPE: *Lillo 319*, Argentina. Probably near Tucumán (P).

S. muelleri Bitt., Repert. Sp. Nov. 12:450. 1913. TYPE: *Müller s.n.*, Brazil. Rio Grande do Sul (W, plants grown in Vienna from seed collected by F. Müeller).

S. chacoense Bitt. var. *angustisectum* Hassl., Ann. Conserv. & Jard. Bot. Geneve 20:186–187, Fig. 4a. 1917. TYPE COLLECTION: *Hassler 470*, Paraguay. Villeta (G); and *Hassler 2805*, Paraguay. Gran Chaco, Santa Elisa (G, lectotype of *S. chacoense*, proposed by Correll).

S. chacoense Bitt. var. *latisectum* Hassl. f. *glabrescens* Hassl., Ann. Conserv. & Jard. Bot. Geneve 20:186, Fig. 4b. 1917. TYPE: *Hassler 326*, Paraguay. Cordillera de Altos (G).

S. chacoense Bitt. var. *latisectum* Hassl. f. *plurijugum* Hassl., Ann. Conserv. & Jard. Bot. Geneve 20:187, Fig. 4d. 1917, *nom. nud.* Based on *Hassler 11849*, Paraguay. Lago Ipacarai (G, GH, US).

S. gibberulosum Juz. et Buk., Rev. Argentina Agron. 3:225. 1936. TYPE: *Millan G-1183*, Argentina. Manfredi, Cordoba (WIR-2877; plants grown in Leningrad from tubers sent by R. Millan).

S. parodi Juz. et Buk., Rev. Argentina Agron. 3:226. 1936. TYPE: *Millan G-980*, Argentina. Río Chico, Tucumán (WIR=K-1603).

S. garciae Juz. et Buk., Rev. Argentina Agron. 3:227. 1936. TYPE: *Garcia s.n.*, Argentina. Near Cordoba (WIR=K-883; plants grown near Leningrad from tubers sent by R. Millan under No. 1201).

S. emmeae Juz. et Buk., Izv. Akad. Nauk U.S.S.R. 2:321. 1937. TYPE: *Knappe s.n.*, possibly from Argentina (LE, not seen; plants grown near Leningrad from tubers sent by the Latvian collector P. T. Knappe, under the name *S. caldasii* var. *glabrescens*.

S. dolichostigma Buk., in Vavilov, Theor. Bases Pl. Breed. 3:72. 1937, *nom. nud.* Based on *Knappe s.n.*, possibly from Argentina (LE, not seen; plants grown near Leningrad from tubers collected originally by P. T. Knappe).

S. knappei Juz. et Buk., Izv. Akad. Nauk. U.S.S.R. 2:322. 1937. TYPE: *Knappe s.n.*, Argentina. Salta (LE, WIR=2882; plants grown near Leningrad from tubers collected originally by P. T. Knappe).

S. schickii Juz. et Buk., Izv. Akad. Nauk. U.S.S.R. 2:324. 1937. TYPE: *Schick s.n.*, Argentina. Siambon, Tucumán (LE, not seen; WIR=2879; plants grown near Leningrad from tubers collected by R. Schick).

S. horovitzii Buk., Rev. Argentina Agron. 4:238. 1937. TYPE: *Horovitz 1029*, Argentina. Potrero Diaz, Salta (WIR, not seen; plants grown near Leningrad from tubers and seeds collected originally by R. Millan, No. G-1047).

S. laplaticum Buk., Rev. Argentina Agron. 4:238. 1937. TYPE: *Millan G-972*, Argentina. Vicinity of Buenos Aires (WIR=K-597; plants grown near Leningrad from tubers collected by R. Millan).

S. boergerii Buk., Rev. Argentina Agron. 4:239. 1937. TYPE: *Henry s.n.*, Uruguay. Agraciada, 15 km from Mercedes, near Arroyo Daca, Soriano (WIR=K-1070; plants grown near Leningrad from material collected by T. Henry).

S. horovitzii Buk. var. *glabristylum* Hawkes, Bull. Imp. Bur. Pl. Breed. &

Genet., Cambridge (June):19. 1944, *nom. nud.* Based in part on *Balls 5921*, Argentina. El Potrerillo, San Antonio, near Jujuy (CPC, F, K, US).

S. saltense Hawkes, Bull. Imp. Bur. Pl. Breed. & Genet., Cambridge (June):19, 113. 1944 (not *S. saltense* (Bitt.) Morton. Contrib. U.S. Nat. Herb. 29:63, 1944). TYPE: *Balls 5935*, Argentina. Near Salta, Quebrada de San Lorenzo, Dept. Salta (CPC).

S. jujuyense Hawkes, Bull. Imp. Bur. Pl. Breed. & Genet., Cambridge (June):19, 114. 1944. TYPE: *Balls 5921, p.p.*, Argentina. El Potrerillo, San Antonio, near Jujuy (CPC).

S. horovitzii Buk. var. *multijugum* Hawkes, Bull. Imp. Bur. Pl. Breed. & Genet., Cambridge (June):19, 114. 1944. TYPE: *Balls 5921, p.p.*, Argentina. El Potrerillo, San Antonio, near Jujuy (CPC, F, K, US).

S. chacoense subsp. *subtilius* (Bitt.) Hawkes, Scottish Pl. Breed. Station, Ann. Rept. 61. 1956. TYPE: *Lillo 319*, Argentina. Probably near Tucumán (P).

S. cuevoanum Cárd., Bol. Soc. Peruana Bot. 5:36, Pl. III-E, Figs. 1-4, Pl. V-C. 1956. TYPE: *Cárdenas 5073*, Bolivia. Santa Rosa de Cuevo, Dept. Chuquisaca, Prov. Calvo (CA).

S. caipipendense Cárd., Bol. Soc. Peruana Bot. 5:35, Pl. III-D, Figs. 1-4, Pl. V-B. 1956. TYPE: *Cárdenas 5072*, Bolivia. Near Caipipendi, Dept. Santa Cruz, Prov. Cordillera, (HNF).

S. arnezii Cárd., Bol. Soc. Peruana Bot. 5:37, Pl. III-F, Figs. 1-6. 1956. TYPE: *Cárdenas 5076*, Bolivia. Between Padilla and Tomina, Dept. Chuquisaca, Prov. Tomina (HNF).

S. chacoense Bitt. f. *gibberulosum* (Juz. et Buk.) Corr., Wrightia 2:172. 1961. TYPE: *Millan 2734*, Argentina. Monte Lena, Catamarca (WIR; authentic material of *S. gibberulosum*, plants grown near Leningrad).

S. chacoense Bitt. f. *caipipendense* (Cárd.) Corr., Wrightia 2:172. 1961. TYPE: *Cárdenas 5072*, Bolivia. Near Caipipendi, Dept. Santa Cruz, Prov. Cordillera (HNF).

S. muelleri Bitt. f. *densipilosum* (Bitt.) Corr., Wrightia 2:173. 1961. TYPE: *Smith, Reitz, and Caldato 9516*, Brazil. Fazenda Campo San Vicente, Chapeco, Santa Caterina (US).

S. limense Corr., Wrightia 2:188. 1961. TYPE: *Soukup 3555*, Peru (F, OCH, US; see Appendix, Ochoa 1962).

S. chacoense subsp. *muelleri* (Bitt.) Hawkes et Hjerting, Scottish Pl. Breed. Station, Ann. Rept. 103. 1963. TYPE: *Müeller s.n.*, Brazil. Rio Grande do Sul (W; plants grown in Vienna from seed collected by F. Müeller).

Plants light green, occasionally forming rosettes, or more frequently subrosette-like, erect, or spreading, and glabrous or sparsely pubescent over all parts. Stems angular, usually light green, or occasionally irregularly pigmented, (10-20-)35-70(-150) cm tall by 3-5(-10) mm in diameter at base,

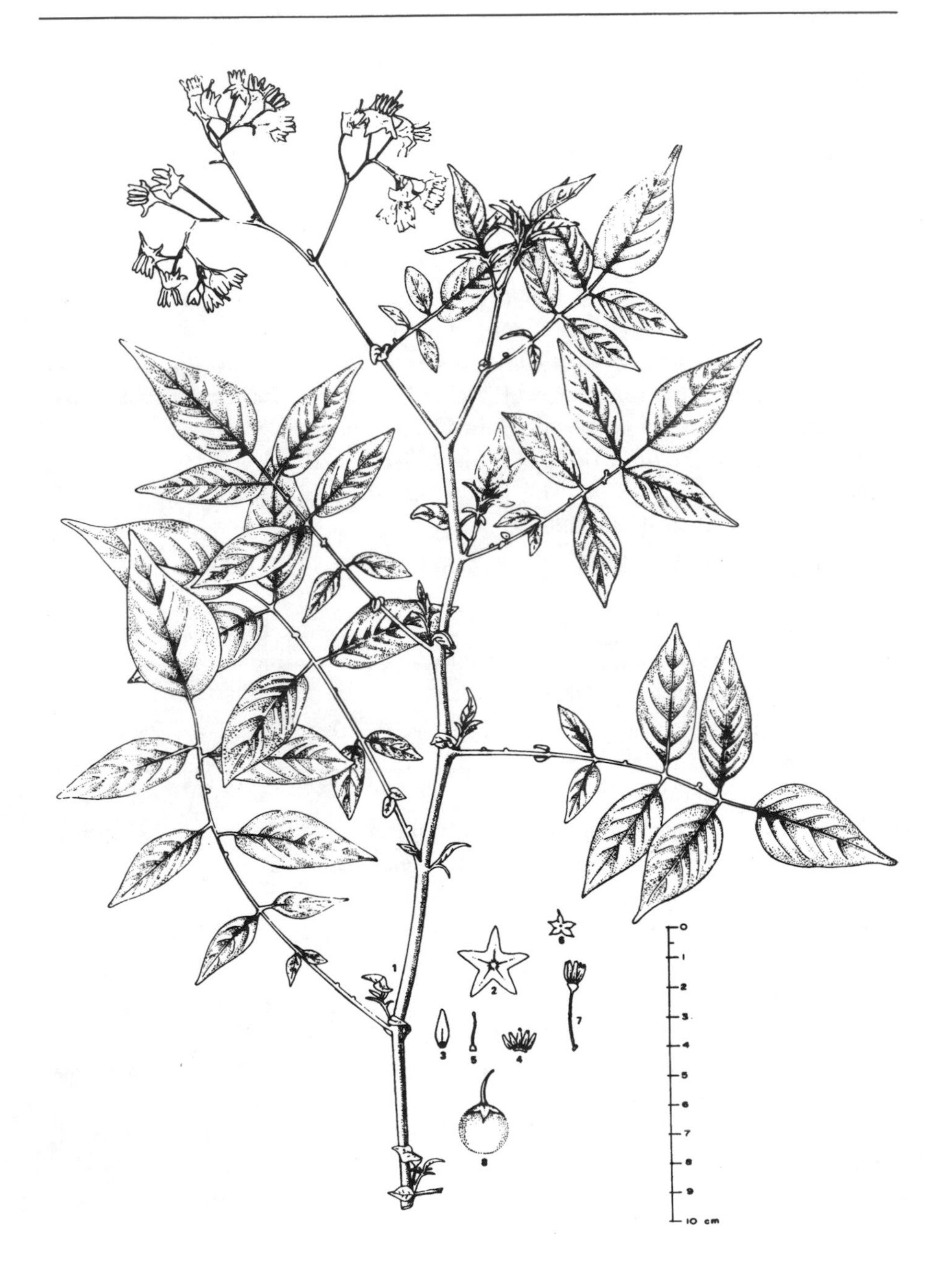

Figure 9. – *Solanum chacoense* (*Ochoa 12026*). 1. Upper part of flowering plant. 2. Corolla. 3. Petal. 4. Stamens, dorsal view. 5. Pistil. 6. Calyx. 7. Pedicel and stamens. 8. Fruit. All ×½.

simple or branched, the internodes appearing short at base and much longer toward the midpoint or upper two-thirds of the winged stem; stem wings straight or rarely undulate, 1-3 mm wide. Stolons 1 m or more long; tubers round to ovoid, 1-2(-3.5) cm long, white or yellowish-white. Leaves imparipinnate, (4-)6.5-15.5(-35) cm long by (2.7-)4.4-11(-15.5) cm wide. Leaflets 4-5(-6) paired, broadly ovate to ovate lanceolate or elliptic-lanceolate to oblong, generally glabrescent above with short hairs, or more rarely densely pilose, the apex shortly acuminate and the base obliquely obtuse to rounded or cordate, sessile or subsessile, or borne on petiolules (1-)3-6(-20) mm long. Terminal leaflets usually of the same size as the laterals, (2-)3.5-7.5(-9.5) cm long by (1-)2-3(-4.6) cm wide. Interjected leaflets usually lacking, or at most 1-3(-10). Pseudostipular leaves small, semiovate or falcate, to 10 mm long and 5 mm wide. Inflorescence cymose, glabrous, or rarely pilose and glandular. Peduncles 8-12(-20) cm long, provided with 35-40 flowers. Pedicels (1-)1.5-2(-3.5) cm long, increasing in diameter towards the calyx, articulated generally in the lower one-third of the stalk or slightly above center. Calyx usually very small, glabrous, 3-4(-5.5) mm long, the lobes broadly ovate and short, 2-3.5 mm long and narrowly ovate-triangular and ending in an extremely short

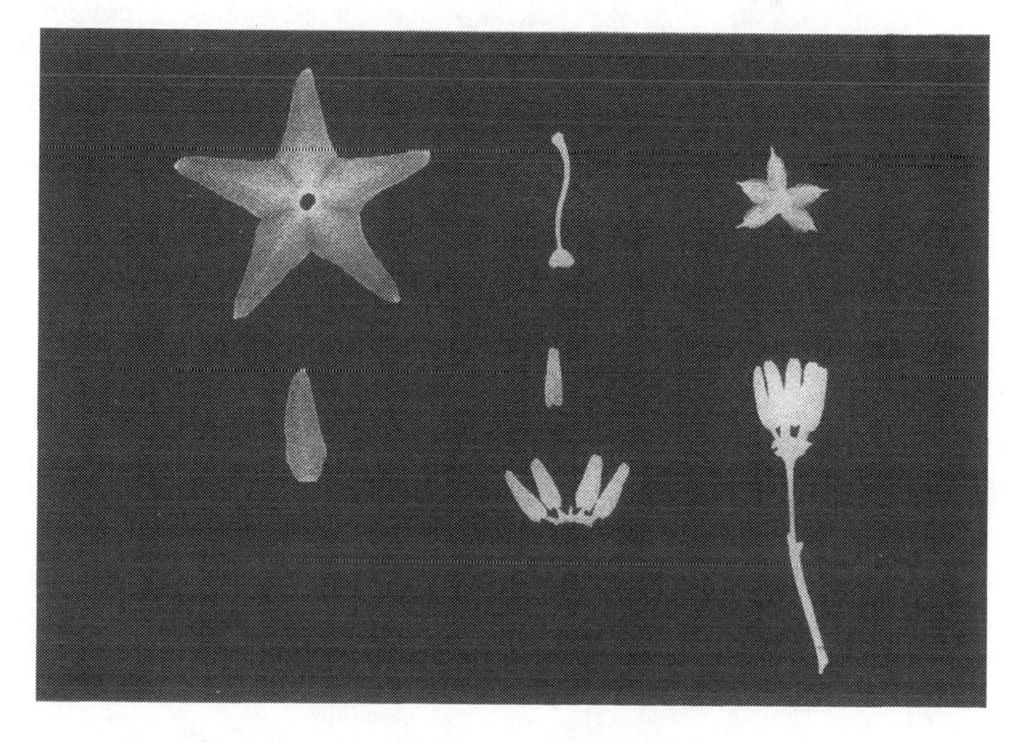

Figure 10. – Floral dissection of *Solanum chacoense* (*Ochoa 12026*), ×1.

apiculate apex, the latter about 0.3-0.6 mm long. Corolla generally stellate, at times substellate or rarely rotate-pentagonal, white, or more rarely grayish-white with pale lilac acumens, (1.7-)2.5-3 cm in diameter, the lobes spreading to reflexed, and usually narrowly triangular-lanceolate 7-12 mm long by 4-7 mm wide at base, or rarely with broad triangular-lanceolate lobes, 9-11 mm long and 8-13 mm broad. Anthers 4-5(-6) mm long by 1.6-1.8 mm wide, borne on glabrous, 1-2 mm long, white-hyaline filaments. Style 9-13 mm long, exerted 7 mm from corolla, usually curved in the upper one-third of its length, glabrous or papillose below; stigma capitate to ovoid and strongly clavellate to bifurcate. Fruit globose to pyriform or ovoid, light green or green-speckled with white spots, or occasionally with 1-2 dark purple vertical stripes.

Chromosome number: 2n=2x=24, 2n=3x=36.

Type: PARAGUAY. Department Concepción, Santa Elisa near Concepción. Lat. 23°10′ S, 57°40′ W. March, 1903; *E. Hassler 2805* (type collection of *S. guaraniticum* var. *angustisectum* Hassl., designated by Correll as lectotype of *S. chacoense* Bitt., G; isotypes, K, P).

The enormous variation and geographic range of *S. chacoense* has been responsible, in part, for the many synonyms published of this species, especially those appearing during the early decades of the twentieth century. This species was discovered in 1911 by the Swiss physician and botanist, Emile Hassler, a one-time resident of Paraguay, who named it *S. guaraniticum*. Unfortunately, this name had already been used almost a century before by Saint Hilaire to designate a nontuberous species of *Solanum*. Georg Bitter, who noted Hassler's error in 1912, renamed the species *S. chacoense*. However, Hassler, who later recognized his own error, tried vainly to recoup priority for this species by giving it a new name, *S. bitteri*, but it was already too late to rename the species by the time of his publication. Later, Hassler transferred all of his subspecific categories under *S. guaraniticum* to *S. chacoense*. More recently, proposals to subdivide this species into lesser entities by Brücher (1956, 1974), Correll (1962), Hawkes (1956a, 1963), and Hawkes and Hjerting (1969) have added more taxonomic confusion with regard to the naming of this plant.

In addition to the above problems, neither Hassler nor Bitter designated a type specimen of either *S. guaraniticum* or *S. chacoense*. The latter species remained without typification for more than fifty years until Correll (1962) designated a lectotype from material of the original Hassler collection (No. 2805), which was found divided among herbaria in Geneva, London, and Paris. Later, Hawkes and Hjerting designated a syntype of *S. chacoense* from another specimen of Hassler's deposited in the museum of Geneva, which had the same number as Correll's lectotype.

Our knowledge of *S. chacoense* today stems largely from collections made in Paraguay rather than Bolivia; however, there exist few in-depth studies on this species. Although Brücher (1974) claims to have re-collected *S. chacoense* in the two localities in Paraguay where Hassler (Nos. 2805 and 326) made his original collections (Santa Elisa and Cordillera de Altos), Brücher's plants have not been seen in any of the large herbaria or museums that have been visited by the author. In connection with the author's botanical explorations of the lowland regions of South America, including the countries of Brazil and Uruguay, he also visited Paraguay in 1983. The localities visited, where *S. chacoense* was previously collected by others, included Villeta, Lago Ipacarai near Asunción, Caacupe, Caaguazu, Coronel Oviedo, and El Alto Parana. Although time considerations prevented him from exploring either the Cordillera de Altos or the Serranias de Amambay, considerable insight was nevertheless obtained with regard to the overall geographic patterns of variation of *S. chacoense*.

Of all the various wild potato species known today, *S. chacoense* has probably the greatest potential for use in potato improvement programs. During the past forty years, many attempts have been made to incorporate the germplasm of this species into various European and North American potato varieties. Sleesman (1940) reports that several species, including *S. chacoense*, *S. commersonii* from Argentina, and *S. polyadenium* from Mexico, have resistance to *Empoasca fabae*. Adams (1946) reports that this species is little attacked by the aphid, *Myzus persicae*, which is known to transmit a viral disease to the cultivated potato. On the other hand, Stelzner (1943), Schaper (1953) and Torka (1948, 1950), working in Germany, discovered that plants of *S. chacoense* are seldom damaged by the Colorado potato beetle (*Leptinotarsa decemlineata*). According to Ross and Baerecke (1950) and Ross (1958), this species also shows some resistance to potato viruses PLRV and PVY. Ross (1958) also indicates that *S. chacoense* has resistance to *Synchytrium endobioticum* and *Corynebacterium sepedonicum*. Moreover, Lebedeva et al. (1978) report that diploids and induced polyploids of this species are resistant to potato virus PVX.

Tests performed at CIP in Lima, Peru, suggest that *S. chacoense* has resistance to a broad spectrum of diseases. Thus, this species appears to have a high level of resistance to bacterial wilt, caused by *Pseudomonas solanacearum*, a serious disease of the potato worldwide (Intl. Pot. Ctr., 1977; P. Schmiediche, 1984, personal communication). This species also shows resistance to various types of nematodes (Mendoza and Jatala, 1978; Intl. Pot. Ctr., 1975, 1978), *Globodera pallida* (*ibid*, 1977), *M. incognita*, *M. arenaria*, *M. javanica*, and *M. hapla* (*ibid*, 1983). *Solanum chacoense* or some of its hybrids is also resistant to *Phthorimaea operculella* (*ibid*, 1980) and *L. huidobrensis* (*ibid*, 1982), and, as pointed out earlier, to *M. persicae* and *E. fabae* (*ibid*, 1978).

Preliminary results of evaluations performed at CIP on recently collected

Paraguayan samples of *S. chacoense* indicate that this species may be resistant to at least two virus complexes, PLRV and PVY (L. Salazar, 1984, personal communication).

Affinities

Solanum chacoense is related to *S. muelleri*. However, it is distinguished from the latter species by the shape of its leaves, its narrower and longer leaflets, its basely articulated pedicels, and by its calyx, which retains a different shape and has longer acumens.

This species is also related to *S. huancabambense* (Ochoa, 1962). The latter species, which is endemic to northern Peru and thus geographically isolated from *S. chacoense*, constitutes the closest known link between series Commersoniana and series Tuberosa. Although this species differs in several respects from *S. chacoense*, the two nevertheless share a number of characteristics in common. For example, both have light-green leaves and stems, white flowers, and similarly shaped leaves, leaflets, and fruits. However, *S. huancabambense* differs from *S. chacoense* in being glandular-pubescent throughout, and by having a differently formed calyx and corolla. The position of the pedicel articulation in *S. chacoense* is also different from the first.

Habitat and Distribution

As pointed out by Hawkes and Hjerting (1969), this species occupies a wide assortment of different habitats throughout its extensive range. It occurs typically along arroyos, borders of irrigation channels and canals, sandy-clay river banks, and in abandoned lands or cultivated fields of cotton and maize, as a weed in gardens and orchards, and in thickets of *Crotalaria* and *Cleome*, where it is frequently associated with *Physalis*, *Convolvulus*, and various species of grasses. It also occurs under the shade of *Alnus* and *Podocarpus*.

Solanum chacoense is rare in Bolivia, but commoner southward where it forms extensive colonies in the lowland regions of Uruguay, Paraguay, and Brazil. Much of our present day knowledge of the distribution and ecology of this species is based on collections and field notes made by the author in Paraguay; for this reason, Paraguay has been addressed more specifically in the following paragraphs.

Contrary to what has been generally reported, this species grows in regions of very high rainfall in Paraguay. In the Cordillera de Altos, for example, one of the classical collecting localities of Hassler (located east of the Paraguay River), the average precipitation exceeds 1000 mm per year. Likewise, the region west of the Paraguay River receives a large ammount of precipitation, but not quite as much as the eastern part. Thus, this western region, after a distance of about a 100 kilometers to the north, gradually fades into a vast desert land inhabited by forests of xerophytic trees and spiny plants. No species of tuber-bearing *Solanum* are known to occur here. On the other hand, the region of the eastern side of the Paraguay River, which includes the Departments of Guaira, Gaazapu, and Itapua, appears to represent a major region of concentration for *Solanum* species.

Numerous large colonies of *S. chacoense* are found along the forested banks of the Paraguay River in the vicinity of Villeta, and in the vicinity of Lake Ipacarai, near Asunción. In these areas, the plants typically grow in the organically rich soil that occurs under the shade of large palms, such as *Arecastrum romanzoffianum*.

The growing season for *S. chacoense* in Paraguay corresponds to the coldest months of the year, May through September. During this season, temperatures may drop down to 5° C due to the periodic occurrence of cold fronts that blow in from the Antarctic. Skies during this stormy period may be cloudy for extended periods.

Finally, it is worth mentioning that *S. chacoense* has been introduced widely outside of its natural range. In these cases, the species has shown an extraordinary ability to adapt and become established in new and varied environments. Thus, in the garden of the Colegio Salesiano in the District of Breña, Lima, this species has been established for more than 40 years as a weed derived originally from material imported from Argentina for horticultural studies (Soukup, personal communication). Moreover, herbarium collections made from living material of this species (*Soukup 3555*) grown in Lima and later deposited in the herbarium of the Smithsonian Institution (US) and in the Chicago Natural History Museum (F) form the base of the species *S. limense* Correll (Ochoa, 1962). And finally, this species has been naturalized for many years in the Himalayas of Simla in northern India. Here, the species was introduced originally for studies on experimental potato breeding (M. D. Upadhya, 1985, personal communication).

This species is called *Papa Ra*, in Guarani, Paraguay (=*Para Papa*), or *Papa de Zorro* in Bolivia.

Solanum chacoense is widely distributed in nature, extending from southern Bolivia (where this species is found at its northern limits) and northern Argentina through southwest Uruguay and southern Paraguay to southern Brazil. Its altitudinal limits range from near sea level to more than 2000 m.

Specimens Examined

Department Chuquisaca

Province Calvo: Santa Rosa de Cuevo, 800 m alt., among bushes in humid sandy soil. Plants 20–30 cm height. Flowers white greenish or yellowish stellate, rather dialipetalous. November, 1954; *Cárdenas 5073* (CA, HNF).

Province Tomina: Between Padilla and Tomina, 2200 m alt., on a sandy slope. February, 1949; *Cárdenas 5076* (HNF).

Department Santa Cruz

Province Cordillera: Near Caipipendi [between Lagunillas and Charagua], 1100 m alt., n.v. *Ahuara Papa* or *Papa de Zorro*. March, 1949; *Cárdenas 5072* (HNF-00027).

Province Valle Grande: Playa Grande, 1800 m alt., along the route from Quebrada Seca to Ariruma. March 24, 1978; *Ochoa 12026* (CIP, OCH).

Crossability

According to the author's studies of *S. chacoense*, this plant is self-incompatible. Moreover, the author has found that this species is easily used in crosses with other wild species with which it is known to occur sympatrically. Thus, hybrids between this species and either *S. berthaultii* or *S. tarijense* are fairly easy to obtain. This species can also be readily crossed with many varieties of cultivated diploid potato, including *S. phureja* and *S. goniocalyx*, but not with diploid wild clones of *S. litusinum*.

Descriptions of some of the more successful crosses made at CIP follow:

Solanum chacoense × *S. berthaultii*. – Crosses between *S. chacoense* (12026) and *S. berthaultii* (12029) are normally 100% successful, yielding an average of 200 fertile seeds per berry. Though first generation hybrids from the above cross are very fertile, these tend to be phenotypically highly diverse. Thus, some hybrid plants tend to resemble *S. chacoense* in the shape of their leaves and flowers. These hybrids have blue to white or whitish flowers that range in shape from stellate to pentagonal. On the other hand, other individuals of the same population present a composite picture of the two species. Thus, some have the leaf shape and dissection of *S. berthaultii*, but are extremely variable in pubescence, ranging from densely pilose to glabrescent. The flowers and inflorescences of these plants resemble those of *S. chacoense*, but the corollas are generally pigmented with purple. However, in different segregates of this same cross, the corolla color may vary from white to purplish. All have a calyx similar to that of *S. berthaultii*, and pedicels articulated above center or in the upper one-third of the stalk.

Solanum chacoense × *S. tarijense*. – Crosses of *S. chacoense* (12026) and *S. tarijense* (12000) are also generally very successful, yielding an average of 125 fertile seeds per berry. These hybrids have leaves that strongly resemble those of *S. chacoense*, but are highly variable in indumentum. Corolla shape of these hybrids varies from the typically stellate condition, with long triangular-lanceolate acumens, to substellate, pentagonal, and rotate (the latter lobed and with short acumens). Flower color varies from white or pure creamy white, to purple, violet, or blue. All have a calyx similar to that of *S. tarijense*, and a pedicel articulated in the upper one-third of the stalk. In connection with these hybrids, it is of some interest that one of the collections by Correll et al. (No. 620-A) from the vicinity of Quiroga, Province of Campero, Department of

Cochabamba, Bolivia, appears to be a natural cross of *S. chacoense* and *S. tarijense*. The above plants, which were deposited in the Lundell Herbarium (LL) and identified originally by Correll as *S. chacoense*, combine the characteristics already noted for our experimentally made hybrids.

Solanum chacoense × *S. litusinum*. – Crosses between *S. chacoense* (12026) and *S. litusinum* (12027) were successful in about 14% of the crosses. These yielded an average of 45 fertile seeds per berry. The inflorescence and form and shape of the leaflets of the F_1 plants obtained in the above cross resembled those of *S. litusinum*, but were less pilose than those of the latter species and with longer petiolules. The hybrids had pedicels articulated near center, or slightly below center. Corolla shape was predominantly stellate, and flower color varied from pure white and light purple to dark purple. All had long-lanceolate anthers, up to 9 mm in length, and calyces similar to those of *S. litusinum*.

4. **Solanum flavoviridens** Ochoa, Am. Pot. Jour. 57:387-390. 1980.

Figs. 11-14.

Plants yellowish-green, vigorous, densely pilose, and strongly glandular. Stem rigid, to 1 m tall and to 15 mm in diameter at base, broadly straight-winged, the wings 1.5-2.5 mm wide, invested with simple, multicellular 1-3 mm-long silvery-white hairs and abundant short hairs terminating in tetralobular glands, and long hairs with simple glands. Stolons 1.5 m or more long and 3-5 mm in diameter, pure white or white with light-purple stripes; tubers round to ovate, to 4 cm long, white or white-tinged with purple-pink, bud eyes white, densely pubescent at base. Leaves imparipinnate, 12-20(-40) cm long by 5-10(-21.5) cm wide, yellow-green, densely pubescent and glandular-sticky on both sides, provided with 3-4 pairs of leaflets and 6-10 interjected leaflets, and borne on petioles 4-7.5 cm long, the latter densely pubescent and glandular, as in the case of the rachis. Leaflets elliptic-lanceolate to ovate-lanceolate, the margins occasionally unequal at base, shortly pilose and finely denticulate. Terminal leaflet more broadly elliptic-lanceolate and larger than the laterals, (2.8-)7-9(-11) cm long by (1.5-)3-4.5(-6.5) cm wide, with obtuse apex and subcordate base. Lateral leaflets gradualy decreasing in size toward the base, the upper pair 5-6.5(-10.5) cm long by 2.5-3(-4.8) cm wide, with obtuse or subpointed apex, the base obliquely rounded and shortly petiolulate. Interjected leaflets small, the largest 12 mm long by 9 mm wide, broadly elliptic-lanceolate, sessile, the smaller ones 2-3 mm long, suborbicular, usually decurrent about the rachis. Pseudostipular leaves partially clasping the stem, broadly falcate, to 18 mm long by 10 mm wide. Inflorescence cymose, 5-7-flowered, borne on short, thick peduncles, 3-5(-10) cm-long by 3 mm in diameter at base, the latter densely pilose and glandular, as in the case of

Figure 11. – *Solanum flavoviridens* (*Ochoa and Salas 11900*). 1. Upper part of flowering plant. 2. Corolla. 3. Petal. 4. Stamens, dorsal view. 5. Pistil. 6. Calyx. 7. Pedicel and stamens. All ×½.

Figure 12. – Leaf, stem, and tubers of a greenhouse plant of *Solanum flavoviridens* grown at the CIP facility near Huancayo (*Ochoa and Salas 11900*). All ×½.

pedicels and calyx. Pedicels articulated slightly above midpoint, the upper part 5-6 mm long and the lower 7-10 mm long. Calyx 7-8(-10) mm long, symmetrical, the lobes narrowly elliptic-lanceolate and gradually attenuate towards the apex. Corolla pentagonal or substellate, 3-3.5(-4) cm in diameter, white, with yellowish-green nectar guides, the lobes hood-shaped and pilose at apex, 9-11(-14) mm long by 17-18(-20) mm wide. Anthers narrowly lanceolate, 5-5.5(-6.5) mm long, 1.2 mm wide at base, light yellow, borne on 0.5-0.8 (-1.8) mm long, glabrous, white-hyalinous filaments. Style 11-12(-13) mm long, exerted 2-3 mm, occasionally sparsely or sometimes densely papillose along the basal two-thirds of its length; stigma small, shortly capitate, slightly broader than apex of style. Fruit globose, light green.

Chromosome number: 2n=3x=36.

Figure 13. – Terminal branch of *Solanum flavoviridens* (*Ochoa and Salas 11900*), ×½.

Type: BOLIVIA. Department La Paz, Province Saavedra, some 10 km east of the road Carijana-Camata, 1800 m alt. March, 1978; *C. Ochoa and A. Salas 11900* (holotype, OCH; isotype, CIP).

As in the case of *S. berthaultii* (*see* Ochoa, 1980a), the strongly glandular pubescence of *S. flavoviridens* may be of potential value in potato improvement programs aimed at the control of harmful insects. Such glandular hairs may be of value in warding off attacks by *M. persicae* (Gibson, 1971, 1974), *Leptinotarsa decemlineata* (Gibson, 1976) and thrips, including *Tetranichus urticae, Empoasca* sp., and *Liriomyza* sp. (Intl. Pot. Ctr., 1977, 1978).

Affinities

Solanum flavoviridens has some characteristics in common with *S. yungasense*, including habit, leaflet shape pedicel articulation, and fruit shape and color. However, it differs markedly from the latter species in leaf dissection, in the shape, size, and color of its corolla and calyx, and in the glandular pubescence and yellowish-green color of its leaves.

Through its dense indumentum and special hair types, *S. flavoviridens* also

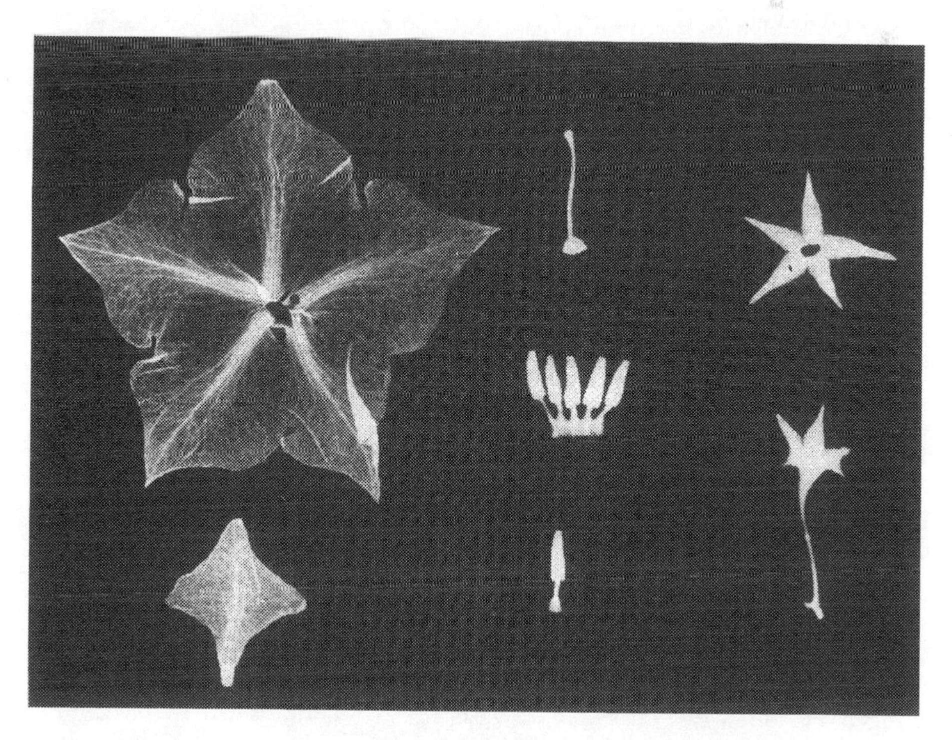

Figure 14. – Floral dissection of *Solanum flavoviridens* (*Ochoa and Salas 11900*), ×1.5.

shows a relationship to *S. berthaultii*. Both species are invested with three kinds of hairs: a multicellular-simple variety; a long, glandular-tipped form; and a short, glandular, tetralobed type. Both have similar shaped corollas, but the color of the first is white and the second is usually purple or lilac.

In the author's opinion, this plant may be a hybrid formed in crosses between *S. yungasense* and some other densely glandular-pubescent species such as *S. berthaultii*. The known, present day range of *S. yungasense*, however, lies more than 80 km beyond the range of *S. flavoviridens*, while the distance separating this species and *S. berthaultii* is even greater. As pointed out below (see *S. yungasense* and *S. microdontum*), gene flow can occur over a broad area of a species range. With respect to the origin of *S. flavoviridens*, this may be a case of introgressive hybridization involving other plants of this series from northwest Bolivia. Only additional field work can resolve this question.

Habitat and Distribution

This species is known only from the Yunga or Subyunga Region of Camata, Province of Saavedra, Department of La Paz, where it grows between 1600 and 1800 m altitude. In this humid-tropical zone, it occurs along forest margins, in open fields and thickets of woody shrubs, and in abandoned, weedy orange orchards, where it grows under invading wild trees. It is found usually in organically rich soils.

Specimens Examined

Department La Paz

Province Larecaja: Umani-Vilaque, 2250 m alt., some 10 km (air distance) southwest of Munaypata-Sorata, near Chahuarani River, in thickets. Vulg. name: *Ckipa Chocke* ('wild potato'). Chromosome number: 2x=3n=36. February 16, 1984; *Ochoa and Salas 15460* (CIP, OCH).

Province Saavedra: Approximately 10 km E of the Carijana-Camata road, across the river from Camata, between 1600 and 1800 m alt., in open fields and among woody shrubs and weeds in old abandoned orange orchard. March, 1978; *Ochoa and Salas 11900* (holotype, OCH; isotype, CIP).

5. Solanum litusinum Ochoa. Phytologia 48:229-230; 232, illustr. 1984.
FIGS. 15-17; MAP 5; PLATE IV.

Plants sparsely pilose. Stem to 70 cm or more tall and up to 1 cm in diameter at base, erect and simple or branched and decumbent, narrowly straight-winged and slightly pigmented towards the basal one-third of its length. Stolons more than 1 m long; tubers marked with lenticels, round to ovate, 2-3 cm long, white. Leaves imparipinnate and little dissected, (7-)17-24 cm long by (4.5-)11-16.5 cm wide, with 2-3(-4) pairs of leaflets and often 2(-3) small interjected leaflets, sessile, borne on petioles 2.5-3 cm long. Leaflets narrowly elliptic-lanceolate to ovoid-lanceolate, irregularly and minutely denticulate, finely and sparsely pilose on the margins and on both the upper and lower surfaces. Hairs of two types, simple and 2-3-cellular or short and glandular-tetralobed at apex (the latter type more densely distributed on the petiolules, petioles, and leaf axils). Terminal leaflets slightly larger and broader than the laterals, (3.5-)5.5-9 cm long by 1.8-2.5(-4) cm wide, with pointed and subacuminate apices and rounded bases. Lateral leaflets 3.5-4.5 cm long by 1.6-2 cm wide, the first and second pairs nearly equal in size, provided with pointed apices and rounded or obliquely rounded bases and borne on petiolules to 10 mm long. Pseudostipular leaves subfalcate or falcate, usually small, 1.5-3.5 (-10) mm long by 1-2(-6) mm wide. Inflorescence cymose to cymose-paniculate, up to 18-flowered, borne on peduncles 5-8(-12) cm long and to 2

Figure 15. – *Solanum litusinum* (*Ochoa 12027*). 1. Upper part of flowering plant. 2. Corolla. 3. Petal. 4. Stamens, dorsal view. 5. Pistil. 6. Calyx. 7. Pedicel and stamens. All ×1.

mm in diameter at base, light green, pilose and glandular, as in the case of the pedicels and calyx. Pedicel articulated above midpoint or along the upper one-third of the stalk, the upper part 7-9(-11) mm long and the lower 12-15 (-30) mm long. Calyx 6-7(-8) mm long, the lobes elliptic-lanceolate, abruptly narrowed at apex and provided with short, pointed acumens 1.5-2 mm long. Corolla stellate, (2.5-)3-3.5 cm in diameter, purple, nectar guides yellowish-green or grayish-white, and bearing triangular-lanceolate lobes (11-)12-13(-14) mm long by (8-)9-10(-11) mm wide. Staminal column cylindrical-conical, asymmetric; anthers narrowly lanceolate, 7-8 mm long and 1.8 mm wide at base, borne on glabrous 0.5-1.5 mm-long, white-hyalinous filaments. Style 11-12.5 mm long, exerted 2.5-3(-3.5) mm, densely papillose along the lower one-third of its length; stigma small, oval-capitate. Fruit globose to ovoid, light green and white verrucose except at base.

Chromosome number: 2n=2x=24.

Type: BOLIVIA. Department Santa Cruz, Province Valle Grande, along the route Quebrada Seca to Ariruma, near the border of Province Florida. March, 1978; *C. Ochoa 12027* (holotype, OCH; isotype, CIP).

Figure 16. – Floral dissection of *Solanum litusinum* (*Ochoa 12027*), ×1.

Affinities

This species resembles *S. berthaultii* in having purple or lilac corollas, short glandular hairs, and light green, globose to ovoid fruits. However, it differs from the latter species in leaf shape and dissection and by having a strongly stellate corolla, the latter bearing acumens more sharply-pointed than that known for any other Bolivian wild potato species. It also lacks the long glandular hairs of *S. berthaultii*.

Because the features of *S. litusinum* recombine the flower shape and pubescence of *S. tarijense* with the flower color of *S. berthaultii*, this proposed species may be a hybrid between the two. However, still another possibility is that it is a hybrid between *S. alandiae* and *S. chacoense*. The latter origin is suggested by artificial crosses made at CIP.

Habitat and Distribution

This species occurs in arid regions. It grows commonly in the shade of spiny shrubs and columnar cacti such as *Cereus* and *Neoraimondia*, or between colonies of *Opuntia* in poor, gravelly soil or on the moist, sandy-clay banks of rivers. This species is very rare, being known today only from its type locality in the Province of Valle Grande in southeast Bolivia, where it occurs along the mouth and lower reaches of the Quebrada Ariruma to an altitude of 1600 m.

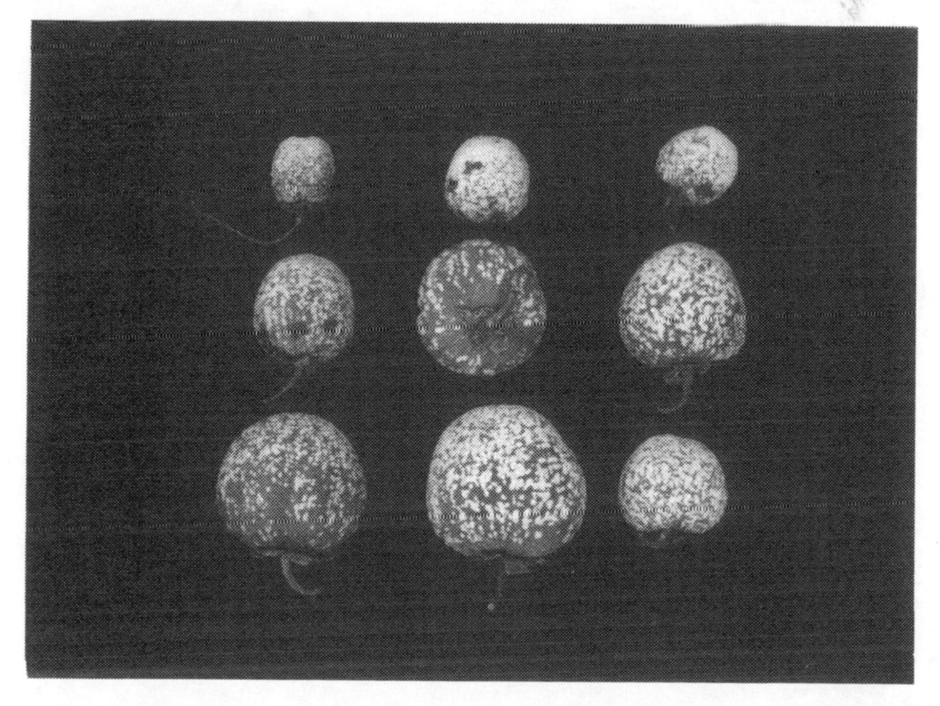

Figure 17. – Berries of *Solanum litusinum* (*Ochoa 12027*), ×1.

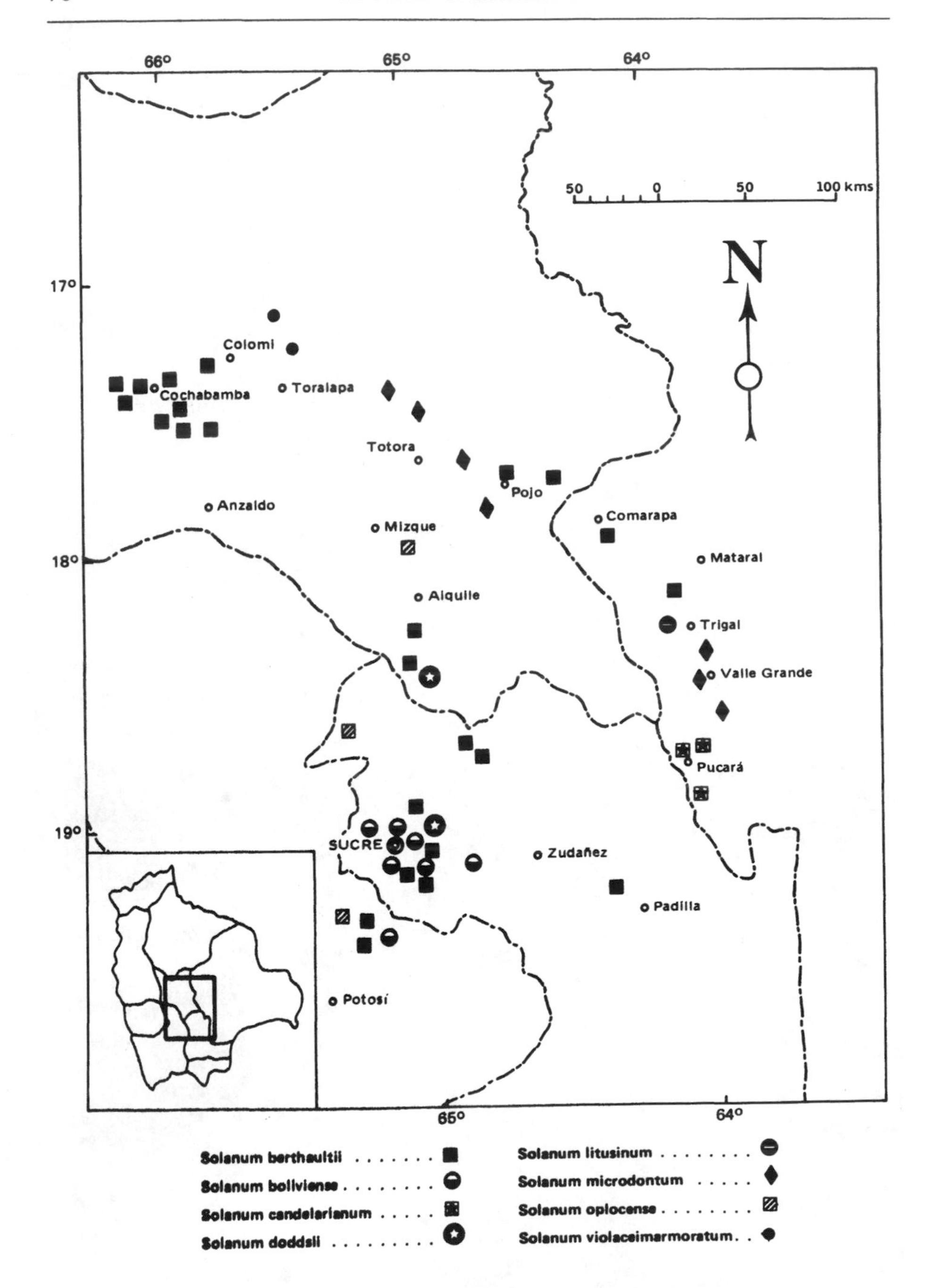

Map 5. – Distribution of wild potato species in Bolivia, showing *Solanum litusinum* (⊖).

Specimens Examined

Department Santa Cruz

Province Valle Grande: La Playa, 1600 m alt. [not 2600 m as erroneously reported on the specimen], along the ancient horse trail of Quebrada Seca to Ariruma. March, 1978; *Ochoa 12027* (holotype, OCH; isotype, CIP).

Crossability

The results of a preliminary crossing program involving this plant and three allopatric species, *S. alandiae*, *S. candolleanum*, and *S. tarijense*, and one sympatric species, *S. chacoense*, indicate there exists a wide range of relationships within this group. Crosses involving the first three species were difficult to make. Though many pollinizations were made, these resulted in few fruits containing relatively few seeds. On the other hand, 60% of the crosses made between this species and *S. tarijense* were successful. These yielded, on the average, 100 fertile seeds per berry.

Solanum litusinum × *S. alandiae*.– The F_1 hybrids from crosses between *S. litusinum* (12027) and *S. alandiae* (12016) resemble the first species very strongly in leaf shape. These typically have 2-3(-4) pairs of leaflets and few interjected leaflets. Various shades of purple were represented among the flower colors of these hybrids, all of which had rotate-pentagonal corollas. Hybrids obtained from crosses of *S. litusinum* (12027) with still another introduction of *S. alandiae* (12014) resemble those obtained from the first cross except for their greater leaf dissection. These plants typically had many leaflets and interjected leaflets, as in the case of *S. litusinum*, but differed from the latter in being more sparsely pilose and having subsessile to shortly petiolulate leaflets.

Solanum litusinum (12027) × *S. tarijense* (12001).– Hybrids derived from this cross were observed to have the dominant leaf shape and pubescence of *S. tarijense* and the long petiolules of *S. litusinum*. All had stellate corollas that varied in color from creamy-white to light lilac. Pedicel articulation was slightly above center or in the upper one-third of the stalk.

Solanum litusinum (12027) × *S. chacoense* (12026).– Crosses between these two species yielded plants with leaves similar to *S. litusinum*. The uniformly stellate corolla of these plants varied in color from pure creamy-white to white-tinged with pale lilac. Their floral peduncles held numerous flowers, the latter borne on pedicels articulated at various positions. The floral calyces of these plants appeared little different from those of *S. litusinum*.

Solanum litusinum (12027) × *S. candolleanum* (11913).– Crosses between these two allopatric species yielded small plants, about 20 cm tall, with short internodes. All had angular, narrowly winged stems, 2-3 mm in diameter. Leaf dissection and leaflet shape was that of *S. litusinum*. The short floral peduncles of these plants (about 2 cm in length) bore 3-5 light-lilac flowers, the latter about 2.5 cm in diameter and with white acumens.

6. Solanum tarijense Hawkes, Bull. Imp. Bur. Pl. Breed. & Genet., Cambridge (June):18, 20, 114-115, Fig. 3. 1944. FIGS. 18-23; PLATE V.

S. trigalense Cárd., Bol. Soc. Peruana Bot. 5:41-42, Pl. III-G, Figs. 1-3, Pl. V-D. 1956. TYPE: *Cárdenas 5069*, Bolivia. Between Trigal and Mataral, Dept. Santa Cruz, Prov. Valle Grande, March, 1955 (HNF-00081, WIS).

S. zudanense Cárd., Bol. Soc. Peruana Bot. 5:31-32, Pl. III-A, Figs. 1-3. 1956. TYPE: *Cárdenas 5077*, Bolivia. Between Zudañez and Tomina, Prov. Zudañez, February, 1949 (LL-photo).

S. berthaultii Hawkes f. *zudanense* (Cárd.) Corr., Wrightia 2:184. 1961. TYPE: *Cárdenas 5077*, Bolivia. Between Zudañez and Tomina, Prov. Zudañez, February, 1949 (LL-photo).

Plants light green, more or less pubescent, and glandular throughout, with a strong and disagreeable odor. Stems 60 cm tall, angular, light green, simple or branched, provided with short, sparse, glandular hairs and very narrow, straight or sinuous wings, these decurrent on the petioles. Stolons 80 cm or more long; tubers ovoid to oval-compressed or round, 2-3(-4.5) cm long, skin white or white with dark-mauve patches. Leaves numerous, imparipinnate, light green, more or less pilose, with glandular trichomes on the rachis and petiolules, (3-)8-30(-39.5) cm long by (2-)3.8-17(-28) cm wide, provided with (3-)4-5 pairs of leaflets and 5-15 interjected leaflets of 2-3 different sizes, the latter borne on sessile or very short petioles. Leaflets elliptic-lanceolate, oval-elliptic to elliptic or ovate-lanceolate, the margins usually entire, or very finely denticulate or slightly undulate, the apex obtuse to slightly acuminate, and the base obliquely rounded to subcordate, or cordate, borne on 15 mm-long petiolules and occasionally subtended by 1-2 small, interjected leaflets. Terminal leaflet nearly same size as the upper laterals, 3-7(-10) cm long by 1.5-3.5(-4.5) cm wide. Pseudostipular leaves semiovate, falcate, 8-12(-15) mm long by 5-6(-8) mm wide. Inflorescence cymose, 8-18(-25)-flowered, the densely glandular peduncle 10-14(-18) cm long, and sparsely and finely pilose. Pedicels 18-25(-30) cm long, recurved near upper end, articulated about the midpoint. Calyx 6-12 mm long, glandular-pilose, the ovate-lanceolate or narrowly elliptic lanceolate lobes narrowed at apex, and terminating in acumens to 5 mm long. Corolla stellate, to 3.5 cm in diameter, white or creamy white, strongly reflexed, the lobes deeply triangular-lanceolate to ovate triangular. Anthers narrowly lanceolate, to 7 mm long and 1.7 mm wide at base, borne on glabrous 1.5-2.5 mm-long filaments. Style 12-13 mm long, long-exerted, densely papillose along the lower two thirds of its length; stigma ovoid to claviform. Fruit globose to ovoid, to 2.5 cm long, light green, more or less white-spotted or verrucose.

Chromosome number: $2n=2x=24$.

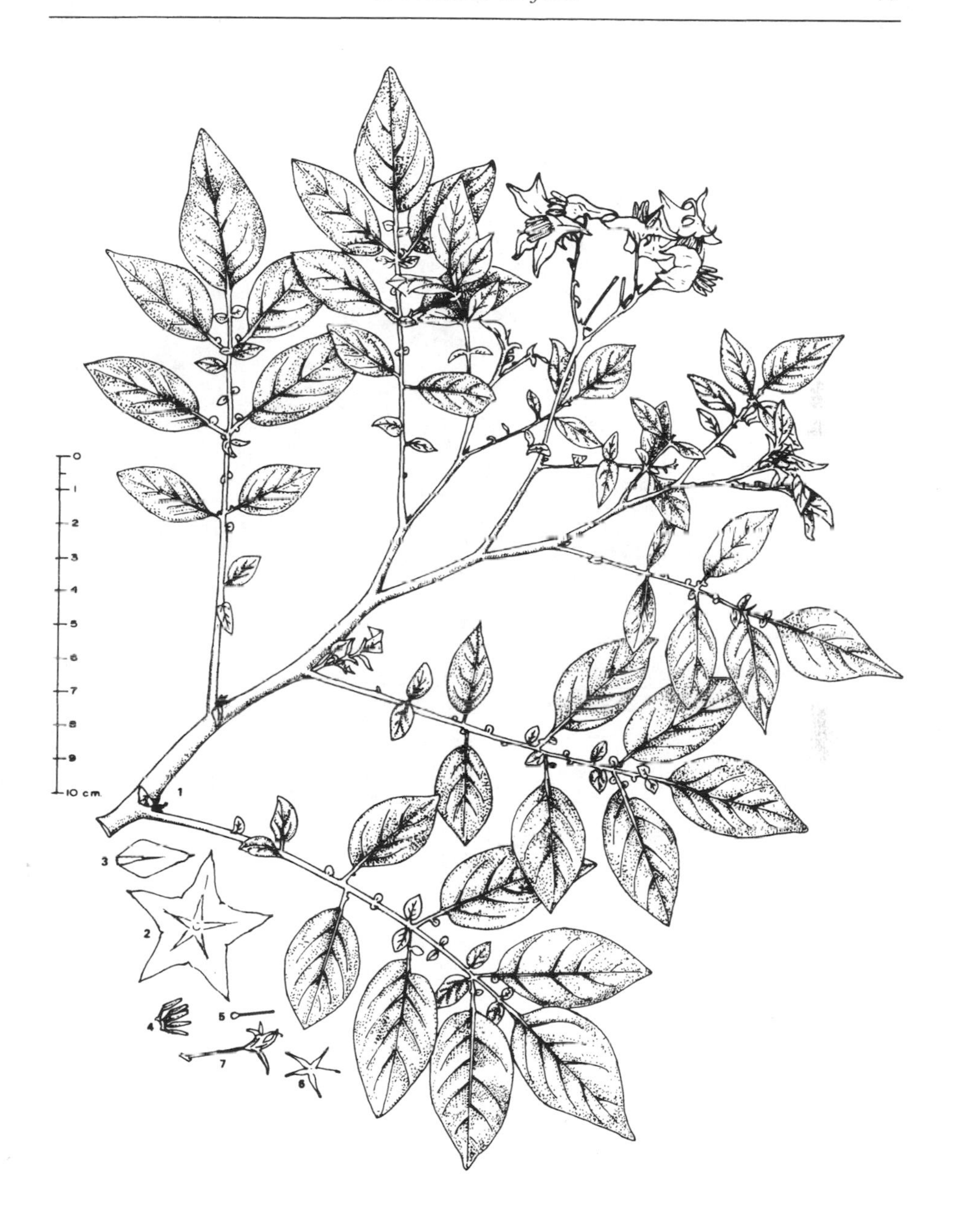

Figure 18. – *Solanum tarijense* (*Ochoa 12000*). 1. Upper part of flowering plant. 2. Corolla. 3. Petal. 4. Stamens, dorsal view. 5. Pistil. 6. Calyx. 7. Pedicel and calyx. All ×½.

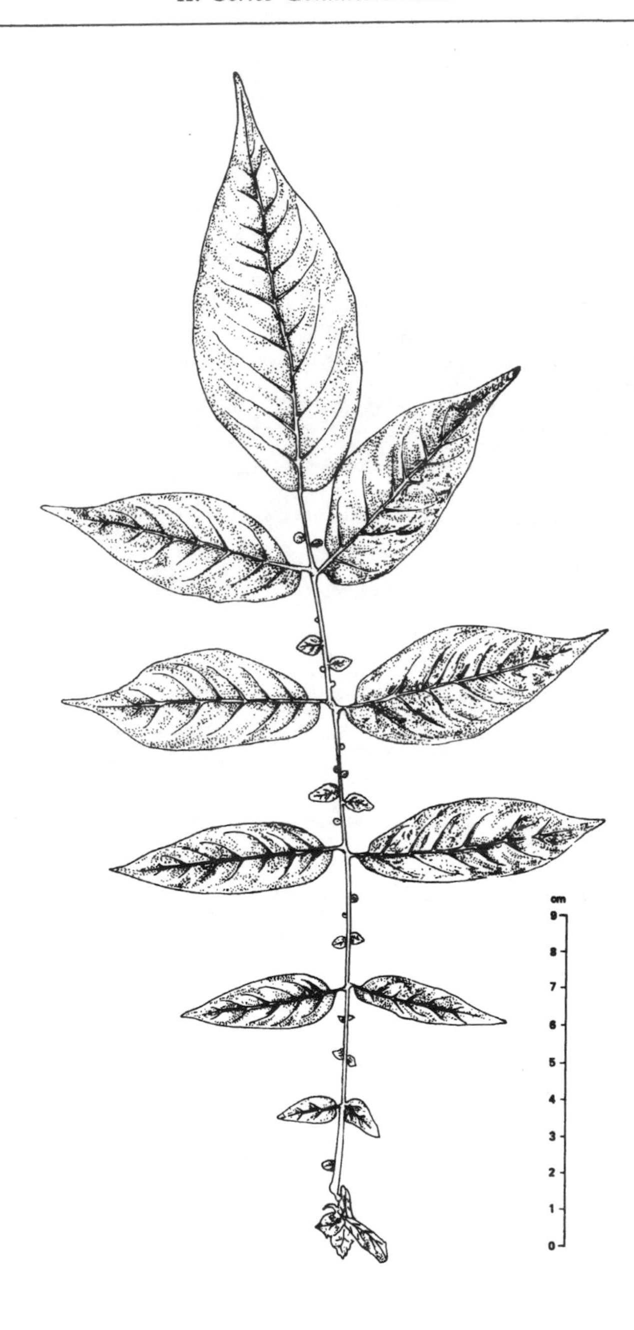

Figure 19. – Leaf of *Solanum tarijense* (*Ochoa and Salas 15529*), ca. ×1.

Type: BOLIVIA. Department and Province Tarija, above lower bank of river a little above Tarija, 2050 m alt. February 26, 1939; *E. K. Balls, W. B. Gourlay, and J. G. Hawkes 6093* (lectotype, designated by Hawkes K; isotypes, CPC, UC, US).

The author studied the type specimen of *S. trigalense* (*Cárdenas 5069*; HNF-00081) at the Herbario Nacional Forestal de Bolivia. In his opinion, the pubescence, leaf, and leaflet shape, and creamy-white stellate corolla of this plant place it clearly as a variant of *S. tarijense.*

The phototype of *S. zudanense* (*Cárdenas 5077*) was examined by the author at the Lundell Herbarium, Renner, Texas. He concluded from a study of this photograph that the plant described by Cárdenas (1956) was a mere variant of *S. tarijense*. His concept of this species was further reinforced by the characters cited by Correll (1962), who, however, classifies *S. zudanense* as a form of *S. berthaultii*.

It is important to remember that Cárdenas' collection numbers do not run sequentially. Thus, the type specimen of *S. zudanense*, which was collected in February 1949, is numbered 5077, while the type specimen of *S. trigalense*, which was collected twelve years later in April 1961, is numbered 5069. To

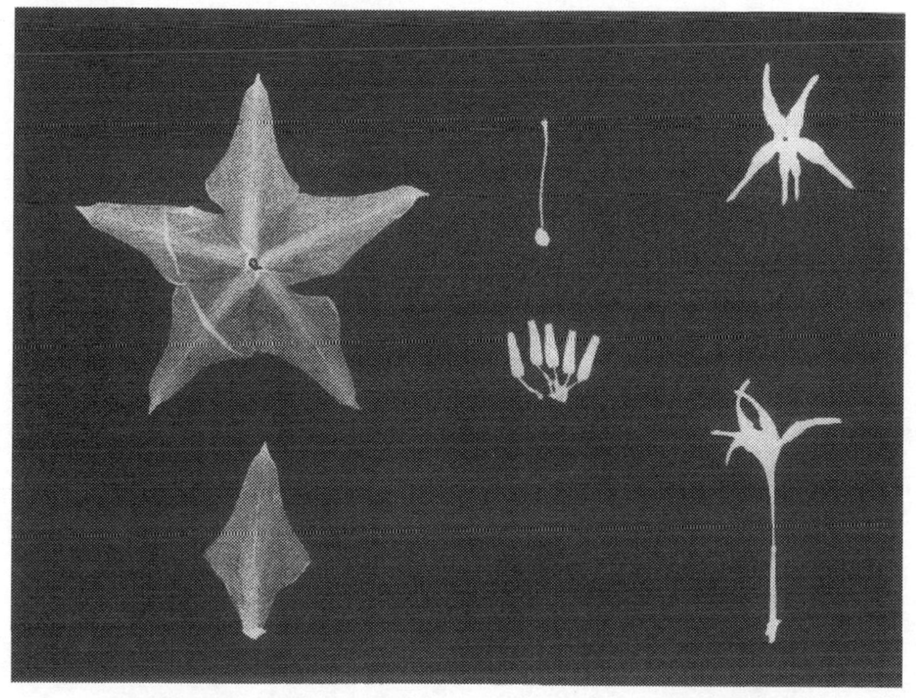

Figure 20. – Floral dissection of *Solanum tarijense* (*Ochoa 12000*), ×1.

add to this problem, we note that *Cárdenas 4503* (= *S. berthaultii*) was collected in February 1949, while *Cárdenas 5070* (= *S. vallegrandense*, type collection) and *Cárdenas 5071* (= *S. vallegrandense* var. *pojoense*, type collection) were collected in March 1955.

Affinities

This species, being glandular pubescent throughout, shows some relationship to *S. berthaultii*. The latter plant, however, is the more densely glandular and pubescent of the two.

Solanum tarijense is also allied to *S. chacoense*. This relationship shows up in the dissection of its leaves, the characteristics of its indumentum, and in the form of its calyx. These features also serve to differentiate these two species taxonomically.

According to Hawkes (1963), *S. berthaultii* may be a natural hybrid of *S. tarijense* and some other blue-flowered species of the series Tuberosa such as *S. sparsipilum*.

Habitat and Distribution

Solanum tarijense occurs preferentially in low rainfall areas. It is found in poor, gravelly or sandy-clay soils in regions dominated by *Acacia*, *Prosopis*, and caespitose and columnar cacti. This species is common in the lower valleys and levels of the phytogeographic formation known as 'Prepuna.'

Figure 21. – Floral dissection of *Solanum tarijense* (*Ochoa and Salas 15598*), ×1.

The Bolivian distribution of *S. tarijense* runs approximately from the Department of Cochabamba to the Departments of Chuquisaca, Santa Cruz, and Tarija. In northern Argentina, it occurs in the Provinces of Jujuy and Salta. Its altitudinal limits run between 2000 and 3000 m.

Specimens Examined

Department Cochabamba

Province Campero: Forest slope about 25 km from Aiquile, 3000 m alt., on road to Sucre. Plants glandular-pubescent; flowers white to lilac, stellate; fruits ovoid, white-spotted and streaked, 2 cm long. February 16, 1960; *Correll, Dodds, Brücher, and Paxman B619* (K, LL, S, US). 129 km from Sucre on road to Aiquile (20 km from Aiquile), 2100 m alt. Summer-green xerophilous forest, shade of thorny bushes. In bud; flowers white. January 29, 1971; *Hawkes, Hjerting, Cribb, and Huamán 4300* (CIP). Road branching westwards at 17 km from Aiquile towards Sucre, at Novillero, 2100 m alt., 1.7 km from main road. Lat. 18°17′ S, 65°13′ W. On edge of maize field in thorn hedge. Plants 75 cm tall, short glands only; flowers stellate, white to palest blue; berries globular, white-verrucose. February 22, 1980; *Hawkes, Hjerting, and Aviles 6438* (CIP). Palca, 2000 m alt., 13 km south of Aiquile (air distance),

Figure 22. – Berries of *Solanum tarijense* (*Ochoa 12005*), ×1.

18°15′ S and 65°12′ W, km 130 on the Totora Sucre road, smooth rock ledges just N of Palca, in soil-filled crevices, with *Acacia*, *Piptadenia*, *Cassia*, *Croton*, *Solanum radicans*, and *Cerastium*. April 11, 1963; *Ugent and Cárdenas 4870* (WIS).

Province Mizque: 18 km from Aiquile on road to Mizque, 2200 m alt. Lat. 18°5′ S and 65°9′ W. In thorn hedges dividing maize fields and by roadside. No long glands; white stellate flowers; white-spotted berries. February 23, 1980; *Hawkes, Hjerting, and Aviles 6451* (CIP).

Department Chuquisaca

Province Azero: Chaco, 900 m alt., eastern Andes, between Monteagudo and Lagunillas, in a petroleum district. *Cárdenas 3515, p.p.* (CPC, grown at Cambridge, England, 1951).

Province Oropeza: Tambockasa, 2850 m alt., from village to 0.5 km past, 16 km from Sucre on Zudañez. On sandstone slopes among rocks, thorn scrub, and cacti. March 4, 1971; *Hjerting, Cribb, and Huamán 4538* (CIP). Near Yotala, 2350 m alt., on road from Sucre to Yotala. On sandy-rocky land,

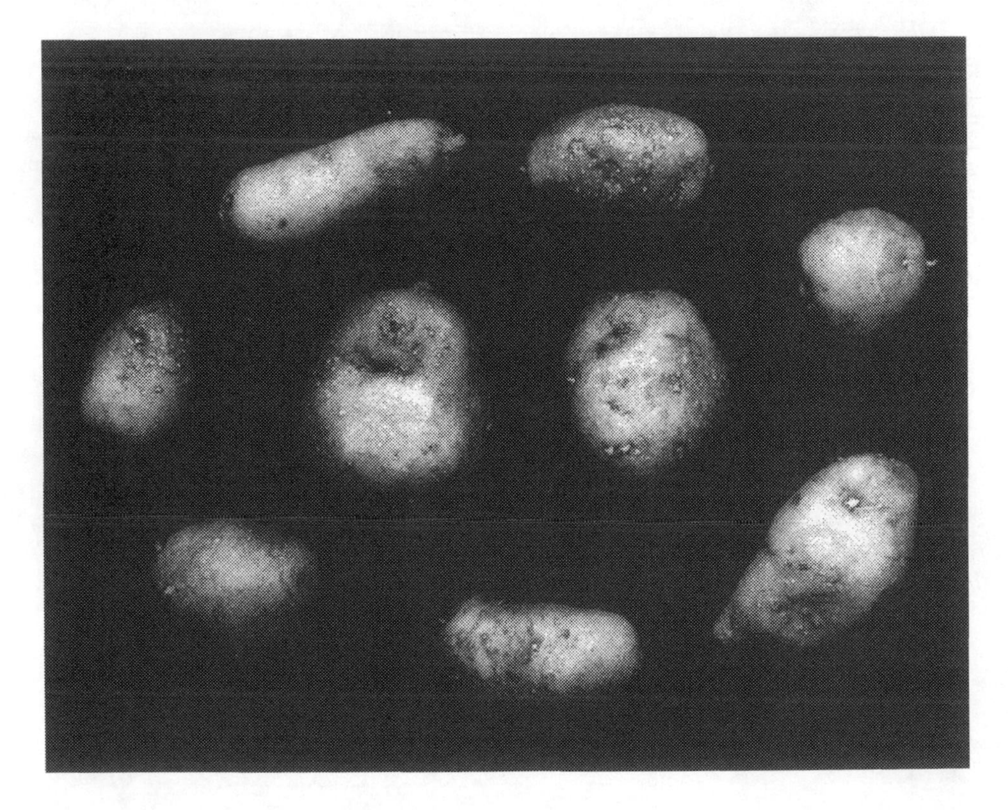

Figure 23. – Tubers of *Solanum tarijense* (*Ochoa 11994*), ×1.

among spiny shrubs and cactus, arid region. Flowers white. March 2, 1984; *Ochoa and Cagigao 15583* (CIP, OCH). La Lecheria, 1800 m alt., near Hueco. In a forested, subxerophytic small valley or dale, among acacias and columnar cacti and various herbs such as *Salvia*, *Commelina*, *Briza*, *Chenopodium* and composites. On rocky soil. Plants up to 1 m tall. Chromosome number 2n=2x=24. April 1, 1984; *Ochoa 15584* (CIP, F, GH, OCH, US). Km 28 on the road Sucre-Aiquile, 1950 m alt., about 15 km ENE of Sucre (air distance), 19°1′ S and 65°7′ W; in soil-filled crevices on a rocky slope with *Solanum zudanense*, *S. gandarillasii*, non-tuber-bearing *Solanum* sp., *Cleome aculeata* subsp. *cordobensis*, *Polygonum*, and *Cerastium*. April 11, 1963; *Ugent and Cárdenas 4885* (WIS), *4886* (US), *4889* (OCH).

Province H. Siles: Road from Sucre to Potosi at km 25 from Sucre, 2700 m alt. Thorn scrub. Plants to 50 cm, with dense glands and hairs; flowers white. January 30, 1971; *Hawkes, Hjerting, Cribb, and Huamán 4304* (CIP).

Province Zudañez: Among shrubs in humid soil along the road from Zudañez to Tomina, 2000 m alt. February, 1949; *Cárdenas 5077* (LL photo). Between Sunchu Tambo and La Quebrada (near Jatun Ckasa), 2600 m alt. Dry region with xerophytic vegetation. Plants robust, very pubescent, glandular. Flowers white, stellate. Common name *Lluttu papa*. Chromosome number: 2n=2x=24. March 24, 1978; *Ochoa 12005* (CIP, F, OCH, US).

Department Potosi

Province Saavedra: 83 km from Sucre on road to Potosi, just below Retiro, 2750 m alt. Amidst *Opuntia* by side of maize field. Plant 30 cm; flower white; no berries yet. January 30, 1971; *Hawkes, Hjerting, Cribb, and Huamán 4307* (CIP).

Department Santa Cruz

Province Caballero: At El Hueco, 1800 m alt., 2 km NE of Quebrada Seca. In forests of shrubs and trees, and with columnar cacti up to 4 m tall. February 29, 1984; *Ochoa and Salas 15561* (CIP, OCH, US).

Province Valle Grande: Near Mataral, 2100 m alt., km 350 on the road to Santa Cruz, on rocky slopes, in the shade. Plants to 60 cm tall; flowers yellowish. April, 1961; *Cárdenas 5069* (HNF-00081, WIS). Near Abra del Arrayanal, 1800 m alt., between forest trees and bushes in a dry region, in humid soil. Infrequent. February 27, 1984; *Ochoa and Salas 15529* (CIP, NY, OCH, US). Between Abra del Arrayanal and Trigal, 1600 m alt., on the road from Mataral to Valle Grande. Could be considered the topotype of the species, as the type locality 'cerca de Trigal' is situated nearby. April 1, 1984; *Ochoa and Salas 15592* (CIP, NY, OCH, US). Near Trigal, 1800 m alt., on the Valle Grande road between Mataral and Trigal. Very infrequent, specimens sterile, very poor; could be considered as topotypes. April 1, 1984; *Ochoa and*

Salas 15593 (CIP, OCH, US). Two km NNW of Quebrada Seca, 1900 m alt., between large colonies of cacti and in the shade of algaroba. April 1, 1984; *Ochoa and Salas 15596* (CIP, OCH, US). Near water tank El Tajo, 1800 m alt., approximately 1 km N of Quebrada Seca on road to Comarapa. In dry, hard lateritic soil, in shade of *huarango Acacia* trees. April 1, 1984; *Ochoa and Salas 15598* (CIP, OCH, US).

Department Tarija

Province Cercado: 33 km from Tarija on road to Entre Rios, 2350 m alt., under trees in damp flush and among sandstone rocks by roadside. March 13, 1971; *Hjerting, Cribb, and Huamán 4667* (CIP). 41 km from Tarija on road to Entre Rios, 2325 m alt., by farm in dry area. Very pentagonal flower. March 13, 1971; *Hjerting, Cribb, and Huamán 4673* (CIP).

Province Mendez: Tarija, 2050 m alt., above lower bank of river, a little above Tarija. In light, sandy loam among other herbs under spiny leguminous shrubs. Appears to be restricted entirely to this locality. February 26, 1939; *Balls, Gourlay, and Hawkes 6093* (lectotype, K, *fide* Hawkes; isotypes, CPC, UC, US). Tolomosa, 1920 m alt., near Tarija. Chromosome number: $2n=2x=24$. March 21, 1978; *Ochoa 11993* (CIP, OCH, US). Abra del Portillo, 1890 m alt., road to Bermejo. Xerophytic vegetation, common name *Papa de Chancho*. Chromosome number: $2n=2x=24$. March 21, 1978; *Ochoa 11994* (CIP, OCH). Cerro Las Canteras, 2200 m alt., in very poor gravelly soil. Subxerophytic region. In full fruit. Common name *Papa de Zorro*. Chromosome number: $2n=2x=24$. March 21, 1978; *Ochoa 12000* (CIP, F, OCH, US). La Tabladita, 1950 m alt., on road Tarija-Bellavista, among acacia trees. Common name *Papa de Zorro*. Chromosome number: $2n=2x=24$. March 21, 1978; *Ochoa 12001* (CIP, F, OCH, US).

Natural hybrids: Some hybrids of *S. tarijense* and *S. berthaultii* were uncovered during the author's Bolivian studies. These are listed below under the Departments of Cochabamba and Potosi.

Department Cochabamba

Province Campero: 21 km from Aiquile on road to Mizque, 2180 m alt. Lat. 18°04′ S, long. 65°10′ W. In a spiny hedge in thorn scrub region with cacti. Plant to 1 m high; berries white-spotted. February 23, 1980; *Hawkes, Hjerting, and Aviles 6453* (CIP).

Province Mizque: In the vicinity of Huara Huara, 2200 m alt., near Ichut'tirana, some 7 km to the east of the highway branch, Aiquile-Totora. March 25, 1978; *Ochoa 12009* (CIP, OCH).

Department Potosi

Province Saavedra: About 50 km from Sucre on the road to Potosi, on brushy slope about stone wall. Flowers blue-lavender with white bands along the inner side. February 17, 1960. *Correll, Dodds, Brücher, and Paxman B628* (LL).

Crossability

Natural crosses.– In addition to natural hybrids of *S. tarijense* and *S. berthaultii* cited above, the author has seen several Argentine collections by Sleumer (*3786, 4000, 4002, 4024*) in the Lundell Herbarium at Renner, Texas. Although these were labeled as *S. tarijense*, they actually represent natural hybrids of this species with diploid plants of *S. oplocense*, or with *S. leptophyes*.

Other Argentine hybrids of this species have been identified by Hawkes and Hjerting (1969). These include crosses between *S. tarijense* and *S. chacoense* and *S. tarijense* and *S. microdontum*, as well as intermediate hybrids and introgressed forms arising in crosses between *S. tarijense* and either *S. vidaurrei*, *S. vernei*, *S. gourlayii*, or *S. spegazzini*

Artificial crosses.– Agronomically, the value of *S. tarijense* lies in its resistance to *Heterodera rostochiensis* (Rothacker and Stelter, 1961) or *Globodera pallida* (Intl. Pot. Ctr., 1978). It is also resistant to such insect pests as *Tetranychus urticae*, *Thrips tabaci*, *Polyphagotarsonemus latus*, *Liriomyza huidobrensis* (*ibid*, 1977), and *Epitrix harilana rubia* (*ibid*, 1979a). It is possible that new and improved varieties of the cultivated potato, incorporating the germplasm of this species, may result from future potato breeding programs.

According to the breeding studies made at CIP, *S. tarijense* is a self-incompatible species. In order to establish crossability and dominance of morphological characteristics in *S. tarijense* and its hybrids, numerous crossses involving this species and other wild forms of either allopatric or sympatric origin were made. Species used in these crosses include *S. alandiae* (12012, 12014, 12016, 12017), *S. berthaultii* (12029), *S. infundibuliforme* (11977), *S. litusinum* (12027), *S. oplocense*-4x (12010), *S. oplocense*-6x (11972), and *S. sparsipilum* (12028, 12030). All, including *S. tarijense* (11993, 11994, 12000, 12001), occur from central to southeast Bolivia. Occasional Peruvian species such as *S. coelestispetalum* (13690, 13715) and *S. marinasense* (13722, 13740, 13748) were also used. Crosses involving the latter two Peruvian species were inefficient, resulting in very few seeds per fruit. On the other hand, crosses between *S. tarijense* and *S. marinasense* were much easier to obtain and yielded, on the average, up to 150 viable seeds per berry.

The characteristics of the F_1 hybrids of *S. tarijense* and the above mentioned Bolivian species are as follows:

Solanum tarijense × *S. alandiae*.– When used as a mother plant in crosses with *S. tarijense*, *S. alandiae* yielded 100 or more seeds per berry. Seed viability was close to 100%; in contrast, the reciprocal cross averaged only 40%. Regardless of the direction in which the cross was made, however, the hybrid populations exhibited the dominant characteristics of *S. alandiae*. These were manifest especially in the shape and dissection of the leaves. The form and color of the flowers of these plants exhibited greater variation, ranging from rotate-pentagonal and pale lilac with whitish acumens to substellate or stellate, and solid white to creamy white. In individuals with substellate or stellate corollas, the articulation of the pedicel was observed invariably to be below the stalk midpoint. One exception to the above pattern of dominance was observed in crosses between *S. tarijense* (11994) and *S. alandiae* (12016). In this cross, the leaves of the hybrids were more like those of *S. tarijense*. However, these plants produced light purple, substellate corollas similar in design to those of *S. litusinum*, with pedicels articulated above center.

Solanum tarijense × *S. litusinum*.– The F_1 populations of these two species are fairly easy to obtain. Mother plants yielded up to 170 viable seeds per berry. Segregates from this cross are similar in appearance to *S. litusinum* in habit and in the form and dissection of their leaves. These have, at most, 2-3 pairs of lateral leaflets. The white flowers of these plants, which are stellate and barely 2.5 cm in diameter, are borne on pedicels articulated near the upper one-third of their stalk. All lack interjected leaflets.

Solanum tarijense × *S. berthaultii*.– As in the case of the above cross involving *S. litusinum*, an F_1 population of these two species was easily obtained. Mother plants used in this cross yielded up to 200 seeds per berry, the highest average seed number obtained in a cross involving *S. tarijense* and any other wild species. The hybrids, in this case, exhibited the dominant morphological characteristics of *S. tarijense*.

Solanum tarijense × *S. sparsipilum*.– First generation plants of this parentage were difficult to obtain. Less than 40% of the pollinations resulted in fruits, and the latter had an average of only 20 viable seeds per berry. Hybrids from this cross have some traits that are intermediate between their parents. Thus, the large corolla of these hybrid plants varies from rotate-pentagonal to pentagonal, and from light lilac-blue to dark purple. In exceptional instances, the hybrids' corollas are found to be substellate and light lilac, or whitish. However, the anthers and calyces of these plants are similar to those of *S. tarijense*. In all segregates, the articulation of the pedicel is variable, ranging from below center to just above midpoint. Moreover, all are more or less sparsely pubescent, being covered with simple, short hairs and very short glandular hairs.

Solanum tarijense × *S. infundibuliforme*.– Fewer seeds are produced when *S. tarijense* is used as a mother plant in this cross than in the reciprocal combination. Thus, when used as a mother plant, the above combination yields berries in about 50% of the pollinations, with an average of about 20 seeds per berry. On the other hand, the reciprocal cross, which is much more efficient, yields up to 130 seeds per berry. The F_1 segregates of the above cross are interesting in that they trend morphologically toward *S. vidaurrei* and *S. leptophyes*. All have delicate stems, 40-50 cm tall, somewhat thickened above but slender and sinuous at base. However, the dominant characteristics of these plants are toward *S. vidaurrei*. All have the rough-pubescent leaves, leaf dissection, and leaflet shape of *S. vidaurrei*. The upper pair of lateral leaflets, which are sparsely pilose and entire-margined, are decurrent on the rachis. Floral articulation occurs in the upper one-third of the pedicel. The inflorescence and floral details are also like those of *S. vidaurrei*. Conceivably, the latter species may represent a stabilized diploid hybrid formed in a complex series of crosses between *S. tarijense*, *S. infundibuliforme*, and *S. leptophyes*.

Solanum tarijense × *S. oplocense*.– All experimental crosses involving this diploid species and either the tetraploid or hexaploid forms of *S. oplocense* were unsuccessful.

7. Solanum yungasense Hawkes, Ann. & Mag. Nat. Hist., ser. 12, 7:697, Figs. 5-6. 1954. FIGS. 24-27; PLATE VI.

Plants light green, sparsely pubescent to glabrous, loosely ascending or decumbent, to 2 m tall. Stem simple or branched, 5-7(-15) mm in diameter at base, with 1-3 mm-wide wings, the latter straight along stem center or upper one-third of the stem, but strongly sinuous at stem base. Stolons up to 1 m long and 2-3(-5) mm in diameter; tubers white, round to oval, 3-4 cm long, provided with abundant lenticels; sprouts densely pubescent, white, and tinged with pale lilac at base and apex. Leaves imparipinnate, (10-)14-21(-35) cm long by (5-)6.5-10(-17) cm wide, rarely pilose to glabrescent, and with few glandular hairs, somewhat rugose, with 4-6 pairs of lateral leaflets and 0-6 interjected leaflets. Lateral leaflets lanceolate or narrowly elliptic lanceolate-acuminate, the apex narrowly attenuate and sharply pointed, and the base subattenuate or round to obliquely rounded, on petiolules 2-3 mm long. Terminal leaflets markedly acuminate, ovate lanceolate, slightly larger than the laterals, 4-7.5(-9) cm long by 1.5-3(-5) cm wide. Pseudostipular leaves falcate, up to 1 cm long. Inflorescence cymose to cymose-paniculate, lax, the peduncle 6-10(-15) cm long, furcate, glabrescent as in the case of the pedicels, many-flowered, with 18-25(-35) blooms. Pedicels 15-20 mm long, articulated usually about the midpoint, the upper part 6-7(-8) mm long and the lower 6-9(-11) mm long. Calyx small, 5.5 mm long, sparsely pilose and glandular, the broad lobes elliptic-lanceolate, nearly round at apex and abruptly narrowed to the 1.5 mm-long acumens. Corolla deeply stellate, 2.5-3 cm in diameter, reflexed, white or yellowish white, with yellow, stellate nectar guides, the lobes triangular lanceolate, 7-9(-11) mm long by 4-5 mm wide at base.

Figure 24. – *Solanum yungasense* (*Hawkes, Hjerting, Cribb, and Huamán 4168*; CIP). Upper part of flowering plant, ca. ×½.

Staminal column doliform, the anthers light yellow, narrowly lanceolate, 5.5-6 mm long, on cordate bases and borne on narrow, 1.5-2 mm-long, glabrous filaments. Style 9 mm long, barely exerted 2 mm, sparsely pilose at base, the upper portion glabrous; stigma capitate to claviform. Fruit globose, light green, white-speckled, 1.5 cm in diameter.

Chromosome number: $2n=2x=24$ and $2n=3x=36$.

Type: BOLIVIA. Department La Paz, Province Nor Yungas, Milluhuaya, 1300 m alt. December, 1917; *O. Buchtien 617* (holotype, K; isotypes, BM, C, GH, M, NY, S, US).

As for many other species that grow in subtropical and humid habitats, *S. yungasense* displays enormous environmental plasticity. Under open-field conditions, plants of this species typically develop short stems (40-50 cm tall) with laterally extended branches and leaves. Plants of this same species, on the other hand, when growing under shrubs in dense shade, develop thicker stems to 2 m tall that bear many branches and larger leaves.

Other morphological modifications of this species are associated with ploidy level. Triploid plants ($2n=3x=36$), for example, are generally more robust

Figure 25. – Living collection of *Solanum yungasense* (*Ochoa and Salas 14842*).

than diploids (2n=2x=24). Also, the stems of the former are thicker and are provided with broader and more sinuous wings than the latter. Moreover, while both diploid and triploid plants have similar leaf dissection, the leaflets of triploids tend to be more elliptic-lanceolate and somewhat more acuminate than diploids, while the flowers are larger with more broadly triangular-lanceolate lobes. Although the flowers of both are occasionally six-merous, this condition tends to occur more frequently among triploids.

Affinities

Solanum yungasense is closely related to *S. chacoense*. Both species have a similar calyx and corolla. Although specimens of *S. yungasense* are occasionally confused in the herbarium with *S. violaceimarmoratum*, the latter species has a violet, rotate corolla. Moreover, the two differ in fruit shape, leaf shape, and dissection. Thus, *S. violaceimarmoratum* has been treated in this work under a different series (*see* Conicibaccata).

Habitat and Distribution

Solanum yungasense grows between 1100 and 1900 m altitude in the subtropical region of Bolivia, called *Yunga*, located principally in the northwestern sector of the country in the Provinces of Nor Yungas and Sud Yungas (Department La Paz). Its distribution extends from southeastern Peru to the watersheds of the eastern slopes of the Bolivian Ceja de Montaña, which in the Department

Figure 26. – Floral dissection of *Solanum yungasense* (*Ochoa and Salas 14842*), ×1.

of Puno, Peru, includes the lower parts of the Valle de Sandia and the Alto Inambari (1500-1800 m alt.). According to Correll (1962), this plant also grows in northwestern Argentina in the vicinity of San Luis, Province of Salta (*Filipovich No. 381*). However, the voucher specimen for this locality was unavailable for examination by the author.

Two Peruvian collections of *S. yungasense*, San Juan del Oro, *Ochoa and Salas 14842*, and Churiuma, *Ochoa and Salas 15041*, both in the Department of Puno, are of some importance, as they extend the known geographic distribution of this series and the two known chromosomal forms of this species. While the center of origin of this plant remains unknown, it is possible that it has migrated in the past from southeastern Peru to northern Argentina.

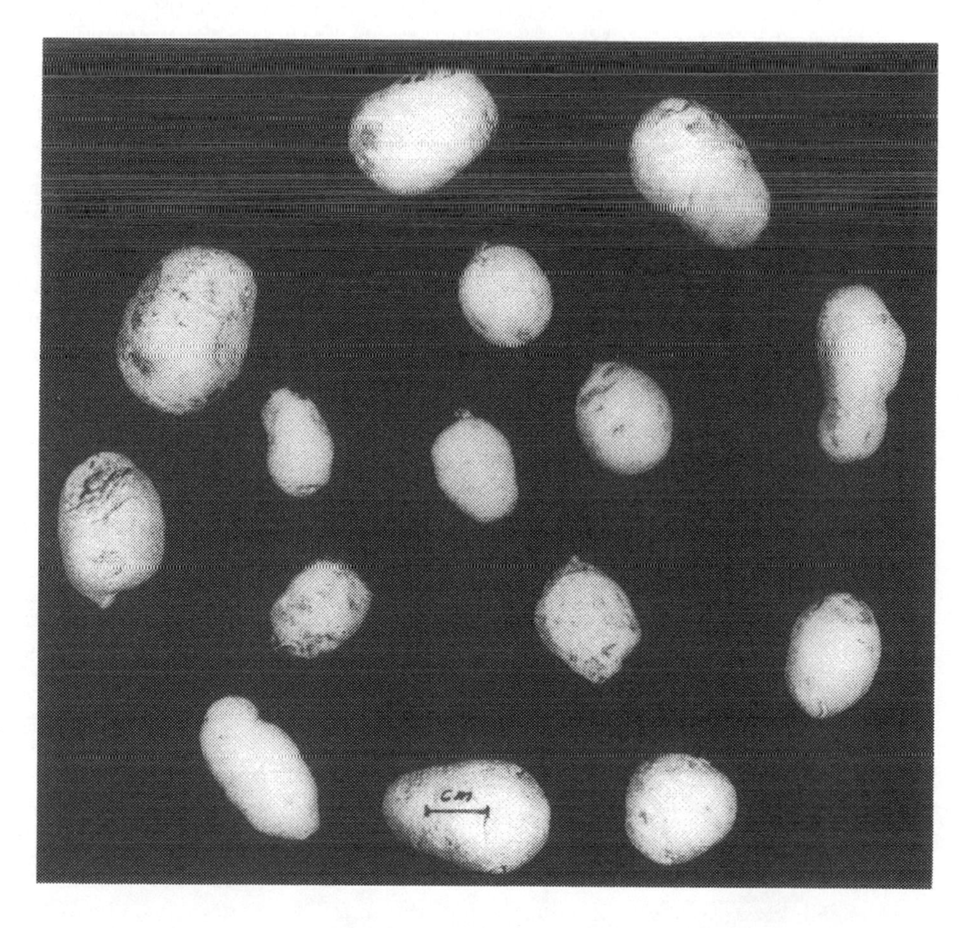

Figure 27. – Tubers of *Solanum yungasense* (*Ochoa and Salas 14842*), ×1.

Specimens Examined

Department La Paz

Province Larecaja: Tipuani, 1400 m alt., Hacienda Simaco on the road to Tipuani. Subtropical region. February, 1920; *Buchtien 5547* (GH, MO, US).

Province Nor Yungas: Milluhuaya, 1300 m alt. Subtropical region. To 1.5 m high. December, 1917; *Buchtien 617* (holotype K; isotypes, BM, C, GM, M, NY, S, US). Polo-Polo, 1100 m alt., near Coroico. October-November, 1912; *Buchtien 4693* (US).

Province Sud Yungas: El Chaco, 1900 m alt. December 1, 1920; *Asplund 1204* (UPS). Puente Villa, 1400 m alt. *Corro s.n.* (CPC-1775, from a plant grown at Cambridge, England, 1948). Puente Villa, 1250 m alt., on the road La Paz-Chulumani. In bushy places, and under the shade of trees, near a river and hanging bridge. Plants 60 cm, flowers cream. January 17, 1971; *Hawkes, Hjerting, Cribb, and Huamán 4168* (CIP). On left bank of Rio Uchumachi, 1500 m alt., on road Coroico-Chulamani, tropical region, between stalks of sugar cane. Plants more than 1 m tall. Infrequent. Flowers small, creamy white. March, 1985; *Ochoa 16150* (OCH).

Crossability

S. yungasense is a self-incompatible species. Moreover, according to Hawkes (1954), this species will not cross with *S. commersonii* nor *S. tarijense*. Nor will it cross easily with *S. chacoense*, to which it is seemingly allied. Hawkes also reports that crosses between *S. yungasense* and different clones of *S. chacoense* are successful in only one case out of nine. The results of these experiments suggest that we are dealing with very distinct species.

Crosses between *S. yungasense* and the diploids *S. vidaurrei* and *S. tarijense* yielded fruits containing varied numbers of seeds. But crosses involving triploid *S. yungasense* and the two diploid species mentioned above yielded fruits containing only 1-2 seeds per berry.

Crosses between *S. yungasense* and *S. phureja* yielded only parthenocarpic fruits.

It is interesting to note that while the fertility of the pollen of diploid *S. yungasense* ranked between 70-78%, the value of triploid *S. yungasense* pollen was surprisingly high for a plant having three sets of chromosomes, falling between 45-50%.

Solanum yungasense × *S. vidaurrei*.– Crosses between triploid *S. yungasense* and diploid *S. vidaurrei* yielded F_1 individuals that were more or less uniform with respect to habit, leaf dissection, and corolla size and color. All appeared to have more dominant characters of *S. vidaurrei* than of *S. yungasense*, including slender, wingless, sparsely pilose stems, 50-60 cm tall. The short, broad leaves of these hybrids were observed to be pubescent and provided with 4-5 pairs of leaflets and several interjected leaflets. The leaflets themselves were shortly petiolulate and broadly elliptic-lanceolate, and had semisharp to obtuse apices and rounded bases. All had small, rotate, light blue to sky blue corollas, these borne on pedicels articulated at or near the midpoint, or in the upper one-third of the stalk. The densely pilose, 10 mm-long calyx (very similar to that of *S. vidaurrei*) was provided with narrowly pointed acumens, these up to 5.5 mm long.

III. SERIES CUNEOALATA

CUNEOALATA Hawkes, Bull. Imp. Bur. Pl. Breed & Genet., Cambridge 118. 1944.

Plants small, spreading, rarely forming rosettes. Leaves imparipinnate, without interjected leaflets. Lateral leaflets narrowly lanceolate, with cuneiform wings, the latter narrowly decurrent about the rachises of each pair. Pedicel articulated at, near, or above center. Corolla rotate-pentagonal to substellate, purple, violet, or white. Fruit globose.

Chromosome number: 2n=2x=24.

Distribution

Central and southern Bolivia, northeastern Chile, northern Peru, and northwestern Argentina, between 2500 and 4200 m alt.

Although Hawkes (1944) initially proposed various other species within the series Cuneoalata, he later (1956a) reduced their number as a result of a series of transfers to *S. infundibuliforme*. Similarly, Brücher (1957a) considered many of Hawkes' species to be no more than variants of *S. infundibuliforme*, including *S. xerophyllum*. Thus, according to Hawkes (1944), Brücher (1957a), and Hawkes and Hjerting (1969), the series Cuneoalata is believed to be constituted at the present time only by its type species, *S. infundibuliforme*. A second species endemic to northern Peru, *S. anamatophilum*, described later by the author (Ochoa, 1964), will be treated in a volume dedicated to the potatoes of Peru.

8. Solanum infundibuliforme Philippi, Anal. Mus. Nac. Chile, 2 ed., Bot. 65. 1891. FIGS. 28-30; MAP 6; PLATE VII.

S. infundibuliforme var. *angustepinnatum* Bitt., Repert. Sp. Nov. 11:388-389. 1912. TYPE: *Philippi s.n.*, Bolivia. Calcalaguai, near Chilean border, March, 1855 (destroyed, US-photo type collection of *S. infundibuliforme* and, in part, of var. *angustepinnatum* ex Herb. B, not seen).

S. platypterum Hawkes, Bull. Imp. Bur. Pl. Breed. & Genet., Cambridge (June):35, 118, Figs. 18-19. 1944. TYPE: *Balls, Gourlay, and Hawkes 6126*, Bolivia. Quebrada Honda, near Villazon, Dept. Tarija [Prov. Aviles], February 27, 1939 (CPC, K, US).

S. microphyllum Hawkes, Bull. Imp. Bur. Pl. Breed. & Genet., Cambridge (June):35, 36, 118, Fig. 20. 1944 (not *S. microphyllum* Dun., 1813) TYPE: *Balls 5980*, Argentina, February 15, 1939 (CPC, UC, US).

S. glanduliferum Hawkes, Bull. Imp. Bur. Pl. Breed. & Genet., Cambridge

(June):35, 37, 118-119, Pl. IIa, Fig. 21. 1944. TYPE: *Balls, Gourlay, and Hawkes 5956*, Argentina, February 10, 1939 (CPC, F, K, US).

S. pinnatifidum Cárd, Rev. Agric. Cochabamba 2(2):33-34, illustr. 1944 (not *S. pinnatifidum* Plate, 1797; not Ruiz and Pavon, 1799; not Roth, 1821). TYPE: *Gandarillas 57*, Bolivia. Anzaldo, Dept. Cochabama, Prov. Tarata, January, 1942 (CA, GH).

S. xerophyllum Hawkes, Jour. Linn. Soc. Bot. 53:108. 1945. TYPE: *Gandarillas 57*, Bolivia. Anzaldo, Dept. Cochabamba, Prov. Tarata, January, 1942 (CA, GH).

S. infundibuliforme var. *albiflorum* Ochoa, Phytologia 46(4):223-224. 1980. TYPE: *Ochoa 11968*, Bolivia. Tajopunta, Dept. Potosi, Prov. Sud Chichas, March 18, 1978 (holotype, OCH).

Plants slender, usually small, 10-25(-40) cm tall, more or less pubescent throughout with short, thick hairs. Stem simple or branched, somewhat flexible, wingless, thin, 1.5-3.5 mm in diameter at the base, erect or decumbent. Stolons long and thin; tubers light gray, rounded to ovoid, 10-20 mm long. Leaves imparipinnate or imparipinnatisect, (3-)5.5(-14.5) cm long by (1-)2.4-4.5(-8) cm wide, lacking interjected leaflets, and provided with 2-4(-5) pairs of lateral leaflets, these borne on petioles 1-2.5(-3) cm long. Terminal leaflet much larger than the laterals, (2-)3-6(-7) cm long by (0.3-)0.8-1.8(-2.5) cm wide, elliptic-lanceolate to rhombic-lanceolate or linear-lanceolate, the apex obtuse and the base attenuate and narrowly decurrent. Lateral leaflets to 4.5 cm long and 1.8 cm wide, narrowly elliptic-lanceolate or lanceolate to linear-lanceolate or linear, the apex obtuse and the base strongly decurrent upon the rachis along the basiscopic side, resembling a broad, triangular wing along the upper part and a narrow wedge below. Pseudostipular leaves, when present, narrowly falcate to linear falcate, small, 7-10(-15) mm long by 1.5-3 mm wide. Inflorescence cymose, (1-)3-5(-10)-flowered, the peduncle 3-5(-8) cm long, sparsely short-pilose as in the pedicels and calyx. Pedicel articulation located more or less toward the center or upper third of the stalk, the upper part 3-6(-8) mm long, and the lower 9-20 mm long. Calyx 5-8.5(-9) mm long, the lobes ovate-lanceolate with narrow, 2-5 mm-long pointed acumens. Corolla rotate to rotate-pentagonal or substellate, 1.2-1.6(-2.7) cm in diameter, the lobes 5-9 mm long including the acumens by 8-12 mm wide, dark blue to light lilac, or pure white with a slight tinge of purple in the interpetalar zone. Anthers triangular-lanceolate, 3.5-4.5 mm long, borne on filaments 1-2 mm long. Style 6.5-9 mm long, papillose along the basal half; stigma globose to ovoid and broader than the apex of the style. Fruit globose, 1.5 cm in diameter, uniformly green or speckled with white spots.

Chromosome number: 2n=2x=24.

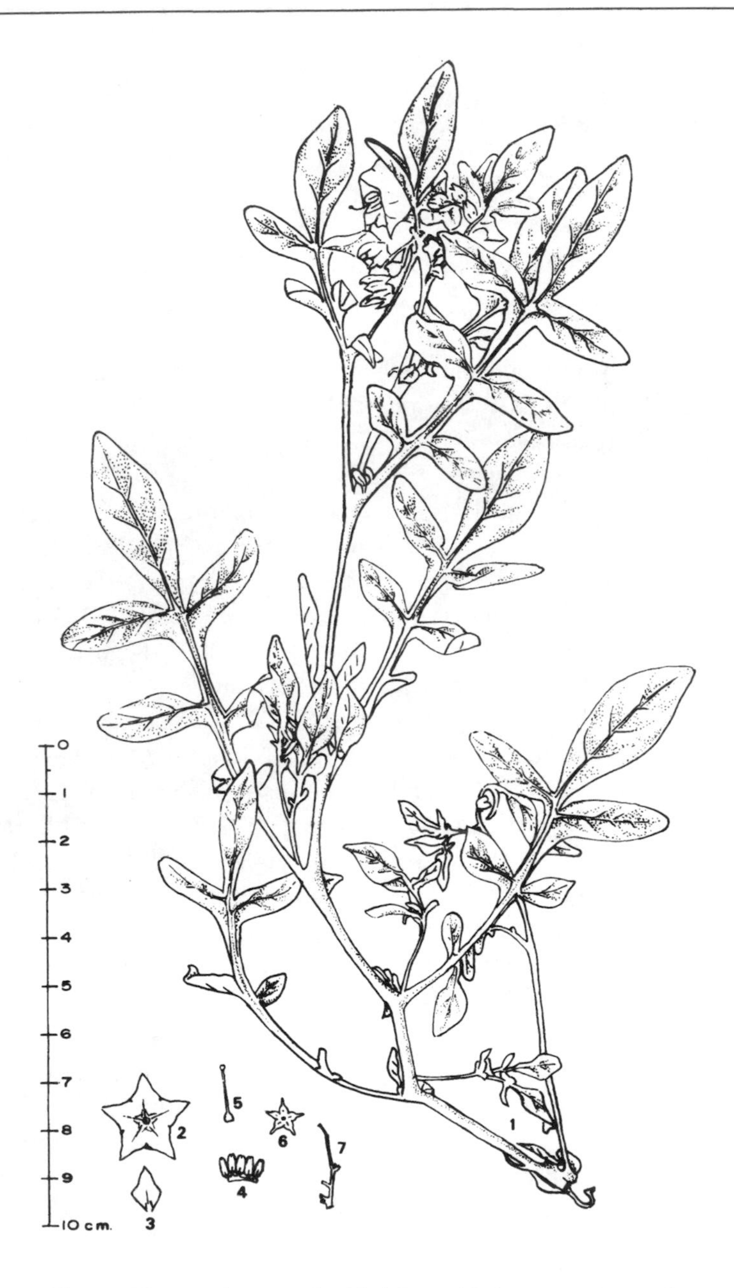

Figure 28. – *Solanum infundibuliforme* (*Ochoa 11968*). 1. Flowering plant, about ×½. 2. Corolla. 3. Petal. 4. Stamens, dorsal view. 5. Pistil. 6. Calyx. 7. Pedicel and articulation. All ×1.

Type: CHILE. Province Tarapaca, Calcalhuay (now in Bolivia, Department Potosi, near Chilean border). March, 1855 [or March, 1856?]; *R.A. Phillipi s.n.* (lectotype, SGO, accession No. 55467, designated by Hawkes and Hjerting).

As noted previously by Brücher (1957a), Correll (1962), and Hawkes and Hjerting (1969), *S. infundibuliforme* is extremely variable. Brücher, who illustrated the leaf and leaflet variation of this species, drew upon only a single source area, the upper valley of Humahuaca (3700 m alt.), near Iturbe in northern Argentina. For this reason, it may be that all of the collections examined by Brücher were merely ecotypes of the same species. The latter conclusion springs in part from the author's own independent examinations of living collections of this species grown under uniform environmental field conditions at the U.S.D.A. Potato Introduction Station, Sturgeon Bay, Wisconsin. These investigations, moreover, suggest that further subdivisions of this species are unwarranted.

Affinities

This plant is closely related to *S. anamatophilum*, a species from northern Peru (Ochoa, 1964). Both have cuneiform and peculiarly winged leaflets. However, *S. anamatophilum* may be distinguished from *S. infundibuliforme* by its woody or subwoody stem, by its more strongly dissected leaves, and by certain floral characteristics.

Figure 29. – Floral dissection of *Solanum infundibuliforme* (*Ochoa 11942*), ×1.

Habitat

This plant grows in extremely inhospitable regions characterized by low rainfall and great variations in temperature, such as occur in the southeastern stretches of the altiplano, and in the arid and semiarid Punas of the Province of Sud Lipez, Department of Potosi, near the northern border zone of Chile. In the Cordillera of Mochara, Province of Sud Chichas (Department Potosi), colonies of very small plants of this species (10-20 cm tall) occur between scattered bunch grasses. In this very high region (3800-4200 m), the plant is associated with *S. acaule* and *S. megistacrolobum*. It is also found in the Prepuna and in the lower, arid desert valleys of the Department of Cochabamba, as well as in the southeast of the country in the Department of Tarija. Here, it occurs mainly between 2500 and 2900 m altitude. In the latter zone, it frequently occurs in association with spiny plants such as cacti and algaroba. Similarly, this species has been collected quite frequently in the dry inter-Andean valleys of the Provinces of Salta and Jujuy in northern Argentina.

Distribution

This species is restricted to the border triangle-region formed by northern Chile, central and southern Bolivia, and northwestern Argentina. Its altitudinal limits lie between 2500 and 4200 m.

Specimens Examined

Department Cochabamba

Province Arani: Toralapa, 3700 m alt., on a slope under the shade of shrubs. January, 1946, *Cárdenas 3683* (HNF, CPC, LL).

Province Carrasco: On road to Santa Cruz, 100 km from Cochabamba, among *Hechtia* on rocky slopes. February 15, 1960; *Correll, Dodds, Brücher, and Paxman B614* (LL).

Province Quillacollo: In clumps of bushes on mountain slopes near San Miguel, above Liriuni. February 14, 1960; *Correll, Dodds, Brücher, and Paxman B610* (LL).

Province Tarata: Anzaldo, 1 km west of railroad station, 3000 m alt. January, 1942; *Gandarillas 57* (HNF, GH). One km west of Anzaldo railroad station, Cochabamba-Villa Villa, 3200 m alt., flowers light bluish-violet, common along the borders of maize fields. January 8, 1943; *Cutler, Cárdenas, and Gandarillas 7600* (HNF, LL, MO).

Department Chuquisaca

Province Sud Cinti: Near km 70 from Villa Abecia on road to Las Carreras. February 22, 1960; *Correll, Dodds, Brücher, and Paxman B650-B651* (LL, US).

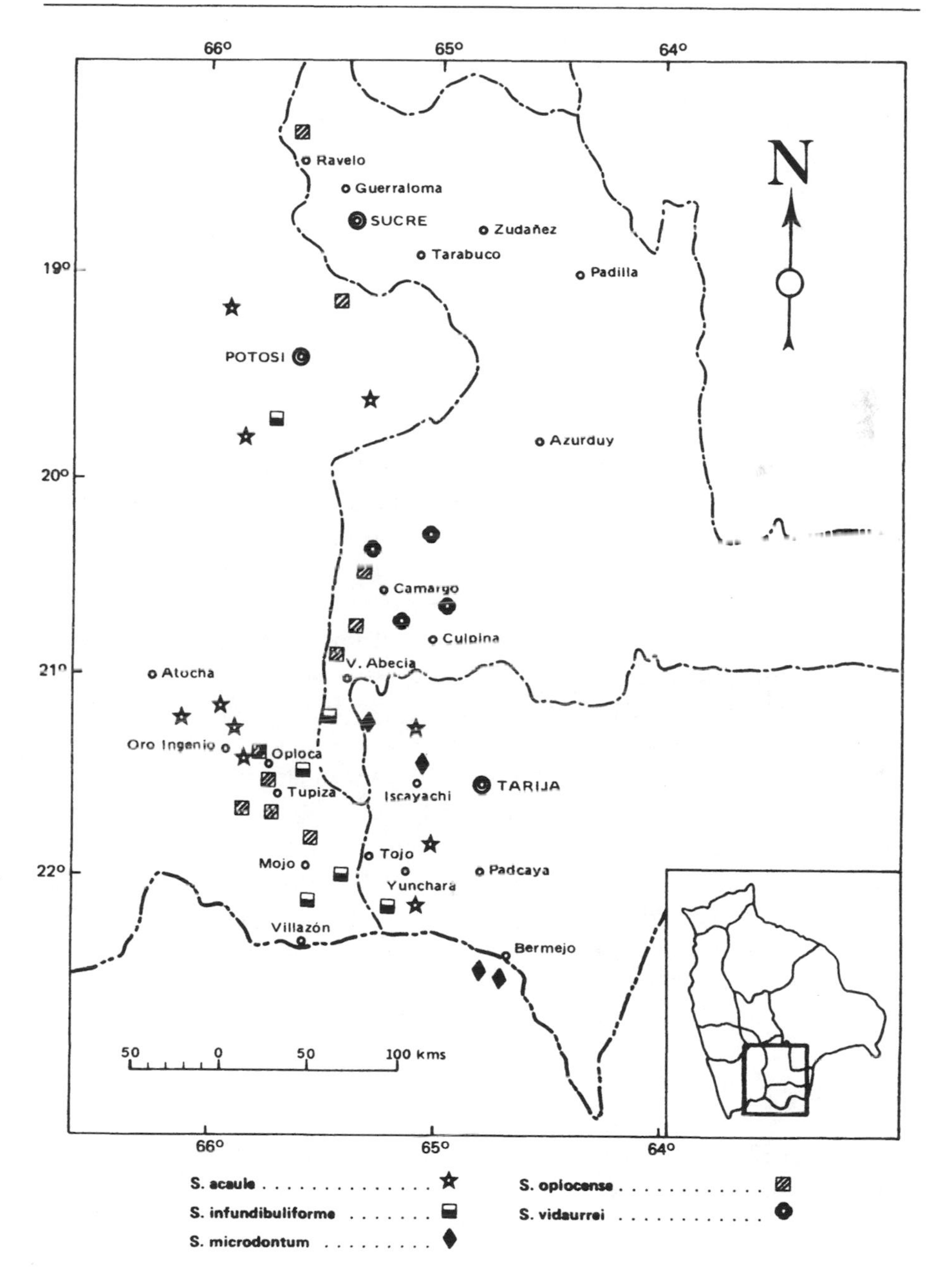

Map 6. – Distribution of wild potato species in southern Bolivia, showing *Solanum infundibuliforme* (▬).

Department Potosi

Province Linares: Near Cucho Ingenio, some 45 km from Potosi on road to Camargo. March, 1952; *Cárdenas s.n.* (HNF-0038). On rocky, dry slopes of canyon among cacti, 35 km from Potosi on road to Camargo, 3500 m alt. February 20, 1960; *Correll, Dodds, Brücher, and Paxman B637-B638* (K, LL, S). Near Cuchu Ingenio, 3600 m alt., beneath bushes on rocky slope. 'Plants 20-40 cm tall, flowers blue.' April 6, 1962; *Cárdenas 6119* (HNF, WIS). 35 km from Potosi on road to Camargo, 3800 m alt. Dry valley with columnar cacti, *Mutisia, Adesmia*, etc., more or less prepuna (c.f. Tilcara). Flowers substellate, purple. February 1, 1971; *Hawkes, Hjerting, Cribb, and Huamán 4321* (CIP). Near Mollepunco, 3880 m alt., on the road from Potosi to Camargo, among stones and cacti, associated with *S. megistacrolobum*. Chromosome number: 2n=2x=24. March 14, 1978; *Ochoa 11942* (CIP, OCH).

Province Omiste: Quebrada Honda, 3000 m alt., near Villazon, on loose shaley stones. February 27, 1939; *Balls, Gourlay, and Hawkes 6126 (CPC, K, US).*

Province Sud Chichas: Along the road from Impora to Tupiza, 3800 m alt., among straw in Puna formation, descending from Cordillera de Impora to Tupiza. Plants 15 cm tall. Berries globose, green, densely white-speckled. Chromosome number: 2n=2x=24. March 17, 1978; *Ochoa 11966* (CIP, OCH). Tajopunta, 4000 m alt. Chromosome number: 2n=2x=24. March 18, 1978; *Ochoa 11968* (OCH, CIP). Near Mojo, 3450 m alt., under the shade of *Acacia* sp. or *Chulque*. Chromosome number: 2n=2x=24. March 20, 1978; *Ochoa 11973* (CIP, OCH). One km from Mojo, on the route Mojo-Yunchara, 3400 alt. Chromosome number: 2n=2x=24. March 20, 1978; *Ochoa 11974* (CIP, OCH). Cuesta de San Miguel, 3500 m alt., on road to Tojo via Yunchara and Iscayachi. Chromosome number: 2n=2x=24. March 20, 1978; *Ochoa 11977* (CIP, OCH).

Province Quijarro: Porco-Uyuni, 3700 m alt. March 24, 1921; *Asplund 3110* (UPS).

Crossability

Solanum infundibuliforme is well adapted for life in arid regions. The small, delicate stems of this plant, together with its poorly dissected leaves, linear leaflets, and highly reduced leaf areas are all suggestive of plant forms that thrive typically under xerophytic conditions. For these reasons, this species would appear to be ideal for use in horticultural improvement programs aimed at incorporating drought resistance into the cultivated potato (Brücher, 1957a; Hawkes and Hjerting, 1969). However, aside from its use in some purely experimental crosses, commercial varieties incorporating the germplasm of this species are currently not to be found in the market.

Crosses between *S. infundibuliforme* and a number of wild and cultivated

South American and Mexican potato species have been made for the purposes of cytological and taxonomic investigations (Choudhury, 1944; Hawkes and Hjerting, 1969). These crosses have involved such diverse plants as *S. chacoense*, *S. tarijense*, *S. microdontum*, *S. phureja*, *S. goniocalyx*, *S. megistacrolobum*, *S. gourlayi*, *S. spegazzinii*, *S. vernei*, and *S. demissum*. As might be expected from the number of species involved, the results of this crossing program were varied. However, with respect to the improvement of the cultivated potato, one multiple cross involving this species and *S. tuberosum* deserves mentioning. A cross obtained by Hawkes and Hjerting (1969) between the Mexican species, *S. demissum*, and an F_1 of the following two South American species *S. infundibuliforme* and *S. chacoense*, were used in further hybridizations with *S. tuberosum*. Although the results of this program have not found application in commercial breeding programs, they do demonstrate how easy it is to obtain multiple exchanges of germplasm in this group of genetically closely related forms.

The results of breeding studies made at CIP suggest that this species is essentially self-incompatible.

Solanum infundibuliforme (11977) × *S. boliviense* (11935, 11938).– Plants derived from reciprocal crosses between these species yielded many berries, these having an average of 275 well-formed and viable seeds per fruit. The various F_1 populations obtained from this cross showed a general resemblance to *S. infundibuliforme*. The dominant characteristics of the latter species appeared in the poorly dissected leaves and small size of the hybrid plants. All

Figure 30. – Floral dissection of *Solanum infundibuliforme* (*Ochoa 11966*), ×1.

had 1-2 pairs of lateral leaflets, these ranging in decurrency from narrow to broad at the base. Terminal leaflets of these hybrids were larger than the laterals and their shapes varied from rhombic-lanceolate or narrowly ovate-lanceolate to elliptic-lanceolate. Flower color of the hybrids varied from very dark purple to light purple. All had a rotate-pentagonal corolla, averaging about 3 cm in diameter. Some features of these hybrids were obviously derived from an influx of genes from *S. boliviense*.

Solanum infundibuliforme (11973) × *S. brevicaule* (11934).– Reciprocal hybrids between these species were easily obtained. Moreover, these yielded an average of 150 viable seeds per fruit. The F_1 segregates of this cross, no matter in which direction they were made, had the dominant leaf shapes and dissection patterns of *S. brevicaule*. However, leaflet decurrency in these hybrids varied from slightly to strongly winged. Some plants bore shortly petiolulate, nondecurrent leaflets. All developed a rotate, pentagonal corolla 3 cm or more in diameter similar to that of *S. brevicaule*. Flower color was mostly violet, but some individuals trended toward dark purple.

Solanum infundibuliforme × *S. tarijense*.– The F_1 hybrids of this cross were generally of greater stature than those of the previous cross, these tending to have the dominant characteristics of *S. tarijense*. However, all had flexible or slightly flexible stems and were provided with leaves having 0-4 interjected leaflets and 4-5 pairs of long-lanceolate leaflets, the latter bearing obtuse apices, and either rounded or asymmetrically rounded bases. In these hybrids, only the upper pair of leaflets was observed to be basally winged; the leaf decurrency of all varied from very narrow to broad and cuneiform. The remaining leaflets were either sessile or shortly petiolulate. All had long peduncles, similar to those of *S. tarijense*, but flower color varied from very pale lilac to dark purple. In different plants the small corolla (2-2.5 cm in diameter) varied in shape from pentagonal to substellate. The anthers of these flowers ranged from 5.5-6 mm in length. Pedicel articulation was in the upper one-third of the stalk.

Solanum infundibuliforme × *S. anamatophilum*.– Crosses between these two species were difficult to make. Notwithstanding the many morphological similarities that are found between the two, the fruits that were obtained from crosses between *S. infundibuliforme* and *S. anamatophilum* contained only 1-3 seeds per berry. Possibly, as explained earlier with respect to the crossability of *S. acaule*, the observed sterility in this instance was due to differences in the numerical balance of the endosperm (Johnston and Hanneman, 1982).

IV. SERIES CIRCAEIFOLIA

CIRCAEIFOLIA Hawkes, Ann. & Mag. Nat. Hist., ser. 12, 7:702. 1954.

Plants less than 1 m tall, spreading or weakly ascending, glabrous or sparsely pubescent. Leaves small, simple or few-dissected. Articulation of the pedicel above center or in the upper one-third of the stalk. Corolla stellate or substellate, white, small, usually less than 2 cm in diameter. Fruit long and narrowly elliptic-lanceolate, the apex acuminate to subacuminate.

Chromosome number: 2n=2x=24.

Distribution

This endemic Bolivian series occurs from the Yungas of the Department of La Paz, the mountains of Choro-Ayopaya, and the valley of the Caine River in the Department of Cochabamba to the southeast Province of Valle Grande in the Department of Santa Cruz. Plants of this series grow in the moist forests of the Yunga and Prepuna formations, between 1800-3700 m altitude.

At present, the series Circaeifolia is known to have only three species: *S. capsicibaccatum* Cárd., *S. circaeifolium* Bitt. (Hawkes, 1956a), and *S. soestii* Hawkes and Hjerting. Recently, Ochoa (1982, 1984a) and Hawkes and Hjerting (1983) have proposed additional subspecific taxa for this series. However, in the present work these have all been reduced to synonomy under *S. circaeifolium*.

Key to Species

1. Leaves imparipinnate, 3-4 jugate, never simple .... 10. *S. soestii*.
1. Leaves simple, or (1-)2-3(-4) jugate.
   2. Leaves typically simple, essentially glabrous or glabrescent .... 9. *S. circaeifolium*.
   2. Leaves typically imparipinnate or rarely with simple leaves, more or less pubescent .... 9a. *S. circaeifolium* var. *capsicibaccatum*.

9. Solanum circaeifolium Bitter, Repert Sp. Nov. 11:385-386. 1912.

FIGS. 31-35; MAP 7; PLATE VIII.

Plants 30-50(-80) cm tall, delicate, glabrous or glabrescent. Stem light green and irregularly pigmented or speckled with lighter spots, spreading or somewhat reclined, slender and slightly flexible, 2.5-3(-5) mm in diameter at base, cylindrical, simple or laterally branched. Stolons long and slender; tubers white, small, 0.5-1(-2) cm long, ovate and rounded or elongated and subcylindric, with white flesh. Leaves essentially simple or with 1(-2) pairs of usually very reduced lateral leaflets, glabrescent or shiny and glabrous above, somewhat pubescent on the veins and duller below. Terminal leaflets much larger than the laterals (when present), (3-)5-7(-9.5) cm long by (1.5-)2.5-4(-5) cm wide, broadly ovate to ovate-lanceolate to triangular-lanceolate, or from narrowly elliptic-lanceolate to broadly elliptic-lanceolate, the apex subobtuse or acuminate to abruptly pointed, the base somewhat cordate or broadly cuneate. Lateral leaflets, when present, elliptic to ovate-elliptic, usually small, (0.5-)0.8-1.5(-2.5) cm long by (0.2-)0.4-0.6(-0.8) cm wide, the apex obtuse, acuminate or pointed, and the shortly petiolulate or sessile base, rounded. Pseudostipular leaves falcate, small, 4-7(-12) mm long by 2.5-3.5(-5) mm wide. Inflorescence cymose, 3-5(-10)-flowered, the peduncle short, 1.5-4(-7.7) cm long, slender, 1-1.5 mm in diameter at the base. Pedicels 15-25 mm long, articulated above center or toward the upper third of the stalk, glabrous or very sparsely pilose as in the case of the calyx. Calyx 5-6(-7.5) mm long, the lobes narrowly triangular-lanceolate with pointed, linear acumens. Corolla stellate, 1.5-2 cm in diameter, white or creamy-white with yellow or greenish-yellow nectar guides, the narrowly to broadly triangular-lanceolate lobes 8-10 mm long by 4.5-6 mm wide. Anthers 4-4.5 mm long, borne on less than 1 mm-long glabrous filaments. Style very slender, 7.5-8 mm long, exerted 2.5 mm, densely papillose and frequently pubescent with short hairs except along the upper one-third of its length; stigma slightly broader than the apex of the style. Fruit long and narrowly elliptic-conical with a pointed apex, up to 3.5 cm long and 1.5 cm in diameter, light green, with one or two vertical light brown stripes.

Chromosome number: 2n=2x=24.

Type: BOLIVIA. Department La Paz, Province Larecaja, Cerro Iminapi, 2650 m alt., near Sorata. March, 1860; *G. Mandon 400* (lectotype, here designated NY; isotypes, BM, F, G, K, P, S, W).

Affinities

As suggested by its delicate habit and its very similar corolla, *S. circaeifolium* (including its variety *capsicibaccatum*) is closely related to *S. soestii*. However, the two may be distinguished from one another by differences in leaf dissection and leaflet shape.

Habitat and Distribution

This species is distributed from the Provinces of Larecaja and Inquisivi in northwestern Bolivia, through the central region of the Department of Cochabamba to the Serranias de Valle Grande in the Department of Santa Cruz in the southeast. It occurs between 1800-3700 m altitude. Variety *circaeifolium* is most common in the northwest provinces; whereas var. *capsicibaccatum* is more abundant in the Departments of Cochabamba and Santa Cruz.

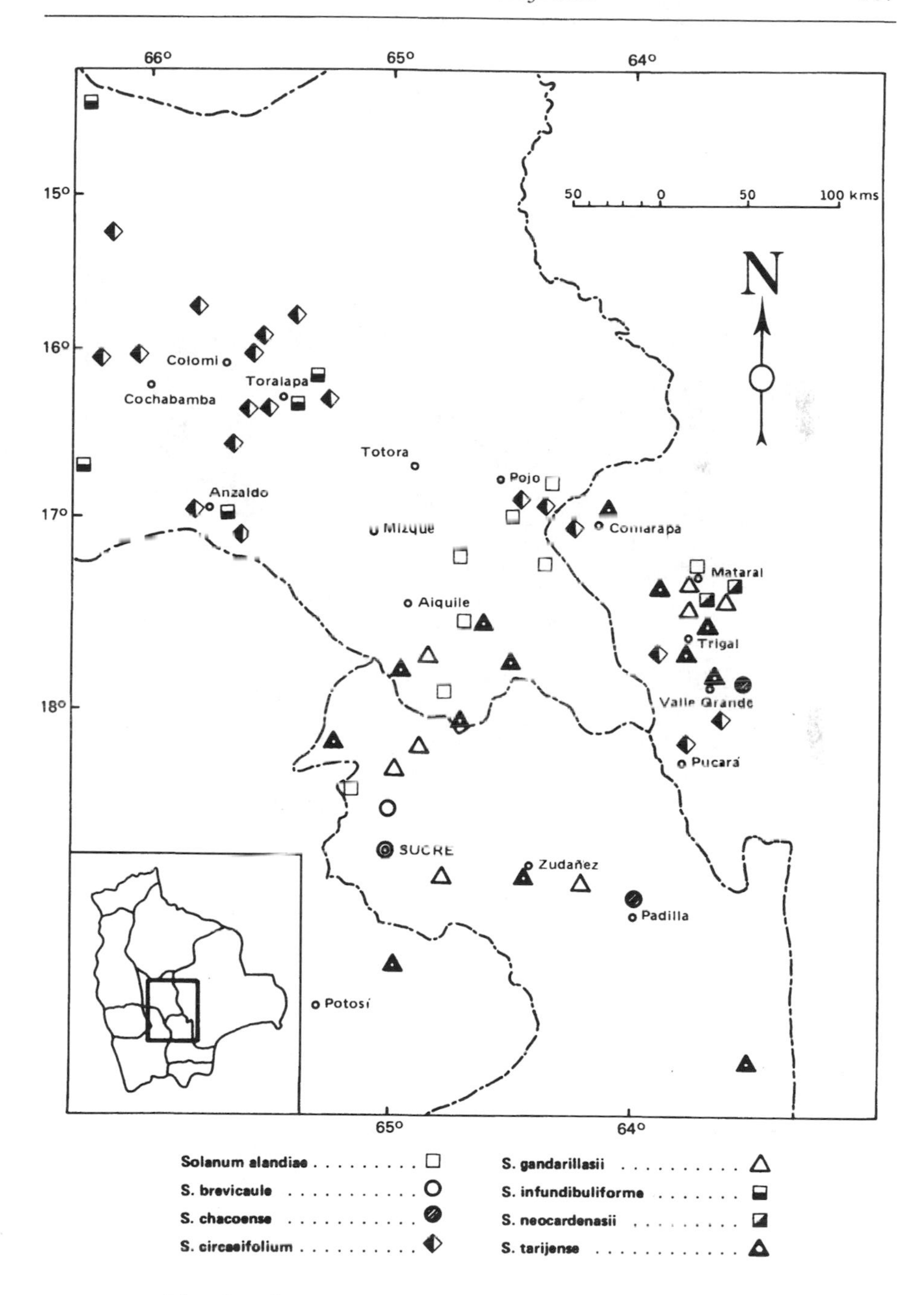

Map 7. – Range patterns of wild potato species in central Bolivia, showing *Solanum circaeifolium* (◆).

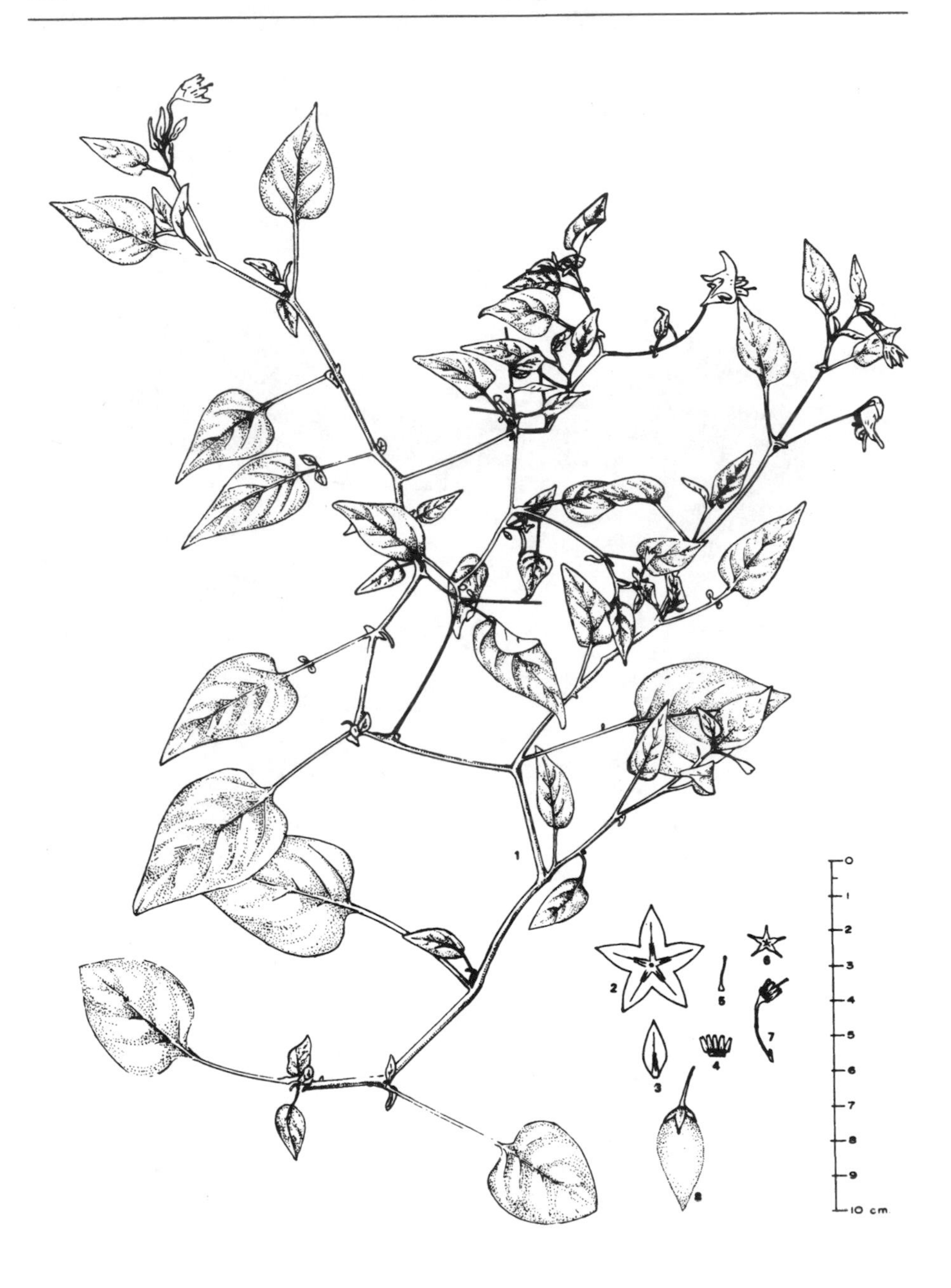

Figure 31. – *Solanum circaeifolium* var. *circaeifolium* (*Ochoa* and *Salas 11909*). 1. Center and upper portion of flowering plant. 2. Corolla. 3. Petal. 4. Stamens, dorsal view. 5. Pistil. 6. Calyx. 7. Pedicel and stamens. 8. Fruit. All ×½.

This species grows typically in the region of the Ceja de Yungas, where it grows in moist soil under the shade of such woody shrubs as *Berberis* and *Barnadesia*, or along forest margins where it grows with *Polylepis*, *Alnus*, and *Cassia*, or in maize fields. It is also found in drier regions such as the Caine River and Cochabamba Valleys. In the latter place, it grows in very dry, stony or sandy-clay soils in the shade of *Schinus molle*.

Figure 32. – Floral dissection of *Solanum circaeifolium* var. *circaeifolium* (*Ochoa and Salas 11806*), ×1.

Figure 33. – Floral dissection of *Solanum circaeifolium* var. *circaeifolium* (*Ochoa and Salas 11909*), ×1.

Specimens Examined

Department La Paz

Province Larecaja: Gran Poder, 3050 m alt., small gold mine about 40 km north of Sorata, in narrow wooded valley below Tacacoma. June 4, 1950; *Brooke 6428* (BM). Cerro Imanapi, 2560 m alt., near Sorata, under bushes in loose, moist, stony soil. 'Plant 30-40 cm tall, flowers white.' February, 1946; *Cárdenas 3639* (CPC, GH, HNF, LL, US, topotypes). Mina Gran Poder, Tacacoma, 3000 m alt., among bushes in humid soil. 'Plants 30-60 cm tall, flowers white.' April, 1956; *Cárdenas 5515* (HNF, LL). Under bushes in moist rocky ravine on Mt. Iminapi near Sorata, 2800 m alt., 1959; *Gandarillas, Rimpau, and Ross 614* (MAX, topotype). 11 km from Sorata on road to Tacacoma, 3500 m alt. Under bushes by roadside. Very glabrous leaves. March 31, 1971; *Hjerting, Cribb, and Huamán 5012* (CIP). Cerro Iminapi, 2650 m alt., near Sorata, among bushes. March, 1860; *Mandon 400* (BM, F-fragment, G, K, NY, P, S, W, type collection). Cerro Iminapi, 2800 m alt., near Sorata. January, 1949; *Ochoa 673* (Ochoa, topotype). Tarapampa, 2500 m alt., about 4 km from Sorata, on the Sorata-Tacacoma horse trail. December, 22, 1977; *Ochoa and Salas 11796* (CIP, OCH). Huallpapampa, 2500 m alt., near Sorata, on the Sorata-Tacacoma horse trail. Chromosome number:

Figure 34. – Type locality of *Solanum circaeifolium*. View of the town of Sorata, Province Larecaja, Department La Paz, at 2700 m altitude.

Figure 35. – Mountain slope (Cerro Iminapi) near Sorata, habitat of *Solanum circaeifolium*.

2n=2x=24. December 22, 1977; *Ochoa and Salas 11806* (CIP, OCH). Atahuallpani, 2700 m alt., near Sorata. Fruits long conical and with a very pointed apex, light green with 1-2 dark violet stripes. Chromosome number 2n=2x=24. March, 1972; *Ochoa and Salas 11909* (CIP, OCH).

Province Inquisivi: Inquisivi-Circuata, 3820 m alt., 1980; *Soest, Okada, and Alarcón S.O.A.-27* (CIP). Unknown locality, 1891; *M. Bang 2509* (NY).

9a. Solanum circaeifolium var. **capsicibaccatum** (Cárd.) Ochoa comb. nov. FIGS. 36-42.

S. capsicibaccatum Cárd., Rev. Agr. Cochabamba 2:35-36, Fig. p. 37. 1944. TYPE: *Gandarillas 60*, Bolivia. Between Huaira Pata and Molle Pujru, Valley of Rio Caine, Dept. Cochabamba, Prov. Tarata, March, 1942 (CA, GH).

S. circaeifolium f. *lobatum* Corr., Wrightia 2:171. 1961. TYPE: *Brooke 6158*, Bolivia. Lagunillas, behind Choro in the St. Elena Valley, Dept. Cochabamba, Prov. Ayopaya, March 7, 1950 (BM).

S. capsicibaccatum var. *latifoliolatum* Ochoa, Phytologia 50:181-182. 1982. TYPE: *Ochoa 11915*, Bolivia. Rio Seco, Dept. La Paz, Prov. Inquisivi, March, 1978 (holotype, OCH; isotypes, CIP, US).

S. circaeifolium subsp. *quimense* Hawkes and Hjerting, Bot. Jour. Linn. Soc. 86:406-408, Figs. 1-2. 1983. TYPE: *Hawkes, Aviles, and Hoopes 6733*, Bolivia. Ichupampa, Dept. Cochabamba, Prov. Ayopaya, March 11, 1980 (holotype, K; isotypes, C, Herb. H. G. Hawkes, not seen).

S. circaeifolium var. *latifoliolatum* (Ochoa) Ochoa, Phytologia 55:19. 1984. TYPE: *Ochoa 11915*, Bolivia. Rio Seco, Dept. La Paz, Prov. Inquisivi, March, 1978 (holotype, OCH; isotypes, CIP, US).

Leaves more or less pubescent including the margins, usually dull-surfaced, normally imparipinnate and provided with (1-)2-3(-4) pairs of leaflets, or very rarely simple-leaved. Terminal leaflet narrowly lanceolate to ovate-lanceolate or broadly elliptic-lanceolate, 5-9 cm long by 3-6 cm wide. Lateral leaflets usually large, occasionally more than half the size of the terminal leaflet, narrowly elliptic-lanceolate, the apex pointed and the base rounded, sessile or borne on short petiolules. Corolla stellate to substellate or strongly pentagonal, 2.5-3(3.5) cm in diameter, the lobes from narrowly triangular-lanceolate and 8-10 mm long by 5-7 mm wide to broadly triangular-lanceolate or pointed ovate-triangular and 12-16 mm long by 9-17 mm wide. Anthers 5-6 mm long. Style 8.5-10 mm long.

In the original diagnosis of *S. capsicibaccatum*, Cárdenas mentions that the type of this species is housed in his herbarium. The type specimen, *Gandarillas*

Figure 36. – *Solanum circaeifolium* var. *capsicibaccatum* (*Ugent 4713*), × ½.

Figure 37. – *Solanum circaeifolium* var. *capsicibaccatum* (*Ochoa 11915*). Flowering plant. 1. Fruit. 2. Corolla. 3. Petal. 4. Stamens, dorsal view. 5. Pistil. 6. Calyx. 7. Pedicel and calyx. All × ½.

60, was collected in March 1942 between Huaira Pata and Molle Pujro in the Province of Tarata (Department Cochabamba).

Cárdenas also cited a second collection of this taxon from the original type locality in his original description (*Cárdenas and Gandarillas 3600*, April, 1944). Although plants from the original collection are no longer to be seen in the Cárdenas Herbarium, other duplicates of this same collection can be viewed at the Lundell Herbarium of the Texas Research Foundation (LL) and the Herbarium of the Commonwealth Potato Collection at Hertford, England (CPC).

According to Alandia (1951), this species shows some resistance to race 'D' (or 'O') of potato late blight disease, *Phytophthora infestans*.

Chromosome number: 2n=2x=24.

Type: BOLIVIA. Department Cochabamba, Province Tarata, between Huayra Pata and Molle Pujro, Valley of Rio Caine, 2800 m alt. March, 1942; *H. Gandarillas 60* (CA, GH).

Specimens Examined

Department Cochabamba

Province Ayopaya: Choro, 3050 m alt., above the Cocapata River, about 160 km northwest of Cochabamba, across the Tunari range, among rocks on steep slope. January 18, 1950; *Brooke 3051* (BM). Choro, 3050 m alt., 66°30′ W and 17°0′ S, above the Cocapata River about 160 km northwest of Cochabamba, across the Tunari range, wedged between rocks on brushy slope. January 18, 1950; *Brooke 3052* (BM). Lagunillas, 2897 m alt., over the mountain range behind Choro in the St. Elena Valley, under bushes on thicketed slope. March 7, 1950; *Brooke 6158* (BM). Near Choro-Ayopaya, under bushes, 3900 m alt. March, 1946; *Cárdenas 3694* (CPC, LL). Near Independencia, by path, Ichupampa, lat. 17°07′ S and long. 66°56′ W, 3000 m alt., among shrubs at edge of natural forest and in maize field (swidden). Locally abundant, to 30 cm high; flowering; with some fruits, spindle-shaped and sharp-pointed. March 11, 1980; *Hawkes, Hjerting, and Aviles 6581* (CIP, from specimen grown at Sturgeon Bay, Wisconsin, USA, under the field No. 9205.01, 1983). 16 km from Independencia on road to Challa, 3220 m alt., lat. 17°12′ S, long. 66°53′ W. Under partial shade of bushes by road. Corolla broad stellate, white. March 12, 1980; *Hawkes, Hjerting, and Aviles 6588* (CIP, from specimen grown at Sturgeon Bay, Wisconsin, USA, No. 9206.2, 1983). Puente San Miguel, 3800 m alt., 17°10′ S and 66°25′ W, above Liriuni on the road Cochabamba-Vizcachas, about 25 km NW of Cochabamba (air distance), on rocky slope with *S. brevicaule*, *S. acaule*, *Oenothera*, *Lupinus*, *Chenopodium*, *Salvia*, *Oxalis*, *Calceolaria* and non-tuber-bearing *Solanum* sp. April 9, 1963; *Ugent 4713, 4722* (OCH); *4716* (US); *4720* (F); *4733* (GH); *4709-11* (K); *4701, 4704-07, 4712, 4719, 4725, 4728-29, 4730, 4731-32, 4734 4735* (WIS).

Province Carrasco: In moist cloud-forest zone between Pojo and Comarapa, 2700 m alt. March 20, 1959; *Cárdenas, Rimpau, and Ross 500* (MAX). In moist cloud-forest zone, near Pojo, 3000 m alt., 110 km from Cochabamba. March 20, 1959; *Cárdenas, Rimpau, and Ross 504a* (MAX). On rocky slope, above km 75 on road to Santa Cruz. February 15, 1960; *Correll, Dodds, Brücher, and Paxman B613* (LL). 212 km from Cochabamba on road to Santa Cruz, 2900 m alt., in Ceja de Monte, new road on damp bushy slope. February 9, 1971; *Hawkes, Hjerting, Cribb, and Huamán 4396* (CIP). 204 km from Cochabamba on road to Santa Cruz, on steep slope below road, lat. 17°46′ S and long. 64°41′ W, 2650 m alt. In deep shade of small trees and shrubs in rich deep leaf mould; plants 10–30 cm; flowers white, stellate; fruits pale green, conical slightly compressed, veined with darker green, especially at the edges, sharp-pointed. March 2, 1980; *Hawkes, Hjerting, and Aviles 6532* (CIP, plants grown at Huancayo, Peru, from seeds obtained at Sturgeon Bay, Wisconsin, USA). Sunchal, 2650 m alt., near km 210 of the highway Cochabamba-Santa Cruz, very close to the Sunchal rivulet, along margins of moist forests under the shade of Cedro and Aliso, and among *Cortaderia* and *Oxalis*. Chromosome number: 2n=2x=24. February 29, 1984; *Ochoa and Salas 15567* (CIP, F, GH, NY, OCH, US).

Figure 38. – Floral dissection of *Solanum circaeifolium* var. *capsicibaccatum* (*Ochoa 11915*), ×1.5.

Province Cercado: Taquiña brewery, on mountain slope a few km from Cochabamba, 2745 m alt., in ditch. January 19, 1949; *Brooke 3013* (BM).

Province Quillacollo: Humid soil under bushes, Liriuni, 2700 m alt. Plants 20-40 cm tall, flowers white. February, 1947; *Cárdenas 5520* (HNF, LL) [*S. capsicibaccatum* var. *liriunianum* Card. is indicated on the HNF specimen, the name, however, is not published]. On mountain slopes near San Miguel, above Liriuni. Flowers creamy white, stellate. February 14, 1960; *Correll, Dodds, Brücher, and Paxman B609* (K, LL, MO, NY, S, U, UC, US). Liriuni. February, 1947; *Cárdenas s.n.* (LL). Liriuni, 4000 m alt., about 8 km NW of Quillacollo. Flowers white. February 21, 1947; *Cutler and Cárdenas 9059* (F). About km 56.5 from Cochabamba on the road to Oruro, 3200 m alt. Edge of barley field among rocks and bushes. Two lateral leaflets; creamy white flowers. February 24, 1971; *Hjerting, Cribb, and Huamán 4486* (CIP). 21 km from Quillacollo on road to Morochata, 3550 m alt., on scree below road. White-cream stellate flowers. March 21, 1971; *Hjerting, Cribb, and Huamán 4748* (CIP). Vicinity of Liriuni, 2600 m alt., near Cochabamba. January, 1949, *Ochoa 672* (OCH).

Province Tarata: Mollepujru. January, 1949; *Cárdenas s.n.* (HNF, LL, topotypes of *S. capsicibaccatum*). Halfway down between Huaira Pata and Molle Pujru, 2800 m alt., valley of the Rio Caine, in stony dry places in shade of *Schinus molle* trees. March, 1942; *Gandarillas 60* (CA, GH, isotype of *S. capsicibaccatum* Card.). Molle Pujru (Caine River), 2800 m alt., on dry stony soil, under 'molle' trees. Plants 20-40 cm tall; flowers white. April, 1944; *Cárdenas 3600* (CPC, LL, paratypes of *S. capsicibaccatum*).

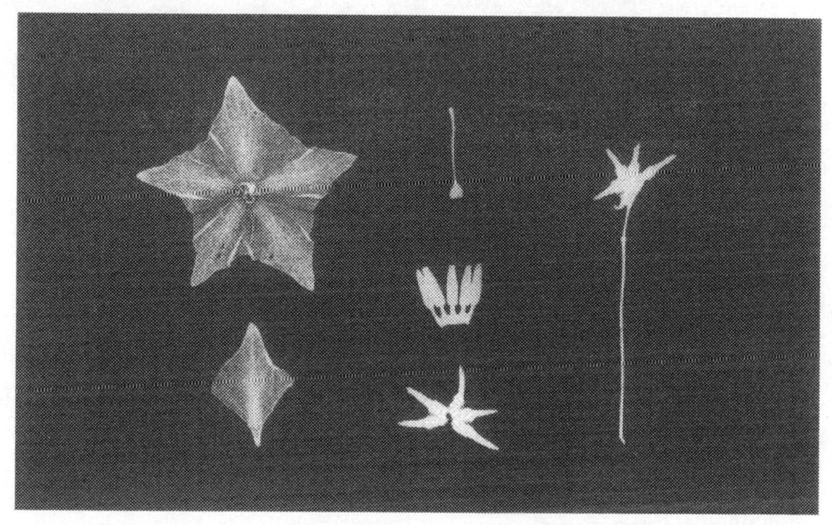

Figure 39. – Floral dissection of *Solanum circaeifolium* var. *capsicibaccatum* (*Ochoa 12027A*), ×1.5.

Department La Paz

Province Inquisivi: Quime, 3050 m alt. [10,000 ft., according to the collector], a small town 160 km from Oruro on the road to Eucalyptus, along the Tres Cruces Pass, Quimsacruz Mountains, moist area. April 7, 1949; *Brooke 3037* (BM). Near Rio Seco, 2900 m alt., vicinity of Quime. Chromosome number 2n=2x=24. March, 1978; *Ochoa 11915* (CIP, OCH). Jañuccacca, 2700 m alt., about 6 km from Quime on the road from Quime to Inquisivi, among large stones and moist thickets of shrubs, principally of the genus *Barnadesia*. Rare. February 23, 1984; *Ochoa and Salas 15489* (CIP, OCH, US). Licoma-Inquisivi, 9 km, 2650 m alt.; 1980; *Soest, Okada, and Alarcón (S.O.A.) 31* (CIP). Licoma-Inquisivi, 17 km, 2860 m alt.; 1980; *Soest, Okada, and Alarcón* (S.O.A.) *32* (CIP).

Department Potosi

Province Bilbao: Ignacita, near Mollevilque, 2500 m alt. In rocky soil. Plants 20 cm tall, flowers white. February, 1960; *Cárdenas 6115* (WIS).

Department Santa Cruz

Province Valle Grande: Near Pucara, 2600 m alt., among bushes in moist soil. Plants 40-60 cm tall; flowers white. March, 1955; *Cárdenas 5514* (LL, type of

Figure 40. – Fruits of *Solanum circaeifolium* var. *capsicibaccatum* (*Ochoa and Salas 15567*), ×1.

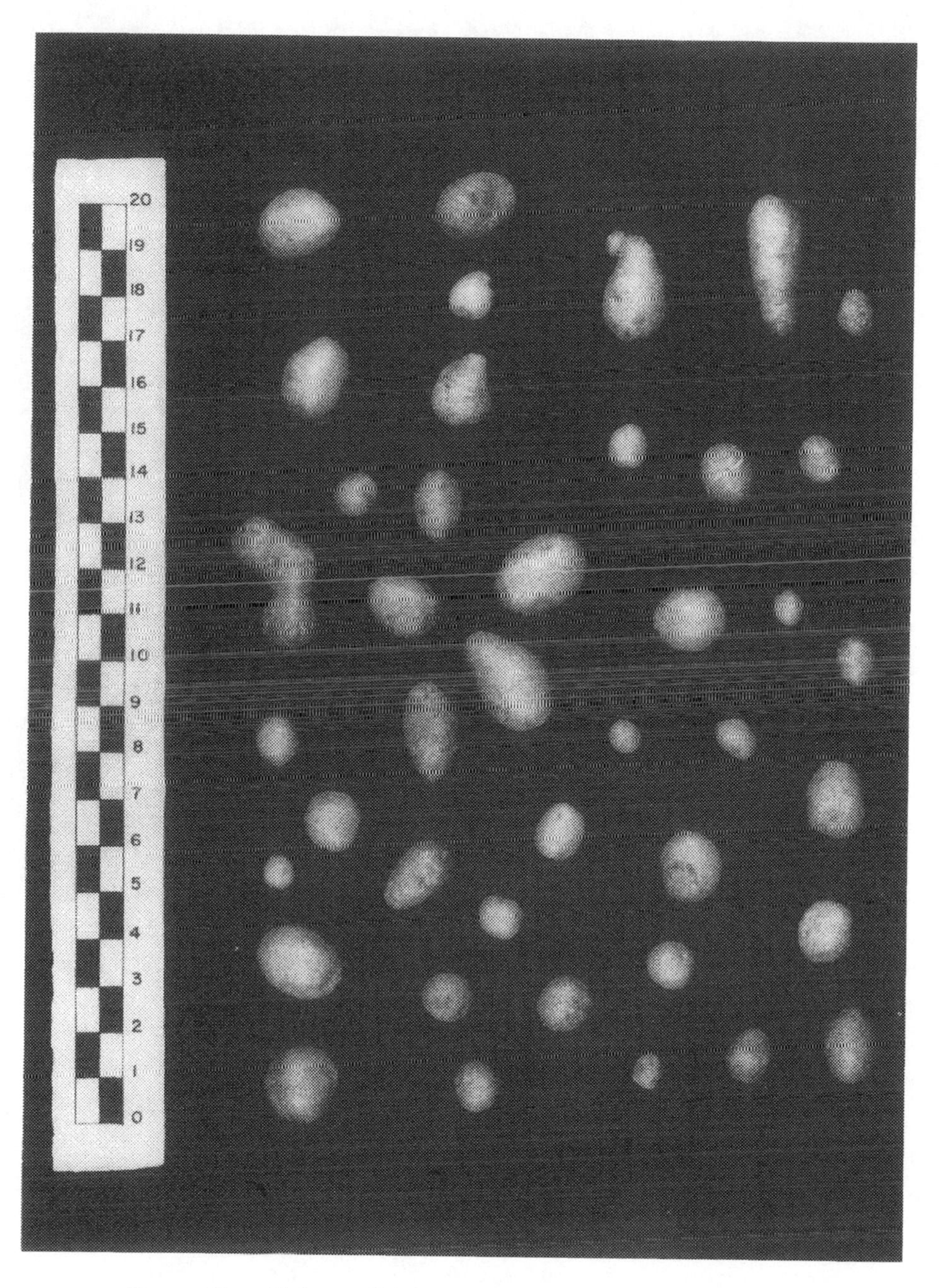

Figure 41. – Tubers of *Solanum circaeifolium* var. *capsicibaccatum* (*Ochoa 11915*), × ½.

S. capsicibaccatum var. *macrophylum* Càrd., never published). Between Higuera and Valle Grande, 2600 m alt., among bushes in sandy soil. March, 1955; *Cárdenas 5519* (LL; HNF, annotated by Cárdenas as *S. capsicibaccatum* var. *glabrescens* Càrd., but never published). On the road from Quebrada Seca to Ariruma, 2200 m alt. March, 1978; *Ochoa 12027A* (CIP, OCH). About 19 km from Valle Grande on road to Pucara, 2650 m alt., lat. 18°36′ S and long. 64°7′ W, bushy places in Ceja formations, growing with No. *HHA 6513* (=*S. microdontum*). Corolla white stellate. March 1, 1980; *Hawkes, Hjerting, and Aviles 6510* (CIP, plants grown at Huancayo, Peru, from seeds obtained at Sturgeon Bay, Wisconsin, USA).

Indefinite: WIS-1730, plant grown at CIP from seed originally obtained from the IR-1 Project, Sturgeon Bay, Wisconsin, with the determination of *S. capsicibaccatum* but without further information.

Crossability

As in the case of the last species, *S. circaeifolium* is self-incompatible. Reciprocal crosses made at CIP between different introductions of *S. circaeifolium* var. *circaeifolium* were 100% successful, yielding on the average 120 viable seeds per fruit.

Similarly, reciprocal crosses betwen different clones of *S. circaeifolium* var. *capsicibaccatum* were also successful, though yielding, on the average, only 80 fertile seeds per berry. Significantly, a large measure of success was also obtained in intervarietal reciprocal crosses involving two introductions of the above variety (11806 and 11909) and *S. circaeifolium* var. *capsicibaccatum* (Wisconsin 1730). The latter series of crosses were particularly successful when *S. circaeifolium* var. *circaeifolium* was used as a mother plant, yielding, in some cases, as many as 130 fertile seeds per fruit. On the other hand, as little as 50 seeds per fruit were obtained when this last variety was used as the male parent.

Interspecific crosses involving *S. circaeifolium* (11806 and 11909) and *S. infundibuliforme* (11927), *S. lepthophyes* (13558), *S. litusinum* (12027), *S. megistacrolobum* (11914), and *S. sparsipilum* (13571 and 13718) were unsuccessful, possibly due to differences involving the numerical balance of the endosperm. It should be noted that the EBN value of *S. circaeifolium* is 1; whereas the remaining species have an EBN value of 2 (Johnston and Hanneman, 1982).

Plants of this series share elongated, pointed fruits with those of Conicibaccata. For this reason, they would appear to be closely related. However, the results of numerous experimental crosses made at CIP between *S. circaeifolium* and various non-Bolivian representatives of series Conicibaccata were largely unsuccessful (Intl. Pot. Ctr., 1985).

Species evaluated in the above interspecific crossing program included *S. limbaniense* (14405), *S. santolallae* (13640), and *S. ayacuchense* (13188), all from southern Peru; *S. rhomboideilanceolatum* (11869) from central Peru; *S. humectophilum* (13247) from northern Peru; and *S. albornozii* (11007) from Ecuador.

Only one extra-territorial species, *S. lignicaule* (11316, 11617, 13584), a Peruvian species regarded by the author as a member of the series Conicibaccata, having an EBN value of 1, yielded partial success in interspecific crosses with *S. circaeifolium*. In the latter case, up to 15 seeds per berry were obtained, but only when *S. lignicaule* was used as the male parent.

On the other hand, crosses involving *S. circaeifolium* and *S. violaceimarmoratum* (11901), a Bolivian species of series Conicibaccata, yielded few berries, averaging 6 viable seeds per fruit, and only when *S. violaceimarmoratum* was used as the mother plant. Similarly, crosses involving *S. circaeifolium* and *S. chomatophilum* (13199, 13205, 13206), when the latter was used as a female parent, yielded few viable seeds. Contrary to the case of *S. circaeifolium*, with an EBN value of 1, the other two species, *S. violaceimarmoratum* and *S. chomatophylum*, have an EBN value of 2. In summary, the results of the above series of crosses would appear to support the treatment of these last two potato groups as separate series.

Although *S. circaeifolium* is difficult to cross with species of other series, it is possible, nonetheless, to occasionally obtain hybrids of this species. For exam-

Figure 42. – Habitat of *Solanum circaeifolium* var. *capsicibaccatum*, in the vicinity of Jañuccacca, 2700 m altitude, between the towns of Quime and Inquisivi, Department La Paz.

ple, several collections of *S. bukasovii* (11058, 11841, 11843, 11859), a Peruvian species of the series Tuberosa, were successfully crossed with *S. circaeifolium* var. *capsicibaccatum* (11915), the latter used as the male parent. Pistillate plants yielded abundant berries, averaging 40 viable seeds per fruit. Segregates from this cross inherited resistance to pathotype P4A of *Globodera pallida* from both parents, and resistance to frost from *S. bukasovii* (Intl. Pot. Ctr., 1984a). Significantly, these hybrids can be easily crossed with the cultivated diploid species *S. phureja*, or with haploids of *S. tuberosum*. Resistance to various pathogenic forms of *G. pallida* may be incorporated in this way.

Other reciprocal combinations were also of interest to this study. Thus, crosses involving different introductions of *S. circaeifolium* var. *circaeifolium* (11806 or 11909 by 11976) always yielded plants having sparsely pilose leaves. The latter characteristic, in this case, appeared to be a dominant trait of the male line. However, regardless of the direction in which this cross was made, other characteristics appeared to be basically unchanged. Thus, the various offspring from this cross were provided with sinuously winged stems that ranged up to 30 cm tall, and leaves that varied from simple to 1-2 jugate. The ovate-lanceolate terminal leaflets of these plants were provided with a pointed or acuminate apex and an asymmetrically rounded or subcordate base. In summary, these plants resembled, in their various morphological characteristics, the classical picture of *S. circaeifolium* var. *circaeifolium*.

First generation hybrid populations derived from reciprocal intervarietal crosses of *S. circaeifolium* var. *capsicibaccatum* (Wisconsin 1730) by *S. circaeifolium* var. *circaeifolium* (11806) are notably different in the relative dissection and pubescence of their leaves and in the respective shapes of their corollas. Thus, the leaves of these plants varied in dissection from simple to 1-2(-4)-jugate and in texture and pubescence from glabrescent and shiny to more or less pilose and dull. The ovate-lanceolate, terminal leaflets of these plants were occasionally broader than the lateral pairs of leaflets, the latter elliptic-lanceolate in shape and provided with acuminate apices. The flowers of these plants varied from narrowly and deeply stellate, to broadly and triangularly stellate.

Some of the specimens examined and treated in the present study represent natural hybrids of *S. circaeifolium* var. *circaeifolium* and *S. circaeifolioum* var. *capsicibaccatum*. Included among the latter group are collections by *Brooke (6158)*, *Ochoa (11915, 12027)*, and *Hawkes, Hjerting, and Aviles (6581, 6588)*. These collections compare favorably to the intervarietal hybrids produced under experimental conditions at CIP.

Natural intervarietal hybrids of *S. circaeifolium* show much transgressive variation. These commonly exceed either parent in such character measurements as leaf and leaflet length and width, and corolla diameter. Thus, in the natural crosses listed above, the terminal leaflets varied from 8.5 to 12 cm in length, and from 5.6 to 8 cm in width. These plants are also unusual in having more highly dissected leaves, these commonly having 3 to 4 pairs of leaflets rather than the customary 1 or 2.

Artificially made hybrids of *S. circaeifolium* var. *circaeifolium* and naturally occurring hybrids of this variety are unique in that their range of variation in leaf and floral characteristics transcend that which is found in nature. Thus, no matter the direction in which the artificial cross is made, F_1 hybrids (or technically, backcrosses) of *S. circaeifolium* var. *circaeifolium* (11806) and the natural intervarietal hybrid (11915) develop poorly dissected leaves (0-2 jugate) and unusually large terminal leaflets. The latter may range up to 14 cm long and up to 8 cm wide. The flowers of these plants are often 3.5 cm in diameter.

Artificial crosses between *S. circaeifolium* var. *capsicibaccatum* (11915) and *S. commersonii* (URU 3), two species with an EBN of 1, yielded fruits containing abundant seed. These experimental crosses may thus be considered as 100% successful. The F_1 hybrids derived from the above series of pollinations displayed primarily the dominant characteristics of *S. circaeifolium*. Thus, these segregates ranged up to 30 cm tall. All had slender, very sinuous, cylindrical stems. The latter were decumbent and branching, and less than 3 mm in diameter. All had 1-2 pairs of dull, sparsely pilose leaflets, these much smaller than the ovate-lanceolate terminal leaflets. None had interjected leaflets. The pedicels of these plants were articulated toward the upper one-third of the stalk. Unlike their parents, which bore stellate flowers, these hybrids developed rotate-pentagonal corollas. The latter had 3 mm-long lobes and 6 mm-long acumens.

10. Solanum soestii Hawkes and Hjerting, Bot. Jour. Linn. Soc. 86(4):406, 409-410, Fig. 3. 1983. FIGS. 43-46.

Plants light green, small, 8-20(-35) cm tall, delicate and graceful, glabrous or glabrescent. Stem erect, 1.5-2.5 mm in diameter, cylindrical, simple or few-branched. Stolons 30-50 cm long, slender, 1.5 mm in diameter; tubers white, round, small, 10-15 mm in diameter. Leaves imparipinnate, (4-)6-10(-14.5) cm long by (2-)3.5-6(-11) cm wide, generally lacking interjected leaflets, provided with 2-3(-4) pairs of lateral leaflets and borne on petioles 1.5-2.5(-4) cm long. Lateral and terminal leaflets narrowly lanceolate to almost linear, borne on petiolules 1-3(-6) mm long, or at times sessile or subsessile. Terminal leaflet (2-)2.5-3.5(-6.5) cm long by (0.6-)1-1.3(-2.4) cm wide, the apex pointed or acuminate and the base rounded. Upper lateral leaflets (1-)2-3(-4.5) cm long by (0.3-)0.6-0.8(-1.6) cm wide, the apex pointed and the base rounded and conspicuously oblique. Pseudostipular leaves small and narrowly subfalcate to subauricular, 2.5-5 mm long by 1.5-2.5 mm wide. Inflorescence cymose,

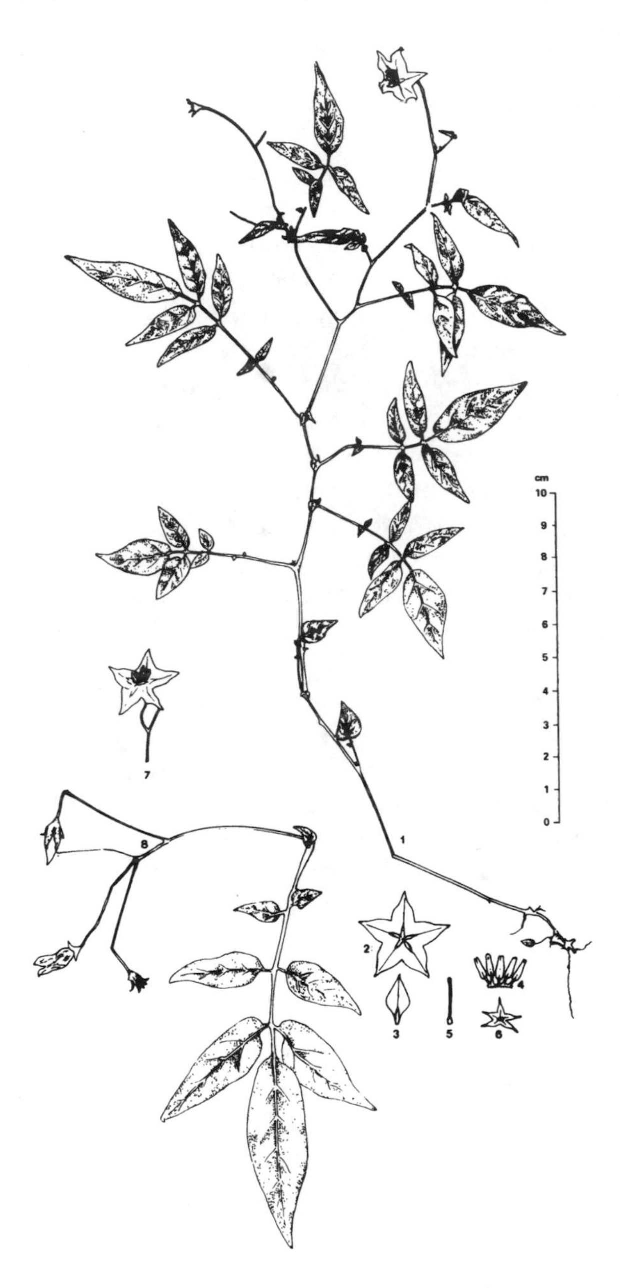

Figure 43. – *Solanum soestii* (*Ochoa and Salas 15502*). 1. Flowering plant. 2. Corolla. 3. Petal. 4. Stamens, dorsal view. 5. Pistil. 6. Calyx. 7. Flower. 8. Fruit, inflorescence and leaf. All × ½.

2-4-flowered. Peduncle 1.5-3(-5) cm long, very slender and almost filiform. Pedicel articulation located toward the upper one-third of the stalk, the upper part 5-7 mm long and the lower 15-25 mm long. Calyx 4-4.5 mm long, glabrous, light green, the lobes very narrowly lanceolate and with attenuate apices or shortly linear. Corolla stellate, 15-20 mm in diameter, creamy white, with yellow nectar guides. Anthers 3.5-4 mm long, elliptic-lanceolate, borne on filaments 1-1.5 mm long. Style 7 mm long, exerted 1.5 mm, papillose along its basal two-thirds; stigma light green, small, capitate, cleft. Fruits light green, long and narrowly elliptic-conical and with pointed apex, 2.5 cm long by less than 1 cm in diameter.

Chromosome number: 2n=2x=24.

Type: BOLIVIA. Department La Paz, Province Inquisivi, 2900 m alt., 4.5 km from Quime on road to Inquisivi. March 14, 1981; *J. G. Hawkes, I. Aviles, and R. W. Hoopes 6729* (holotype, K; isotypes, C, Herb. J.G. Hawkes, not seen).

Affinities

This species is closely related to *S. circaeifolium*. The two resemble one another quite closely in life form and in their respective floral, fruit, and tuber characteristics. Because of their many common features, *S. soestii* could be considered by some as a mere variety of *S. circaeifolium*. However, as the two differ strongly in the form of their leaves and leaflets, they are maintained here as separate species.

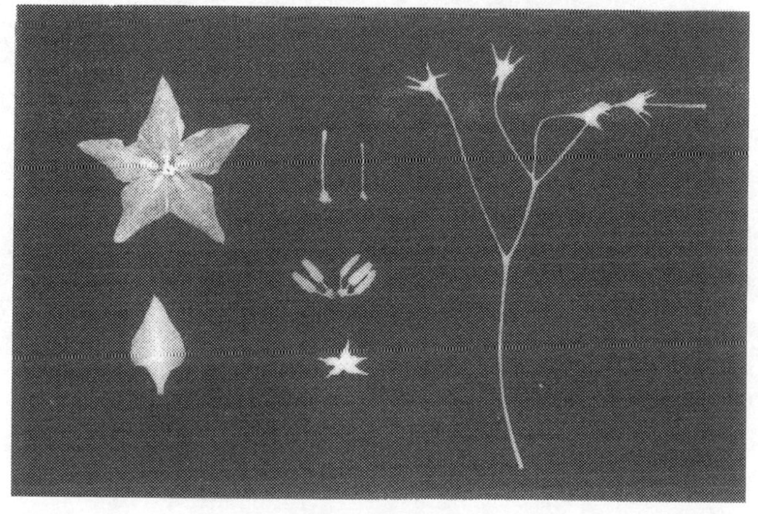

Figure 44. – Floral dissection of *Solanum soestii* (*Ochoa and Salas 15502*, topotype), ×1.

Figure 45. – Floral dissection of *Solanum soestii* (*Ochoa and Salas 15503*), ×1.

Figure 46. – *Solanum soestii*, in its type locality, between the towns of Quime and Inquisivi, 2900 m altitude, Province Inquisivi, Department La Paz.

Habitat and Distribution

This plant occurs in moist, foggy places. It is frequently found in thickets among herbaceous plants, where it is often associated with *Barnadesia polycantha, Salvia haenkei, Begonia bridgesii,* two composite species, *Verbesina semidecurrens* and *Heliopsis bupthalmoides* and a non-tuber-bearing *Solanum* species. Currently, this plant is known only from several locations between Quime and Inquisivi, in the Province of Inquisivi, Department of La Paz, where it grows at altitudes between 2700 and 2900 m.

Specimens Examined

Department La Paz

Province Inquisivi: About 4 to 5 km from Quime on road to Inquisivi, 2850 m alt., on steep talus slope, 30–40 m above road, among bushes and rocks on moist soil, with ferns, begonias, *Salvia, Barnadesia,* non-tuber-bearing *Solanum* sp., several composites and other herbacious plants. February 23, 1984; *Ochoa and Salas 15502* (CIP, F, NY, OCH, US, topotypes). Near Rosanani, 2750 m alt., on the road from Quime to Inquisivi, among shrubs on stony slope above roadside. February 23, 1984; *Ochoa and Salas 15503* (CIP, NY, OCH, US). Rosanani, 2750 m alt., on rode from Quime to Inquisivi, in open fields and near margins of thickets. February 23, 1984; *Ochoa and Salas 15504* (CIP, NY, OCH, US). Between Rosanani and Inquisivi, 2750 m alt., in open fields and along moist thicket margins. Plants large, with well-dissected leaves and large leaflets. February 23, 1984; *Ochoa and Salas 15505* (CIP, F, NY, OCH, US).

V. SERIES CONICIBACCATA

CONICIBACCATA Bitter, Repert. Sp. Nov. 11:381. 1912. Oxycarpa Rydb. Bull. Torrey Bot. Club 51:146, 172. 1924.

Plants erect to erect-descending or spreading, to 3 m or more tall, glabrous or pubescent, tuber-bearing or rarely non-tuber-bearing. Leaves imparipinnate, variously dissected. Pedicels variously articulated above base. Corolla rotate-pentagonal to rotate-substellate, at times appearing 10-lobed, blue, white, lilac, or purple. Fruit long-conical and with pointed apex, or prominently ovoid-conical and with obtuse apex.

Chromosome number: 2n=2x=24, 2n=4x=48, and 2n=6x=72.

Habitat and Distribution

This group extends from Mexico to Bolivia and is one of the more widely distributed series of this section. Plants of this series occur most frequently in the climatically temperate region of the Yunga, or Ceja de Monte. Here, they occur in moist forests at elevations ranging from 1500 to 3000 m. They are also found in the Prepuna formations of the lower valleys and in the high, cold and moist Paramo and Puna at elevations of 4000 m.

Although the series Conicibaccata ranks second in size to series Tuberosa and contains a very large number of species, only the following three occur in Bolivia: *S. bombycinum*, *S. neovavilovii*, and *S. violaceimarmoratum*.

Key to Species

1. Plants puberulent; stem angular, winged; corolla rotate to rotate pentagonal, purple or light blue with white acumens; leaves usually 3-4 jugate.
   2. Pedicels articulated in the upper one-third of the stalk; calyx small, 4-5 mm long, with attenuate lobes; corolla with short acumens, these 1.5-2.5 mm long .... 12. *S. neovavilovii*.
   2. Pedicels articulated below center or occasionally slightly above center; calyx much larger, 6-7 mm long, with abruptly narrowed lobes; corolla with longer acumens, these 4-5.5 mm long .... 13. *S. violaceimarmoratum*.
1. Plants densely pubescent; stem cylindrical, wingless; corolla rotate-pentagonal to substellate, lilac; leaves usually 1-2 jugate .... 11. *S. bombycinum*.

11. Solanum bombycinum Ochoa, Am. Pot. Jour. 60:849-852, Fig. 1. 1983.

Figs. 47-49; Plate IX.

Plants to 60-70(-120) cm tall, densely pubescent. Stem erect, cylindrical, wingless, 6-8(-10) mm in diameter at the base, usually simple, pigmented, or green and densely speckled with dark purple along the lower two-thirds of the stalk, very pilose with short hairs. Stolons 40-50 cm or more long, slender, 2-2.5 mm in diameter; tubers with white buds, round to oval-compressed, very lenticulate, usually large, to 7 cm long and 3.5 cm in diameter, the skin tinged with light purple or dark purple, or whitish when immature. Leaves imparipinnate, densely pilose, (7.5-)12-16(-26) cm long by (5.5-)10-13(-15) cm wide, generally without interjected leaflets or at most only with 1 or 2 very small blades, and provided with 1-2(-3) pairs of lateral leaflets, the latter borne on petioles 7-9(-12) cm long. Juvenile leaves silvery white or straw-yellow, soft velvety or very densely pubescent with short hairs above and densely pilose with short hairs below except for along the nerves, which are invested with long, broad hairs. Terminal leaflets much larger than the laterals, (3.5-) 6.5-8.5(-11) cm long by (3-)4.4-6.5(-9.5) cm wide, broadly ovate-lanceolate to elliptic-lanceolate or occasionally nearly orbicular, the apex strongly acuminate or pointed and the base broadly rounded to subcordate or cordate. Lateral leaflets elliptic-lanceolate, the apex acuminate and the base rounded to subcordate and borne on petioles 2-4 mm long, the upper pair (2-)3.5-7(-8.5) cm long by (1-)2-4(-5) cm wide, the second pair half the size of the first, and the third, when present, sessile, very small, and having an obtusely pointed apex. Pseudostipular leaves small, 6-9(-11) mm long by 4-5(-7) mm wide, obliquely elliptic-lanceolate or broadly subfalcate. Inflorescence cymose to cymose-paniculate, 4-7(-12)-flowered. Peduncle short, 2.5-5(-7.5) cm long and to 2.5 mm in diameter at base, densely covered with very short hairs and some tetralobed glandular hairs. Pedicel pubescent with very short hairs, articulated near center or slightly below center, the upper or distal part 5-7(-9) mm long, and the lower or proximal end 2.5-5(-8) mm long. Calyx pubescent with very short hairs, 6-7(-8) mm long, the membranaceous, narrowly elliptic-lanceolate lobes either attenuate, or bearing abruptly narrowed acumens, the latter either similar or of unequal length. Corolla rotate-pentagonal to substellate, 2.5-2.8 cm in diameter, dark lilac within and light lilac with whitish acumens without, the lobes short, 1.5-2.5 cm long excluding the 5.5-6(-7) mm-long acumens. Anthers 5-5.5(-6) mm long, yellowish-orange with light-brownish spots at base, borne on 1.5-2 mm-long sparsely pilose filaments. Style 8.5-9 mm long, densely papillose along the lower two-thirds of its length; stigma small and oval-capitate. Fruit oval-conical or long-conical and obtusely pointed.

Chromosome number: 2n=4x=48.

Type: BOLIVIA. Department La Paz, Province Franz Tamayo, between Cheke-Chekeni and Chullumayo, 2000 m alt., on the horse trail to Mojos. February, 1983; *C. Ochoa and A. Salas 14964* (holotype, OCH; isotypes, CIP, US).

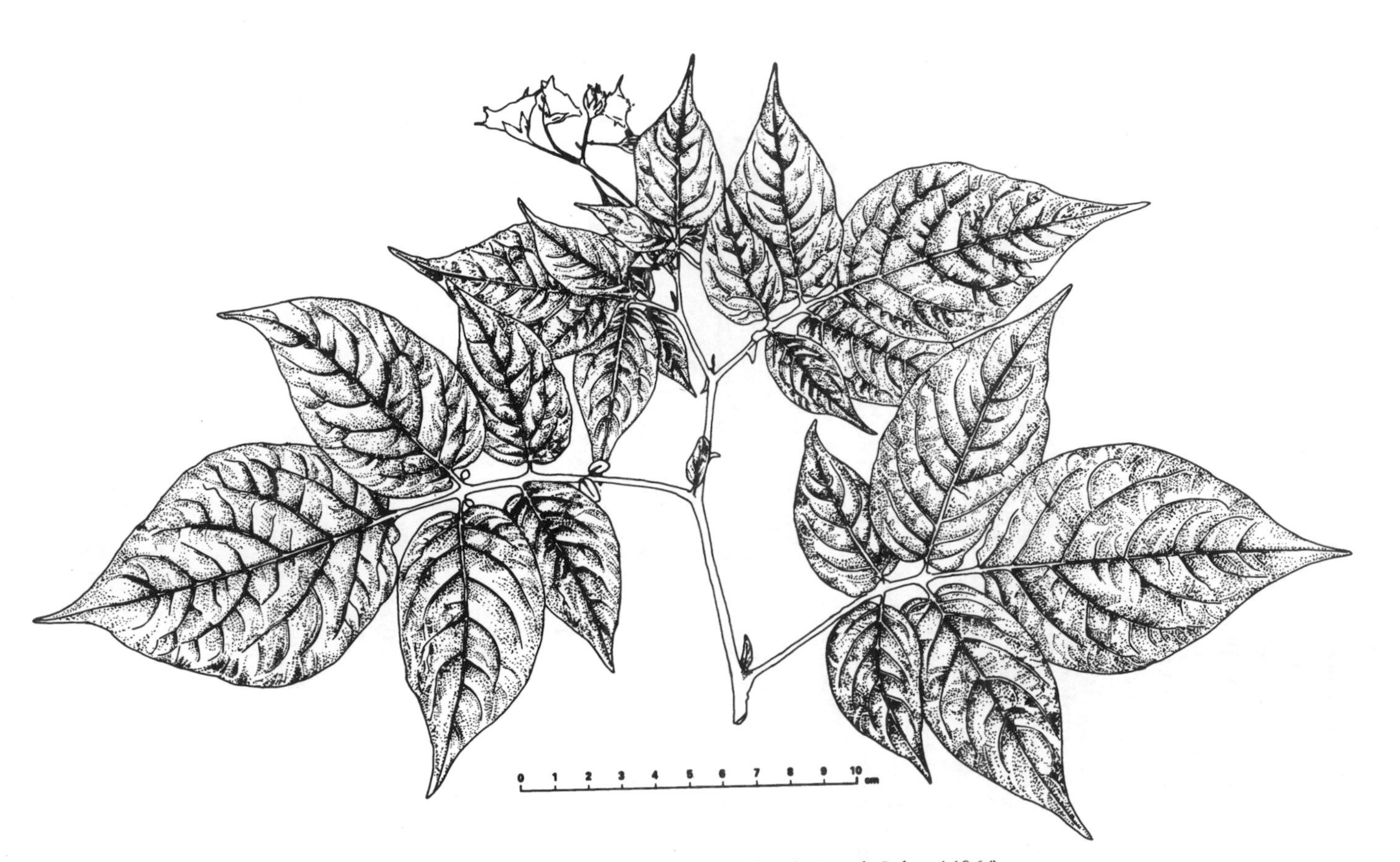

Figure 47. – *Solanum bombycinum* (*Ochoa and Salas 14964*).
Upper part of flowering plant, × ½.

Figure 48. – *Living plant of Solanum bombycinum (Ochoa and Salas 14964).*

It may be of some interest here to note that the flowers of this species are seldom fully opened. As in the case of *S. acaule*, the pedicels of *S. bombycinum* are normally strongly recurved downward in fruit. This suggests that the fruits of *S. bombycinum*, as in the case of the latter cleistogamous species, may be self-planting.

Affinities

This species is closely related to *S. villuspetalum* Vargas. Both species are densely pubescent and have cylindrical stems, poorly dissected leaves, rotate-pentagonal corollas, and pubescent filaments. The two differ, however, with respect to flower number, leaflet shape, and form of the inflorescence. The leaflets also differ with respect to their placement on the rachis. Moreover, the leaves of *S. villuspetalum* are somewhat glossy; whereas those of *S. bombycinum* are lusterless. In *S. bombycinum*, the branches of the inflorescence are somewhat lax, and the flowers, as previously mentioned, are rarely opened even at maturity. Lastly, it is important to mention here that *S. villuspetalum* is diploid (2n=2x=24), while *S. bombycinum* is tetraploid (2n=4x=48).

Habitat and Distribution

This species is known at the present time from only a single locality. It occurs under bushes and trees in the lush, subtropical watershed region of Chullumayu. Here it occurs at an elevation of 2000 m along the horse trail connecting Checke Checkeni to Mojos. Other plants known to be associated with this

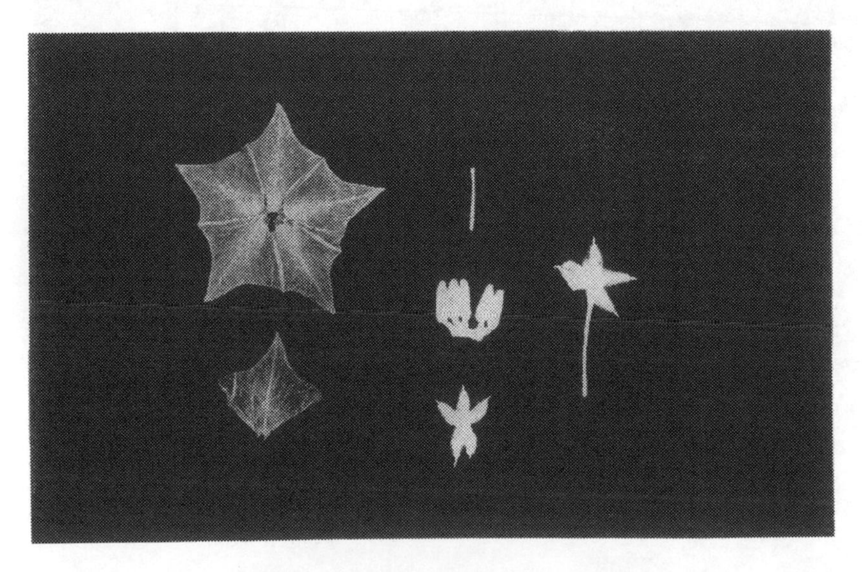

Figure 49. – Floral dissection of *Solanum bombycinum* (*Ochoa and Salas 14964*), ×1.

species include a woody member of the genus *Fuchsia*, herbaceous ferns, tree ferns, *Oxalis*, and several different species of grasses, including the genus *Chusquea*.

Specimens Examined

Department La Paz

Province Franz Tamayo: Between Checke Checkeni and Chullumayu, 2000 m altitude. February, 1983; *Ochoa and Salas 14964* (holotype, OCH; isotypes, CIP, F, NY, US).

Crossability

This self-fertile species crosses easily with *S. tuberosum* subsp. *andigena*. Reciprocal crosses, however, involving this tetraploid species and four diploid taxa, *S. circaeifolium* var. *circaeifolium* (11806), *S. circaeifolium* var. *capsicibaccatum* (WIS 1730), *S. yungasense* (15041), and *S. neovavilovii* (14967), were completely unsuccessful.

Two to three parthenocarpically developed seeds per berry were produced on plants derived from reciprocal crosses of *S. bombycinum* (14964) and *S. limbaniense* (14290). The F_1 hybrids from the latter cross had stems with short, sinuous internodes, and leaves, pedicels, and peduncles that were very similar to those of *S. limbaniense*. The pedicels of these plants were articulated near the center of the stalk. All hybrids bore small, white, rotate corollas, these tinged externally with light blue.

12. **Solanum neovavilovii** Ochoa, Am. Pot. Jour. 60:919-923, Fig. 1. 1983.

FIGS. 50-53; PLATE X.

Plants usually small, (12-)25-35(-70) cm tall, erect. Stem slender, 3-6(-8) mm in diameter, simple or branched and slightly pigmented toward base, straight or slightly flexible, very sparsely pilose and narrowly winged, the wings straight and barely distinguishable. Stolons slender, 1-1.5 mm in diameter and up to 50 cm or more long, white; tubers white, the smaller 1.5-2.5 cm long, rounded or oval to compressed-oval, the larger 3.5-5.5 cm long and 1.5-2 cm in diameter, long subcylindrical and rounded or obtuse at both ends or occasionally rounded at one end and pointed or fusiform at the other, buds white, glabrous. Leaves imparipinnate, small and broad, (7-)12-16(-21) cm long by (5-)7-9(-12.5) cm wide, poorly dissected, 3-4(-5)-jugate and with 0-5(-9) interjected leaflets, the latter, when present, very small, 1-1.5 mm in diameter to 2.5 mm long, orbicular to elliptic and sessile. Leaflets elliptic-lanceolate or narrowly elliptic-lanceolate, subsessile and shortly petiolulate,

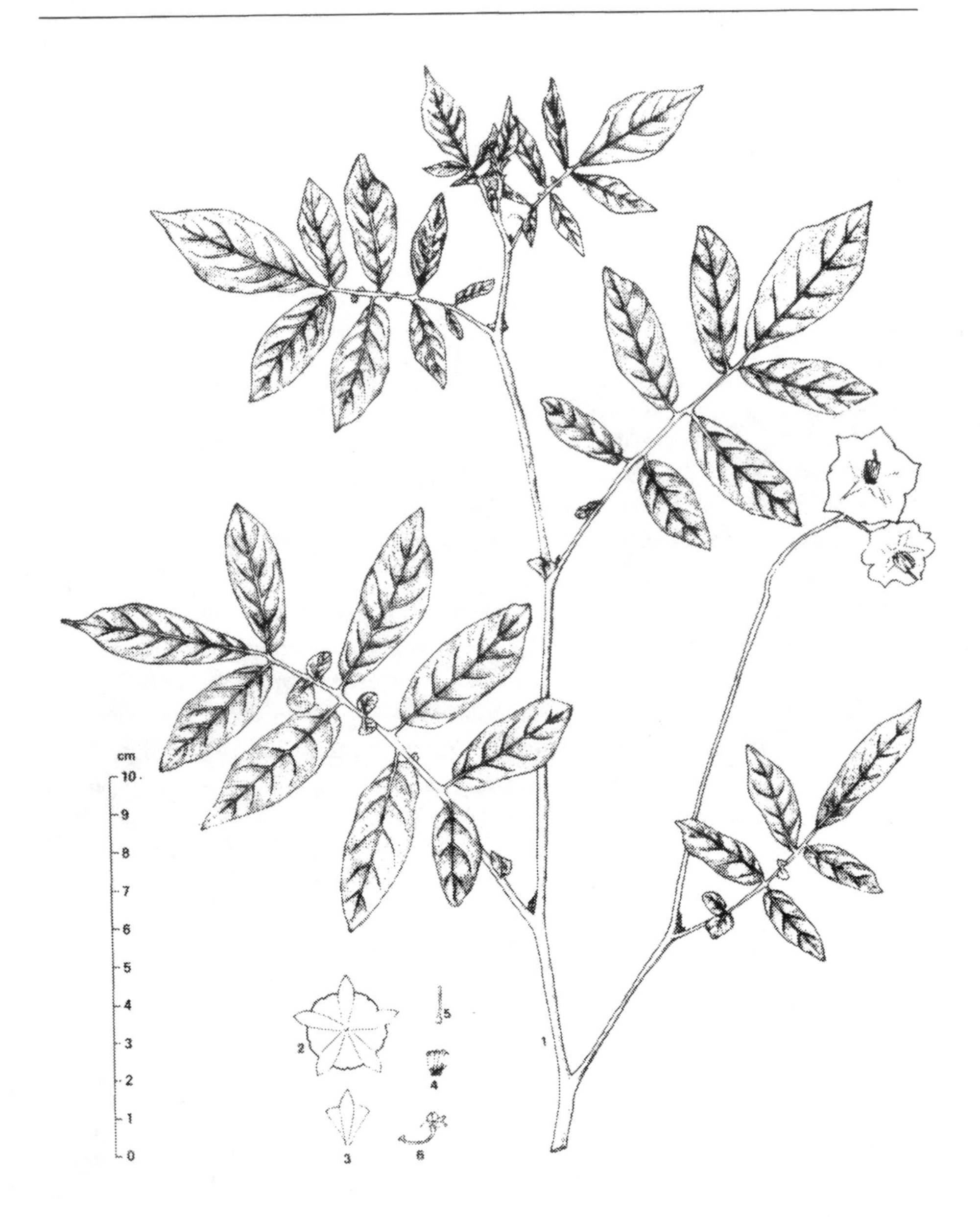

Figure 50. – *Solanum neovavilovii* (*Ochoa and Salas 14962*).
1. Upper part of flowering plant. 2. Corolla. 3. Petal.
4. Stamens. 5. Pistil. 6. Pedicel and calyx.

glabrescent or sparsely pilose below and somewhat more densely pilose above with many short, simple hairs and some tetralobed glandular hairs, the latter especially frequent on the rachis and leaf axiles. Terminal leaflet of equal or slightly smaller size than the upper pair of laterals, the latter (2.5-)4-5.5 cm long by (1.3-)1.5-2 cm wide, and provided with a pointed or shortly acuminate apex and an attenuate base. First pair of lateral leaflets (1.5-)3-4.5 cm long by (0.8-)1.3-1.6 cm wide and smaller than the second pair, the latter (1.5-)3.3-5.5 cm long by (0.8-)1.7-2.2 cm wide. Inflorescence cymose, 2-4-flowered. Peduncle slender, sparsely pilose, (2-)3.5-6(-15) cm long and to 1.8 mm in diameter at base. Pedicels articulated above center, the upper part 5-6 mm long and the lower 7-9 mm long, pilose with many short hairs alternating with fewer multicellular and longer hairs. Calyx small, 4.5-5 mm long, symmetrical or asymmetrical, light green and pilose with short, fine, white hairs, the lobes narrowly-lanceolate and attenuate at apex. Flowers light sky blue with white acumens or sky blue with white narrow nectar guides. Corolla rotate, small, 2-2.5(-2.8) cm in diameter, the lobes 1.5-2.5 mm long excluding the very short, 2 mm-long acumens. Anthers narrowly lanceolate 4-5 mm long, light yellow, borne on glabrous, 0.5-0.7 mm-long, white-hyalinous filaments. Style short and very slender, 6.5-7 mm long, densely papillose along the lower two-thirds of its length; stigma light green, very small, scarcely wider than the apex of the style; ovary ovate-conical. Fruit long-conical with obtuse apex, light green, 3 cm long by 1.7-1.8 cm in diameter.

Chromosome number: 2n=2x=24.

Type: BOLIVIA. Department La Paz, Province Franz Tamayo, Mayupampa, 3000 m alt., east of Quecara. February, 1983; *C. Ochoa and A. Salas 14961* (holotype, OCH; isotype, CIP, US).

Affinities

Solanum neovavilovii is somewhat related to *S. limbaniense*, a species from southeast Peru. The two are similar in habit, pedicel articulation, and fruit shape. Both also have a rotate corolla with short lobes. However, the two differ markedly in corolla size and color, anther size and shape, style length, and calyx shape and pubescence, as well as in the dissection of the leaves.

The original description of *S. neovavilovii* was based upon the type and paratype collections (*Ochoa and Salas 14961, 14969*). At the time of its publication, it was thought that this species had some affinity to the Peruvian *S. bukasovii* and the Bolivian *S. leptophyes*. However, more recent studies of this plant, based on collections gathered from its native habitats, as well as of accessions of this species that were grown under uniform environmental conditions at CIP, have failed to support the above theory.

Figure 51. – Terminal branch of *Solanum neovavilovii* (*Ochoa and Salas 14961*).

Habitat and Distribution

This species grows mostly in the Ceja de Yunga of Bolivia, a cool-misty, cloudy region characterized by high rainfall. Here, it grows between 2500 and 3200 m in forested quebradas and on elevated rock formations, the latter covered largely with mosses and succulent plants. Geographically, the range of this species extends from Mayupampa and Chimpainioc (east of Quecara) to Jatum Pampa and Sayhuanimayo in the vicinity of Puina (northeast of Cordillera de Apolobamba, Province Franz Tamayo, Department La Paz).

Specimens Examined

Department La Paz

Province Franz Tamayo: Mayupampa, 3000 m alt., about 4 km E of Quecara, on road to Mojos, among shrubs and forest margins. Chromosome number: 2n=2x=24. February 6, 1983; *Ochoa and Salas 14961* (holotype, OCH; isotypes, CIP, US) and *Ochoa and Salas 14962* (topotypes, OCH, CIP, US). Between Quecara and Chullumayo, 3000 m alt., on rock formations covered with soil and mosses, in open field. February 6, 1983; *Ochoa and Salas 14967* (CIP, OCH, US) and *Ochoa and Salas 14968* (CIP, OCH). Chimpainioc, near Mayupampa, 2800 m alt., about 7 km E of Quecara. Chromosome number: 2n=2x=24. February 6, 1983; *Ochoa and Salas 14969* (OCH, paratype). Sachapampa, 3200 m alt., near Jatun Sencca, about 5 km from Puina on road to Sayhuanimayo, along the right bank of the Rio Puina, in a 'roso de monte' or an agricultural clearing within a forest. Chromosome number: 2n=2x=24.

Figure 52. – Floral dissection of *Solanum neovavilovii* (*Ochoa and Salas 14962*), ×1.

February 9, 1983; *Ochoa and Salas 14993* (CIP, OCH, US). Near Ayamachay, 2500 m alt., on the road from Puina to Sayhuanimayo, along the right bank of the Rio Puina, about 16 km from Puina. Chromosome number: 2n=2x=24. February 9, 1983; *Ochoa and Salas 14994* (OCH).

Crossability

Varied success was obtained in artificial reciprocal crosses involving this self-incompatible species and other wild, tuber-bearing diploid species of *Solanum*. All crosses involving *S. bombycinum* (14964), no matter in which direction they were made, were unsuccessful. On the other hand, crosses of *S. neovavilovii* (14993) with *S. limbaniense* (14290 and 15079) resulted in almost 40% successful crosses with fruits containing an average of 60 viable seeds.

Similarly, crosses between *S. leptophyes* (13720) and *S. neovavilovii* (14994) were only partially successful, these yielding, on the average, 60 seeds per berry. Crosses involving *S. yungasense* (15041) and *S. neovavilovii*, on the other hand, yielded an average of 70 seeds per berry, while those involving *S. candolleanum* (15002) yielded 130 seeds per berry.

Figure 53. – Tubers of *Solanum neovavilovii*, ×1.

A surprising amount of success was achieved in crosses involving *S. neovavilovii* and *S. raquialatum* (13958), a species of northern Peru. Eighty percent success was obtained in crosses between these two species when *S. raquialatum* was used as a mother plant. In this case, an average of 300 viable seeds per fruit were obtained. As these two species are geographically isolated and also very different from one another morphologically, they have been treated under different taxonomic series (*see* Ochoa, 1962).

First generation hybrids of *S. neovavilovii* (14994) and *S. limbaniense* (14288) or *S. limbaniense* (15079) and *S. neovavilovii* (14993) exhibited some dominant characteristics of each parent. These segregates were similar to *S. neovavilovii* in leaf dissection and leaflet shape, but had the floral characters and pigmented stems of *S. limbaniense*.

Solanum leptophyes (13720) × *S. neovavilovii* (14994).– Crosses between these two species yielded plants that were very similar in habit, leaf shape, and dissection to *S. leptophyes*. However, the flowers of the hybrids retained characteristics of both parents. The pedicels, which were unlike those of either parent, were articulated at their midpoints or below center.

Solanum neovavilovii (14933) × *S. candolleanum* (15002).– Hybrid plants of this cross expressed the dominant floral and indumentum traits of *S. candolleanum*. The traits of the latter were especially evident in the size and pilosity of the calyx, the size and color of the corolla (violet to purple), and in the general pubescence of the plant. On the other hand, these hybrids resembled *S. neovavilovii* in the shape of their leaflets, in their less vigorous and slender stems, and in their smaller pseudostipular leaflets.

Solanum neovavilovii (14993) × *S. yungasense* (15041).– Except for certain corolla characteristics, this cross yielded plants having the dominant leaf and floral traits of *S. neovavilovii*. The rotate to rotate-pentagonal corollas of these hybrids varied in color from white to blue-violet. All had white acumens.

Solanum neovavilovii (14993) × *S. raquialatum* (13958).– Plants from this cross had leaves and purple flowers closely resembling the first species.

13. Solanum violaceimarmoratum Bitter, Repert. Sp. Nov. 11:389. 1912.

FIGS. 54–56.

S. violaceimarmoratum var. *papillosum* Hawkes, Bull. Imp. Bur. Pl. Breed. & Genet., Cambridge (June):12, 14, 113. 1944. TYPE: *Balls, Gourlay, and Hawkes 6275*, Bolivia. Track from Colomi to Incachaca below Llanta Aduana, Dept. Cochabamba, Prov. Chapare, March 16, 1939 (CPC, K).

Plants vigorous, erect. Stem to 3 m tall and to 15 mm or more in diameter at base, strongly purple-spotted, puberulent or essentially glabrous, usually

Figure 54. – *Solanum violaceimarmoratum* (*Ochoa 11908*). 1. Upper part of flowering plant. 2. Corolla. 3. Petal. 4. Stamens, dorsal view. 5. Pistil. 6. Calyx. 7. Pedicels and stamens. 8. Fruit. All×½.

thick, simple or branched, ascending, with broad, straight or sinuous green wings along the lower half, and narrow, straight wings above. Stolons more than 1 m long; tubers yellowish-white, ovate to rounded, 5-6 cm long by 3.5 cm in diameter. Leaves imparipinnate, (8.5-)17.5-32.5 cm long by (5.5-)11-19 cm wide, with 3-4(-5) pairs of leaflets and 0-5(-7) interjected leaflets and borne on long petioles (2.5-)5-7 cm long, sparsely pubescent above, with short, slender, or broad glossy hairs, and more densely pubescent below. Leaflets ovate to ovate-lanceolate or elliptic-lanceolate to broadly lanceolate, the apex attenuate or abruptly pointed or acuminate, and the base obliquely rounded to subcordate, borne on petiolules 8-12(-17) mm long. Terminal leaflets much broader and somewhat longer than the laterals, (3.5-)6.5-12 cm long by (1.8-)2.5-6 cm wide. Upper lateral leaflets (2.8-)5-9 cm long by (1.5-)2-4 cm wide, sessile or shortly petiolulate. Pseudostipular leaves ovate or falcate, 1.2-1.8(-2.5) cm long by 0.7-0.9(-1.2) cm wide. Inflorescence cymose, 10-14 (-18)-flowered. Peduncle pigmented and puberulent, 10-12(-18) cm long and to 2.5 mm in diameter at base. Pedicels puberulent and glandular, articulated near or somewhat below center, the lower part 4-6(-9) mm long and the upper 6-8(-11) mm long, and gradually enlarging in diameter toward calyx. Calyx symmetrical to asymmetrical, pigmented, 7(-8.5) mm long, glabrous or glabrescent, the narrowly lanceolate lobes abruptly narrowed and tipped with

Figure 55. – Living plant of *Solanum violaceimarmoratum* (*Ochoa 11908*).

pointed acumens to 1.5 mm long. Corolla rotate to rotate-pentagonal, usually small, 2.5-2.8(-3.5) cm in diameter, dark purple with black nectar guides, or occasionally light blue with yellow nectar guides, and colored light purple with white acumens outwardly, the lobes short, barely 2.5 mm long and tipped with conspicuous, or prominent, acumens to 5.5 mm long. Staminal cone cylindrical-conical; anthers 5-6(-7) mm long, very narrowly lanceolate and borne on glabrous or sparsely pilose 1-2 mm-long filaments. Style 8-9 (-10) mm long, exerted 2.5 mm, glabrous or papillose, or occasionally puberulent on lower half; stigma small, ovate-capitate. Fruit dark green, obtusely pointed, ellipsoid-conical to long-conical, to 3 cm long by 1.7 cm in diameter.
Chromosome number: 2n=2x=24.

Type: BOLIVIA. Department La Paz, Province Nor Yungas, Unduavi, 3300 m alt. February 12, 1907; *O. Buchtien 764* (holotype, US).

Although *S. violaceimarmoratum* occasionally bears pubescent filaments and styles, these traits are not characteristic of the species as a whole. Normally, the styles of this species are papillose and the filaments are glabrous. However, all variations of these traits can be found in plants drawn from a single population.

It is of some interest to note that plant collections of *S. maglia* from Chile and *S. laxissimum* and *S. santolallae* from Peru have been occasionally mislabeled as *S. violaceimarmoratum*.

Affinities

This species, which is not closely related to any other wild tuber-bearing species of *Solanum*, is endemic to Bolivia. Features that are unique to this species include its flower color, cylindrical stems, pubescence, leaf shape and dissection.

This plant bears a superficial resemblance to *S. urubambae* and *S. villuspetalum*. Both have pubescent filaments, long petioles, and corollas with long, prominent acumens. And yet, this plant would not appear to be very closely related to these two Peruvian species, nor, for that matter, to *S. buesii*, another Peruvian plant with which it has been occasionally confused.

Habitat and Distribution

Solanum violaceimarmoratum occurs along the moist, tropical or subtropical eastern slopes of the Bolivian Andes between 3000 and 3400 m, just beneath the Prepuna zone, in a region commonly referred to as either the Ceja de Yunga or Ceja de Montaña. Here, it is found mostly in loose, moist soils along river banks or in rocky or sandy-clay soils. It occurs also at lower elevations, as for example near Sacramento Alto (above Coroico) in the Province of Nor

Yungas, Department of La Paz, where it grows between 2200 and 2000 m altitude; and near Incachaca in the Province of Chapare, Department of Cochabamba, where it is found at 2200 m. At the Incachaca locality this species is associated with tree ferns, herbaceous ferns, orchids, bamboos, fuchsias, and woody species of *Solanum* and *Monnina*.

It is distributed 120 km from Mataro and Caspichaca, near Chulina in the Province of Saavedra, Department of La Paz, to Unduavi (Province Nor Yungas), its type locality, and from Unduavi to the Province of Chapare (Department Cochabamba), in central Bolivia. The altitudinal limits of this species lie between 2000 and 3500 m.

Specimens Examined

Department Cochabamba

Province Ayopaya: Lagunillas, 2800 m alt., over the mountain range behind Choro in the St. Elena Valley, thicket. March 5, 1950; *Brooke 6172* (BM).

Province Chapare: Incachaca, 2300 m alt. February 28, 1929; *Steinback 9488* (GH, K). Incachaca, 2700 m alt., slender shrub, 1 m tall, in moist soil. April, 1938; *Cárdenas 689* (HNF, LL, US). Trail from Colomi to Incachaca, below

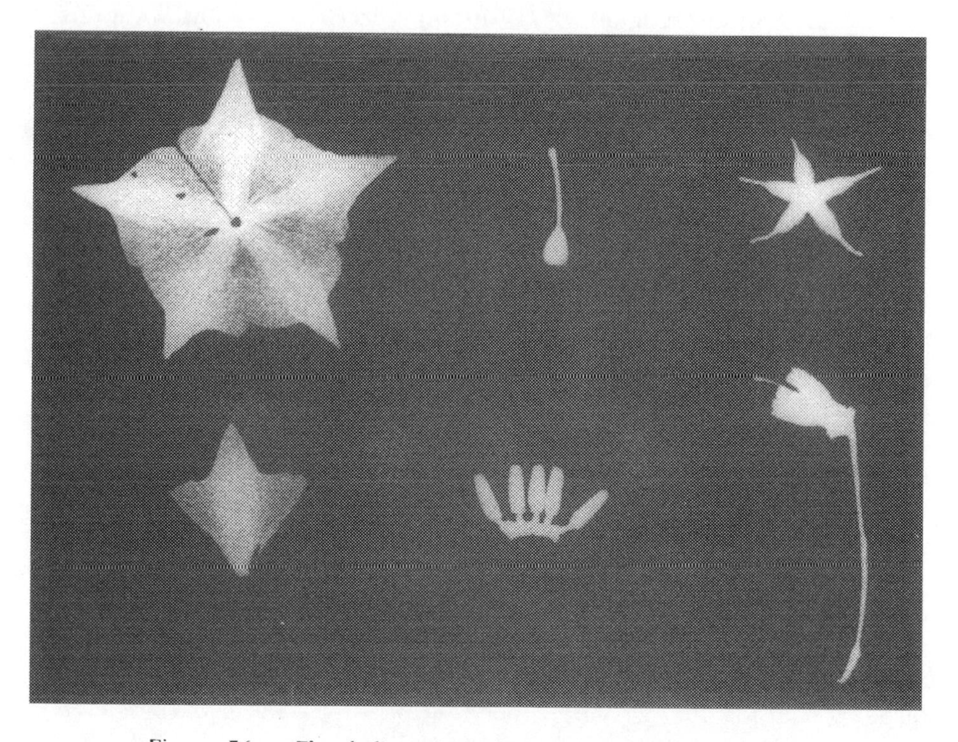

Figure 56. – Floral dissection of *Solanum violaceimarmoratum* (*Ochoa 11908*), ×1.

Aduana Llanta, 3500 m alt. March 16, 1939; *Balls, Gourlay, and Hawkes 6275* (CPC, K). Llanta Aduana, near Colomi, 3000 m alt., among shrubs on right side of trail to Incachaca. Herb up to 2.4 m tall, flowers purple, berries conical-elongated, tubers large. February-March, 1946; *Cárdenas 3637* (CPC, F, GH, LL, US). Incachaca, 2350 m alt., small power station about 128 km NW of Cochabamba, river bank. August 17, 1950; *Brooke 6716* (BM). Km 103 of road from Cochabamba to Chapare, 3000 m alt., on moist, rocky soil. Plants 30-50 cm tall, flowers blue. March, 1961; *Cárdenas 6116* (HNF No. 00095, WIS; determined as *S. chaparense* by Cárdenas). Incachaca, near Llanta Aduana, 2000 m alt., among shrubs in moist soil. Plants 0.8-3 m tall, flowers blue. March, 1961; *Cárdenas 6118* (HNF, WIS). 104 km on road from Cochabamba to Chapare and Villa Tunari, 3025 m alt., 20 m before Hotel, also for 0.5 km further on. On open bank above road in Ceja de Montaña. Flowers blue violet (one plant with paler lilac flowers) pentagonal-rotate. Fruit conical. February 17, 1971; *Hjerting, Cribb, and Huamán 4436* (CIP). Km 100 on Cochabamba to Tunari road, 3150 m alt. In Ceja de Montaña by roadside; in natural vegetation, not as a weed. Colony up to 1.5 high. February 18, 1971; *Hjerting, Cribb, and Huamán 4474* (CIP). Cochabamba-Aguirre-Todos Santos road, between Comercocha and Palmar, 3100 m alt., about 70 km SW of Todos Santos (air dist.), at km 104, 17°10′ S and 65°40′ W. Mountainous forest with many tree ferns, bamboo, orchids, ericads, melastomes, *Fuchsia*, *Epilobium*, *Monnina*, *Passiflora*, and both tuberous and nontuberous *Solanum* spp. April 19, 1963; *Ugent 5014, 5016, 5017* (WIS), *5018* (US), *5019* (F), *5022* (K), *5024* (OCH), *5026, 5028, 5030, 5031, 5032* (WIS).

Department La Paz

Province Nor Yungas: Unduavi, 3300 m alt., in forests. February 12, 1907; *Buchtien 764* (US). Unduavi, 3000 m alt., among bushes along right side of trail to Yungas. Herbs 0.8 to 1.2 m tall, moist soil. Flowers purple. February, 1946; *Cárdenas 3638* (CPC, GH, LL, US, topotype of *S. violaceimarmoratum*). Road from Unduavi to Coroico, 3275 m alt., about 6 km from Unduavi, on disturbed slope below roadside, and in the Ceja formation above roadside. Flowers stellate or pentagonal, violet or violet with white acumens; fruit round or elongated. April 3, 1971; *Hjerting, Cribb, and Huamán 5040* (CIP). Road from Unduavi to Coroico, 3225 m alt., about 10 km from Unduavi, flowers striped violet. April 3, 1971; *Hjerting, Cribb, and Huamán 5042* (CIP). Unduavi, 3350 m alt., 44 km from La Paz. 1980; *S.O.A.-7A and S.O.A.-7E* (CIP) [collected, perhaps, by Soest, Okada, and Alarcón]. Road from Unduavi to Coroico, 3350 m alt., 20 km E of Unduavi, wild potato growing in fill along roadside; flowers purple; only one plant with one tuber was found. February 23, 1954; *Smith and Torrico 183-H-T* (LL). Sacramento Alto, 2800–

2900 m alt., above Coroico. Chromosome number: 2n=2x=24. March 7, 1978; *Ochoa 11908* (CIP, OCH, US).

Province Saavedra: Caspichaca, 3500 m alt., about 5 km before Chullina, between Mataro and Chullina. Chromosome number: 2n=2x=24. March 4, 1978; *Ochoa and Salas 11901* (CIP, OCH, US).

Province Sud Yungas: Chulumani, 2400 m alt. February 5 to April 13, 1980; *Soest, Okada, and Alarcón S.O.A.-15* (CIP).

Indefinite: La Paz, 3600 m alt. March, 1914; *Buchtien s.n.* (G, GH, NY), *Bang 2519 p.p.* (NY). Santa Rosa, 2000 m alt. April 14, 1892; *Kuntze s.n.* (NY). April 13-20, 1892; *Kuntze s.n.* (NY).

Crossability

Reciprocal crosses made by the author between *S. violaceimarmoratum* (11901), a self-incompatible species, and *S. circaeifolium* var. *capsicibaccatum* (11915) were largely negative. However, in 30% of the cases, berries with parthenocarpic seeds were obtained. Similarly, reciprocal crosses involving *S. circaeifolium* var. *circaeifolium* (11806) were also very difficult to achieve, yielding only 6 viable seeds per fruit. These were successful only in cases when *S. violaceimarmoratum* was used as the mother plant.

Crosses involving *S. violaceimarmoratum* and *S. commersonii* (series Commersoniana) were unsuccessful, possibly due to differences between these species involving the EBN. However, crosses involving this species and other plants of series Conicibaccata yielded varying results, depending upon which species was used, as well as in which direction the cross was made. Little or no success was obtained when this plant was crossed with a diploid introduction of *S. santolallae* (13640). An average of only 4-8 seeds per berry was obtained. On the other hand, reciprocal crosses involving the following three other diploid species of series Conicibaccata were very successful: *S. laxissimum* (11855), *S. limbaniense* (14291 and 14405), and *S. rhomboideilanceolatum* (11869).

Crosses made with *S. chomatophilum* (series Conicibaccata) and *S. commersonii* (series Commersoniana) were negative. However, crosses with *S. humectophilum* (series Conicibaccata) yielded an average of 8 seeds per berry, but only when the latter species was used as the mother plant.

Solanum violaceimarmoratum × *S. circaeifolium*.– Plants obtained from crosses between these two species tended to be morphologically more similar to *S. violaceimarmoratum*, particularly resembling this species in corolla shape and color, leaf dissection, and vigor. The very large calyx of these plants was borne on centrally articulated pedicels ranging from 5 to 6 cm long.

In contrast to what was found above, all other crosses involving *S. violaceimarmoratum* and other species of this section were more similar to the second parent.

Solanum violaceimarmoratum × *S. laxissimum*.– The hybrid derived from this cross appeared to resemble *S. laxissimum* in leaf and leaflet shape, and in having small petiolules, but some ranged to nearly 20 mm in length. All had small, dark purple to light lilac flowers.

Solanum violaceimarmoratum × *S. limbaniense*.– As in the case of the above cross, the hybrid plants derived from this combination were more similar to the second species in leaf characteristics. However, their rotate to strongly pentagonal flowers varied in color from whitish to light lilac and purple.

Solanum violaceimarmoratum × *S. rhomboideilanceolatum*.– The F_1 plants of this cross had the dominant leaflet shape of *S. rhomboideilanceolatum*. However, their leaves varied in dissection from 2-jugate without interjects to 5-jugate with several interjects; while their flowers, which were borne on long-peduncled inflorescences, varied in color from dark purple to light lilac. The dense pilosity of the leaves of these hybrids was a character of these plants unlike that of either parent.

Solanum violaceimarmoratum × *S. santolallae*.– Plants of this cross had leaves and leaflets that were similar to those of *S. santolallae*. However, the corolla shape and color of these hybrids were dominant features that were inherited from *S. violaceimarmoratum*.

VI. SERIES MEGISTACROLOBA

MEGISTACROLOBA Cárdenas and Hawkes, Jour. Linn. Soc. Bot. 53:93. 1945.

Plants stoloniferous and tuber-bearing. Stems short and rosette-like or occasionally elongated. Leaves imparipinnate to imparipinnatisect or occasionally simple, usually without interjected leaflets. Terminal leaflets conspicuously much larger than the laterals. Lateral leaflets, when present, broadly decurrent upon the rachis. Peduncle usually short. Pedicel articulation well marked, above midpoint and often just below calyx. Corolla rotate, pentagonal, or broadly stellate, light lilac to very dark purple. Fruit globose to ovoid.

Chromosome number: 2n=2x=24.

This series is closely related to Cuneoalata. Both have species having decurrent leaflets, though they are much less broadly decurrent in Cuneoalata. Moreover, this series appears to be related in part to Acaulia. Thus, some species of Megistacroloba are acaulescent or have a short floral peduncle or rosette-like habit.

The type species of this series, *S. megistacrolobum*, was proposed by Bitter in 1912. Later, Bukasov (1955a) proposed a new series, Alticolae, in which he included the Bolivian and northern Argentine species that were formerly placed in Megistacroloba. This left only two Peruvian species in the original series, *S. hawkesii* and *S. raphanifolium*.

The above concept of series Megistacroloba was also one shared by Brücher (1959b). However, his classification was revised a short time later by Bukasov, this time in collaboration with Kameraz (Bukasov and Kameraz, 1959). The net effect of these changes was to reestablish the original concept of series Megistacroloba as proposed by Cárdenas and Hawkes.

Series Megistacroloba, along with all other taxonomic series of potatoes, was monographed in detail by Correll (1962). A brief treatment was also published by Hawkes (1963). Six years later, Hawkes and Hjerting (1969) published a treatment of series Megistacroloba as part of their own work on the potatoes of Argentina, Brazil, Paraguay, and Uruguay. In this monograph, the two wide-ranging Bolivian species, *S. megistacrolobum* and *S. boliviense*, are treated in greater detail.

Major differences are found in the treatment of this series by Hawkes (1963) and Correll (1962). Thus, while Hawkes upholds *S. ellipsifolium*, a Bolivian species, Correll places it in synonomy under *S. toralapanum* var. *subintegrifolium*. Moreover, Correll treats the Peruvian diploid species *S. dolichocremastrum* as a member of series Megistacroloba, while Hawkes classifies both this plant and the Colombian *S. flahaultii* as synonyms of the Ecuadorian *S. paucijugum*, the latter of series Conicibaccata. In the author's opinion, the latter two tetraploids, as well as the diploid *S. dolichocremastrum*, should be maintained as separate species.

Similarly, Correll classified *S. boliviense* in series Transaequatorialia, while Hawkes placed it in Megistacroloba. Both authors, however, agree that *S. toralapanum* should be classified as a species of series Megistacroloba, and that *S. decurrentilobum* should be treated as a synonym of *S. toralapanum*.

Recently, Hawkes and Hjerting (1985a) have proposed *S. astleyi* as a new Bolivian wild species of series Megistacroloba. However, herbarium material of this new taxon has not been available to the author. Nevertheless, he has concluded from a study of the type photo of this collection (*Hondelman, Astley, and Moreira 208*) and its original diagnosis (Hawkes and Hjerting, 1985a) that it should be classified probably as an ecotype of *S. boliviense*.

Distribution

The author believes that series Megistacroloba consists of one-named hybrid and seven species. The members of this series and their respective distributions are as follows: *S. hastiforme* (northern Peru); *S. sogarandinum* (northern Peru); *S. dolichocremastrum* (central Peru); *S. raphanifolium* (=*S. hawkesii*, southern Peru); *S. megistacrolobum* (southern Peru, Bolivia, and northwest Argentina); *S. boliviense* (Bolivia and northern Argentina); *S. sancta-rosae* (northern Argentina); and *S.* × *brucheri* (Argentina).

The two Bolivian species of this series, *S. boliviense* and *S. megistacrolobum* (including var. *toralapanum*), extend from southern Peru across Bolivia to northwestern Argentina. They occur between 2000 and 4300 m altitude.

Key to Species

1. Leaves usually simple, or with 1-2 pairs of very small, nondecurrent or rarely decurrent leaflets; calyx to 6 mm long, with ovate-lanceolate lobes and poorly delimited acumens; corolla rotate or rotate-pentagonal, dark purple .... 14. *S. boliviense*.

1. Leaves simple or with 1-3(-5) pairs of decurrent leaflets; calyx to 9 mm long, with ovate-triangular lobes and generally well-delimited acumens; corolla substellate or rotate, lilac, violet, or purple.

   2. Leaves simple, broadly elliptic to oblanceolate, or if imparipinnate with one to several small, or highly reduced and narrowly decurrent leaflets .... 15. *S. megistacrolobum* var. *megistacrolobum*.

   2. Leaves simple, broadly triangular-ovate, or if irregularly imparipinnate to imparipinnate with one to several small but not highly reduced decurrent leaflets .... 15a. *S. megistacrolobum* var. *toralapanum*.

14. Solanum boliviense Dunal. DC. *Prodr.* 13(1):43. 1852.

FIGS. 57-60; PLATE XI.

Plants rosette-like to postrate-extending or ascending to decumbent, sparsely pubescent with short hairs. Stems to 40 cm tall, sparsely pilose with short hairs, angular, usually wingless, simple or branched. Stolons long and slender; tubers small, 1-2 cm long, round to ovoid, yellowish-white. Leaves usually simple, more rarely with one or two pairs of very small lateral leaflets, sparsely pubescent above with stout, white, adpressed hairs, and glabrescent below or pubescent only on the veins and veinlets. Leaf blade or terminal leaflet (when present) extremely large, 4.5-8(-12) cm long by 2-4(-5.8) cm wide, elliptic-lanceolate, elliptic or ovate to obovate, with an obtuse or pointed apex and a rounded or very narrowly cuneate base. Lateral leaflets approximately 2 cm long by 1 cm wide, borne on 1.5-3(-4) cm-long petiolules. Pseudostipular leaves very small, less than 10 mm long, arcuate or semilunate. Inflorescence cymose, few-flowered (3-7), borne on 2-4(-5) cm-long peduncles. Pedicels articulated in the upper one-third of the stalk, the upper part 4-8 mm long, and the lower 25 mm long. Calyx 4-5 mm long, the lobes ovate-lanceolate and tipped with 1 mm-long pointed acumens. Corolla rotate to pentagonal, (2-) 2.4-2.8(-3.2) cm in diameter, dark bluish purple with grayish nectar guides and with usually well-delimited shoulder lobes, the latter 7-11 mm long (including the 2.5 mm-long acumens) by 10-15 mm wide. Anthers 5-5.5 mm long, narrowly lanceolate, noncordate at base, borne on 1-2 mm-long filaments. Style 9-10 mm long, exerted 2.5-3 mm, sparsely pilose below center; stigma capitate, cleft. Fruit globose or occasionally slightly compressed or foreshortened, to 2.5 cm in diameter, uniformly dark green or occasionally dark green and irregularly speckled with white dots.

Chromosome number: 2n=2x=24.

Type: BOLIVIA. Department Chuquisaca; *A. D'Orbigny 1170* [without date and locality, but probably collected in the vicinities of today's town Sucre, in February, 1833] (holotype, here designated P; isotypes, W, MPC).

The holotype of *S. boliviense*, collected by D'Orbigny under the number *1170*, is deposited in the Natural History Museum of Paris. Isotypes are housed in the Natural History Museums of Montpellier and Vienna. Although the date and exact locality of this collection were not indicated on the original labels, it is possible that these were collected in the vicinity of Chuquisaca (now Sucre) between January and February of 1833. D'Orbigny, it should be noted, was at that time detained in Chuquisaca awaiting an interview with Don Andres Santa Cruz, President of the Republic (D'Orbigny, 1945).

One germplasm accession of *S. boliviense*, examined by the author at CIP

(OCH 11937), is unusual in that the upper parts of the stems, peduncles, and pedicels are densely covered with very short, simple hairs and short glandular, tetralobular type A hairs. Moreover, the stems and petioles of these plants are narrowly grooved and wingless, and the pseudostipular leaves are obsolete. However, the leaves of this accession, as in the case of D'Orbigny's holotype, are sparsely pilose on the upper surfaces and margins of the leaflets and pubescent on the veinlets below.

Affinities

This species is very closely related to *S. megistacrolobum*. It differs principally from that species in habit, leaf dissection, terminal leaflet shape, pubescence, and corolla size and color.

Habitat and Distribution

This species extends from the Department of Chuquisaca, Bolivia, to the Province of Salta, Argentina. It grows preferentially in poor, stony soils between 2600 and 3700 m alt. Within this zone, which is not entirely frost-free, it grows primarily in open fields, among grasses and rocks, along the margins of cultivated fields, in barley fields on brushy slopes, and in thickets of *Polylepis*. It is associated occasionally with *S. megistacrolobum*, *S. leptophyes*, and *S. brevicaule*.

Specimens Examined

Department Chuquisaca

Province Oropeza: Chuquisaca (*fide* Bitter), without date and exact locality; *D'Orbigny 1170* (MPU, P, W, type collection). On hills above Sucre, dry slopes among stones, 2600 m alt. March 6, 1939; *Balls, Gourlay, and Hawkes 6147* (CPC). Guerraloma, near Sucre, 3019 m alt. Growing in open, exposed places in poor, stony, dry soil. Stem short, slightly branched, the branches creeping 2-3 cm along ground. Leaves simple, about 3-5 cm long, ovate. Flowers few. Corolla to 2.5 cm in diameter, blue-purple. Berry dark green, globular. March 8, 1939; *Balls, Gourlay, and Hawkes 6188* (CPC, K, US). Guerraloma, near Sucre, 3000 m alt. Growing in fields with potato and bean crops. Native name: *Qquita papa*. March 8, 1939; *Balls, Gourlay, and Hawkes 6190* (BM, CPC, US). 20 km S of Sucre. February, 1949; *Cárdenas s.n.* (LL). Punilla, about 15 km NW of Sucre (air distance) on the Sucre-Ravelo road, 18°55′ S and 65°25′ W. Margins of weedy wheat field, with *Solanum boliviense*, *S. pachytrichum*, nontuberous *Solanum* sp., *Setaria plumosa*, *Calceolaria* and *Veronica*. April 12, 1963; *Ugent and Cárdenas 4896, 4897-4901, 4902-4908* (WIS). Villa Maria, 2850 m alt., 19°1′ S and 65°10′ W, a small farm about 10 km NE of Sucre (reached by jeep but not serviced by road). Margins of a

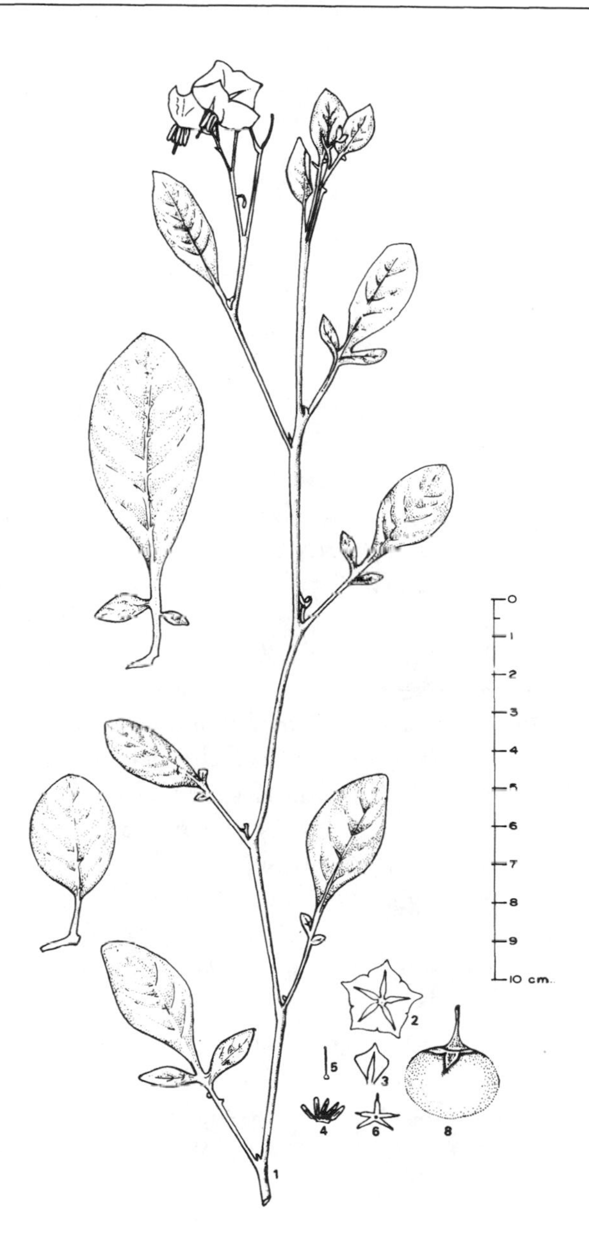

Figure 57. – *Solanum boliviense* (*Ochoa 11929*). 1. Upper part of flowering plant. 2. Corolla. 3. Petal. 4. Stamens. 5. Pistil. 6. Calyx. 7. Leaves. 8. Fruit. All ×½.

weedy wheat field and surrounding thickets, with *Solanum boliviense*, *S. pachytrichum*, *Tagetes*, *Cerastium*, *Medicago*, *Rumex*, and *Brassica*. April 12, 1963; *Ugent and Cárdenas 4928, 4929* (WIS). Guerraloma, 16 km from Sucre on road to Ravelo, 3275 m alt., by roadside among stones and bushes in a wetter area. Rotate deep violet flowers. March 5, 1971; *Hjerting, Cribb, and Huamán 4553* (CIP). Vicinity of Guerraloma, 2770 m alt. Chromosome number: 2n=2x =24. March 11, 1978; *Ochoa 11929* (CIP, OCH). Ckuchu Tambo, 2940 m alt., on road from Sucre to Guerraloma. Flowers very dark purple, corolla pentagonal. Chromosome number: 2n=2x=24. March 12, 1978; *Ochoa 11933* (CIP, OCH) and *Ochoa 11933A* (OCH). Between Ckuchu Tambo and Guerraloma, 3000 m alt. Chromosome number: 2n=2x=24. March 13, 1978; *Ochoa 11935* (CIP, OCH). Mountain ascent, between Sucre and Guerraloma, 2900 m alt. Chromosome number: 2n=2x=24. March 13, 1978; *Ochoa 11937* (CIP, OCH). Near Guerraloma, 3100 m alt. Chromosome number: 2n=2x=24. March 13, 1978; *Ochoa 11938* (CIP, OCH).

Province Sud Cinti: Vicinity of Ingahuasi, 3120 m alt., among barley fields. Fruit globose, slightly compressed toward distal part. Chromosome number: 2n=2x=24. March 16, 1978; *Ochoa 11954* (CIP, F, NY, OCH, US).

Province Yamparaez: Near Tarabuco. February, 1949; *Cárdenas s.n.* (LL). Siclla, 3600 m alt., on hill among *Cactaceae* and *Portulacaceae*. March 6, 1959; *Alandia and Rimpau 344* (MAX). East of Siclla, 3600 m alt.; 1959; *Ross, EBS 1795* (from plants grown at Sturgeon Bay, Wisconsin, USA). Siclla, 3600 m alt.; 1959; *Ross, EBS 1847* (from plants grown at Sturgeon Bay, Wisconsin, USA). 20 km SSE of Sucre on the Sucre-Yamparaez road, 2800 m alt., 19°5′ S and 65°15′ W. Weedy previously cultivated slope (now fallow) with *Solanum boliviense*, *S. radicans*, *Chenopodium*, *Tagetes*, *Verbena*, *Setaria*, and *Plantago*. April 12, 1963; *Ugent and Cárdenas 4956-4960, 4961-4963, 4964-4967* (WIS), *4968-4972* (US), and *4973-4975* (F). Km 48 on road from Sucre to Zudañez, 3100 m alt., by roadside at edge of barley field. Habit of *S. boliviense* but flower of *S. megistacrolobum*. March 4, 1971; *Hjerting, Cribb, and Huamán 4542* (CIP). Abra, between Cordillera de Tarabuco and Mandinga, 3380 m alt. Flowers purple, pentagonal, among thickets of *Polylepis*. March 12, 1978; *Ochoa 11930* (CIP, OCH).

Department Potosi

Province Saavedra: Open fields about 80 km from Potosi on road to Sucre; flowers dark purple, rotate-stellate; plants spreading; fruits dark green, orbicular, 2 cm in diameter. February 18, 1960; *Correll, Dodds, Brücher, and Paxman B633, B634* (F, G, K, LL, MO, NY, S, U, UC, US). Pampas de Lequesana, 3700 m alt., near Betanzos, in open field on rocky soil. Plant provided with blue flowers. April, 1963; *Cárdenas 6121* (WIS).

Figure 58. – Living plant of *Solanum boliviense* (*Ochoa 11933*).

Crossability

Of the various crosses involving this species and other wild and cultivated Bolivian potatoes made at CIP, some were more successful than others. Seed set, for example, was noticeably less in hybrids of *S. boliviense* and polyploid species than in crosses involving wild diploids.

Among the more successful hybrid combinations produced at CIP, averaging 100 to 200 fertile seeds per fruit, were the following: *S. boliviense* (11935 and 11937) × *S. alandiae* (12016 and 12017); *S. boliviense* (11937) × *S. tarijense* (11994); *S. vidaurrei* (12003) × *S. boliviense* (11937); *S. boliviense* (11935 and 11938) × *S. infundibuliforme* (11973 and 11977); *S. brevicaule* (11934) × *S. boliviense* (11937); and *S. candolleanum* (14958 and 14959) × *S. boliviense* (11937).

In contrast to the above, only relatively few seeds (15-30) were produced in crosses between *S. boliviense* (11935) and *S. sparsipilum* (13644) or *S. boliviense* (11935) and *S. soukupii* (15543). Moreover, only 35% of the crosses made between *S. goniocalyx* f. *amarilla* and *S. boliviense* (11935) were successful. In the latter case, an average of only 25 fertile seeds per fruit was obtained.

Though many fruits were produced in crosses between *S. boliviense* (11935 and 11937) and tetraploid *S. acaule* (12037 and 15095), these on the average contained relatively few seeds (40-50). On the other hand, while 20 crosses were attempted between *S. boliviense* (11937) and tetraploid *S. oplocense* (12010A), all resulted in failure.

Crosses involving Peruvian diploid species gave equally varied results. An average of over 200 seeds per fruit, for example, was obtained in crosses

Figure 59. – Floral dissection of *Solanum boliviense* (*Ochoa 11929*), ×1.

between *S. boliviense* (11929) and *S. abancayense* (13600); whereas an average of only 5-14 seeds per fruit was obtained in crosses between *S. boliviense* (11929) and *S. bukasovii* (13115 and 13862). Crosses between *S. coelestispetalum* (13591) and *S. boliviense* (11935) all resulted in failure.

Finally, the progeny of *S. boliviense* and two Peruvian species, *S. bukasovii* or *S. soukupii*, more closely resemble the Peruvian parents in habit and leaf characteristics.

The more successful Bolivian crosses are described in the following sections.

Solanum boliviense (11938, 11935) × *S. infundibuliforme* (11973, 11977).– Plants of this parentage develop uniformly dark purple, rotate-pentagonal to pentagonal flowers borne on pedicels articulated about their midpoints, or in the upper one-third of their stalks. All lack interjected leaflets and have terminal leaflets that are much larger and more broadly elliptic-lanceolate than their laterals. These strongly resemble *S. infundibuliforme* in their delicate habit and in having broadly decurrent and poorly dissected, normally 2-3 jugate leaves.

Solanum boliviense × *S. alandiae*.– The F_1 hybrids of *S. boliviense* (11937) and *S. alandiae* (12016) develop vigorous, branching stems. Similar to *S. boliviense*, these plants grow 30-40 cm tall. However, their leaves and floral characteristics are more like *S. alandiae*. All have poorly dissected leaves, lacking interjected leaflets and bearing 1 to 2 pairs of subsessile, lateral leaflets, which are much smaller than the rhombic-lanceolate to ovate-lanceolate terminal leaflets. The terminals are more or less pubescent and have pointed apices and slightly attenuate bases. The hybrids have rotate to rotate-pentagonal flowers, borne

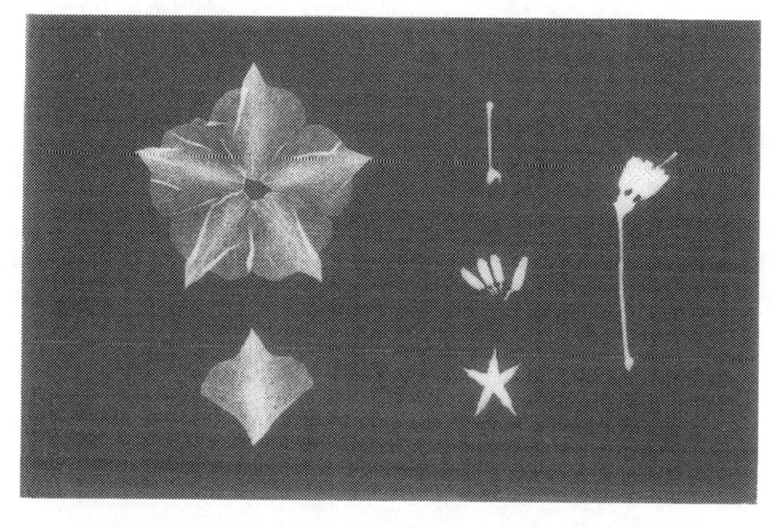

Figure 60. – Floral dissection of *Solanum boliviense* (*Ochoa 11933*), ×1.

on 25-35 mm-long pedicels that are articulated in the upper one-third of their stalks. These hybrids have 6-7 mm-long anthers and 3-3.5 cm-broad corollas that are colored in various shades of lilac. Hybrids of *S. boliviense* (11937) and *S. alandiae* (12017), on the other hand, unlike the cross involving the previously cited accession of *S. alandiae*, are much more similar to *S. boliviense* in leaf characteristics and in having stems to 15 cm tall. These hybrids bear simple leaves or imparipinnate leaves with only a single pair of very small lateral leaflets; however, the floral characteristics of these hybrids resemble those of *S. alandiae*.

Solanum boliviense (11937) × *S. tarijense* (11994).– Plants of this parentage are similar to *S. boliviense* in habit and leaf form. However, with respect to the pubescence and light-green or yellowish-green color of their leaves and the white or creamy-white color of their flowers, they are much more like *S. tarijense*.

Solanum boliviense (11935) × *S. acaule* (12037).– Hybrids of this cross, while more vigorous (40-50 cm tall) and branched than either parent, tend to be more like *S. boliviense* in leaf and floral characteristics. The leaves of these hybrids are provided with 1 to 2 pairs of lateral leaflets and very large terminal leaflets. All bear purplish, 3 cm-broad flowers borne on pedicels that are clearly articulated centrally or within the upper one-third of their stalks. The majority are sterile.

15. Solanum megistacrolobum Bitter, Repert. Sp. Nov. 10:536. 1912.

FIGS. 61-70; MAP 8.

S. alticola Bitt., Repert. Spec. Nov. 12:5-6. 1913. TYPE: *Mandon 398*, Bolivia. Near Mt. Sorata, Dept. La Paz, Prov Larecaja, 1860 (P). *S. alticolum* var. *xanthotrichum* Hawkes, Bull. Imp. Bur. Pl. Breed. & Genet., Cambridge (June):42, 120, Fig. 26. 1944. TYPE: *Balls 5986, p.p. and 5981, p.p.*, Argentina. San Gregorio, 4000 m alt., above Tilcara, Prov. Jujuy (CPC). *S. tilcarense* Hawkes, Bull Imp. Bur. Pl. Breed. & Genet., Cambridge (June):41, 119-120, Pl. II(b), Fig. 25. 1944. TYPE: *Balls 5986, p.p.*, Argentina. San Gregorio, 4000 m alt., above Tilcara, Prov. Jujuy (CPC).

Plants small, rosette-like or branched, with weakly decumbent, abbreviated stems. Stems (1-)2-35 cm tall by 3-5 mm in diameter at base, glabrescent or more or less sparsely pilose with brilliant white hairs. Stolons 70-90 cm or more long; tubers small, 1-2 cm long, round to ovoid, occasionally moniliform, slightly compressed, yellowish-white to brownish-yellow. Leaves (3-) 7-12(-20) cm long, simple, or if imparipinnate, with 1-4(-5) pairs of lateral

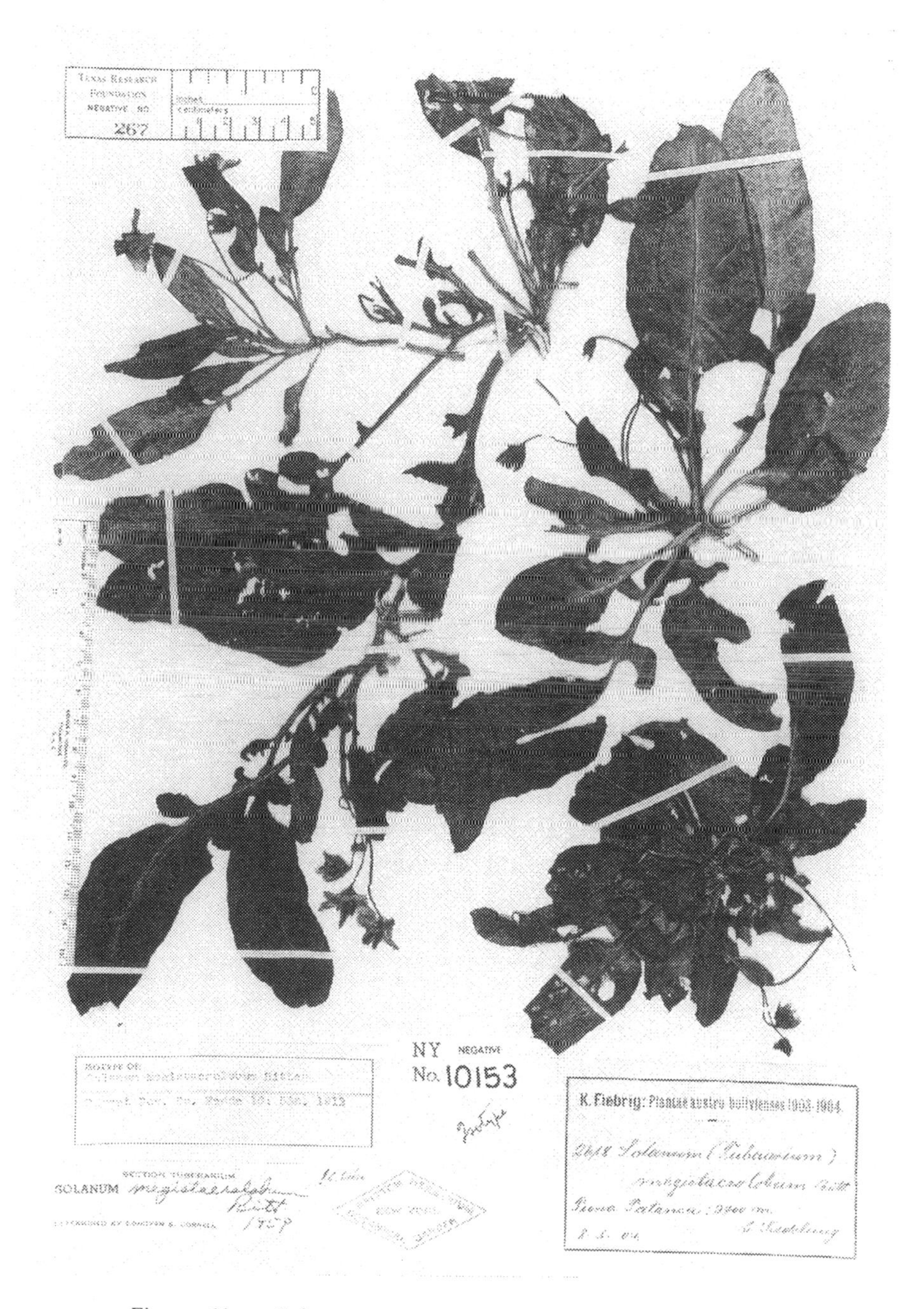

Figure 61. – *Solanum megistacrolobum* (*Fiebrig 2618*; W, photo, type collection). Photo: courtesy of the Chicago Field Museum.

leaflets and without interjected leaflets, sparsely invested with short, thick adpressed hairs on the margins and upper surfaces of the leaflets and with few long conspicuous hairs on the rachis and larger veins below, but more densely covered with shorter hairs on the veinlets. Terminal leaflet, or blades in the case of simple leaves, (2-)4-7(-12) cm long by (1-)2.5-4(-6) cm wide, exceedingly polymorphic, ranging from suborbicular to elliptic, broadly elliptic or oval-elliptic to lanceolate and oblanceolate, or from long-ovate to long-oblong or occasionally rhombic, the apex varying from broadly rounded to obtuse or rarely apiculate, and the base from broadly rounded to cuneate, or in rosette-like plants, gradually decurrent upon the petiolule or truncate. Lateral leaflets 1-3.5 cm long by 0.5-2.5 cm wide, much smaller than the terminals and gradually decreasing in size toward the leaf base, usually ovate to oblong, more or less narrowly decurrent basiscopically on the rachis of the leaf, the apex rounded, obtuse or rarely pointed. Pseudostipular leaves diminutive or absent. Inflorescence lax, few-flowered (1-8), borne on short peduncles, 3-5 cm long (or absent in dwarf or rosette-like plants). Pedicels sparsely pilose and somewhat glandular, articulated 4-12 mm below calyx and of two sizes, either short (2-3 cm) or long (8-10 cm). Calyx 4-6(-9) mm long, regular or occasionally asymmetric, the pointed or shortly acuminate, clearly defined lobes varying from ovate-triangular to ovate-lanceolate. Corolla 2.5-3(-3.5) cm in diameter, broadly substellate to pentagonal (rarely hexagonal or heptagonal), frequently lilac but varying to dark purple, the lobes much broader than long and with poorly delimited, externally pubescent acumens. Anthers 4-5 mm long by 1.8 mm broad, frequently irregular and occasionally somewhat lobulate at base, borne on 1-1.5 mm-long broadly triangular filaments. Style 7-10 mm long, glabrous or very sparsely pilose along the basal one-third of its length; stigma shortly capitate or entirely clavate to cleft or strongly channeled, bilobed, or nearly bifid. Fruit 1.5-2(-2.5) cm in diameter, globose to oval-pyriform, or ovoid to occasionally slightly laterally compressed, green or dark green.

Chromosome number: $2n=2x=24$.

Type: BOLIVIA. Department Tarija, Province Aviles, Puna Patanca, 3700 m alt. January 8, 1904; *K. Fiebrig 2618, p.p.* (lectotype, designated by Hawkes and Hjerting, S; BM, F-fragment and phototype, G, GOET, HBG, LD, LL-fragment, M, NY, SI, U, W, Z).

Although the holotype of *S. megistacrolobum* was destroyed during World War II in a bombing raid of Berlin that set fire to the Dahlem Museum, a large number of isotypes of this species exist and are preserved in various herbaria throughout the world. Perhaps more duplicate types exist for this species than for any other tuber-bearing *Solanum*. The author has examined more than 40

Figure 62. – *Solanum megistacrolobum* (*Fiebrig 2618*; photo, type collection). Photo: courtesy of the New York Botanical Garden Museum.

isotypes of *S. megistacrolobum* that were distributed in 15 world herbaria. The majority of these differed from the originally described plant (or holotype) in habit, pubescence, and leaf decurrency and dissection.

Among the more extreme forms noted among the isotype collections of this species are some that are housed in the Munich (M) and Stockholm (S) museums. However, there are two specimens of this collection in Göttingen, Germany (GOET), mounted on one sheet, that compare favorably with the original diagnosis.

One of Fiebrig's type collections in Göttingen is probably a hybrid of *S. megistacrolobum* and *S. acaule* (c.f. Correll, 1962). This plant was collected on the same date and in the same locality as *Fiebrig 2618*. It is similar to still another collection of Fiebrig's (No. 3429) housed in the museum at Lund, Sweden (LD). Moreover, this specimen is very similar to *Brücher 546* (LL).

In part, difficulties in the classification of *S. megistacrolobum* can be traced to its great environmental plasticity. Studies of this plant at CIP suggest that it is especially affected by local variations in light, moisture, and soil (*see* Figs. 68–69 for *S. megistacrolobum*, and Figs. 73, 75–76 for var. *toralapanum*). However, the species as a whole is also genotypically highly variable, and its various ecotypes tend to intergrade in nature.

To a certain extent, *S. megistacrolobum* has been affected by the introgression of genes from other species. Such is the case with respect to certain populations of this plant from northwestern Argentina. Here, the species has been reported to cross naturally with *S. acaule* and *S. infundibuliforme* (Brücher, 1959b; Correll, 1962; Hawkes, 1963; Hawkes and Hjerting, 1969).

Brücher (1959b) has studied and illustrated the enormous range of leaf variation that occurs in wild populations of this species from Tres Cruces, Argentina (3700 m alt.). He suggests that the unusual leaf aroma of this species may be due to the presence of glandular hairs; however, not all plants of this species are as equally glandular or odoriferous.

The author has noted the frequent occurrence of immature plants among the many type specimens, general herbarium collections, and germplasm accessions of *S. megistacrolobum* and other species of this group that he has examined. Such plants frequently have only floral buds and juvenile leaves. Because the juvenile leaves tend to be simple and markedly different in shape and dissection from those of the adult plants, the identification of immature forms is often difficult.

Affinities

Solanum megistacrolobum appears to be closely related to several other species of series Megistacroloba. Its leaves, in particular, resemble those of the Argentine species *S. sancta-rosae*. Both have similar-appearing terminal leaflets and lateral leaflets that are decurrent.

The rosette-like habit of this species is shared by the central Bolivian and northern Argentine species *S. boliviense*, to which it may also be related. This species differs from *S. boliviense*, however, in the shape and size of its corolla, in its sinuous, weakly decumbent stems, and in the form and decurrency of its leaflets.

Caulescent plants of *S. megistacrolobum* resemble *S. sogarandinum*, a northern Peruvian species of series Megistacroloba. Both have similar-appearing leaves and decurrent lateral leaflets. The two differ, however, in the number of flowers borne on the stem and in several other important ways including the branching pattern of the inflorescence, the size, shape, and color of the corolla and the calyx, the length of the peduncles and pedicels, and the shape of the anther column.

This species also forms a connecting link to series Acaulia. In particular, the species *S. acaule* is very similar in habit to rosette-forming plants of *S.*

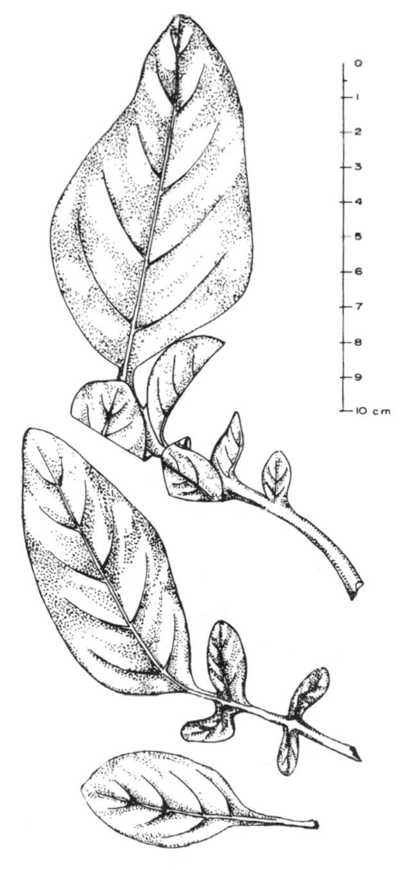

Figure 63. – Leaves of *Solanum megistacrolobum* (*Fiebrig 2618*; NY), ×1.

megistacrolobum. Solanum acaule, however, differs from the latter species in having rotate corollas and in often having the articulation of the pedicels marked by only a ring of pigment.

Habitat and Distribution

Solanum megistacrolobum is endemic to the high, cold, moist Puna and Prepuna formations of the central Andes, where it occurs at elevations ranging from 2700-4000 m, or occasionally as high as 4200 m. It extends along high plateaus from Cailloma, La Raya, and Puno in southern Peru, to Jujuy and Salta in northwestern Argentina.

In Bolivia, it occurs commonly in the north in the Department of La Paz (Provinces Larecaja, Los Andes, Muñecas, and Omasuyos). In this sector of the country it is found frequently in close association with *S. acaule*. In central Bolivia, it occurs primarily in the Departments of Oruro (Province Poopó), Cochabamba, and Chuquisaca, where it is often associated with *S. infundibuliforme*. In southern Bolivia, it extends from the Department of Potosi (Provinces Frias, Saavedra, and Sud Chichas) to the Department of Tarija (Province Aviles and Punas de Patanca). In the latter two departments it frequently grows with *S. leptophyes* and *S. boliviense*.

This species grows frequently between bunch grasses in open fields, in thickets of *Polylepis*, or along the borders of cultivated fields, cattle corrals, and shepherd's huts in rich, organic soil. When growing in fields it forms colonies of rosette-like plants; however, individuals sprouting from rock crevices develop elongated stems.

Specimens Examined

Department Cochabamba

Province Arani: Rumi-Rumi, 3200 m alt., about 91 km from Cochabamba on the Santa Cruz road. Among stones under small trees of *Buddleja* sp. February 26, 1984; *Ochoa and Salas 15521* (CIP, OCH). Toralapa, 3700 m alt., old farm house; *Cárdenas s.n.* (HNF-0087).

Province Ayopaya: Jancko Cala, 3900 m alt., 17°10′ S and 66°25′ W, above Puente San Miguel and Liriuni on the Cochabamba Vizcacha road, about 25 km NNW of Cochabamba (air dist.). Weedy cultivated potato field with *S. decurrentilobum*, *Oxalis tuberosum*, *Capsella*, *Lupinus*, and nontuberous *Solanum* sp. April 9, 1963; *Ugent 4804* (K). Near Hacienda Sailapata, 3600 m alt., 16°50′ S and 66°25′ W, about 65 km NNW of Cochabamba (air dist.). High, cold, wet grassland, roadside and banks of small stream, with *S. ureyi* (Papa Aporoma), *Stipa*, *Cerastium*, and many cushion plants. April 9, 1963; *Ugent 4838-39* (K).

Province Chapare: Colomi, near Cochabamba, damp sandy soil, 3400 m alt. March 15, 1939; *Balls, Gourlay, and Hawkes 6241* (CPC).

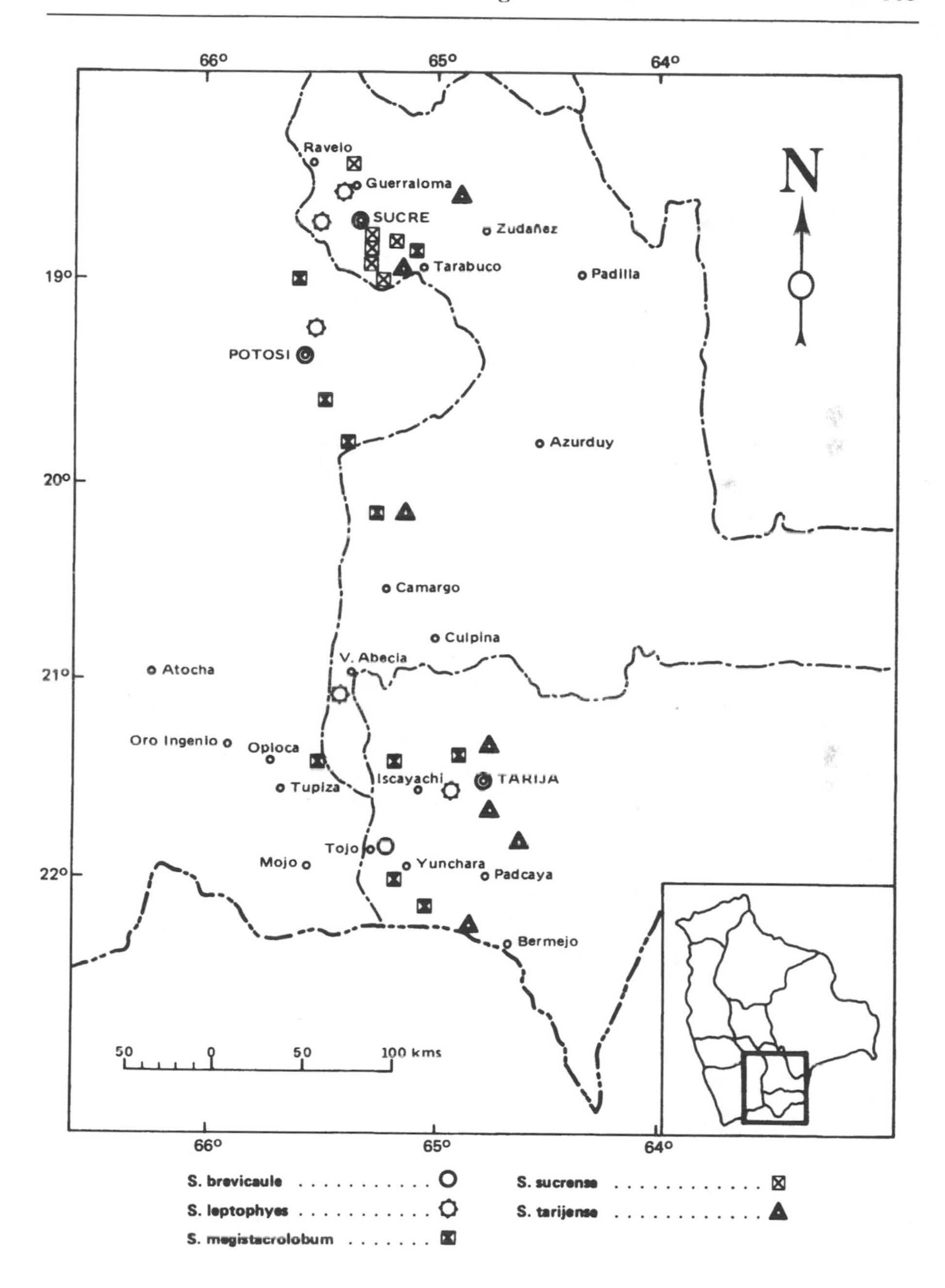

Map 8. – Distribution of wild potato species in southern Bolivia, showing *Solanum megistacrolobum* (■).

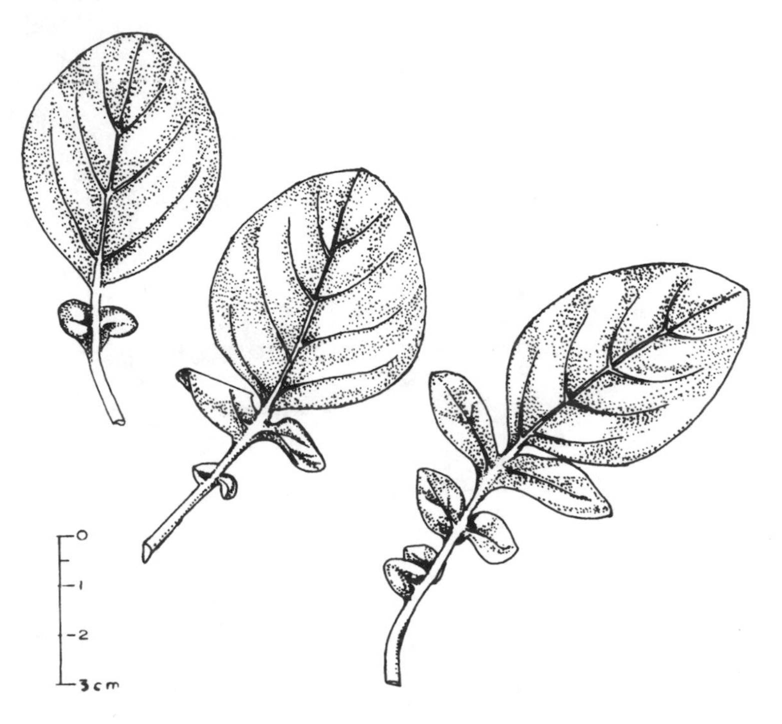

Figure 64. – Leaves of *Solanum megistacrolobum* (*Correll et al. B603*; US), ×1.

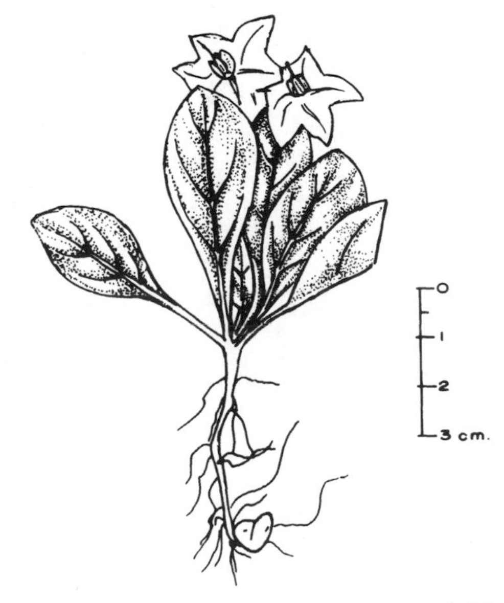

Figure 65. – Flowering plant of *Solanum megistacrolobum* (*Shepard 244*; US), ×1.

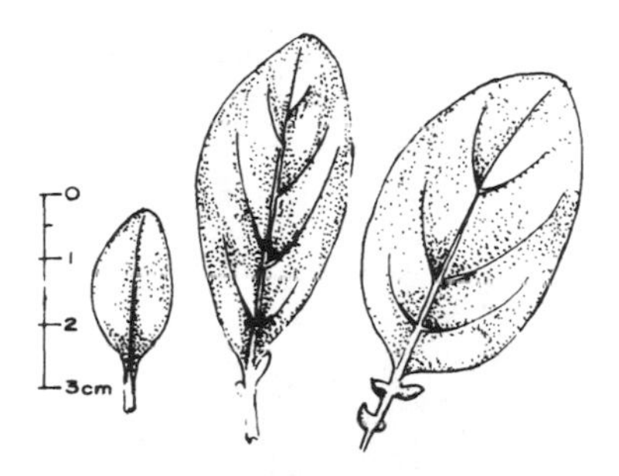

Figure 66. – Leaves of *Solanum megistacrolobum* (*Cárdenas 397*; US), ×1.

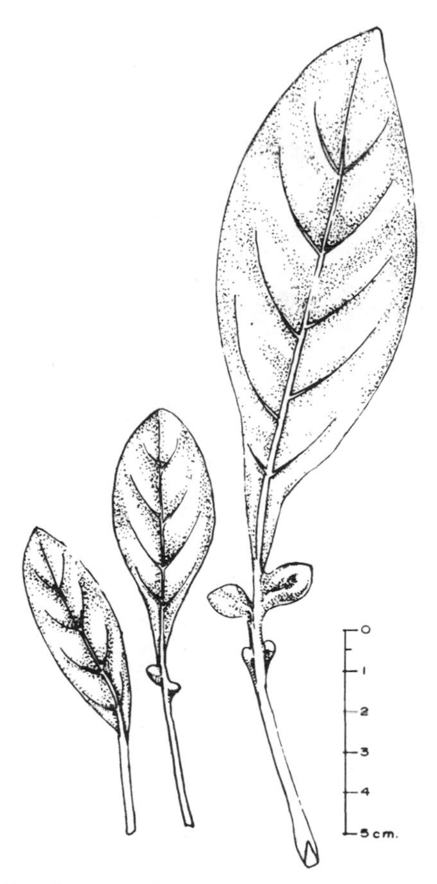

Figure 67. – Leaves of *Solanum megistacrolobum* (*Correll et al. B652*; US), ×1.

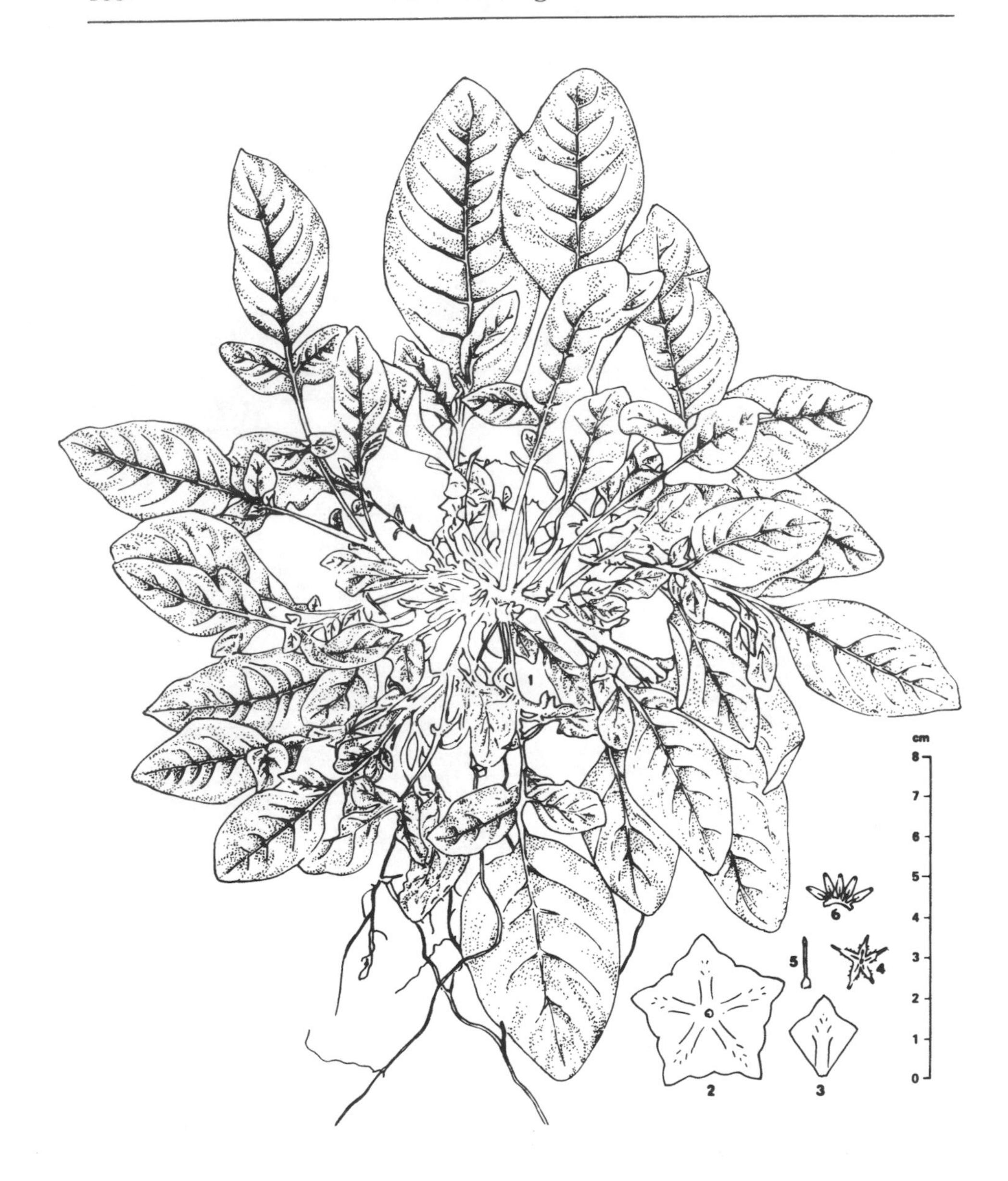

Figure 68. – Plant of *Solanum megistacrolobum* (*Ochoa 11981*; topotype). 1. Plant. 2. Corolla. 3. Petal. 4. Calyx. 5. Pistil. 6. Stamens, dorsal view.

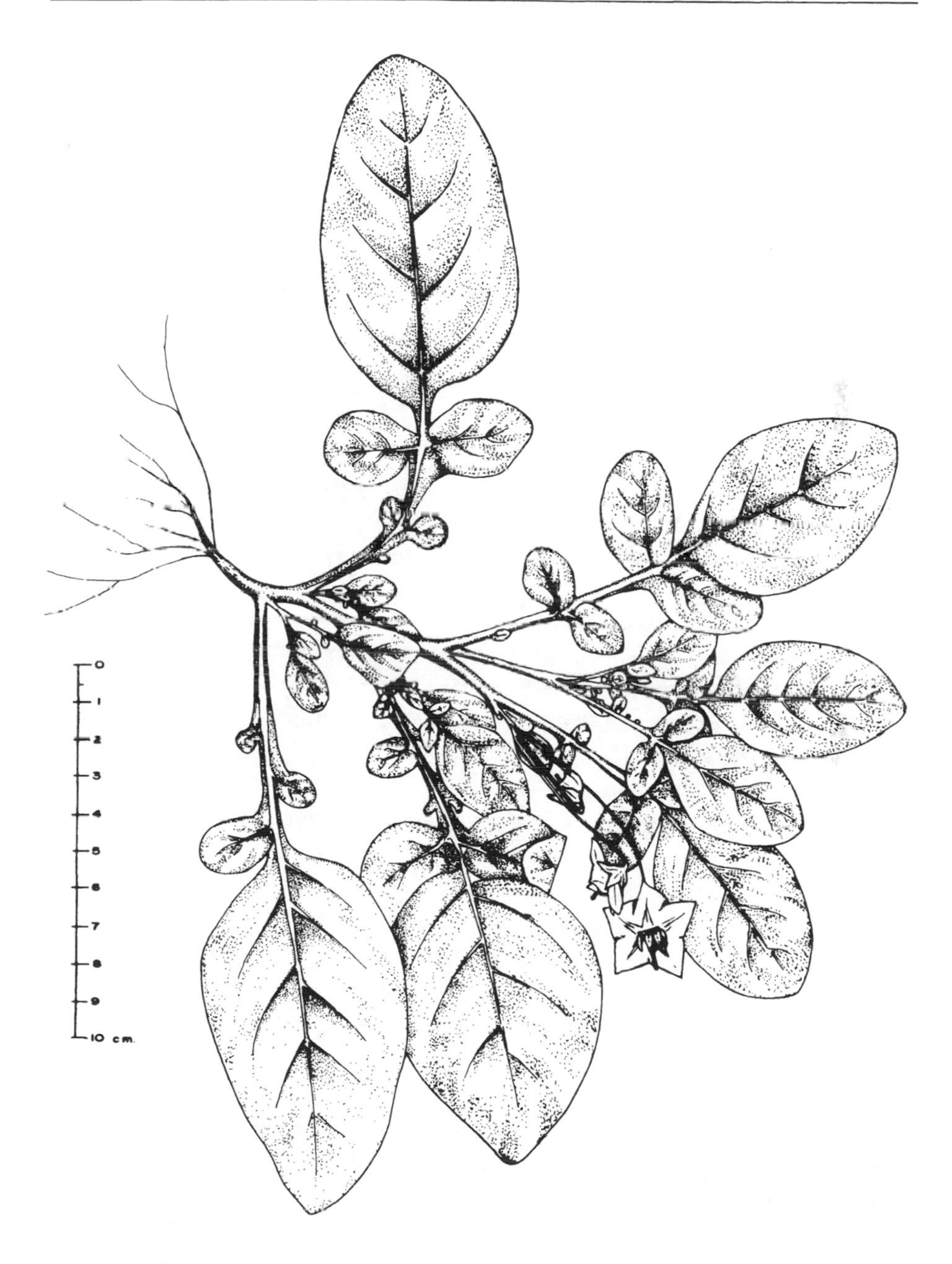

Figure 69. – Flowering plant of *Solanum megistacrolobum* (*Ochoa 11981*), grown at the CIP experiment station, Huancayo, Peru.

Department Chuquisaca

Province Nor Cinti: Sivingamayu. February, 1949; *Cárdenas 4501* (HNF, LL). Sivingamayu, among stones and about potato fields, 3300-3500 m alt. February 28, 1959; *Ross, Alandia, and Rimpau 148-150* (MAX). 126 km on road from Potosi to Camargo, 3500 m alt., near Padcoyo (just north). On edges of field only. February 1, 1971; *Hawkes, Hjerting, Cribb, and Huamán 4323* (CIP).

Province Yamparaez: 5 km W of Tarabuco, 3500 m alt., in sandstone among *Polylepis*. March 5, 1959; *Alandia and Rimpau 333, 335* (MAX).

Department La Paz

Province Ingavi: Hills above Guaqui. January 24, 1903; *Hill 350* (K). Guaqui, 3900 m alt. January 31, 1921; *Asplund 2223* (LL, UPS). About 1 km E of Desaguadero, 3850 m alt., on the Desaguadero-La Paz road. Rocky, weedy, muddy bank along the roadside. March 28, 1963; *Ugent 4557 and 4558* (WIS).

Province Laracaja: Near Mt. Sorata. Circa 1860; *Mandon 398* (P, type of *S. alticolum*).

Province Murillo: Chuquiaguillo to Incachaca, 4100 m alt. December 24, 1920; *Asplund 1914* (UPS).

Province Pacajes: Corocoro, 4000 m alt. February 14, 1921; *Asplund 2413* (UPS). Corocoro, 4100 m alt. February 16, 1921; *Asplund 2461* (UPS). Ulloma, 3800 m alt. February 22, 1921; *Asplund 2571* (UPS). Rosario, 3950 m alt., in sandy places. February 14, 1921; *Shepard 244* (US).

Department Oruro

Province Cercado: Hacienda Huancaroma, 3800 m alt., near Eucalyptus. February 19-24, 1934; *Hammarlund 127* (F, S). On rocky slope N of Eucalyptus, on road to Oruro; flowers lavender, rotate-stellate. February 11, 1960; *Correll, Dodds, Brücher, and Paxman B602 and B603* (K, LL, NY, S, US).

Province Poopó: Between Pasña and Cieneguilla, 4000 m alt., on road from Oruro to Potosi. March, 1978; *Ochoa 11918* (CIP, OCH).

Department Potosi

Province Tomas Frias: Miraflores. February, 1949; *Cárdenas s.n.* (HNF, LL). Near Potosi, 4000 m alt. February, 1949; *Cárdenas s. n.* (HNF, LL). Near Potosi, 3950 m alt. March 3, 1959; *Ross 201* (MAX). Lecherias of Potosi, 3950 m alt., stony hill. March 3, 1959; *Ross and Vidaurre 195* (MAX). On rocky dry slopes of canyon, 3500 m alt., among cacti, 35 km from Potosi on road to Camargo. February 20, 1960; *Correll, Dodds, Brücher, and Paxman B639* (LL, MO, NY, U, US). Below Potosi on road to Lecherias, 3950 m alt., weed of cereal field in nearby grassland. Flat rosettes in grassland, semirosettes in cereal field. A few berries. January 31, 1971; *Hawkes, Hjerting, Cribb, and Huamán 4317*, mixed with two leaves of var. *toralapanum* (CIP). Road to Lecherias, 0.75

km below Potosi, 3950 m alt., by wall. March 7, 1971; *Hjerting, Cribb, and Huamán 4590* (CIP).

Province Linares: In sandy loam, 4000 m alt. February, 1933; *Cárdenas 397* (US). Alccatuyo, 3500 m alt., between Belen and Cuchu Ingenio. February 24, 1953; *Petersen, Hjerting, and Reche 1028* (OCH).

Between Hornos and Cucho Ingenio, 3600 m alt., near Lalava. February, 1953; *Ochoa 2078* (CIP, OCH). About 7 km from Cucho Ingenio, 3550 m alt., near Potosi, under spreading bushes. March 1, 1959; *Alandia, Rimpau, and Ross 180* (MAX).

Department Tarija

Province Aviles: Puna near Patancas, 'bei Siedelung,' 3700 m alt. January 8, 1904; *Fiebrig 2618* (BM, F-fragment, G, GOET, HBG, LD, LL, M, NY, S, SI, U, W, Z, type collection of *S. megistacrolobum*). Between Copacabana and the serrania of Tajsara, 3800 m alt. February, 1953; *Ochoa 2076* (CIP, OCH) Between Abra de Tincuya and Quebrada Honda, 4000 m alt. February, 1953; *Petersen, Hjerting, and Reche 995* (OCH). Yunchara, Bachufer (?), 3500 m alt. February 26, 1959; *Ross, Alandia, and Rimpau 129* (MAX). Plateau, between Cerro Tajsara and Patanca, 3700 m alt., with *Stipa ichu* and *Solanum acaule*. Chromosome number: 2n=2x=24. March 19, 1978; *Ochoa 11981* (CIP, OCH, topotypes). Punas de Patanca, 3750 m alt., 14 km SE of Copacabana. Associated with *Solanum acaule.*, between bunches of *Stipa ichu*. Chromosome number: 2n=2x=24. March 21, 1978; *Ochoa 11985* (CIP, OCH, topotypes). Punas

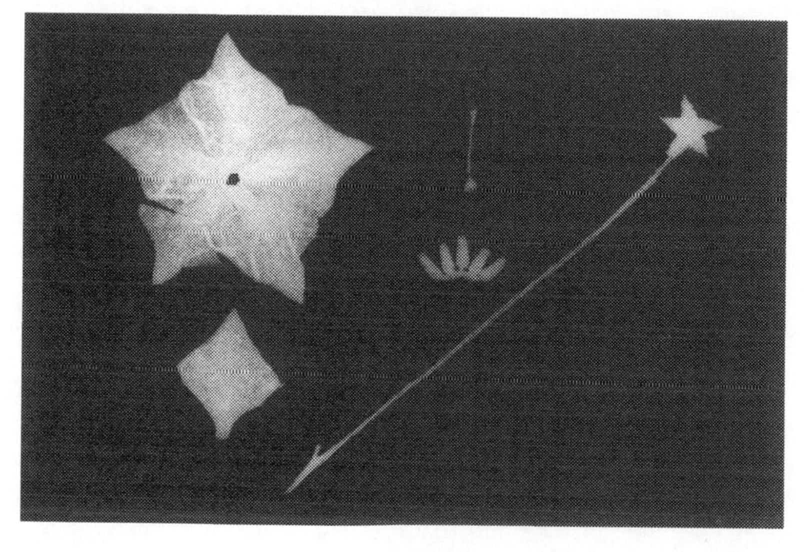

Figure 70. – Floral dissection of *Solanum megistacrolobum* (*Ochoa 11981*, topotype), ×1.

de Patanca, 3780 m alt., 10 km NE of Copacabana. Associated with *Solanum acaule* and *Stipa ichu*. Chromosome number: 2n=2x=24. March 21, 1978; *Ochoa 11987* (CIP, OCH, topotypes). Punas de Patanca, 3780 m alt., 10 km NE of Copacabana. Chromosome number: 2n=2x=24. March 21, 1978; *Ochoa 11988* (CIP, OCH, topotypes).

Province Mendez: Sama, 3500 m alt., between Tarija and Villazon, in sandy loam among stones at base of walls. February 27, 1939; *Balls, Gourlay, and Hawkes B6100* (CPC, E, K, US). About rock walls along stream between Las Carreras and Iscayachi. February 22, 1960; *Correll, Dodds, Brücher, and Paxman B652* (K, LL, NY, S, US). About walls and on piles of rock, Iscayachi. February 22, 1960; *Correll, Dodds, Brücher, and Paxman B654* (F, G, LL, US). Between Iscayachi and Tarija, 3700 m alt. March, 1978; *Ochoa 11991* (OCH).

Indefinite: Titicaca; *Pentland s.n.* (K). Potosi (?); *Vidaurre s.n.*

15a. Solanum megistacrolobum var. toralapanum (Cárdenas and Hawkes) Ochoa comb. nov.

FIGS. 71-76; PLATE XII.

S. decurrentilobum Cárd. and Hawkes, Jour. Linn. Soc., Bot. 53:97-98, Fig. 3. 1945. TYPE: *Cárdenas 3503*, Bolivia. Near Tiraque, Hacienda Toralapa, Dept. Cochabamba, Prov. Arani, February, 1944 (CPC).

S. toralapanum Cárd. and Hawkes, Jour. Linn. Soc. Bot. 53:98-99, Fig. 4. 1945. TYPE: *Cárdenas 3504*, Bolivia. Tiraque, Hacienda Toralapa, Dept. Cochabamba, Prov. Arani, February, 1944 (CPC).

S. toralapanum var. *subintegrifolium* Cárd. and Hawkes, Jour. Linn. Soc. Bot. 53:99-100, Fig. 5. 1945. TYPE: *Cárdenas 3505, p.p.*, Bolivia. Tiraque, Hacienda Toralapa, Dept. Cochabamba, Prov. Arani, February, 1944 (CPC).

S. ellipsifolium Cárd. and Hawkes, Jour. Linn. Soc. 53:100-101, Fig. 6. 1945. TYPE: *Cárdenas 3505, p.p.*, Bolivia. Tiraque, Hacienda Toralapa, Dept. Cochabamba, Prov. Arani, February, 1944 (CPC).

S. ureyi Cárd., Bol. Soc. Peruana Bot. 5(1-3):32-33, Pl. III-B, Fig. 1, Pl. IV-D. 1956. TYPE: *Cárdenas 3262*, Bolivia. Hacienda Sailapata, Dept. Cochabamba, Prov. Ayopaya, November, 1935 (holotype, CA; isotype, US).

Plants usually densely pilose. Leaves (3-)6-15(-25) cm long, simple, or irregularly imparipinnate to imparipinnatisect or pinnately lobed, provided with 1-3(-4) pairs of lateral leaflets. Terminal leaflets, or blades in the case of simple leaves, (2.5-)5-9(-16) cm long by (1.5-)3.5-4.5(-6.5) cm wide, elliptic, obovate to broadly oblanceolate to obovate-suborbicular, tapered at base. Lateral leaflets much smaller than the terminals, 4-5 cm long, obliquely triangular-ovate, the apex obtuse and the base sessile and broadly decurrent

Figure 71. – *Solanum megistacrolobum* var. *toralapanum* (*Correll et al. B605A*; LL), ×½.

basiscopically upon the rachis, appearing broadly triangularly winged. Peduncles to 8 cm long, densely pilose as in the case of the pedicels and calyx. Pedicels short, 1-3 cm long. Corolla usually rotate, rarely subpentagonal, violet to dark violet, or occasionally light purple or lilac, (2.5-)3-3.8 cm in diameter, with short acumens. Anthers 4.5-5.5 mm long.

As in *S. megistacrolobum* var. *megistacrolobum*, the great majority of plants of var. *toralapanum* examined by the author are also densely invested with short tetraglandular hairs (e.g., see *Ochoa 11960, 11963, 11964*). These glands impart a strong, disagreeable odor to the leaves.

The author examined a series of 18 specimens of wild potato (mounted on 7 sheets) collected from the type locality of *S. ureyi*, in the vicinity of Sihuancani, near Sailapata, Department of Cochabamba (*Ugent 4840-41, 4842-44, 4845, 4848-49, 4851-52, 4854-58, 4859-61*). All represent, in his opinion, variants of *S. megistacrolobum* var. *toralapanum*.

Moreover, a series of 8 specimens of wild potato (on 3 sheets) collected from

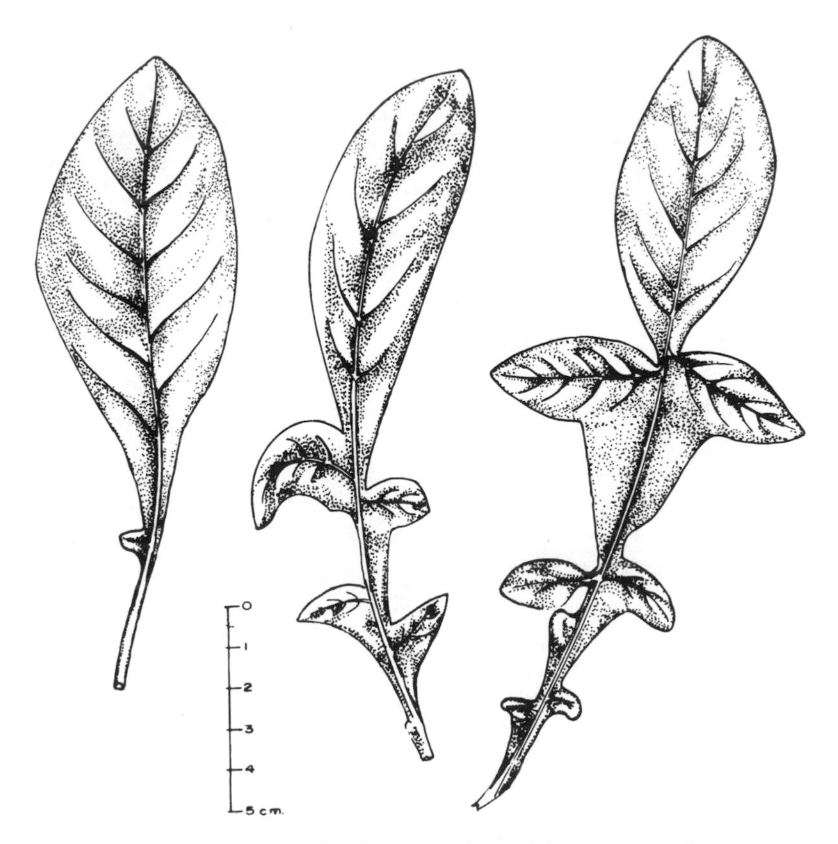

Figure 72. – Leaves of *Solanum megistacrolobum* var. *toralapanum* (*Correll et al. B605*; LL), ×1.

the type locality of *S. toralapanum* (*Ugent 5070-72, 5077-80, 5081-82*) are similar to *S. ellipsifolium*, the latter regarded by the author as a synonym of *S. megistacrolobum* var. *toralapanum*.

A densely pilose plant with simple leaves and rotate corolla that was collected by the author from near Huañacta in the Province of Inquisivi, Department of La Paz, is very close to *S. ureyi* (*see* Fig. 73, *Ochoa-11914*, drawn from original collection, and Fig. 75, *Ochoa-11914*, grown at CIP-Huancayo, Peru). This specimen compares very favorably with the photo and drawing of

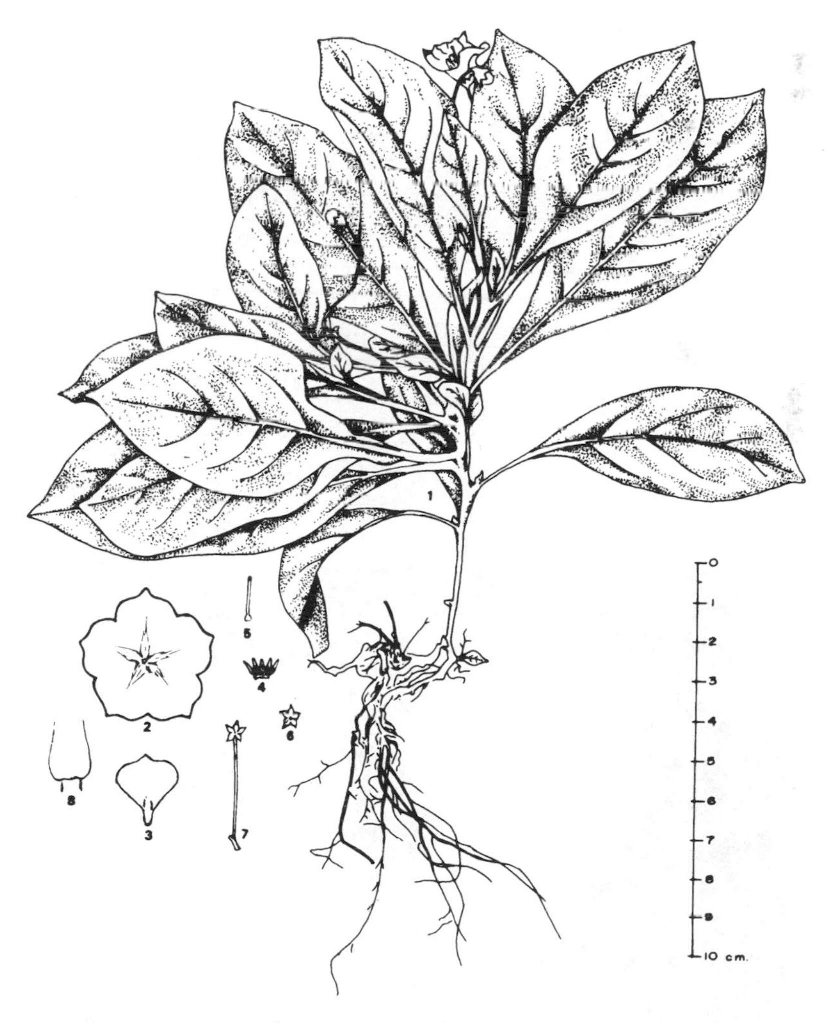

Figure 73. – *Solanum megistacrolobum* var. *toralapanum* (*Ochoa 11914*). 1. Flowering plant. 2. Corolla. 3. Petal. 4. Stamens, dorsal view. 5. Pistil. 6. Calyx. 7. Pedicel and calyx. All ×1. 8. Anther, basal one third, ×½.

the holotype of *S. ureyi* by Cárdenas (1956), and the drawing and floral dissection of this same species by Correll (1962). Another plant of this same collection (*see* Fig. 76, *Ochoa-11914*, grown at CIP-Huancayo), however, has leaves that are more typical of *S. megistacrolobum* var. *toralapanum*.

Chromosome number: 2n=2x=24.

Type: BOLIVIA. Department Cochabamba, Province Arani, near Tiraque, Hacienda Toralapa, 3700 m alt. February, 1944; *M. Cárdenas 3504* (CPC).

Specimens Examined

Department Cochabamba

Province Arani: Near Tiraque, 3700 m alt., Hacienda Toralapa, in moist cultivated soil. February, 1944; *Cárdenas 3503* (CPC, type collection of *S. decurrentilobum*). Near Tiraque, 3700 m alt., Hacienda Toralapa. February, 1944; *Cárdenas 3504* (CPC, type collection of *S. toralapanum*). Hacienda Toralapa, 3700 m alt., near Tiraque, moist soil near cultivated area. February, 1944; *Cárdenas 3505 p.p.* (CPC, type collections of *S. ellipsifolium* and *S. toralapanum* var. *subintegrifolium*). Toralapa, 3700 m alt., near an old hacienda-house, in soil covered with grass, among stones. January, 1946; *Cárdenas 3682* (CPC, HNF). Toralapa, 3700 m alt., near Aliso woods. In sandy soil, 10 cm tall, blue flowers, n.v. *Hampatu Papa*. March, 1946; *Cárdenas 3681* (HNF) [note: the wrong date sequence compared with *Cárdenas 3682*]. Near Tiraque. March,

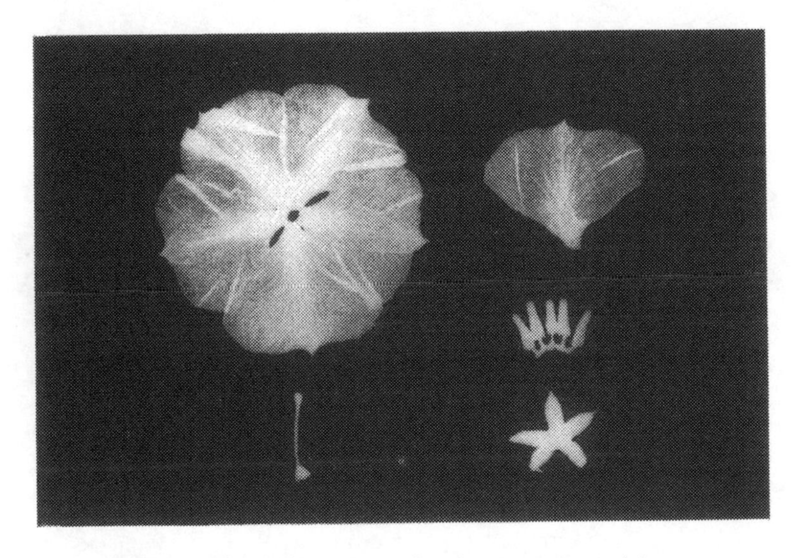

Figure 74. – Floral dissection of *Solanum megistacrolobum* var. *toralapanum* (*Ochoa 11914*), ×1.

1946; *Cárdenas s.n.* (CPC, herbarium specimen from a plant grown at Cambridge, England, 1951). On rocky grassy slope near Koari, 3800 m alt., on the Cochabamba-Santa Cruz road; flowers dark reddish purple. February 13, 1960; *Correll, Dodds, Brücher, Cárdenas, and Paxman B605-A* (K, LL, NY, S, US). On rocky grassy slope near Koari, 3800 m alt., on the Cochabamba-Santa Cruz road; flowers dark reddish purple. February, 13, 1960; *Correll, Dodds, Brücher, Cárdenas, and Paxman B605-B* (K, LL, NY, S, US). Toralapa, 3700 m alt., 15 km E of Tiraque (air dist.), km 75 on the Cochabamba-Comarapa highway, 17°25′ S and 65°35′ W. Rocky grassland, with *Stipa*, *Solanum toralapanum*, *Brassica*, and *Cerastium*. April 20, 1963; *Ugent 5070-5072* (OCH), *5077-5082* (WIS), *5087-5089, 5091-5093* (WIS, topotype of *S. toralapanum*), *5086* (F), *5090* (US). Hacienda Toralapa, 3500 m alt., turning of the Cochabamba to Santa Cruz road between km 71 and 72, in fallow (first year, with some cereals). Rosettes with only the inflorescence above ground surface, some with elliptic leaves, others with several leaf lobe pairs. January 26, 1971;

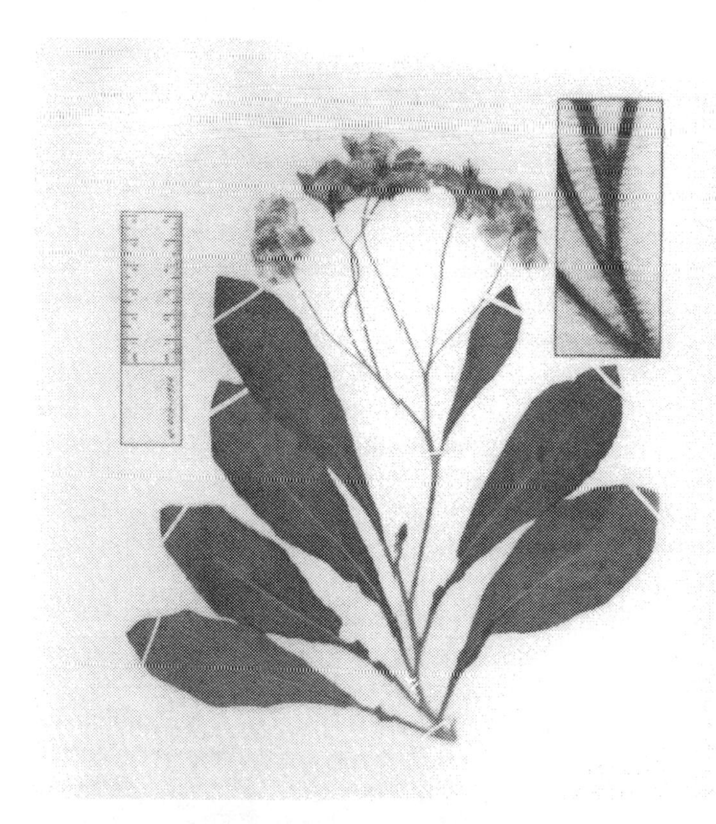

Figure 75. – Flowering plant of *Solanum megistacrolobum* var. *toralapanum*. Simple leaves (*Ochoa 11914*), ×1. Plant grown at CIP, Huancayo, Peru.

Hawkes, Hjerting, Cribb, and Huamán 4247 (CIP, topotype of *S. toralapanum*). Toralapa, 3600 m alt., in fields of Experimental Station. April 20, 1978; *Ochoa and Jatala 12098* (OCH).

Province Arque: Aguas Calientes Station, 3250 m alt., about halfway between Oruro and Cochabamba. January 30, 1949; *Brooke 3011a* (BM). About 71 km from Cochabamba on old road to Oruro, 3550 m alt., 17°38′ S and 66°29′ W, on scree below road; also occurs in grass and among bushes in the same general area. Rosette plants with more or less subsimple leaves, very hairy; flowers deep blue-violet. March 9, 1980; *Hawkes, Hjerting, and Aviles 6562* (CIP).

Province Ayopaya: Wet grassy slope, Hacienda Sailapata, 3600 m alt., on grassy slope. Plants 20-30 cm height, flowers blue-violet. November, 1935; *Cárdenas 3262* (holotype of *S. ureyi*, CA; isotype, US). Near Morochata, 3000 m alt., Hacienda Piusilla. February, 1944; *Cárdenas 3502* (CPC, paratype of *S. decurrentilobum*). Near Choro-Ayopaya, 4000 m alt., grassy wet place. March, 1946; *Cárdenas 3691* (CPC). Janko Cala, 3900 m alt., above Puente San Miguel and Liriuni on the Cochabamba Vizcachas road, about 25 km NNW of Cochabamba (air dist.), weedy cultivated potato field with *Solanum decurrentilobum, Oxalis tuberosum, Capsella, Lupinus* and nontuberous *Solanum* sp. April 9, 1963; *Ugent 4792-1,2,3,4,5, 4793* (WIS), *4795-4797, 4799, 4800, 4805, 4826-4828* (OCH), *4798* (F), *4806, 4808, 4809-1,2,3, 4810, 4813* (WIS), *4817-18* (WIS), *4819* (US), *4820, 4829* (WIS). Sihuincani, 3600 m alt., near Hacienda Sailapata, about 65 km NNW of Cochabamba (air dist.), 16°50′ S and 66°25′ W. High, cold, wet grassland, roadside and banks of small stream, with *Solanum ureyi* (Papa Aporoma), *Stipa, Cerastium*, and many cushion plants. April 9, 1963; *Ugent 4840-41, 4842-44, 4859-61* (WIS, topotypes of *S. ureyi* Card.), *4845* (US), *4848-49* (F), *4851-52* (WIS), and *4854-58* (OCH). Suncani on road to Cuatro Cruces, 10 km from Hacienda Sailapata. April 9, 1963; *Ugent 4853* (WIS). 27 km from Kami on road from Cruce de Challa to Independencia, 3730 m alt., Abra de Ornoni. 17°15′ S and 66°54′ W. Sheltered under ichu grass and with very abundant manure. Very large plants and rotate flowers. Berries dark brownish with white spots. March 10, 1980; *Hawkes, Hjerting, and Aviles 6572* (CIP). 24 km from Independencia on road to Challa, 3730 m alt., 17°14′ S and 66°54′ W. In open shaley ground by side of road in wet puna formation. Plants mostly in rosettes. Flowers deep rich purple. March 12, 1980; *Hawkes, Hjerting, and Aviles 6589* (CIP).

Province Carrasco: Jatun Pino, 3300 m alt., near Monte Punko, stony soil. Plants 10-20 cm tall, flowers blue. January, 1960; *Cárdenas 6112* (HNF, WIS).

Province Chapare: Near Colomi, 3700 m alt., sandy and bushy soil. February, 1946; *Cárdenas 3680* (CPC, HNF). Colomi, 3700 m alt., 30 miles ENE of Cochabamba, damp rocky slope. January 7, 1949; *Brooke 3002* (BM, F). Colomi. January, 1949; *Cárdenas s.n.* (HNF, LL). Colomi, 3400 m alt. January,

1949; *Cárdenas 4515* (CPC). Colomi, 3400 m alt. January, 1949; *Cárdenas 4516* (CPC). Choro, 3050 m alt., above the Copata River about 100 miles NW of Cochabamba, above the Tunari range. Open slope. January 18, 1950; *Brooke 3050* (BM). Choro, 3350 m alt., above the Cocapata River about 100 miles NW of Cochabamba, across the Tunari range, in short grass. January 22, 1950; *Brooke 3053* (BM, F, U). Lagunillas, 3050 m alt. On an open grassy slope. March 4, 1950; *Brooke 6185* (F, NY). 35 km from Tranca de Sacaba on road to Palca, 3760 m alt., 17°14′ S and 66°3′ W, amongst stones in puna vegetation. March 15, 1980; *Hawkes, Hjerting, and Aviles 6613* (CIP, OCH, from plants grown at Sturgeon Bay, Wisconsin, USA). About 40 km from Tranca de Sacaba along road to Palca, 3730 m alt., 0.5 km from Palca, 17°13′ S and 66°3′ W, wayside amongst stones. No flowers. March 15, 1980; *Hawkes, Hjerting, and Aviles 6622* (CIP).

Province Quillacollo: Janko Cala, 3900 m alt., above Puente San Miguel and Liriuni on the Cochabamba-Vizcachas road, about 25 km NNW of Cocha-

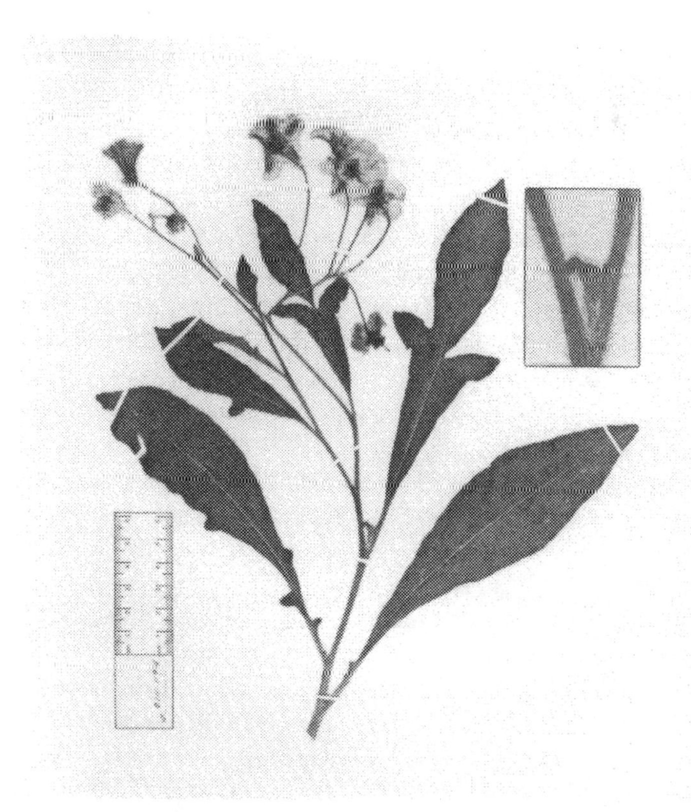

Figure 76. – Flowering plant of *Solanum megistacrolobum* var. *toralapanum*, divided leaves (*Ochoa 11914*), ×1. Plant grown at CIP, Huancayo, Peru.

bamba (air dist.). Weedy cultivated potato field with *Solanum decurrentilobum*, *Oxalis tuberosum*, *Capsella*, *Lupinus*, and nontuberous *Solanum* sp. April 9, 1963; *Ugent 4822* (WIS). 28.3 km on road from Quillacollo to Morachata, 4000 m alt., between Janco Cala and Puente San Miguel, growing in stony ground amongst rocks by roadside, very abundant. Rosettes, stems absent; peduncles erect; leaves more or less simple, some lobed; flowers deep rich purple. January 27, 1971; *Hawkes, Hjerting, Cribb, and Huamán 4270* (CIP).

Indefinite: Ventilla, 4000 m alt. February, 1955; *H. and O. Brücher 571* (LL).

Department Chuquisaca

Province Cinti: January, 1846; *Weddell s.n.* (P).

Province Oropeza: 29 km from Sucre on road to Ravelo, 3300 m alt., growing among bushes in wet valley in quiet luxuriant vegetation. By roadside in grassy slopes. March 5, 1971; *Hjerting, Cribb, and Huamán 4556* (CIP).

Department La Paz

Province Aroma: Road from Panduro to Quime, 4040 m alt., 76 km. February, 5-April 13, 1980; *Soest, Okada, and Alarcón (S.O.A.) 19* (CIP).

Province Inquisivi: Near Huañacta, on road from Tres Cruces to Quime, 4 km down slope from Mt. Abra to Quime. Chromosome number: 2n=2x=24. March, 1978; *Ochoa 11914* (CIP, OCH). Between Tres Cruces and Rio Seco, 3500 m alt., descending to Quime. March, 1978; *Ochoa 11916* (CIP, OCH). Millaque, 3400 m alt., about 5 km N of Quime, along margins of cultivated potato fields. February 24, 1984; *Ochoa and Salas 15518* (CIP, OCH, US).

Province Loayza: Viloco, 4500 m alt., large tin mine about 100 miles from Oruro via Eucalyptus and Cacsata, near Araca, on earth slope in cloud district. March 21, 1949; *Brooke 3033* (BM).

Department Oruro

Province Cercado: Hacienda Huancaroma, 3800 m alt., near Eucalyptus. February 19-24, 1934; *Hammarlund 129* (GH, NY, S).

Department Potosi

Province Saavedra: Colavi, 3800 m alt., in sandy place. April, 1933; *Cárdenas 358* (US).

Province Sud Chichas: Paso de Machu Cruz, 4250 m alt., between Impora and Tupiza, Cordillera de Mochara. Upper part of the pedicel very short, calyx-lobes with obtuse apex. Chromosome number: 2n=2x=24. March 17, 1978; *Ochoa 11960* (CIP, OCH). Descending from Machu Cruz, Cordillera de Mochara, 4210 m alt., between Impora and Tupiza. March 17, 1978; *Ochoa 11962, 11964* (OCH). Near Paso de Machu Cruz, Cordillera de Mochara, 4210 m alt., between Impora and Tupiza. Associated with *Solanum acaule*. Fruits globose, pure green, upper part of the pedicel 10 mm long, calyx symmetrical with lobes shortly acuminate, pointed. Chromosome number: 2n=2x=24. March 17, 1978; *Ochoa 11963* (CIP, OCH).

Department Tarija

Province Mendez: 32 km from Tarija on road to Iscayashi, 3000 m alt., Cuesta de Sama. By roadside in sandy soil. March 14, 1971; *Hjerting, Cribb, and Huamán 4695* (from plants grown at Sturgeon Bay, Wisconsin, USA). Iscayachi, 4300 m alt., without date of collection; *Brücher s.n.* (LL, herbarium specimen from plant cultivated in Mendoza, Argentina).

Indefinite: Bolivia, without date and place of collection; *Pflanz 94* (US) [only known photo of a specimen originally housed in the Berlin-Dahlem Museum, which was destroyed during World War II].

Crossability

Natural hybrids.– *Solanum megistacrolobum* is a self-incompatible species (Pandey, 1960). The author collected a series of plants in the vicinity of Mollepunco (Department Potosi, Province Frias; *OCH 11941*, CIP, OCH) that may represent natural crosses between this species and *S. infundibuliforme*. These putative hybrids were found on a rocky hillside with plants of *S. infundibuliforme* and globular and columnar cacti. Similar to *S. infundibuliforme*, these plants had 2 cm-broad flowers and very small (1-1.5 cm-broad) green and white-speckled, globose fruits. In leaf shape and dissection, however, these plants resembled *S. megistacrolobum* var. *toralapanum*. Like their parental species, the hybrids were also rosette-like to subrosette-like in habit and had light purple, rotate corollas.

Still other collections classified by the author under *S. megistacrolobum* var. *toralapanum* may be of intravarietal origin. Some of these had pentagonal, lilac corollas similar to those observed in artificial crosses between *S. megistacrolobum* var. *megistacrolobum* and var. *toralapanum*.

Artificial hybrids.– According to Dunnet (1959), *S. megistacrolobum* may be valuable in breeding programs aimed at the improvement of the cultivated potato. Studies conducted by this researcher at the Duddington Plant Breeding Station in Scotland show that this species has some resistance to the nematode *Heterodera rostochiensis* (=*Globodera pallida*). More recently, resistance to pathotypes P4A and P5A of *Globodera* has been found in *S. megistacrolobum* as well as in *S. capsicibaccatum* and *S. sparsipilum* (Intl. Pot. Ctr., 1980, 1981, 1982). It is possible that tolerance to the previously mentioned pest could be transferred genetically to the tetraploid cultivated potato by means of an appropriate series of crosses. In this respect, the observations of Quinn et al. (1974) on the formation of unreduced gametes in this species are of some interest. In this connection, it is of interest to mention that up to 5% of the pollen grains produced by two accessions of this species, examined by the author (OCH 11963, 11988), were unreduced.

This species also shows some resistance to frost. Brücher (1959b), for example, reports that at least some populations of *S. megistacrolobum* are capable of surviving in the mountains of Tilcara, Argentina, at elevations ranging up to 4300 m. In these areas, the plants are subjected to repeated nightly frosts. In some instances, they may be actually covered by a blanket of snow for several days. Brücher also indicated that *S. alticola* (=*S. megistacrolobum*) is capable of withstanding temperatures of −10° C in Argentina.

Frost-resistant populations of *S. megistacrolobum* occur in other areas of the Andes. According to observations made by the author, *S. megistacrolobum* is capable of withstanding temperature extremes of −5° C at the Camacani Experimental Station near Puno, Peru, the latter situated at an elevation of 3800 m. *Solanum acaule*, at this same location, survives temperature extremes of −7° C.

A varying degree of success was met with at CIP in crosses involving *S. megistacrolobum*. Hybrids of this species and tetraploid *S. acaule*, for example, were easily made. These yielded an average of 100 fertile seeds per fruit. Similarly, crosses between *S. megistacrolobum* and diploid cultivars of *S. phureja* and *S. stenototum* were easily accomplished. However, plants of the latter parentage yielded only an average of 20–30 fertile seeds per fruit. Crosses between *S. megistacrolobum* and *S. infundibuliforme*, on the other hand, were more difficult. In the latter case, only very few seeds per fruit were obtained.

The results of the remaining crosses made at CIP are summarized in the following sections.

Diploid crosses yielding between 50 and 120 fertile seeds per fruit:

S. sogarandinum (13324) × *S. megistacrolobum* var. *megistacrolobum* (13544).

S. megistacrolobum var. *toralapanum* (11914) × *S. raphanifolium* (13775).

S. megistacrolobum var. *toralapanum* (11964) × *S. sparsipilum* (12028).

S. megistacrolobum var. *toralapanum* (11914) × *S. circaeifolium* var. *capsicibaccatum* (WIS 1730).

Incompatible diploid and diploid-polyploid crosses: Diploid *Solanum megistacrolobum* var. *toralapanum* (11963) × hexaploid *S. oplocense* (11972).

Diploid *S. megistacrolobum* var. *toralapanum* (11914) × diploid *S. candolleanum* (11805).

Diploid *S. megistacrolobum* var. *toralapanum* (11960) × diploid *S. laxissimum* (11855).

Diploid *S. megistacrolobum* var. *toralapanum* (11914) × triploid *S. yungasense* (14842).

Tetraploid *S. oplocense* (11947) × diploid *S. megistacrolobum* var. *toralapanum* (11916).

Descriptions of some of the more successful crosses made at CIP are as follows:

Solanum megistacrolobum × *S. sparsipilum*.– Hybrid plants of this origin strongly resemble *S. megistacrolobum* in having strongly pentagonal to substellate flowers. Their leaves are also similar to *S. megistacrolobum*, though the terminal leaflet is somewhat smaller. The corollas of these hybrids, on the other hand, are very small, and each peduncle bears numerous flowers. The latter characteristics are more typical of *S. sparsipilum*.

Solanum megistacrolobum × *S. circaeifolium* var. *capsicibaccatum*.– The F_1 plants of this parentage are similar to *S. megistacrolobum* in habit, leaf shape, and pubescence. However, they are more like *S. circaeifolium* var. *capsicibaccatum* in having small, creamy-white flowers. Their corollas, however, are rotate or subrotate and not stellate as in typical *S. circaeifolium* var. *capsicibaccatum*.

Solanum megistacrolobum × *S. bukasovii*.– Plants of this cross resemble *S. bukasovii* in habit and floral characters. Their leaves are also similar in shape and dissection to the latter species, but the individual leaflets are sessile and basiscopically decurrent. The latter characteristics are probably inherited from *S. megistacrolobum*.

Solanum megistacrolobum × *S. chomatophilum*.– Morphologically, hybrids of this parentage are similar to *S. chomatophilum* regardless of the direction in which the cross is made. The floral and foliar features of *S. megistacrolobum* are almost totally masked. However, as in the case of *S. megistacrolobum*, these hybrids are capable of surviving temperatures of −5° C. The latter characteristic of these hybrids may be of value to plant breeders concerned with the improvement of the cultivated potato.

Solanum megistacrolobum × *S. sogarandinum*.– Segregates of this cross differ from one another greatly with respect to plant form and habit. However, their floral characteristics are similar to those of *S. sogarandinum*.

Solanum megistacrolobum × *S. raphanifolium*.– The F_1 hybrids of this cross are similar in habit and foliar characteristics to *S. megistacrolobum*. Only their flowers are intermediate between their parental species.

Solanum megistacrolobum × *S. boliviense*.– Morphologically, these hybrids are similar to *S. boliviense*. All have purple flowers, sinuous stems, and leaves that are similar to *S. boliviense*.

VII. SERIES TUBEROSA

TUBEROSA Rydberg, Bull. Torrey Bot. Club 51:146-147. 1924, *nom. nud.*

Tuberosa (Rydb.) Buk., *sensu stricto*, ex Buk. and Kameraz, Bases of Potato Breeding 18. 1959.
Andigena Buk., ex Buk. and Kameraz, Bases of Potato Breeding 24. 1959.
Transaequatorialia Buk., ex Buk. and Kameraz, Bases of Potato Breeding 21. 1959.
Vaviloviana Buk., ex Buk. and Kameraz, Bases of Potato Breeding 18. 1959.
Andreana Hawkes, Bull. Imp. Bur. Pl. Breed. & Genet., Cambridge (June):50. 1944, *nom. nud.*
Minutifoliola Corr., Texas Res. Found. Contrib. 4:216-218. 1962.

Plants erect. Stems long, thick, branched, and pubescent. Leaves imparipinnate to simple, usually large. Terminal leaflet typically larger than the laterals. Pedicel articulated at or above midpoint. Calyx with long-lanceolate lobes. Corolla rotate-pentagonal. Anthers frequently functional, sometimes abortive or sterile. Stigma large. Fruit globose to subglobose.

Chromosome number: 2n=2x=24, 2n=3x=36, 2n=4x=48, and 2n=5x =60.

Habitat and Distribution

The species of this series extend along the Andes of South America from Venezuela, Colombia, and Ecuador to Peru, Bolivia, Argentina, and Chile and grow from near sea level to 4000 m. This group includes the cultivated potato and its closest relatives.

Key to Species

1. Plants to 40 cm tall.
    2. Plants rosette-like or subrosette-like.
        3. Leaves with 3-4(-5) pairs of leaflets; stem 4-5 mm or more in diameter .... 18. *S. brevicaule.*
        3. Leaves with (-1)2-3 pairs of leaflets, stem usually 1.5-2.5 mm in diameter .... 16. *S. achacachense.*
    2. Plants not forming rosettes.
        4. Leaves with 3-6(-7) pairs of leaflets.
            5. Pedicels articulated centrally or slightly above center; plants nonglandular; flowers lilac .... 23. *S. leptophyes.*
            5. Pedicels articulated below calyx; plants densely glandular; flowers white ... . 25. *S. neocardenasii.*
        4. Leaves with 1-2(-3) pairs of leaflets.
            6. Leaves dark green; flowers dark purple .... 19. *S. candelarianum.*

6. Leaves light green; flowers purple, lilac or white.
   7. Calyx-lobes obtuse; flowers white .... 22. *S. gandarillasii.*
   7. Calyx-lobes pointed.
      8. Calyx-lobes long-acuminate, acumens; linear to linear-subspatulate; flowers white .... 26. *S. okadae.*
      8. Calyx-lobes shortly acuminate; acumens triangular-lanceolate; flowers purple or lilac .... 31. *S. virgultorum.*

1. Plants usually more than 50 cm tall.
   9. Leaves with 1-3 pairs of leaflets or rarely simple.
      10. Flowers purple or lilac, with white or whitish-lilac acumens.
         11. Corolla pentagonal, lilac, with 7-8 mm-long ovate-triangular lobes; fruit globose .... 17. *S. alandiae.*
         11. Corolla rotate, light purple, with 4.5-5 mm-long, whitish, slightly sinuous lobes and white acumens; fruit conical .... 24b. *S. microdontum* var. *montepuncoense.*
      10. Flowers white.
         12. Calyx lobes broadly ovate to abruptly lanceolate, with linear to subspatulate acumens, the latter varying in length and width; corolla pentagonal; plant usually 1-2 m tall; fruit globose, green to almost glaucous.
            13. Leaves simple or rarely with 1-2 pairs of leaflets; terminal leaflet up to 15 cm long and 7 cm broad and much larger than the laterals .... 24a. *S. microdontum* var. *metriophyllum.*
            13. Leaves with 1-2(-3) pairs of leaflets, rarely simple; terminal leaflet smaller, up to 7 cm long by 4.5 cm broad .... 24. *S microdontum* var. *microdontum.*
         12. Calyx lobes narrowly elliptic-lanceolate, with equal size linear to linear-subspatulate acumens; corolla rotate to rotate-pentagonal; plants generally less than 40 cm tall; fruit globose, dark green .... 26. *S. okadae*
   9. Leaves with 3-5 pairs of leaflets, never simple.
      14. Leaflets glabrous, with finely denticulate margins, ovate-elliptic to elliptic-lanceolate .... 21. *S. dodsii.*
      14. Leaflets more or less pubescent, with entire margins.
         15. Leaflets ciliate, the upper pair decurrent upon the rachis.
            16. Pedicels articulated below center or slightly above base; stems 5 mm or more in diameter, sinuous .... 27. *S. oplocense.*
            16. Pedicels articulated about or above center; stems slender, less than 2 mm in diameter, nonsinuous .... 30. *S. vidaurrei.*
         15. Leaflets non ciliate, sessile or petiolulate.
            17. Stem up to 2 cm in diameter; leaflets more or less narrowly elliptic-lanceolate and pointed; fruit 2-3 cm in diameter .... 20. *S. candolleanum.*
            17. Stem 1-1.2 cm in diameter; leaflets ovate-elliptic to ovate or obtuse; fruit 1-1.5 cm in diameter.
               18. Stem sinuous, broadly winged; leaves 3-4 jugate and with 5-10 interjected leaflets; pseudostipular leaves large, up to 20 mm or more in length .... 29. *S. sucrense.*
               18. Stem nonsinuous, narrowly winged; leaves 4-5 jugate and with 15-20 interjected leaflets; pseudostipular leaves small, up to 12 mm long .... 28. *S. sparsipilum.*

16. Solanum achacachense Cárdenas, Bol. Soc. Peruana Bot. 5:30-31, Plate II, Fig. 1, Plate IV, Fig. C. 1956. FIG. 77.

Plants small, subrosette-like at base, sparsely pilose. Stems usually 8-12 cm tall and 1.5-2.5 mm in diameter at base, erect, somewhat flexible, slender, pilose with 1.5-2 mm-long silvery-white hairs. Stolons to 50 cm or more long; tubers small, 5-12 mm in diameter, white, round. Leaves imparipinnate, 3-6 cm long by 2-3 cm broad, with (1-)2-3 pairs of leaflets, sparsely pubescent on both sides and on the rachis and petioles. Terminal leaflets much larger than the laterals, 2-2.7 cm long by 1-1.8 cm wide, ovate-elliptic to elliptic-lanceolate, the apex broadly pointed to obtuse, and the base shortly attenuate to rounded. Lateral leaflets narrowly elliptic-lanceolate, the apex obtuse, and the base rounded to obliquely rounded and sessile, decreasing gradually, or at times abruptly, in size toward the base. Upper lateral leaflets 1.3-1.4 cm long by 0.5-0.6 cm wide, the base narrowly decurrent on the rachis along its basiscopic side. Inflorescence terminal, cymose. Peduncle to 7 cm or more long, darkly pigmented or tinged with blue-violet, sparsely pilose as in the case of the pedicels and calyx, divided into short, 14 mm-long pedicels. Pedicels articulated near or about center or toward the upper one-third of the stalk. Calyx small, 5-5.5 mm long, with narrowly elliptic-lanceolate lobes, the apex attenuate or narrowed to 1-1.5 mm-long acumens. Corola rotate to rotate-pentagonal, usually large, to 3.5 cm or more in diameter, blue-violet to light purple, with yellowish-green nectar guides. Anthers narrowly lanceolate, 6.5-7 mm long, borne on glabrous 1-1.5 mm-long filaments. Style short, 8 mm long, slightly exerted from corolla, papillose along the lower one-third of its base; stigma small, capitate. Fruit globose to ovate, 1.5 cm long, green.

Chromosome number: 2n=2x=24.

Type: BOLIVIA. Department La Paz, Province Omasuyos, between Achacachi and Sorata, near Ilabaya, 3800 [4000] m alt. February, 1946; *M. Cárdenas 3688* (holotype, CA).

Despite its specific name of *achacachense*, this species is not found in the immediate vicinity of the village of Achacachi, which is located on the altiplano along the road from La Paz to Sorata between the villages of Huarina and Huarisata. In the original diagnosis of *S. achacachense*, Cárdenas indicated that its type locality was situated along the highest pass between Achacachi and Sorata. After exploring this area, the author determined that the locality in question is actually Apacheta of Ilabaya. The latter hamlet is situated at 4050 m above a small Indian village called Occo Hualata, which in turn is situated some 7 km from Huarisata along the road to Sorata, or about 30 km from Achacachi. In an earlier publication (Ochoa, 1985), the author erroneously assigned one of his collections of this species (No. 15901) as the neotype of *S. achacachense*.

Affinities

This species appears, at least superficially, to be related to *S. pumilum* from southern Peru. Both are small plants, somewhat rosette-like at base, with pubescent stems and leaves, and both have poorly dissected leaves. This plant also shares some characteristics in common with *S. bukasovii*, a central and southern Peruvian species with more highly dissected leaves. Both are large-flowered, subrosette-like plants with pilose leaves and stems. However, the character associations that are observed here may not be a true reflection of the phylogenetic relationships of these species, but rather merely an expression of traits that are normally associated with plant forms (ecotypes) adapted to growth at higher elevations.

Habitat and Distribution

This species is known presently only from its type locality, Apacheta, situated in the shadow of Pico de Illampu. Here, the plant grows in black, rocky soil at 4000 m elevation in cool, foggy grassland (Puna) dominated by *Stipa ichu* and other bunch-forming species.

Figure 77. – *Solanum achacachense* (*Ochoa 15901*).

Although no other tuber-bearing *Solanum* species are known from Apacheta, two other wild potatoes, *S. acaule* and *S. leptophyes*, are known to occur in the vicinity of Occohualata, a village situated at somewhat lower elevations (3780 m) than Apacheta and along the same road which leads from Huarisata to Sorata. Here, the above species grow in a Puna grassland dominated by *Lupinus microphyllum*, *Stipa*, *Poa*, *Plantago*, *Gnaphalium*, *Lobivia*, and *Cajophora* spp.

Specimens Examined

Department La Paz

Province Omasuyos: Near Ilabaya, 3800 m alt. Between Achacachi and Sorata. On humid grassy, stony soil among rocks. Plants 5-8 cm tall; flowers blue. February, 1946; *Cárdenas 3688* (CA, holotype). Pass between Achacachi and Sorata, 4000 m alt. January, 1985; *Ochoa 15901* (OCH, topotype).

17. Solanum alandiae Cárdenas, Bol. Soc. Peruana de Bot. 5(1):11-12, Plate I(A), Figs. 1-3, pp. 18-19. 1956. FIGS. 78-84; PLATE XIII.

S. torrecillasense Cárd., Bol. Soc. Peruana de Bot. 5(1):15-16. 1956. TYPE: *Cárdenas 5088*, Bolivia. Near Torrecillas (km 247), Dept. Santa Cruz, Prov. Valle Grande, March, 1955 (CA, not seen).

Plants light green, robust, very branched, to 1 m tall. Stems erect to semidecumbent, 10-12 mm in diameter at base, glabrescent with very sparse, short, white hairs, and with broad, straight or sinuous wings, the lower two-thirds slightly pigmented brown. Stolons white, more than 1 m long and 2-3 mm in diameter; tubers white, round to ovate, or ovate-compressed, 3-5 cm long. Leaves imparipinnate, with 2-3(-4) pairs of leaflets and 0-5 interjected leaflets, somewhat longer than broad, (7-)12-19(-28) cm long by (5-)9-15(-23) cm wide, borne on petioles (1-)2-5(-6) cm long, densely and shortly pilose above (i.e., soft to touch) and along the leaf margins, and less sparsely pilose below, except for the densely pubescent nerves. Terminal leaflets larger than the laterals, (3-)5-9(-13) cm long by (1.5-)3.5-7(-8) cm wide, broadly elliptic-lanceolate to ovate-lanceolate, the apex pointed to subacuminate, and the base rounded or asymmetrically rounded or slightly cuneate and very narrowly decurrent on the pigmented rachis. Lateral leaflets elliptic-lanceolate or narrowly elliptic-lanceolate, the apex pointed and the base obliquely rounded or rounded, borne on petiolules 1-3(-6) mm long. Interjected leaflets 2-6(-7) mm long by 1-3.5(-4) mm wide. Pseudostipular leaves large, (12-)20-30(-40) mm long by (7-)10-15(-20) mm wide. Inflorescence cymose or cymose-paniculate,

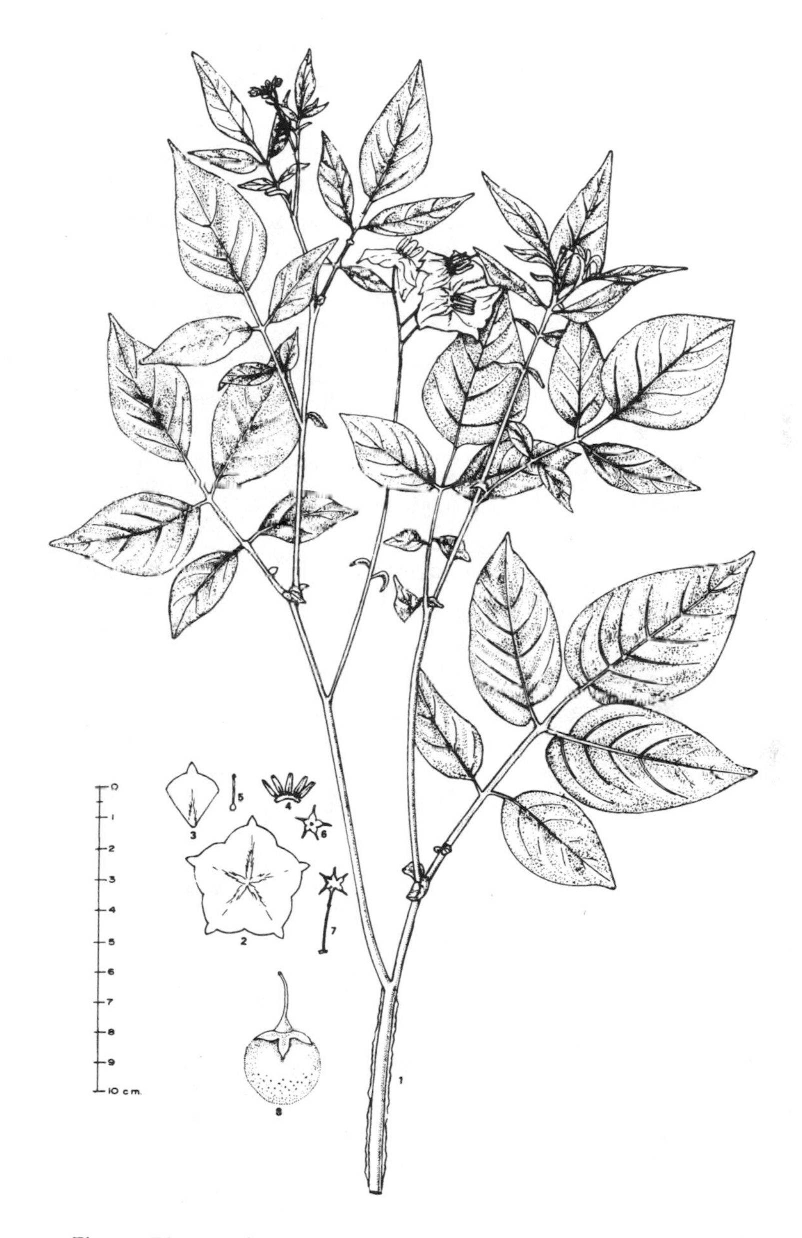

Figure 78. – *Solanum alandiae* (*Ochoa 12012*). 1. Upper part of flowering plant. 2. Corolla. 3. Petal. 4. Stamens, dorsal view. 5. Pistil. 6. Calyx. 7. Pedicel, articulation and calyx. 8. Fruit. All ×½.

terminal, 15-20 or more-flowered, borne on peduncles (3.5-)5-9(-12) cm long, shortly and sparsely pilose as in pedicels and calyx. Pedicel articulated above center or toward the upper one-third of stalk, the upper part 10-12 mm long and the lower 14-16(-27) mm long. Calyx 6-6.5(-7) mm long, light green as in the case of the pedicels, the lobes elliptic-lanceolate and abruptly narrowed to 1.5-2 mm-long pointed acumens. Corolla pentagonal to subpentagonal, 2.6-3(-3.5) cm in diameter, light lilac with greenish-yellow nectar guides within, and dark lilac on the exterior. Anthers large, 7 mm long, cordate at base, borne on glabrous, 0.7-1.2 mm-long white-hyalinous filaments. Style 9-10 mm long, curved, exerted 2 mm, densely papillose below midpoint; stigma small, capitate. Fruit light green, densely white-speckled and verrucose about the midpoint, globose to ovate or occasionally long-conical, 2-2.5 cm long and up to 1.8 cm in diameter at base.

Chromosome number: 2n=2x=24.

Type: BOLIVIA. Department Cochabamba, Province Carrasco, km 140 between Cochabamba and Santa Cruz, 3200 m alt. March, 1955; *M. Cárdenas 5079* (syntype, HNF; isosyntype, LL).

In the Bolivian National Forestry Herbarium of Cochabamba (HNF) are three specimens of this species mounted on two sheets. Each is marked with the accession number 00005. The original labels on these two sheets, both

Figure 79. – Floral dissection of *Solanum alandiae* (*Ochoa 12011*), ×1.

hand-written by Cárdenas, read 'No. 5079 Co-type, 1955.' However, one sheet is annotated 'Km 140 on the road from Cochabamba to Santa Cruz,' while the other is labeled 'Km 144 on the road from Cochabamba to Santa Cruz.' Cárdenas' description would appear to have been made from the sheet that is labeled 'Km 140.' Since Cárdenas did not designate a holotype for this species, these two sheets must necessarily be regarded as syntypes.

Cárdenas has indicated that his type collection [holotype?] of *S. torrecillasense* (No. 5088) is housed in the Martín Cárdenas Herbarium in Cochabamba (CA); however, the author (C.M.O.) did not see this in the CA or HNF herbaria in Cochabamba. This variant of *S. alandiae* was collected at km 247 on the road from Cochabamba to Santa Cruz, in the Department of Santa Cruz, Province of Valle Grande, near the locality of Torrecillas.

The author has examined two collections by Correll et al. (B615 and B618A) that were labeled *S. torrecillasense*. Although it appeared to him, at least at first, that these two collections represented hybrid plants of *S. microdontum* and *S. alandiae*, the results of later field and herbarium studies by the author suggested that they should be included under *S. alandiae*. A thorough search by the author for *S. torrecillasense* at its type locality (km 247 along the road from Cochabamba to Santa Cruz) yielded only collections of *S. alandiae*.

Affinities

This species does not appear to be related to any other tuber-bearing *Solanum* species, except *S. doddsii*.

Figure 80. – Floral dissection of *Solanum alandiae* (*Ochoa 12014*), ×1.

Habitat and Distribution

Solanum alandiae occurs at elevations between 2000 and 2900 m in the Prepuna, or inter-Andean valleys of Bolivia. Here it grows in moist maize fields and in barren ground, as well as along the borders of arroyos, river banks, and thickets. Its range extends from eastern Cochabamba to northern Chuquisaca.

Specimens Examined

Department Cochabamba

Province Campero: About 25 km from Aiquile, 3000 m alt., on brushy slope on the way to Sucre. Flowers light lilac, rotate-stellate. February 16, 1960; *Correll, Dodds, Brücher, and Paxman B618-A* (K, LL, S, US).

Province Carrasco: Km 140 on the road from Cochabamba to Santa Cruz, 3200 m alt., in cultivated field, plant 20–40 cm tall; flowers blue. March, 1955; *Cárdenas 5079* (syntype of *S. alandiae* HNF; isosyntype, LL). From km 144 between Cochabamba and Santa Cruz, 3000 m alt. April, 1957; *Cárdenas PI 243501* (plant grown at Sturgeon Bay, Wisconsin, USA; CIP). Ichupampa, 3000 m alt., between Lluttupampa and La Habana, along road from Cochabamba to Santa Cruz. February 26, 1984; *Ochoa and Salas 15526* (CIP, OCH). La Habana, 2600 m alt., near km 180 between Cochabamba and Santa Cruz, in clay soil among ruins of demolished corral. Dry and stony region. February 26, 1984; *Ochoa and Salas 15527* (CIP, OCH, US). Torrecillas, 2900 m alt., near km 247 of the route Cochabamba-Santa Cruz, between Siberia and

Figure 81. – Floral dissection of *Solanum alandiae* (*Ochoa 12017*), ×1.

Comarapa, among thickets. February 29, 1984; *Ochoa and Salas 15566* (CIP, OCH).

Province Mizque: 192 km from Sucre, 44 km from Aiquile, among rocks on brushy slope. Leaves pubescent; flowers light lilac. February 15, 1960; *Correll, Dodds, Brücher, and Paxman B615* (LL, MO, UC). Km 192 on road from Sucre to Epizana, 2600 m alt., 40 km from Aiquile, below road on grassy slope. Small fruit; blue-lilac pentagonal flower. March 3, 1971; *Hjerting, Cribb, and Huamán 4535* (CIP). Vicinities of La Hoyada, 2300 m alt., between Aiquile and Totora. Light lilac flowers. March 25, 1978; *Ochoa 12011* (CIP, OCH). Above La Hoyada, 2300 m alt., along the road between Aiquile and Totora. March 25, 1978; *Ochoa 12012* (CIP, LE, OCH, P, US). On hill above La Hoyada, 2350 m alt., along the road between Aiquile and Totora. March 25, 1978; *Ochoa 12013* (CIP, OCH, US). Serranias in the vicinity of La Hoyada, 2400 m alt., along the road between Aiquile and Totora. Flowers lilac. March 25, 1978; *Ochoa 12014* (CIP, OCH, US). In the vicinity of La Hoyada, 2300 m alt., along the route Aiquile-Totora-Episana. In cultivated field; flowers lilac. March 25, 1978; *Ochoa 12016* (OCH). Between La Hoyada and Azul Cocha, 2650 m alt., along the road between Aiquile and Totora. March 25, 1978;

Figure 82. – Berries of *Solanum alandiae* (*Ochoa 12012*), ×1.

Ochoa 12017 (CIP, LE, NY,OCH, US). Along the road between Aiquile and Totora, 2750 m alt. March 25, 1978; *Ochoa 12018* (CIP, OCH, US).

Department Chuquisaca

Province Oropeza: Punilla, about 15 km NW of Sucre on the Sucre-Ravelo road, 2600 m alt., 18°55′ S and 65°25′ W. Margins of a weedy wheat field, with *S. boliviense*, *S. pachytrichum*, nontuberous *Solanum* sp., *Setaria plumosa*, *Calceolaria*, and *Veronica*. April 12, 1963; *Ugent and Cárdenas 4909* (WIS).

Department Santa Cruz

Province Caballero: Near El Potrerillo, 2000-2200 m alt., between Abra de Quiñe and Mataral, in dry region, among *huarango* trees and cacti on hard, cracked soil. Plants flowerless, infrequent. February 29, 1984; *Ochoa and Salas 15562* (CIP, OCH).

Crossability

Interspecific crosses made at CIP between *S. alandiae* and such other wild Bolivian diploid (2n=24) potato species as *S. berthaultii*, *S. boliviense*, *S.*

Figure 83. – Berries of *Solanum alandiae* (*Ochoa 12017*), ×1.

brevicaule, *S. candolleanum*, *S. leptophyes*, *S. sparsipilum*, and *S. tarijense* yielded, in all of the above-mentioned cases, a high average number of fertile seeds per fruit.

On the other hand, crosses of *S. alandiae* (12014) with *S. infundibuliforme* (11977), *S. litusinum* (12027) with *S. alandiae* (12016), or *S. microdontum* (12025 with *S. alandiae* (12014 and 12017), yielded only 10 seeds per fruit. However, the reciprocal crosses of these last four species averaged up to 30 viable seeds per fruit.

In crosses of *S. alandiae* (12016 and 12017) with two Bolivian tetraploid (2n=48) species, *S. acaule* (8611, 13554) and *S. oplocense* (11945, 11948), an average of 75 to 30 viable seeds per fruit were obtained, respectively.

Interspecific crosses between *S. alandiae* and various Peruvian species were also made. When artificially fertilized with pollen from such wild diploid species as *S. abancayense*, *S. ambosinum*, *S. huancabambense*, *S. raquialatum*, or *S. raphanifolium*, plants of *S. alandiae* yielded abundant fertile seed. Similarly, crosses between this plant and two cultivated diploid species, *S. goniocalyx* and *S. phureja*, were very successful, yielding, on the average, about 35 viable seeds per berry.

Figure 84. – Habitat of *Solanum alandiae*, vicinity of La Hoyada, 2300 m altitude, between the towns of Aiquile and Totora, Province Mizque, Department Cochabamba.

In the following paragraphs, a summary is given of the salient characteristics of the first generation hybrid plants that were obtained between *S. alandiae* and various wild diploid Bolivian species of potato at CIP.

Solanum alandiae (12018) × *S. sparsipilum* (12028).– With the exception of a tendency toward the development of more numerous interjected leaflets, the plants of this cross were more like those of *S. alandiae* than *S. sparsipilum*. All were provided with long-peduncled, cymose inflorescences that bore numerous strongly pentagonal, dark to light lilac flowers, the latter with conspicuously long, white acumens. The pedicels of these hybrids were articulated slightly above center, the upper part of the stalk being about 15 mm long and the lower 16–18 mm long.

Solanum alandiae (12012) × *S. microdontum* (12025).– Plants of this parentage closely resemble *S. alandiae* in habit, leaf shape, and flower. All have strongly pentagonal flowers of various shades of lilac, and pedicels articulated in the upper one-third of the stalk. However, some individuals of this cross have leaves that are more dissected than those of *S. alandiae*. Moreover, the pubescence of these hybrids varies from slightly pilose to medium pilose, but not densely pilose as in *S. microdontum*.

Solanum alandiae (12017) × *S. candolleanum* (11913).– The hybrids of this cross have terminal leaflets that are almost the same size as their laterals. However, the leaves of these plants are intermediate between their parental forms in dissection, while their petiolule length and leaf and leaflet shapes correspond more closely to *S. candolleanum*. The flower characteristics of these hybrids, however, are similar to those of *S. alandiae*. All have a pentagonal corolla that varies in color from dark to light lilac, and the articulation of the pedicel occurs in the upper one-third of the stalk. In summary, the morphological characteristics of these hybrids are, in general, closer to *S. alandiae* than *S. candolleanum*.

Solanum alandiae (12012) × *S. litusinum* (12027).– Plants of this cross resemble *S. alandiae* very closely in the majority of their characteristics. All have robust stems bearing large, 3–4 jugate leaves that are provided with 2–6 interjected leaflets. The mostly rotate-pentagonal flowers of these hybrids varies in color from purple to light lilac. Pedicels are articulated somewhat above center; the upper part of the stalk being 10–12 mm long and the lower 14–16 mm long. Most are sparsely and shortly pilose.

One exception to the above F_1 hybrid population of plants involved a single specimen that developed a purple, stellate corolla. In this specimen, the pedicel was articulated below the midpoint of the stalk.

In contrast to the above cross, the reciprocal combination yielded mostly plants that bore a greater resemblance to *S. litusinum*. All had purple, rotate-pentagonal flowers, and fine, slender stems bearing small, 2–4 jugate leaves, with 2-5 very small interjected leaflets. The pedicels of these hybrids were articulated above the midpoint.

As far as is known, *S. litusinum* does not occur in the same habitats as *S. alandiae*. However, some features of the former species show some relationship to *S. alandiae*. These include plant habit, form and dissection of the leaves, and pubescence. As pointed out previously under the species *S. litusinum*, this plant may represent a hybrid formed in crosses of either *S. alandiae* by *S. chacoense*, or *S. alandiae* by *S. tarijense*.

Solanum alandiae × *S. chacoense*.– Plants derived from crosses between *S. alandiae* (12012) and *S. chacoense* (12026) share many features in common with *S. alandiae* including stout, robust, broadly winged stems and 3-4 jugate leaves, the latter with elliptic-lanceolate leaflets and up to 12 interjected leaflets. Moreover, these also share with this species a common calyx shape, anther size (8-9 mm) and point of articulation of the pedicel. However, they resemble *S. chacoense* in having light green, 30 cm-long peduncles, and cymose to cymose-paniculate inflorescences that bear 30 or more broadly stellate flowers. The flower color of these hybrids, however, varies from creamy white with a bordering tinge of lilac, to pure pale lilac, dark lilac, or purple, the latter three corollas having white acumens.

Some differences were observed in plants derived from crosses involving *S. chacoense* (12026) and other introductions of *S. alandiae*. Hybrids involving *S. alandiae* (12014) were similar in all respects to the ones obtained above except that these bore rotate-pentagonal flowers, none of which were creamy white. Crosses involving *S. alandiae* (12016), on the other hand, differed from the others in having lilac-purple to purple, large, rotate-pentagonal flowers up to 4 cm in diameter, borne on centrally articulated pedicels. And lastly, as plants obtained from supposed crosses of *S. alandiae* (12017) with *S. chacoense* were mostly identical to *S. alandiae*, a failure in the transfer of *S. chacoense* pollen is indicated. One plant, however, which appeared to represent the cross, had dark purple, rotate-pentagonal flowers borne on pedicels articulated in the upper one-third of the stalk; in these characteristics this plant resembled *S. alandiae*. However, in its small calyx with short, elliptic-lanceolate and apiculate lobes, it resembled *S. chacoense*.

Solanum alandiae × *S. oplocense*.– Though F_1 hybrids between diploid accessions of *S. alandiae* and tetraploid accessions of *S. oplocense* are easy to obtain, the plants resulting from this cross are always sterile, as can be normally expected. All, however, have inherited characteristics derived from both parents. Thus, crosses of diploid *S. alandiae* (12016) with tetraploid *S. oplocense* (11945) are similar to the first species in leaf dissection and more similar to the last in leaflet shape. The upper lateral pair of leaflets and the base of the terminal leaflet of these hybrids tend to be narrowly decurrent on the rachis.

Only rarely are these structures conspicuously decurrent. The purple flowers, which are borne upon cymose and cymose-panniculate inflorescences, are like those of *S. oplocense*. Articulation of the pedicel occurs toward the midpoint of the stalk, the upper part of which ranges between 12-14(-16) mm long. In the cross *S. alandiae* (12017) × *S. oplocense* (11945), the flowers are rotate and light purple. The thinner leaves of these hybrid forms are also more similar to *S. oplocense*. In contrast to the above segregates, the leaves of plants derived from crosses of diploid *S. alandiae* (12017) with tetraploid *S. oplocense* (11948) are much more dissected. They resembled *S. alandiae* in having up to 4 pairs of leaflets, and up to 12 ovate, subsessile interjected leaflets up to 12 mm long. Plants of this cross have dark purple, pentagonal flowers, borne on pedicels articulated in the upper one-third of the stalk. All bear thick, vigorous stems similar to *S. alandiae*.

Solanum alandiae × *S. tarijense*.– All crosses made between these two species were very similar in appearance to *S. alandiae*. In *S. alandiae* (12014) × *S. tarijense* (11994), the plants produced cymose or cymose-panniculate inflorescences, these developing a dozen or more pentagonal, very light sky-blue flowers. The latter were borne on pedicels articulated about their midpoints. Similar results were obtained in crosses of *S. alandiae* with another introduction of *S. tarijense*, No. 12001. However, the latter cross yielded stellate to substellate, creamy-white flowers tinged with lilac along the petal margins. One other combination, *S. alandiae* (12016) × *S. tarijense* (11994), yielded hybrid plants having sky blue flowers with white or whitish acumens and calyces very similar to *S. alandiae*. In these plants, the articulation of the pedicel occurred about the midpoint or in the upper one-third of the stalk. In another cross of *S. alandiae* (12017) with *S. tarijense* (12001), the hybrids developed leaves, inflorescences, and pedicels similar to *S. alandiae*, and light sky blue, strongly pentagonal flowers. The pedicels of these plants were articulated at the midpoint or toward the upper one-third of the stalk. The larger leaves of these hybrids bore up to 14 interjected leaflets, the latter number considerably more than found typically in *S. alandiae*. In one final cross of *S. alandiae* (12018) × *S. tarijense* (12001), the hybrid plants produced leaves similar to *S. tarijense*. These plants bore few interjected leaflets, up to four pairs of lateral leaflets, and sky blue, whitish-tipped corollas that varied in shape from substellate to stellate. Pedicels were articulated above center or in the upper one-third of the stalk. As pointed out previously, the F_1 hybrids of this cross were more similar in appearance to *S. alandiae* than to *S. tarijense*.

Solanum alandiae × *S. berthaultii*.– Hybrids of *S. alandiae* (OCH-12014) × *S. berthaultii* (OCH-12029) have the leaf shape and dissection of *S. alandiae* and the inflorescence of *S. berthaultii*. They vary, however, in corolla shape from substellate to rotate-pentagonal, and in flower color from very light lilac to dark purple. All bore a variable number of flowers borne on pedicels articulated in the upper one-third of the stalk. Their indumentum varied from sparsely to densely pilose. In the reciprocal cross, *S. berthaultii* (12029) × *S. alandiae* (12014), the indumentum was also variable. These hybrids resembled *S. alandiae* in habit as well as in leaf shape and dissection, and *S. berthaultii* in the respective features of their inflorescence. The few to many substellate to rotate-pentagonal flowers of these hybrids, however, varied from very light violet or almost whitish to dark purple. All had the pedicels articulated in the upper one-third of the stalk.

Solanum alandiae (OCH 12014) × *S. leptophyes* (OCH 13720).– The plants of this hybrid population share many features in common with *S. leptophyes*. The great majority are delicate, weak stemmed plants, about 25 cm tall. Most are provided with 4-5 pairs of small, narrowly elliptic-lanceolate leaflets, borne on short petiolules, and a variable number of minute interjected leaflets. Although the corolla of these plants was rotate as in *S. leptophyes*, flower color was as *S. alandiae* in being lilac. All had pedicels articulated slightly above the base. The upper part of the stalk ranged from 10-12 mm long, while the lower ranged from 12-14 mm long. The anthers of these plants, which were about 6 mm long, had cordate bases.

Solanum alandiae (OCH 12014) × *S. infundibuliforme* (OCH 11977).– The F_1 hybrids that developed from the seeds of this cross were very weak and small, with stems approximately 10 cm tall. All withered before flowering. Similar to *S. infundibuliforme*, these plants had 1-2(-3) pairs of lateral leaflets. The upper pair of leaflets was broadly decurrent upon the rachis.

18. Solanum brevicaule Bitter, Repert. Sp. Nov. 11:390, Fig. 166, Plate XX. 1912. FIGS. 85-87.

S. liriunianum Cárd. and Hawkes, Jour. Linn. Soc. Bot. 53:106, Fig. 10. 1945. TYPE: *Cutler 7690*, Bolivia. Dept. Cochabamba, Liriuni, Prov. Quillacollo, February 5, 1943 (F, MO, CPC, US).

S. colominense Cárd., Bol. Soc. Peruana Bot. 5:21, Plate 2A, Fig. 1. 1956. TYPE: *Cárdenas 3686, p.p.*, Bolivia. Colomi-Chapare, Dept. Cochabamba, Prov. Chapare, January, 1946 (HNF; LL, photo).

S. mollepujroense Cárd. and Hawkes, Jour. Linn. Soc. Bot. 53:103-104, Fig. 9. 1945. TYPE: *Cárdenas 3510*, Bolivia. Mollepujru, Dept. Cochabamba, Prov. Tarata, March, 1944 (CPC).

Plants of small stature, densely pilose with white, thick hairs along the stem, peduncles, and rachis, and short fine hairs on the leaves and pedicels. Stems (15-)25-30(-60) cm tall and 4-5(-8) mm in diameter, vigorous, sinuous, simple or branched, occasionally rosette-like or with short basal internodes and longer terminal internodes, pigmented on the axils and nodes, narrowly straight-winged. Stolons long and slender; tubers round to ovate, yellowish-white, small, 1.5-2.5(-4) cm long and 1-2 cm in diameter. Leaves imparipinnate, dark green above and light green below, short and broad, finely and sparsely pilose, 5.5-11(-20) cm long by (2.3-)3-5(-9) cm wide, with 4-5 pairs of leaflets and (3-)5-8(-15) interjected leaflets of diverse sizes, borne on 7-15 mm-long petiolules. Leaflets elliptic, elliptic-lanceolate, or broadly elliptic-lanceolate to ovate, or occasionally elliptic-oblanceolate, the apex obtuse to subacuminate or pointed, and the base rounded to obliquely rounded or broadly cuneate. Terminal leaflets almost of the same size or somewhat larger than the laterals, broadly elliptic-lanceolate, (2.8-)3.5-4(-5.5) cm long by 1.1-2(-4) cm wide. Upper lateral leaflets (1.8-)2.3-3.3(-4) cm long by (0.7-)1-1.8 (-2.5) cm wide, the lower ones decreasing gradually in size towards the leaf base. Pseudostipular leaves broadly falcate, or occasionally slightly clasping the stem, 6-7(-15) mm long by 3-4(-7) mm wide. Inflorescence terminal or lateral, cymose to cymose-paniculate, usually with 3-7(-15) flowers. Peduncle sparsely pilose, slightly pigmented, slender, 5-8(-12) cm long by 2 mm in diameter at base. Pedicels slender, sparsely pilose, articulated in the upper one-third of the stalk or rarely at or below center, the upper part (5-)7-9(12) mm long and the lower 12-14(-20) mm long. Calyx symmetrical or somewhat irregular, sparsely pilose, 6-7 mm long, pigmented, the elliptic-lanceolate lobes narrowed to short, fine acumens. Corolla usually rotate, 2.5-3 cm in diameter, dark purple or blue-violet with yellowish-green or yellow nectar guides, the lobes short, 2-3 mm long excluding the 3-3.5 mm-long acumens. Staminal cone cylindrical-conical; anthers narrowly lanceolate, 5.5-6.5 mm long, distinctly cordate at base, borne on glabrous, 1-2 mm-long white hyalinous filaments. Style 9-10(-11) mm long, exerted 2.5-3 mm, very densely papillose along the lower two-thirds of its length; stigma small, capitate, cleft. Fruit globose to ovoid, green, to 2 cm in diameter.

Chromosome number: 2n=2x=24.

Type: BOLIVIA. Department Cochabamba [Province Cercado], vicinity of Cochabamba, [March?] 1891; *M. Bang 1100* (lectotype, here designated, US; isotypes, BM, E, G, K, M, MO, NY, PH, US, W, WU).

Although the pedicel of *S. brevicaule* is usually articulated at its midpoint or in the upper one-third of its stalk, this character varies in different plants of this species. Thus, occasionally, the pedicels are articulated at or below center. In the latter, the upper part of the stalk may be up to 12 mm long, while the lower may be up to 5 mm long. In *S. bukasovii* and *S. leptophyes*, on the other hand, the position of the pedicel articulation is much more constant.

Solanum brevicaule is also a variable species with respect to leaf dissection,

Figurc 85. – *Solanum brevicaule* (*Ochoa 11934*). 1. Plant. 2. Corolla. 3. Petal. 4. Stamens, dorsal view. 5. Pistil. 6. Calyx. 7. Pedicel and calyx.

vigor, and size of plant. According to a note made by E. K. Balls on one of his herbarium specimens of this species (No. 6242), the above variations may be due to such environmental factors as shade and the relative moisture and richness of the soil.

The author has studied isotypes of *S. liriunianum* Cárdenas and Hawkes that were collected by Hugh Cutler (7690) and deposited in the Missouri Botanical Garden Herbarium, St. Louis, USA. Bearing in mind the great environmental plasticity of this group – as has already been noted for specimens studied by E. K. Balls – and their stable floral characteristics, the author has concluded that *S. liriunianum* should be treated as a synonym of *S. brevicaule*. The above position is further supported by the fact that one of the above collections consists of two very different appearing plants mounted on one sheet. One of the two is typical in appearance to *S. brevicaule*. It is a very small plant, not more than 20 cm tall, having 3-4 pairs of leaflets and sinuous stems, bearing short internodes at the base. The other plant is almost three times taller and more robust than the latter, but not rosette-like at the base. In appearance this taller plant is very similar to the other isotypes of *S. liriunianum* that are housed in the U.S. National Museum (US) and the Commonweath Potato Collection

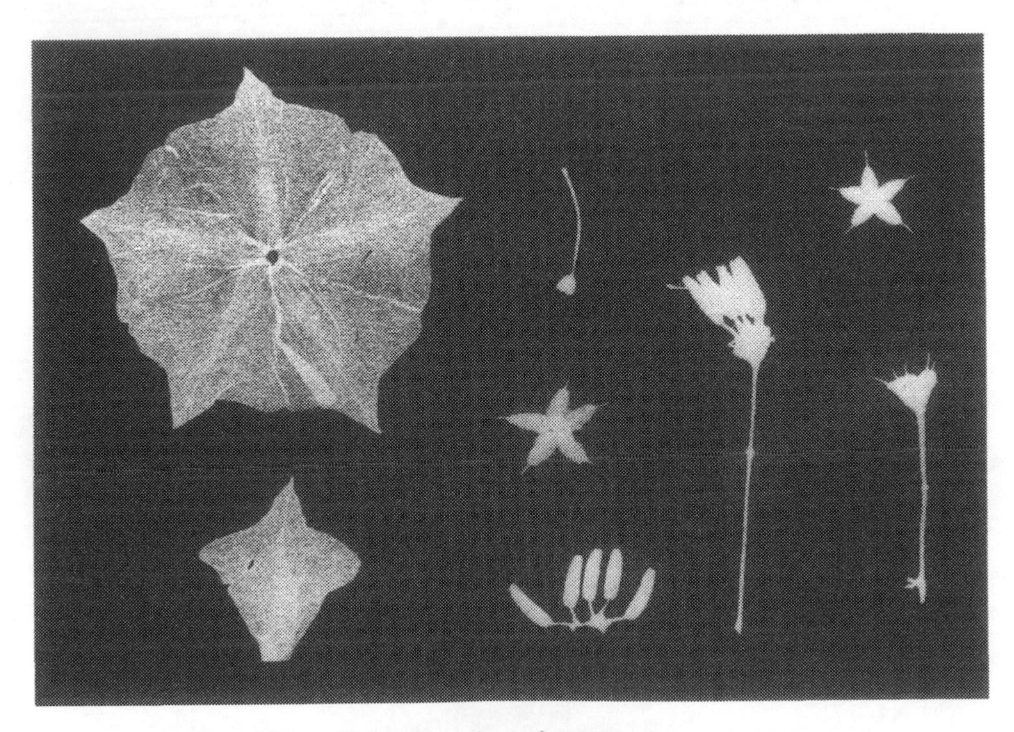

Figure 86. – Floral dissection of *Solanum brevicaule* (*Ochoa 11934*), ca. ×1½.

Figure 87. – *Solanum brevicaule* (isotype of "*S. liriunianum*", *Cutler 7690*; US). Photo: courtesy of the Natural History Museum, Smithsonian Institution.

(CPC). The duplicates of Cutler (No. 7690) that are maintained in the Chicago Natural History Museum (F) under the accession number 1290343, as well as in the Vienna Botanical Garden Herbarium (WU), are similar to the collections of *S. brevicaule* made by Bang (No. 1100) and Ochoa (No. 659).

The plant described as *S. colominense* should probably be considered as no more than a variant of *S. brevicaule*. A phototype of this entity is found in the Lundell Herbarium (LL) of the Texas Research Foundation. The same specimen is also very similar in appearance to the two plants that were collected by Balls (No. 6381) at Huancaroma (near Eucaliptos), Bolivia. These specimens were mounted on one sheet and accessioned by the U.S. National Museum (US) under the number 1779343.

A number of collections of *S. brevicaule* (*Ugent 4652, 4653, 4654, 4655, 4699, 4700*) are more vigorous and better developed than what might be considered as normal for this species. Although these might be considered at first glance to show introgression from some other blue-flowered species, such as *S. sparsipilum*, it is more likely that these unusually large plants have resulted from optimal growth conditions at their respective habitats. Thus, all were said to grow in the shade of shrubs, in organically rich, moist, black soil.

Incidentally, it should be pointed out that two species, *S. brevicaule* and *S. leptophyes*, are included under Cárdenas' collection No. 6114. Both are mounted on the same sheet, dated April, 1958. Cárdenas No. 6113, a collection of *S. candelarianum* dated February, 1961, is not numbered sequentially.

Affinities

This species is related to *S. bukasovii*. Both have similar leaf dissection and habit. As in *S. bukasovii*, the stems of this plant vary considerably in height. Some of the specimens of this plant that are included in Bang's type collection (No. 1100) are no more than 10-15 cm tall. However, the two species are markedly different in other respects. Thus, the flowers of *S. brevicaule* are smaller and are provided with a differently shaped calyx. Moreover, the plant is less pubescent and bears differently shaped leaves, and smaller and differently colored fruits.

The author agrees with Correll (1962) that *S. bukasovii* and *S. brevicaule* should be maintained as separate species. However, unlike Correll, this author sees little or no relationship of this species to *S. rhomboideilanceolatum*, *S. lobbianum* and *S. canasense*. *Solanum rhomboideilanceolatum*, it should be pointed out, is a diploid member of the series Conicibaccata. This particular species is endemic to the eastern flanks of the central Andes, where it occurs in the low, moist, forested zone known as Ceja de Montaña. *Solanum lobbianum*, on the other hand, is a tetraploid species of the series Conicibaccata. It was recently rediscovered by the present author in Colombia. And lastly, *S. canasense*, a diploid plant from southern Peru, will be considered in a forthcoming work by this author as a mere variant of *S. bukasovii*.

Habitat and Distribution

This species occurs primarily on rocky soil in semiarid valleys. It also grows on moist slopes between 2600 and 4000 m altitude. According to Tossi's ecological map, the latter zone lies between 'Montano' and 'Subalpino.' It is distributed from the Provinces of Aroma and Inquisivi in the vicinity of La Paz to the Provinces of Jujuy and Salta in northwestern Argentina. In central Bolivia, it occurs in semitropical valleys in the vicinities of Cochabamba, Sucre, and Tarija. Throughout much of its range, this species occupies habitats similar to those of *S. leptophyes*.

At its higher altitudinal limits, *S. brevicaule* grows in steppe grasslands dominated by *Stipa ichu*. At somewhat lower elevations, as for example, at Puente San Miguel along the road from Cochabamba to Vizcachas near Liriuni (3800 m), it grows in moist, rich soil along with *S. megistacrolobum*, *S. acaule*, *Oenothera*, *Salvia*, *Lupinus*, *Oxalis*, *Calceolaria*, *Begonia*, *Nasturtium*, *Epilobium*, and various genera of ferns. At still lower elevations, as for example at its type locality along the lower slopes of the Tunari range near Cochabamba (alt. 2800 m), it grows with *Prosopis* sp., *Nicotiana glauca*, *Opuntia* sp., *Trichocereus* sp., *S. sparsipilum*, and *S. berthaultii*.

Specimens Examined

Department Cochabamba

Province Cercado: Vicinity of Cochabamba; 1891; *Bang 1100*, type collection of *S. brevicaule* (lectotype, US; isotypes, BM, E, G, K, M, MO, NY, PH, US, W, WU). Vicinity of Cochabamba; 1891; *Bang 1101* (BM, E, G, GH, K, MO, NY, PH, US, W, WIS, Z). In cultivated fields west of Cochabamba, weed. January 3-7, 1924; *Hitchcock 22832* (NA). Near Cochabamba, 2700 m alt., mountain slopes at vicinity of Cerveceria Colon. January 27, 1949; *Ochoa 659* (OCH, US). Above Cochabamba, Cerveceria Colon, 2850 m alt. On waste ground beside path, on edges of fields, etc., common. Calyx regular; flowers blue-violet. February 11, 1971; *Hawkes, Hjerting, Cribb, and Huaman 4408* (CIP).

Province Arque: Aguas Calientes Station, 3324 m alt. [10,900 feet, according to the collector], railway between Oruro and Cochabamba, on barren soils, plants 30 cm tall, branched principally at base. March 11, 1939; *Balls 6202* (US). Aguas Calientes Station, 3250 m alt., about halfway between Oruro and Cochabamba. January 27, 1949; *Brooke 3008, 3010* (BM); January 30, 1949; *Brooke 3011* (BM).

Province Ayopaya: Road from Cochabamba to Morochata, 3800 m alt., in shady places. March, 1946; *Cardenas 3693* (CPC). Puente San Miguel, 3800 m alt., along the Tunari-Cochabamba road, among bushes in humid soil. Plants 10-30 cm tall, flowers blue-violet. February, 1952; *Cardenas 5518* (HNF, LL). Puente San Miguel, 3800 m alt., above Liriuni on the Cochabamba-Vizcachas road, about 25 km NNW of Cochabamba (air dist.), 17°10′ S and 66°25′ W. On rocky slope with with *S. brevicaule*, *S. capcisibaccatum*, *S. acaule*, *Oenothera*, *Lupinus*, *Chenopodium*, *Salvia*, *Oxalis*, *Calceolaria*, and nontuberous *Solanum* species. April 9, 1963; *Ugent 4673*, (US), *4675, 4676, 4677, 4679, 4680, 4681, 4685* (WIS), *4682* (F, WIS), *4687* (OCH), *4688* (K), *4689* (WIS), *4690* (OCH), *4691, 4695* (WIS), *4696* (K, OCH, US, WIS), *4697* (F, US), *4699* (WIS), *4700* (WIS), *4770* (F, US) and *4771* (WIS). Puente San Miguel, along the road Tunari-Quillacollo-Cochabamba; *Cardenas*, without number and date (HNF-0011).

Province Chapare: Quebrada de Colomi, 3200 m alt. [10,500 feet, according to the collector], among stone and under bushes in poor soil, plants small, however larger when growing in humus rich soils; stems 15-30 cm tall; leaves 10-22 cm long and provided with 2-4 pairs of leaflets. March 15, 1939; *Balls 6242* (US). Colomi. January, 1946; *Cardenas 3686 p.p.* (HNF; LL photo). Colomi, 3700 m alt., 30 miles ENE of Cochabamba, among large stones on slope. February 9, 1949; *Brooke 3003* (BM). About 40 km from Tranca de Sacaba, 3730 m alt., along road to Palca, 1/2 km before Palca, 17°13′ S and 66°3′ W. Growing on valley side among stones and *ichu* grass, wet Puna. Flowers 4 cm in diameter, dark violet. March 15, 1980; *Hawkes, Hjerting, and Aviles 6619, 6621* (plants grown at Sturgeon Bay, Wisconsin, USA; CIP).

Province Quillacollo: Liriuni, 3000 m alt., about 18 km NW of Cochabamba, in small damp thicket on hillside. February 5, 1943; *Cutler 7690* (CPC, F, K, MO, US). Near Liriuni, among bushes on mountain slope. February 14, 1960; *Correll, Dodds, Brücher, and Paxman B607, B608* (LL). Near San Miguel, above Liriuni, among bushes on rocky mountain slope. February 14, 1960; *Correll, Dodds, Brücher, and Paxman B611* (LL, US). Liriuni, 3500 m alt., 17°20′ S and 66°20′ W, above Escuela Bellavista, about 17 km NW of Cochabamba (air distance). Roadside thickets with *Galium*, *Tagetes*, *Salvia*, and nontuberous *Solanum* sp. April 9, 1963; *Ugent 4652-1, 4652-3, 4652-4, 4653, 4654* (WIS), *4652-2* (F), *4655* (US, WIS). Jankho Khala, 3900 m alt., 17°10′ S and 66°25′ W, above Puente San Miguel and Liriuni, on the Cochabamba-Vizcacha road, about 25 km NNW of Cochabamba (air distance). In weedy cultivated potato field with *S. decurrentilobum*, *Oxalis tuberosum*, *Capsella*, *Lupinus*, and non-tuberous *Solanum* sp. April 9, 1963; *Ugent 4832* (WIS). Baños de Liriuni, 3100 m alt., 14 km from Quillacollo on road to Morochata. Waste stony place outside baths. Some large plants, some small; with rich purple large flowers and some berries. Plants to 30 cm. 2n=24. January 27, 1971; *Hawkes, Hjerting,*

Cribb, and Huamán 4251 (CIP). 22 km from Quillacollo on road to Morochata, 3650 m alt., by new Puente San Miguel, 17°20′ S and 66°21′ W. In shade of little building and trees. Very dissected leaves, dark blue flowers and round green berries about 2 cm diam. February 25, 1980; *Hawkes, Hjerting, and Aviles 6468* (plants grown at Sturgeon Bay, Wisconsin, USA; CIP). 25.5 km from Quillacollo on road to Morochata, 3850 m alt., below hairpin bend by river. Lat. 17°19′ S and [long.] 66°22′ W. Locally abundant in rich leaf mould under *Polylepis* in partial shade by river. Plants 10–50 cm tall; flowers rich deep blue-purple, to 4 cm diam. Berries round to oval, slightly 2-grooved, to 2 cm long, very abundant. April 2, 1980; *Hawkes, Hjerting, and Aviles 6690* (plants grown at Sturgeon Bay, Wisconsin, USA; CIP). 22 km from Quillacollo on road to Morochata, 3650 m alt., near Puente San Miguel but on old road. Lat. 17°20′ S, long. 66°21′ W. Under shade of *Polylepis, Berberis, Escallonia,* etc. Plants with large flowers and about 2–3 berries per plant. April 3, 1980; *Hawkes, Hjerting, and Aviles 6701* (plants grown at Sturgeon Bay, Wisconsin, USA; CIP).

Province Tarata: Near Anzaldo, 3600 m alt., part of a random collection of 17 plants, growing along the margins of cultivated fields, flowers violet to light violet, darker when growing under the shade. January 8, 1943; *Cutler, Cárdenas, and Gandarillas 7602* (F, US). Near Anzaldo, 3600 m alt., lands of Humberto Gandarillas, flowers light blue-violet, with darker or occasionally white centers. January 9, 1943; *Cutler, Cardenas, and Gandarillas 7623* (F, MO, NY, US). Mollepujro, slope of Rio Caine. March, 1944; *Cardenas 3510* (CPC). Mollepujro. February, 1961; *Cardenas s.n.* (CA).

Department Chuquisaca

Province Nor Cinti: 126 km from Potosi on road to Camargo, 1 km from Padcaya, 3400 m alt. March 29, 1974; *Hawkes, van Harten, and Landeo 5730* (CIP).

Province Oropeza: Guerraloma, 2800 m alt., near Sucre, on rocky soil, plants 30 cm tall, flowers blue. April, 1958; *Cárdenas 6114* (US, a sheet with two plants, only plant No. 1 belongs to *S. brevicaule*). Near Guerraloma, 3100 m alt., some 18 km from Sucre. Associated with *S. boliviense* and several species of grass, on poor rocky soil. Tubers white, small, 1–2 cm in diameter. Chromosome number: 2n=2x=24. March 13, 1978; *Ochoa 11934* (CIP, F, GH, LE, NY, OCH, P, SI, US, WIS). Punilla, near 2600 m alt., 18°55′ S and 65°25′ W, about 15 km NW of Sucre (air distance), on the Sucre-Ravelo road. Margins of a weedy wheat field with *S. boliviense* and *S. pachytrichum* and other nontuberous *Solanum* sp, *Setaria plumosa, Calceolaria,* and *Veronica.* April 12, 1963; *Ugent and Cárdenas 4911* (WIS). Villa Maria, about 2850 m alt., 19°1′ S and 65°10′ W, a small farm about 19 km NE of Sucre. Margins of a weedy wheat field and surrounding thickets, with *S. pachytrichum*, *Tagetes*, *Cerastium*, *Medicago*, *Rumex*, and *Brassica*. April 12, 1963; *Ugent and Cárdenas 4927, 4934* (WIS), *4937* (F).

Province Tomina: In moist places among bushes. December 1845-January 1846; *Weddell 3725* (P).

Department La Paz

Province Aroma: Patacamaya, Luribay, 3980 m alt. February 5-April 13, 1980; *Soest, Okada, and Alarcón (S.O.A.)-56* (CIP).

Province Inquisivi: Canayapu, 2700 m alt., on the road from Quime to Inquisivi, among moist thickets of *Calceolarias*, composites, and grasses. February 23, 1984; *Ochoa and Salas 15497* (CIP, NY, OCH, US).

Department Oruro

Province Cercado: Finca Huancaroma, 3700 m alt. [12,000 ft., according to the collector], near Eucalyptus, among stones at the base of walls. Stems 10-20 cm long; corolla bluish-purple; fruit green or white speckled, slightly ovate, 2.5 cm long. April 1, 1939; *Balls 6381* (K, US). Aguas de Castilla, 4000 m alt., on dry rocky slopes, suburb of Oruro. February 1, 1949; *Brooke 3014 p.p, 3020 p.p* (BM). Aguas de Castilla, 4026 m alt. [13,000 ft. according to the collector], among stones near stream, in dry area. February, 1949; *Brooke 3017* (F).

Department Potosi

Province Frias: 1 km below Potosi on road to Lecherias, 3800 m. March 27, 1974; *Hawkes, van Harten, and Landeo 5703* (CIP).

Department Tarija

Province Aviles: Near edge of fields along the Tojo river, near Tojo. Flowers blue-violet. February 23, 1960; *Correll, Dodds, Brücher, and Paxman B656* (LL).

Crossability

According to screening tests conducted at CIP, *S. brevicaule* shows some resistance to early frost and attacks by *Globodera pallida*. For these reasons, this species could be of value in breeding programs aimed at the improvement of the cultivated potato.

Crosses between *S. brevicaule* and other Bolivian diploid wild species such as *S. alandiae*, *S. infundibuliforme*, *S. leptophyes*, and *S. sparsipilum* were very easily obtained. These yielded abundant fruits and seed. This species also crosses very easily with such diploid allopatric species of southern Peru as *S. coelestispetalum* and *S. raphanifolium*.

The two crosses that were attempted using tetraploid species gave varied results. Thus, many seeds were obtained when *S. brevicaule* was used as the staminate parent in crosses with *S. acaule*. On the other hand, only empty berries were obtained when *S. brevicaule* was used as the pistilate parent in crosses with *S. oplocense*.

Solanum brevicaule (11934) × *S. infundibuliforme* (11973).– Hybrids of this parentage are generally similar in appearance to *S. brevicaule* when the latter is used as a mother plant. However, these hybrids strongly resemble *S. infundibuliforme* in the dissection and shape of their leaves. For example, all have the first and second lateral pairs of leaflets broadly decurrent on the rachis. The dark purple flowers of these plants range up to 3 cm in diameter and bear a rotate-pentagonal shaped corolla. Their pedicels are articulated in the upper one-third of their stalks.

In the reciprocal cross, *S. infundibuliforme* (11973) by *S. brevicaule* (11934), the F_1 segregates are similar to *S. brevicaule* with respect to leaf shape. Some plants of the previously mentioned population have the upper pair of lateral leaflets broadly decurrent on the rachis, while in others the lateral leaflets are shortly petiolulate. Although there is much variability, in general, with respect to decurrency, there was a definite tendency for the hybrid plants to resemble *S. brevicaule*. Most were erect, vigorous plants with simple or occasionally branched, delicate stems ranging to 40 cm tall. While the rotate-pentagonal corollas of these hybrids ranged in color from light to dark purple, the majority were of the lighter variety. All had pedicels articulated in the upper one-third of the stalk.

19. Solanum candelarianum Cárdenas, Bol. Soc. Per. Bot. 5(1):12-13. 1956. Plate 1-B (not *S. candelarianum* Buk., in Bull. Appl. Bot. Genetic & Pl. Breed., Leningrad, Suppl. 47:60, 218, 1930). FIGS. 88-91.

S. avilesii Hawkes and Hjerting, Bot. Jour. Linn. Soc. 86(4):410-412, Fig. 4. 1983. TYPE: *Hawkes, Hjerting, and Aviles 6521*, Bolivia. 28 km from Valle Grande on road to Pucara near top high peak, Dept. Santa Cruz, Prov. Valle Grande, March 1, 1980 (holotype, K, not seen; isotype, Herb. J. G. Hawkes, not seen).

Plants usually small, pilose throughout with white and more or less dense hairs. Stems (10-)20-25(-50) cm tall and 2-3(-5) mm in diameter at base, erect, simple or branched, somewhat pigmented, and provided with very narrow and bearly distinguishable straight wings. Stolons 15-25 cm long by 1-1.5 mm in diameter; tubers round to ovate, small, 10-15 mm in diameter, white. Leaves imparipinnate, without interjected leaflets, 4.5-9 cm long by 3-6 cm wide, and provided with 1-2(-3) pairs of leaflets, the latter borne on petiolules 0.6-1.5(-2.5) cm long. Leaflets dark green and finely pilose above with dense,

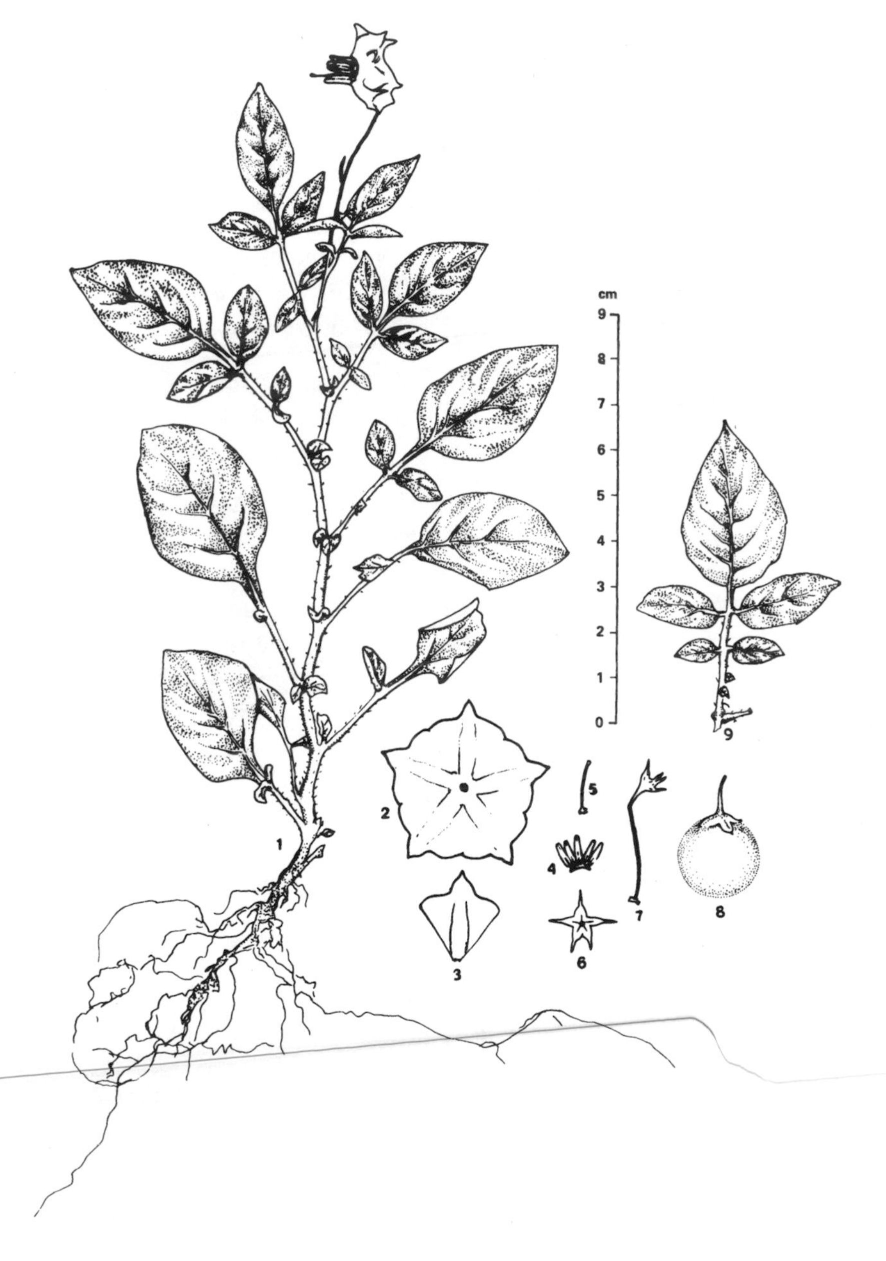

Figure 88. – *Solanum candelarianum* (*Ochoa and Salas 15542*). 1. Flowering plant. 2. Corolla. 3. Petal. 4. Stamens, dorsal view. 5. Pistil. 6. Calyx. 7. Pedicel and calyx. 8. Fruit. 9. Leaf from different specimen. All ×1.

white, velvety hairs, lighter green and somewhat pilose below and shortly pilose along the finely denticulate leaf margins. Terminal leaflets longer and much wider than the laterals, 3-4 cm long by 2-3 cm wide, broadly elliptic-lanceolate to ovate-lanceolate or elliptic-lanceolate, the apex pointed or shortly acuminate and the base rounded to subcordate or slightly attenuate. Lateral leaflets to 0.9-3.5 cm long by 0.6-1.7 cm wide, ovate-oblong to narrowly elliptic, with pointed apices and rounded to oblique bases, subsessile or borne on 1-3 mm-long petiolules. Pseudostipular leaves very small, 3-5(-8) mm long by 1.5-2.5(-3.5) mm wide, broadly subfalcate to obliquely elliptic. Inflorescence cymose, 3-5(-7) flowered. Peduncles to 5 cm long and 1.5-2 mm in diameter at base. Pedicels articulated in the upper one-third of the stalk, the upper part 5-7(-8) mm long and the lower 15-17 mm long. Calyx 5.5-6.5 mm long, symmetrical, dark purple, provided with narrowly elliptic-lanceolate lobes, the latter abruptly narrowed to finely pointed 1.5-2 mm-long acumens. Corolla rotate to rotate-pentagonal or occasionally hexagonal, 2.5-3 cm in diameter, purple to dark purple or blue-lilac with grayish-white nectar guides. Anthers narrowly lanceolate, 5 mm long, cordate at base, borne on glabrous 0.3-0.6 mm-long filaments. Style 8.5-9 mm long, exerted 2 mm, papillose along the basal one-third of its length; stigma small, capitate, light green. Fruit globose to ovoid, 1.5-2 cm long, green.

Chromosome number: 2n=2x=24.

Figure 89. – Floral dissection of *Solanum candelarianum* (*Ochoa and Salas 15542*), ×1.

Type: BOLIVIA. Department Santa Cruz, Province Valle Grande, between Candelaria and Pucara, 2700 m. March, 1955; *M. Cardenas 5080* (CA).

Solanum candelarianum was discovered more than three decades ago by Cárdenas who found it growing in southern Bolivia along roadsides between Candelaria and Pucara in the Province of Valle Grande, Department of Santa Cruz. Unfortunately, the holotype of this species was not, at the time of this investigation, in any of the herbaria that were reviewed by the author, including the National Forestry Herbarium of Cochabamba, Bolivia, where the majority of Cardenas' potato collections were said to be kept. In view of this situation, the author proposed a neotype for *S. candelarianum* (Ochoa, 1985) based upon materials of this species that were collected at the original type locality (*Ochoa and Salas 15542*). However, the author, with the help of Ricardo Salaues (former lecturer of Systematic Botany, University of Cochabamba), has recently located some other type collections of Cardenas. Among these was the type collection of *S. candelarianum* (*Cárdenas 5080*), consisting of two plants mounted on one sheet located in the Herbarium Cardenasianum (CA). Both plants proved to be identical to the collections of this species that were made previously by this author at its type locality.

Also, a Cardenas collection (No. 6113) determined by the author as *S. candelarianum*, but labeled by Cárdenas as '*S. totorensis*,' was also in the National Forestry Herbarium (NHF) of Cochabamba, Bolivia. This collection consists of five small plants ranging between 10–25 cm in height mounted on a

Figure 90. – Floral dissection of *Solanum candelarianum* (*Ochoa and Salas 15543*), ×1.

single sheet. All are simple or few-branched and bear stems with short internodes at base, and dark purple, rotate-pentagonal flowers (the latter 'blue' according to Cárdenas).

Affinities

This species resembles *S. brevicaule* in its stem height and pubescence, as well as in the size, form, and color of its corolla. However, it differs substantially from *S. brevicaule* in the form and dissection of its leaves and in having smaller and darker purple flowers, narrower stems, and pedicels that are articulated toward the upper one-third of their stalks. This species also resembles *S. virgultorum* in habit, pubescence, leaf shape, and dissection.

Habitat and Distribution

This species occurs in temperate or cool and semiarid mountain districts at elevations ranging from 2300 to 3200 m. In this zone, it usually grows in poor and somewhat moist, rocky or sandy-clay soils in open fields where it is often associated with shrubs and other herbaceous plants. It is known only from southern Bolivia in the Provinces of Carrasco (Department Cochabamba) and Valle Grande (Department Santa Cruz).

Figure 91. – Cerro Merma, 2650 m altitude, habitat of *Solanum candelarianum*, a short distance from its type locality, 28 km from the town of Valle Grande on the way to Pucara, Province Valle Grande, Department Santa Cruz.

Specimens Examined

Department Cochabamba

Province Carrasco: Jatun Pino, 3200 m alt., near Monte Punco, 'in humid soil under bushes, plants 10-30 cm tall, flowers blue.' February, 1961; *Cardenas 6113* (HNF, US, WIS).

Department Santa Cruz

Province Valle Grande: Between Candelaria and Pucara, 2700 m alt. On humid slopes. Plants 30-60 cm tall, flowers blue-purple. March, 1955; *Cardenas 5080* (CA). 28 km from Valle Grande on road to Pucara, 2950 m alt., near top of high peak. Lat. 18°38′ S, long. 64°09′ W. Amongst spiny shrubs in rich soil on rock outcrops. Plants about 5-10 cm, flowers blue-purple. March 1, 1980; *Hawkes, Hjerting, and Aviles 6521* (CIP, from plants grown at the Exp. Stn. at Sturgeon Bay, Wisconsin, USA). Some 32 km from Valle Grande on the road towards Pucara, 2500 m alt., among thickets, in moist soil. Plants very small with dark green leaves and purple flowers. Chromosome number: $2n=2x=24$. February 28, 1984; *Ochoa and Salas 15539* (CIP, OCH, US). Between Candelaria and Pucara, 2400-2500 m alt., from Valle Grande on the road toward Pucara. Among large stones and protected by *Agave* sp. Plants small, 15 cm tall, pubescent; fruit globose. February 28, 1984; *Ochoa and Salas 15542* (topotypes, CIP, NY, OCH, US). El Cruce, 2700 m alt., approximately 40 km from Valle Grande on the road toward Pucara. Among thickets, in stony but rich soil. Plants small up to 20 cm tall; flowers purple. February 28, 1984; *Ochoa and Salas 15543* (CIP, OCH). Alto Grande, 2400 m alt., some 26 km from Valle Grande on the road toward Pucara, along margins of potato fields, ollucus (*Ullucus tuberosus*) and gourd (*Cucurbita* sp.). Common name *papa de zorro*. February 28, 1984; *Ochoa and Salas 15553* (CIP, OCH, US).

20. Solanum candolleanum Berthault, Ann. Sci. Agron. and Etrang., Paris, ser. 3, 6(2):184, 185, 190, Plate 1 (p. 292). 1911. FIGS. 92-96.

S. mandonii A. DC., Bibl. Univ., Arch. Sci. Phys. et Nat., ser. 3, 15:438. 1886. TYPE: not known. Not *S. mandonis* van Heurk et Muell., Arg. in Heurk, Obs. Bot., p. 78. 1870.

Plants large, erect, very robust. Stems 1 m or more tall and to 2 cm in diameter at base, simple or branched, thick, somewhat pigmented, pilose with 2-4 mm-long hairs, provided with broad, straight or slightly sinuous wings. Stolons 1-1.5 m long by 3-4 mm in diameter, fleshy; tubers very large, 12-14 cm long by 3.5-4.5 cm in diameter, white, ovoid to compressed-oval, lenticu-

Figure 92. – *Solanum candolleanum* (*Ochoa 11913*). 1. Upper part of flowering plant. 2. Corolla. 3. Petal, 4. Stamens, 5. Pistil. 6. Calyx. 7. Pedicel and calyx. 8. Fruit. All ×½.

late, and provided with shallow eyes. Leaves imparipinnate, strongly dissected, to 35 cm long by 28 cm wide but averaging 20 cm long by 13.5 cm wide, borne on 1.5-3 cm-long petiolules and provided with 4-5(-6) pairs of lateral leaflets and numerous decurrent, sessile interjected leaflets, numbering 27 or more and of two or three different sizes. Leaflets dark green and densely pilose above, and light green or grayish-green and densely pilose below, elliptic-lanceolate or narrowly elliptic-lanceolate, the apex subobtuse or subacuminate, and the base rounded or obliquely rounded, borne on petiolules 1.5-5(-7) mm long. Terminal leaflets about the same size or slightly larger and broader than the upper pair of laterals, 6-8(-12.5) cm long by 2.5-4(-6) cm wide. Upper lateral leaflets 5.5-7(-12) cm long by 2-2.4(-3) cm wide, occasionally narrowly decurrent about the rachis, the latter densely covered with short, glandular and tetralobular hairs. Pseudostipular leaves broadly falcate, obliquely semiovate, 9.5-16(-26) mm long by 4-8(-13) mm wide. Inflorescence terminal or lateral, cymose, 10-15-flowered. Peduncles to 18 cm long, slightly pigmented. Pedicels light green, articulated above center or in the upper one-third of the stalk, 2-3 cm long and more densely pilose than the peduncles, the upper part 6-10 mm long and the lower 25-45(-75) mm long. Calyx pilose, 6-8(-10) mm long, the lobes narrowly lanceolate and pointed or acuminate at tip. Corolla rotate to rotate-pentagonal, 3-3.5 cm in diameter, blue-violet or dark purple, the acumens triangular and pointed and up to 6 mm long. Anthers lanceolate, 6 mm long, distinctly cordate at base, borne on glabrous, 1-1.5 mm-long white-hyalinous filaments. Style 11 mm long, densely papillose along the lower two-thirds of its length; stigma claviforme, cleft. Fruit globose to ovoid, large, to 3.5 cm in diameter, green and flecked with white dots except for the upper third, which is whitish-green.

Chromosome number: 2n=2x=24 and 2n=3x=36.

Type: BOLIVIA. Department La Paz, Province Larecaja, rocky slopes of Illampu, Lancha de Cochipata, vicinity of Sorata, 3500 m alt. November, 1858; *G. Mandon 397* (lectotype, here designated, P; isotypes, BM, G, K, W).

Prior to Berthault's 1911 publication of this species, the original type specimen of this plant (*Mandon 397*) was first labeled *S. tuberosum* L. This classification of Mandon's specimen was later replicated in Baker's (1884), 'A Review of the Tuber-bearing Species of *Solanum*.' And yet another element of confusion was added to the classification of this plant when Berthault (1911) erroneously cited Mandon's original type specimen as No. 997 rather than 397.

In 1886, De Candolle, who recognized the distinctiveness of this species, published the name *S. mandoni*, based on Mandon 397. However, Mandon's original type collection, which is housed in the Paris Herbarium, has a note on it by Berthault which reads, '. . . not the same as *S. mandonis* van Heurck, which

Figure 93. – Living plant of *Solanum candolleanum* (*Ochoa and Salas 11913*), grown at CIP, Huancayo, Peru.

is very different from *S. mandoni* A. DC., 1870, which is based on a collection of Mandon, No. 422.' Thus, van Heurck's name, which was published earlier, has priority. The latter non-tuber-bearing species, it should be noted, is related to *S. albidum*. Because of the above nomenclatorial conflict, Berthault (1911) renamed this species *S. candolleanum*, in honor of the great Swiss botanist, Alphonse De Candolle.

Two species are included under Mandon's collection No. 397 (Accession No. 208) in the Kew Herbarium of London, England. Both were determined originally as *S. tuberosum*. One of these specimens, as indicated by the annotations of Bitter and Correll, is that of *S. candolleanum* Berthault. The other, according to Correll, is *S. tuberosum* (*sensu latu*), although Bitter has labeled it as *S. brachyodon*, a name, however, which remains unpublished. In the author's opinion, *S. brachyodon* is probably a variant of *S. virgultorum* (Bitt.) Cárdenas and Hawkes.

The two other duplicate collections of Mandon 397, which are housed in the Vienna Museum (Accession Nos. 33056 and 33053) are both *S. candolleanum*. However, one of the two (No. 33053) was classified as *S. brachyodon* by Bitter, who noticed the unusual shape of its fruit. But Bitter noted that this Vienna specimen was not as typical as the one at Kew.

Habitat and Distribution

This species is known from the vicinity of Pelechuco and Sorata in the northwestern part of the Department of La Paz, Bolivia. It is also known from the western side of the Cordillera de Apolobamba in the Peruvian Province of Sandia, where it is most frequently found near Sina.

Solanum candolleanum grows in cool, moist, well-drained soils between 3000 and 4000 m altitude. It is frequently found under shrubs or along the forested margins of riverbanks, arroyos, small streams, and quebradas. In the vicinity of Tacacoma (Province Larecaja), it grows in cool, moist, foggy pasture lands. Here it is associated with *Buddleja montana*, *Lupinus* sp., and *Mutisia mandoniana*. In moist quebradas, it grows with *Tropaeolum*, *Calceolaria*, *Oxalis*, and *Bomarea* spp., several non-tuber-bearing species of *Solanum*, and *Salvia dombeyii*. In the vicinity of Sorata, it grows commonly with *Ulupica*, a shrubby species of *Capsicum* which bears small, round pungent fruits used commonly as a condiment.

Affinities

This species does not appear to be very closely related to any other wild potato species. However, it does have some characteristics in common with *S. stenotomum* (2n=24) and *S. tuberosum* subsp. *andigena* (2n=48), two cultivated Andean species with which *S. candolleanum* may play an evolutionary role. All three have similar life forms and also are known to produce abundant berries.

Moreover, the leaves and leaflets of *S. candolleanum* are similar in shape to those of *S. stenotomum*, while the vigor of the plant and its flower color and shape are reminiscent of *S. tuberosum* subsp. *andigena*.

Living plants of this species were grown from seeds and tubers collected by the author from its type locality. These represent the first germplasm collections that have been made for this species since the time of Mandon. Subsequent cytological work on this collection confirmed its diploid status (2n=24; Ochoa, 1956). Other ploidy determinations on material collected by this author in Bolivia in 1977 and 1978 (Nos. 11805, 11814, 11835, 11896, 11897, 11913) were also diploid. However, two other accessions (Nos. 14983 and 14986) were triploid (2n=36).

Specimens Examined

Department La Paz

Province Camacho: Pantini, 3600 m alt., near Italaque. Chromosome number: 2n=2x=24. February 28, 1978; *Ochoa and Salas 11897* (CIP, OCH, US).

Province Franz Tamayo: Unupuquio, 3500 m alt., on the road from Pelechuco to Queara, among stones and near plants of *Puya*. Chromosome

Figure 94. – Floral dissection of *Solanum candolleanum* (*Ochoa and Salas 11805*), ×1.

number: 2n=2x=24. February 5, 1983; *Ochoa and Salas 14955* (CIP, OCH, US). Quellentique, 3200 m alt., some 6 km E of Pelechuco, among shrubs along the right margin of the Pelechuco River. Chromosome number: 2n=2x=24. February 5, 1983; *Ochoa and Salas 14958* (CIP, OCH). Leitique, 3000 m alt., on the road between Pelechuco and Apolo, among isolated forest groves, near the left margin of the Pelechuco River. Chromosome number: 2n=2x=24. February 5, 1983; *Ochoa and Salas 14959* (CIP, NY, OCH, US). Vicinity of Queara, 3440 m alt., among thickets of shrubs and herbs. Flowers dark purple. The leaves apparently more dissected than in diploid specimens. Chromosome number: 2n=3x=36. February 7, 1983; *Ochoa and Salas 14983* (CIP, OCH). Vicinity of Queara, 3440 m alt., among shrubs and weeds. February 7, 1983; *Ochoa and Salas 14984* (CIP, OCH). Vicinity of Huaycckochayoc, 3200 m alt., some 5 km from Puina, between Puina and Queñuapampa, along the left margin the the Puina River near waterfall, in very humid, black soil. With *Blechnum*, *Hieracium*, and *Bidens andicola*, also with several ferns. Flowers very dark purple. Chromosome number: 2n=3x=36. February 7, 1983; *Ochoa and Salas 14986* (CIP, OCH). Jatun Sencca, 3200 m alt., on the Puina-Sayhuanimayo road, right margin of the

Figure 95. – Floral dissection of *Solanum candolleanum* f. *sihuanpampinum* (*Ochoa and Salas 11835*), ×1.

Puina River. Plants more than 1 m tall, stems thick and strongly winged up to 2.5 cm in diameter at the base. Associated with *Bromus pitensis*, *Blechnum*, and *Tropaeolum* sp. Chromosome number: 2n=2x=24. February 9, 1983; *Ochoa and Salas 14990* (CIP, OCH). Toro Toro, 3400 m alt., between Ayamachay and Puina, protected by abundant thickets of *Rubus*. Plants vigorous, more than 1 m tall, stems pigmented or subpigmented, winged, 1.5 cm in diameter at the base. Chromosome number: 2n=2x=24. February 10, 1983; *Ochoa and Salas 15002* (CIP, NY, OCH, US). Pauchinta Cucho, 3500 m alt., above Toro Toro, among thickets of *Rubus*, *Salvia dombeyii*, and *Salpichroa* sp. In very humid soil. February 10, 1983; *Ochoa and Salas 15004* (CIP, OCH). Above Pauchinta Cucho, 3600 m alt., between Pauchinta Cucho and Puina. Chromosome number: 2n=2x=24. February 10, 1983; *Ochoa and Salas 15011, 15012, 15013* (CIP, OCH).

Province Larecaja: Near Sorata, 3500 m alt., rocky slopes of Illampu, Lancha de Cochipata. November, 1858; *Mandon 397* (P, BM, G, K, W). Near Sorata, *Cárdenas s.n.* (CA, topotype). Tacacoma, 3600 m alt., some 20 km air distance NW of Sorata. Abundant in the cemetery of Tacacoma, in loose, humid soil. Plants large and vigorous, more than 1 m tall, stems thick, robust and branched. Tubers large, up to 10 cm long, ovoid to oval flattened. Flowers dark blue. Chromosome number: 2n=2x=24 (Ochoa, 1958). May, 1955;

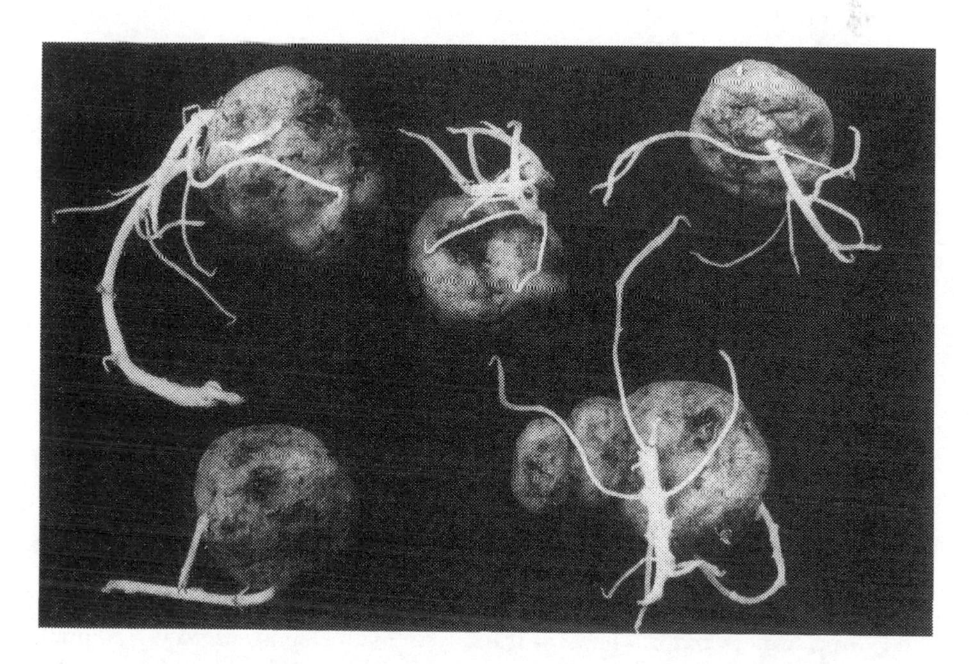

Figure 96. – Tubers of *Solanum candolleanum*, collected at Tacacoma, Province Larecaja, Department La Paz (*Ochoa and Salas 11814*), ×⅓.

Ochoa CO-2274. Pata Pata, 3600 m alt., on the lower slopes of Illampu, above Yamuco. Plants robust, stems subpigmented to light green, broadly winged, flowers light blue. Common name *Gentil Achochil Chocke*. Chromosome number: 2n=2x=24. December 20, 1977; *Ochoa and Salas 11805* (CIP, OCH, P, US, topotypes). Cemetery of Tacacoma, 3600 m alt. Plants robust, very pilose, stems strongly pigmented. Chromosome number: 2n=2x=24. December, 1977; *Ochoa and Salas 11814* (CIP, LE, NY, OCH, P, US). District of Sorata, near the mines of Martinsito, 4000 m alt., on the slopes of Illampu. Plants very vigorous, stems up to 2 cm in diameter at the base, very pilose. Leaves also very pubescent. Chromosome number: 2n=2x=24. March 9, 1978; *Ochoa and Salas 11913* (CIP, F, GH, LE, NY, OCH, P, SI, US). Sorata-Tacacoma, 23 km, 3980 m alt. 1980; *Soest, Okada, and Alarcón S.O.A. 65* (CIP). Umajalanta, 3750 m alt., some 4 km E of Ancoma, between Ancoma and the Yunga zone of Tushuaya. Among *Stipa ichu*, *Cajophora*, and a non-tuber-bearing *Solanum*, on large stones covered by moss. Common name *Ckeña chulpa*, its berry is known locally as *mamunco*. February 17, 1984; *Ochoa and Salas 15463* (CIP, GH, NY, OCH, P, US). Pongopata, 3900 m alt., along the road from Sorata to Ancoma, some 2 km before the junction of Llipichi, in humid soil associated with *Stipa ichu*. Plants tender without flowers, very scarce. February 17, 1984; *Ochoa and Salas 15469* (CIP, OCH). Vicinity of Pongopata, 3900 m alt., between Sorata and Ancoma, near the junction of Llipichi. February 17, 1984; *Ochoa and Salas 15470* (CIP, NY, OCH, US).

Province Muñecas: Moco Moco, 3500 m alt. Chromosome number: 2n=2x=24. February 28, 1978; *Ochoa and Salas 11896* (CIP, NY, OCH, P, US).

20a. **Solanum candolleanum** Berthault f. **sihuanpampinum** Ochoa.

This plant differs from the typical species in many features. These include less vigorous stems and much less pilose peduncles, pedicels, and calyces. The stem of this form is less pigmented and less pilose with shorter hairs and narrower, straight wings. Moreover, the upper part of the pedicel is longer (9-11 mm or more long) and gradually increases in diameter towards the calyx. The calyx is reflexed and strongly recurved with 2-3 mm long, narrowly pointed acumens, while the anthers are smaller, 5-5.5 mm long, and the style is somewhat more exerted. The inflorescence is more or less cymose paniculate. Its dark purple flowers are notably different from the blue flowers of the typical species. The leaves are provided with a larger number of interjected leaflets and 5-6 pairs of leaflets, the latter being narrowly elliptic-lanceolate, strongly dark purple, and long petiolulate. The petiolules of the second and third upper pairs of lateral leaflets are provided always with one interjected leaflet. The pubescence of the leaves differs from the typical species in having dense, short hairs. Fruits to 20 mm in diameter, varying in shape from globose to ovoid and in color from light green and speckled with white dots or dark purple stripes, when young, to dark green tinged with green-lilac at base, when mature.

Specimens Examined

Department La Paz

Province Omasuyos: Sihuanpampa, 3350 m alt., about 4 km above Catalone near the right margin of the Ancoaqui River. Chromosome number: 2n=2x=24. December, 1977; *Ochoa and Salas 11835* (CIP, OCH, US).

Crossability

Since it is important to ascertain the numerical balance of the endosperm in order to attain a better understanding of the results of crossability experiments among the species of this tuber-bearing section of *Solanum*, an effort was made at CIP to determine the EBN and ploidy level for each of the Bolivian species studied, following, as far as possible, the methods outlined by Johnston and Hanneman (1980, 1982), Johnston et al. (1980), Hanneman (1983), and Ehlenfeldt and Hanneman (1984). Our results for *S. candolleanum*, in this case, indicate an EBN value of 2. This explains why we were so easily able to cross certain diploid introductions of *S. candolleanum* (11814, 11896, 11897, 11913, 14958, 15012) with Bolivian diploid species, such as *S. alandiae* (12018), *S. boliviense* (11935), *S. brevicaule* (11934), *S. infundibuliforme* (11977), *S. litusinum* (12027), *S. oplocense* (11927), *S. sparsipilum* (12028, 12030), and *S. vidaurrei* (12003), all of which, according to our independent investigations at CIP, have an EBN value of 2, a value previously reported by Johnston and Hanneman (1980) and Hanneman (1983).

In the above crosses, the average number of viable seeds obtained per fruit varied from 50 to 80. However, an unexpected and unexplained rise in the number of viable seeds that were obtained was realized in crosses involving *S. candolleanum* and such allopatric Peruvian diploid species as *S. ambosinum* (13003 and 12079), *S. bukasovii* (11873, 13166, 13862), *S. coelestispetalum* (13609, 14326) and *S. raphanifolium* (13775, 14296), all of which have an EBN value of 2. In these cases, the average number of viable seeds produced ranged from 250 to 350.

In crosses involving diploid *S. candolleanum* and tetraploid Bolivian introductions of *S. acaule*, both of which have an EBN value of 2, a large number of fruits were obtained. These averaged up to 90 viable seeds per berry. On the other hand, crosses of *S. candolleanum* (11835 11897) and tetraploid *S. sucrense* (11926), the latter with an EBN value of 4, yielded, on the average, no more than 50 viable seeds per fruit.

Two examples of crosses not producing viable seed, involving species with similar ploidy level and EBN value to *S. candolleanum*, were *S. candolleanum* (11896) × *S. microdontum* (12025) and *S. megistacrolobum* (11914) × *S. candolleanum* (11805). On the other hand, crosses that failed for reasons of differences in ploidy level and/or EBN value included the following: *S. candolleanum* (11805) × *S. circaeifolium* (11806; 2x, EBN=1); *S. candolleanum* (11896) × *S. circaeifolium* var. *capsicibaccatum* (WIS 1730; 2x, EBN=1); *S. candolleanum* (15015) × *S. lignicaule* (14416; 2x, EBN=1); and *S. candolleanum* (11835) × *S. oplocense* (11972; 6x, EBN=4).

The following descriptions of F_1 hybrid populations of *S. candolleanum* were drawn from living plants.

Solanum candolleanum × *S. litusinum*.– Although generally easy to obtain, crosses involving different introductions of *S. candolleanum* were found to give very different results. Plants derived from crosses of *S. candolleanum* (11805) with *S. litusinum* (12027) yielded individuals that closely resembled *S. candolleanum* in leaf shape and dissection, as well as in having more robust stems. The white-tipped flowers of these hybrids, however, varied in color from dark purple to very light violet. The pedicels of these plants were articulated at about the midpoint of the stalk. On the other hand, hybrids derived from crosses of *S. candolleanum* (11913) with *S. litusinum* (12027), were more similar in leaf shape and dissection to *S. litusinum*. These produced rotate to rotate-pentagonal flowers that varied in color from dark purple to light blue. In individuals of this cross, the articulation of the pedicel varied in position from below center to the upper one-third of the stalk.

Solanum candolleanum (11835) × *S. sucrense* (11926).– Plants derived from crosses between *S. candolleanum* and tetraploid *S. sucrense* had vigorous stems, large flowers that ranged from 2.8 to 4 cm in diameter, and leaves that were more similar in shape and dissection to the first species than the second. Corolla color varied from light blue, or light indigo to dark purple, while the pedicels were articulated mostly at the midpoints or occasionally in the upper one-third of their 40 mm-long stalks. Because chromosome counts were not made, the ploidy level of these plants remains unknown. However, unlike most triploids – the chromosomal condition which would be expected here – some individuals of this cross developed berries with well-formed seeds.

Solanum candolleanum × *S. sparsipilum*.– Plants derived from different introductions of *S. candolleanum* and *S. sparsipilum* were of dissimilar appearance. Individuals derived from crosses between *S. candolleanum* (OCH 11814) and *S. sparsipilum* (OCH 12028), for example, resembled the first species in the form and dissection of their leaves, which have pointed, elliptic-lanceolate leaflets. As in the case of the two parental forms that were used in this cross, all had a long-peduncled, cymose inflorescence. The pentagonal flowers of these hybrids were up to 4 cm in diameter. Plants derived from crosses of *S. candol-*

leanum (OCH 11897) and *S. sparsipilum* (OCH 12028) were also similar to *S. candolleanum* in leaf shape and dissection. These hybrids produced very light purple flowers similar to those of *S. candolleanum.* Moreover, crosses of *S. candolleanum* (OCH 11897) and *S. sucrense* (OCH 11926) mostly resembled *S. candolleanum.* However, occasional segregates of this hybrid combination developed leaves similar in dissection to *S. sparsipilum.* Their flower color varied from very light violet to dark purple, and their pedicel articulations occurred in the upper one-third of the stalk. And lastly, it should be pointed out here that hybrids formed in crosses between *S. candolleanum* (OCH 11835) and *S. sparsipilum* (OCH 12028) were extremely variable in appearance. Although all had pentagonal corollas, their flowers varied from 2.8 to 4 cm in diameter and from light lilac to dark purple in color. Similarly, these hybrids were variable in leaf dissection. Some had numerous interjected leaflets and as many as 8 pairs of lateral leaflets, whereas others had few interjected leaflets and as few as 4 pairs of lateral leaflets. Leaflet shape ranged from narrowly elliptic-lanceolate and pointed to broadly elliptic-lanceolate and obtuse.

Solanum candolleanum (OCH 11896) × *S. boliviense* (OCH 11935).– Hybrids of this parentage tend to be very similar to *S. candolleanum* with respect to plant size, habit, vigor, and leaf shape and dissection. The intensely dark purple coloration of the corolla, however, is one feature of these plants that appears to be inherited from their second parent. All had a pentagonal corolla ranging up to 3.5 cm in diameter, and their pedicels were articulated in the upper one-third of the stalk.

Solanum candolleanum (OCH 15012) × *S. brevicaule* (OCH 11934).– Plants of this cross had only one feature in common with *S. brevicaule*, and that was stem size, which varied from 20 to 30 cm in height. In all other respects, they tended to be unique, resembling neither parent. Thus, all had poorly dissected, sparsely short-pubescent leaves, these without interjected leaflets. Also, all had 2 or at most 3 pairs of pointed, elliptic-lanceolate lateral leaflets with asymmetrically rounded bases, borne on short petiolules. All hybrid plants had glabrous stems.

21. **Solanum doddsii** Correll, Wrightia 2:186. 1961. FIG. 97.

Plants delicate. Stems 50 to 60 cm tall and up to 7 mm in diameter at base, usually branched, ascending or decumbent by flowering time, lightly pigmented below center, glabrescent with barely distinguishable hairs, and provided with light green, narrow wings. Tubers round to compressed-oval, 2-3 cm long, white, with smooth skin and white flesh. Leaves with 4–5 leaflets and 5-12 interjected leaflets of various sizes. Leaflets elliptic-lanceolate, the apex obtuse and the base obliquely rounded, borne on petiolules to 10 mm long, glabrescent with short, simple, and tetralobular-glandular hairs on both sides and finely pilose and denticulate along margins. Pseudostipular leaves broadly subfalcate, small, 7-10 mm long by 3-5 mm wide. Inflorescence cymose or cymose-paniculate, 6-8-flowered, light green, borne on peduncles 4-6 cm long, puberulent as in the case of pedicels and calyx. Pedicels slender, articulated above center or in the upper one-third of the stalk, the upper part 7-8 mm long and the lower 12-20 mm long. Corolla pentagonal to rotate-pentagonal, 2.8-3.4 cm in diameter, light lilac with yellowish-gray nectar guides, the acumens conspicuous and narrowly triangular lanceolate, 4 mm long by 2.5-3 mm wide at base. Staminal cone cylindrical conical; anthers narrowly lanceolate, cordate at base, 6-6.5 mm long, borne on 3 mm-long filaments. Style to 14 mm long, exerted 3-4 mm, densely papillose along the lower two-thirds of its length; stigma oval, laterally compressed. Fruit globose to ovoid, 1.5 cm long, light green.

Chromosome number: 2n=2x=24.

Type: BOLIVIA. Department Cochabamba, Province Campero, on rocky woody slope at km 192 from Sucre, 44 [km] from Aiquile. February 15, 1960; *D. S. Correll, K. S. Dodds, H. Brücher, and J. G. Paxman B616* (holotype, LL).

Although the leaf margins of this species are very slightly serrate, or occasionally denticulate and pilose, these characteristics are not always conspicuous and easily determined. This is especially true with regard to the lower leaves of the plant. However, the middle and upper leaf margins are ciliate and more strongly serrate.

Affinities

This species is related to *S. alandiae*. Both have large anthers and light violet to lilac flowers. However, the corolla of *S. doddsii* is rotate (not stellate as indicated in the original description), and the anthers of this plant are larger than those of *S. alandiae*. Moreover, its leaves are more dissected than those of the former species and its leaflets are more finely serrate and narrowly lanceolate.

Habitat and Distribution

This species is known only from the following two localities: near Zapatera (2250 m alt.) in the Department of Chuquisaca, and near Aiquile (2600-2800 m alt.) in the Department of Cochabamba. In both localities, this species grows on rocky shrubby slopes and along the margins of moist, humid forest land in Prepuna formations between Sucre and Aiquile.

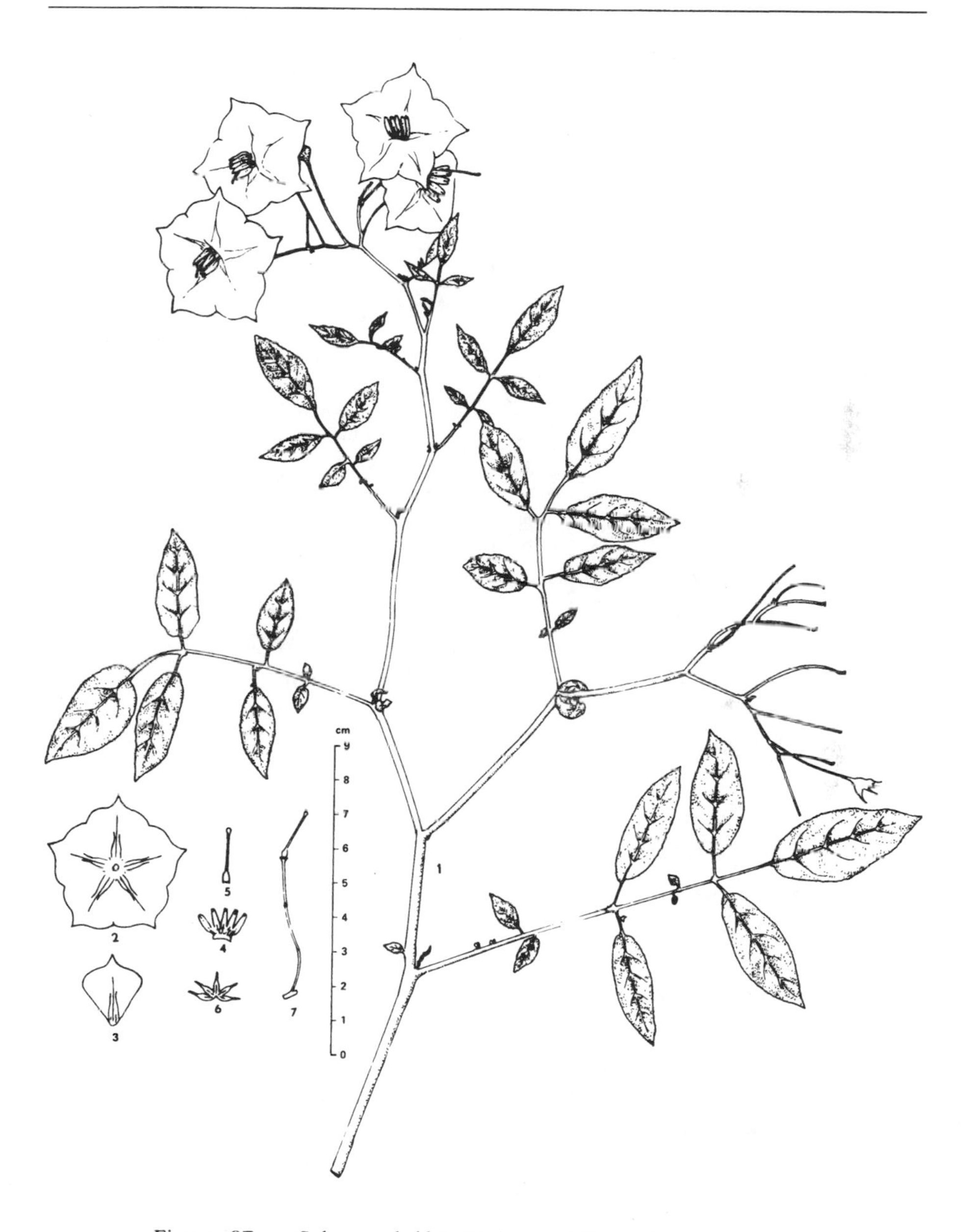

Figure 97. – *Solanum doddsii* (*Ochoa 12004*) 1. Upper part of flowering plant. 2. Corolla. 3. Petal. 4. Stamens. 5. Pistil. 6. Calyx. 7. Pedicel and pistil. All ×1.

Specimens Examined

Department Cochabamba

Province Carrasco: Km 192 on road from Sucre to Epizana, 40 km from Aiquile, 2600 m alt., on scree above road. March 3, 1971; *Hjerting, Cribb, and Huamán 4534* (CIP).

Province Mizque: On rocky wooded slope at km 192 from Sucre, 44 km from Aiquile, leaves essentially glabrous, flowers light lilac. February 15, 1960; *Correll, Dodds, Brücher, and Paxman B616* (holotype, LL).

Department Chuquisaca

Province Oropeza: Above Zapatera, 2250 m alt., along road from Sucre to Aiquile, among large rock piles in pockets of moist soil. Chromosome number: 2n=2x=24. March 24, 1978; *Ochoa 12004* (CIP, OCH).

Crossability

Natural hybrids.– One collection (*Ochoa 15563*) made by the author may be a natural hybrid of *S. doddsii* and *tarijense*. This material was collected at Campanera (1600 m alt.) near km 285 in a xerophytic region between Cochabamba and Santa Cruz in the Province of Caballero, Department of Santa Cruz (CIP, OCH).

The rather striking intermediate characteristics of this plant are described as follows:

Leaves dark green, similar to *S. alandiae* but less pilose and somewhat shiny (especially above), leaves smooth to the touch, poorly divided, with 1-2(-3) pairs of leaflets. Terminal leaflets much larger than the laterals. Stems vigorous, more than 1 m tall, very branched, with abundant inflorescences and flowers. Pedicels 25-30 mm long, articulated about the center, the upper part up to 12 mm long. Calyx 7-8, reflexed, light green with pointed, 2.5 mm long, shortly pilose acumens. Corolla large, up to 4 cm in diameter, stellate, white with a lightly tinged, pale blue spot on the external side, especially on newly opened flowers. Anthers up to 7 mm long.

These plants resemble *S. tarijense* in calyx and corolla shape, as well as in shape and color of their anthers. They resemble *S. doddsii* in having poorly dissected leaves and serrulate leaflet margins.

This particular composite-combination of characters fits very well with what might be expected of a hypothetical cross between these two species.

Lastly, it should be mentioned that the above putative hybrids have oblong to globose, light green fruits. These fruits have a whitish cast from the apex extending about one-third the length of the fruit and are covered almost completely with small white dots (verrucose).

Chromosome number: 2n=2x=24.

22. Solanum gandarillasii Cardenas, Bol. Soc. Peruana Bot. 5(1):16, Plate I(E). 1956. FIGS. 98, 99; PLATE XIV.

Plants low, succulent, easily broken, glabrous or glabrescent, light green. Stems 10-15(-35) cm tall and 5-6 mm in diameter at base, simple or branched, narrowly winged. Roots shallow. Stolons short and delicate, 20-30 cm long by 1.5-2.5 mm in diameter, milky-white; tubers small, 10-15 mm long, globose to ovoid, yellowish-white. Leaves imparipinnate, rarely simple, (7-)12-19(-30) cm long by (4.5-)7.5-15(-23.5) cm wide, shiny above and dull-glabrescent below, with shortly pilose margins, poorly dissected, with 1-3 pairs of lateral leaflets. Interjected leaflets absent. Leaflets broadly elliptic-lanceolate or ovate to ovate-elliptic, the apex broadly obtuse or shortly acuminate, and the base rounded or subcordate, narrowly decurrent upon rachis and borne on 2-4 mm-long petiolules. Terminal leaflets rounded or slightly cuneate at base, slightly larger to much larger than the laterals, 5-9(-13) cm long by 3-5(-9) cm wide. Pseudostipular leaves very narrowly subfalcate, 7-13 mm long by 4-7 mm wide. Inflorescence cymose, 6-8(-12)-flowered. Peduncles slender, to 9 cm long. Pedicels articulated at or above center, the upper part 10-12 mm long and expanding in width apically, the lower 12-16(-30) mm long. Corolla 1.8-2.5 cm in diameter, rotate, white with yellowish-green nectar guides, the acumens short and not well developed. Calyx 5-7(-8) mm long, the obtuse lobes broadly ligulate and strongly reflexed or revolute. Anthers light yellow, 5.5-6 mm long, cordate at base, borne on white-hyalinous, 0.7-1 mm-long filaments. Style 10-11 mm long, densely papillose along the lower two-thirds of its length, exerted 2.5 mm; stigma ovoid, cleft. Fruit globose to ovoid, 1.5 cm long, light green and sparsely flecked with white dots.

Chromosome number: $2n=2x=24$.

Type: BOLIVIA. Department Santa Cruz, Province Florida, near Mataral, 2000 m alt. March, 1955; *M. Cardenas 5068* (lectotype, CA [=HNF-00034] here designated; isotype, LL).

Affinities

This species is not related to any other species. It is distinguished primarily by its small, fragile stems, light green foliage, and shallow roots and stolons, and poor tuberization. The broadly ligulate and reflexed, or revolute calyx lobes of this species are also very unusual and are not known from any other tuber-bearing *Solanum* species.

Habitat and Distribution

This species grows in arid places. It is found usually in the shade of various species of cacti, *Pitcairnia* (Bromeliaceae), *Jatropha* (Euphorbiaceae), and vari-

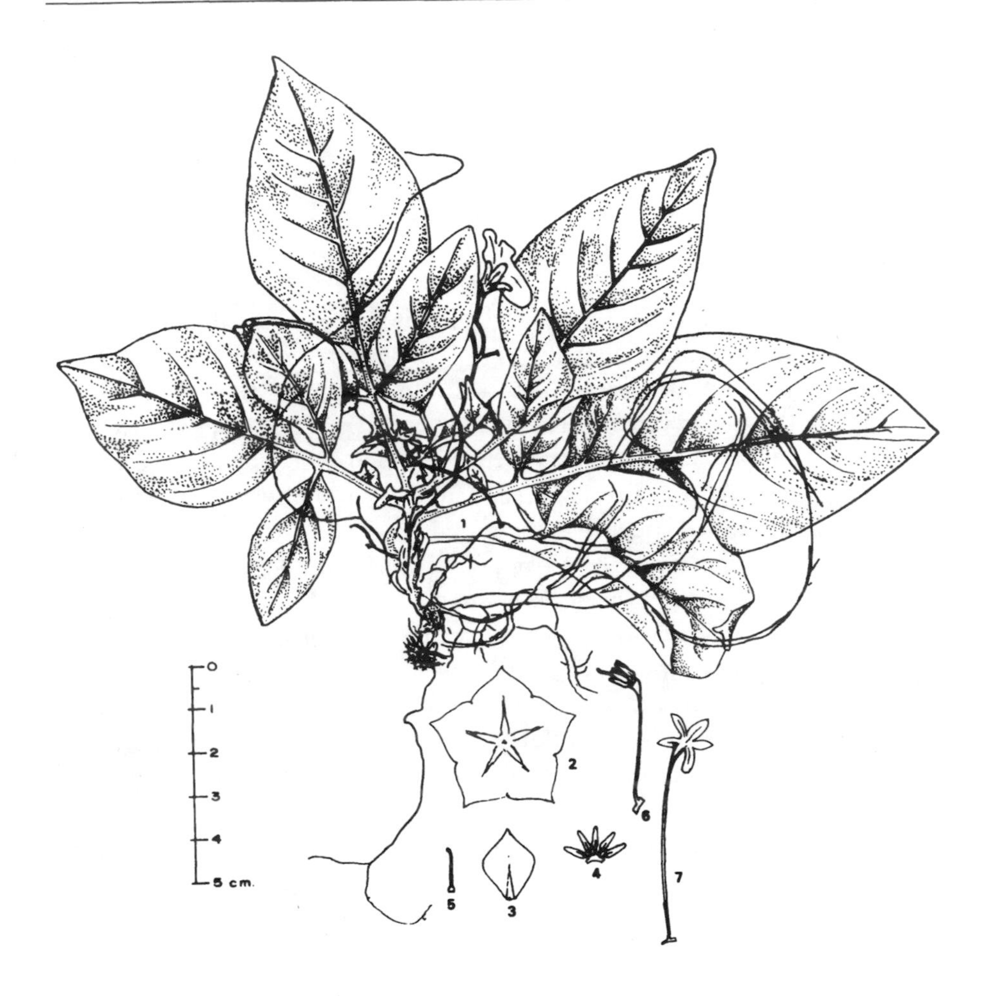

Figure 98. – *Solanum gandarillasii* (*Ochoa 12007*). 1. Flowering plant. 2. Corolla. 3. Petal. 4. Stamens, dorsal view. 5. Pistil. 6. Pedicel and stamens. 7. Pedicel and calyx. All ×1.

ous spiny, non-tuber-bearing *Solanum* species in poor, hard-dry, gravelly clay soils. Its distribution extends from the southern part of the Department of Cochabamba to the northwestern corner of the Department of Santa Cruz, Bolivia. It is found mostly between 1450 and 3000 m altitude.

Specimens Examined

Department Cochabamba

Province Campero: About 40 km S of Aiquile, 2500 m alt., along roadside, under bushes. March 11, 1959; *Ross, Cardenas, Alandia, and Rimpau 259* (MAX). About 20 km S of Aiquile, 2200 m alt., along roadside. March 11, 1959; *Ross, Cardenas, Alandia, and Rimpau 268* (MAX). Forested slope about 25 km from Aiquile on road to Sucre, 3000 m alt. Plants small, flowers white, rotate-pentagonal, petals robust. February 16, 1960; *Correll, Dodds, Brücher, and Paxman B617* (LL). In cactus-acacia-jatropha forest, near Quiroga. Plant low, robust; flowers white, rotate. February 16, 1960; *Correll, Dodds, Brücher, and Paxman B620* (K, LL, S, UC, US), *B620-1* and *B620-2* (LL).

Province Mizque: Loma de Condor Ckecha, 2050 m alt., several km NE of Chacco, on dry stony slopes. Plants small, flowers white, leaves light green, fruits ovoid, green-speckled with sparse white dots. Common name *Alcco Papa* (Papa de Perro). Chromosome number: 2n=2x=24. March 24, 1978; *Ochoa 12007* (OCH). Huara Huara, 2200 m alt., near Ichu Ttirana, on stony slope, at a short distance from other tuber-bearing *Solanum* species as *S.*

Figure 99. – Floral dissection of *Solanum gandarillasii* (*Ochoa 12010*), ×1.

oplocense and *S. berthaultii*. Plants small, flowers white, provided with rotate corollas and reflexed calyces, the latter with broadly ligulate lobes. Chromosome number: 2n=2x=24. March 25, 1978; *Ochoa 12010* (OCH).

Department Chuquisaca

Province Oropeza: Between Sucre and Yamparaez (Tambo Kasa), 2780 m alt., in sandy soil with *Cactaceae*. March 5, 1959; *Alandia and Rimpau 331* (MAX). At km 28 from the Sucre-Aiquile road, 1950 m alt., about 15 km ENE from Sucre (air distance), 19°1′ S and 65°7′ W; in crevices filled with soil in a rocky steep slope, with *S. zudanense*, *S. gandarillasii*, nontuberous *Solanum* sp., *Cleome aculeata* subsp. *cordobensis*, *Polygonum*, and *Cerastium*. April 11, 1963; *Ugent and Cardenas 4891, 4892, 4893* (WIS).

Province Zudanez: About 10 km from Zudanez to Tomina, 2200 m alt. March 8, 1959; *Ross and Cardenas 209* (MAX).

Department Santa Cruz

Province Florida: Near Mataral, 2000 m alt., in clay soil, under shrubs. Plant 10–30 cm tall, flowers rotate, white. March, 1955; *Cárdenas 5068*, type collection of *S. gandarillasii* (lectotype, CA [=HNF-00034], here designated; isotype, LL).

Province Valle Grande: Between the mountains La Tipa and El Estanque, 1700 m alt., some 7 km (air distance) SW of Mataral, under the shade and protection of spiny shrubs and large colonies of cactus, *Opuntia*, *Bromeliaceas*, and *Pitcairnia*, near a small colony of *S. neocardenasii*. February 28, 1984; *Ochoa and Salas 15557* (CIP, NY, OCH, US). Near Mataral, 1450 m alt. Among abundant colonies of diverse genera and species of cacti and under the shade of spiny shrubs. Xerophytic locality. February 28, 1984; *Ochoa and Salas 15558* (CIP, NY, OCH, US, topotypes). El Alto, 1900 m alt., near the mountain pass of the Arrayanal, on road from Mataral to Valle Grande, near plants of *S. tarijense*. April 1, 1984; *Ochoa and Salas 15588* (CIP, NY, OCH, US).

23. Solanum leptophyes Bitter, Repert. Spec. Nov. 12:448. 1913.
FIGS. 100–103.

S. gourlayi Hawkes, Bull. Imp. Bur. Pl. Breed. & Genet., Cambridge (June):43, 120, Fig. 27. 1944. TYPE: *Balls, Gourlay, and Hawkes 5979*, Argentina. San Gregorio, 3650 m above Tilcara (CPC, K).

S. pachytrichum Hawkes, Bull. Imp. Bur. Pl. Breed. & Genet., Cambridge (June):45, 121, Fig. 31. 1944. TYPE: *Balls, Gourlay, and Hawkes 6177*, Bolivia. Near Sucre, Dept. Chuquisaca, Prov. Oropeza, March 8, 1939 (CPC, K, UC).

S. leptophyes f. *gourlayi* (Hawkes) Correll, Wrightia 2:188. 1961. Type: *Balls, Gourlay, and Hawkes 5979,* Argentina. San Gregorio, 3650 m above Tilcara (CPC, K).

Plants small and graceful, pilose. Stem slender and somewhat flexible, 10-15(-30) cm tall by 3-5 mm in diameter, simple or branched from base, erect or decumbent, divided into short internodes at base, very narrowly and inconspicuously winged, lightly pigmented along the lower one-third of its length. Tubers small, 1.5-2 cm long, round to ovoid or oval-compressed, with white skin and flesh. Leaves imparipinnate, long and narrow, variously dissected with 4-5(-7-8) pairs of lateral leaflets and (0-)5-11(-16) interjected leaflets. Leaflets more or less pilose on both sides, usually small, (2-)3-4(-6) cm long by (0.8-)1.5-2(2-4) cm wide, elliptic-lanceolate or narrowly elliptic-lanceolate to very narrowly elliptic-lanceolate, the upper pairs nearly the same size as the terminal leaflets and occasionally decurrent upon rachis, and the

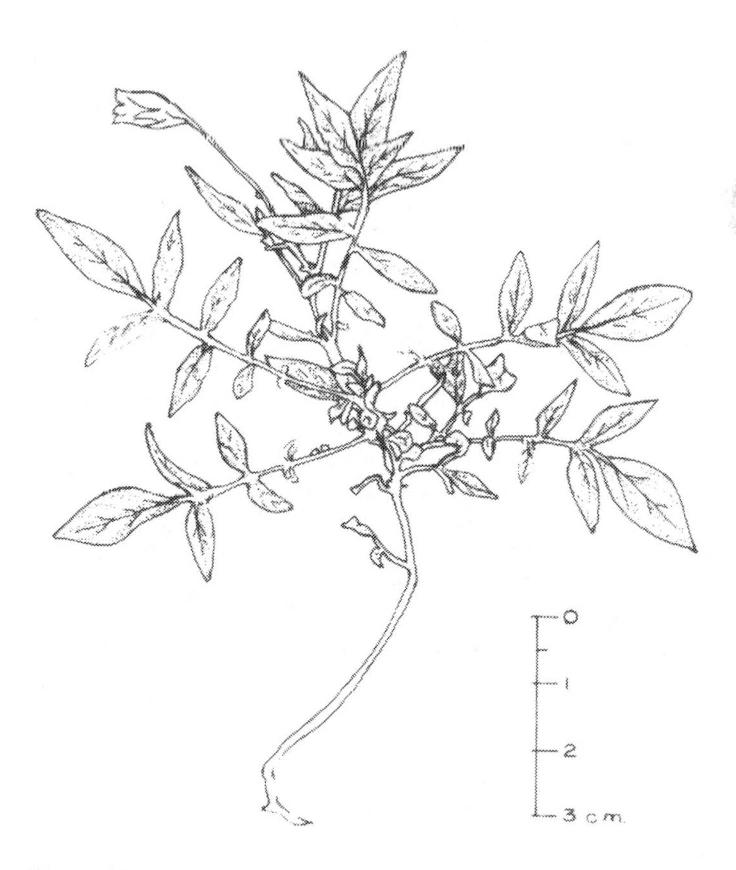

Figure 100. – *Solanum leptophyes* (*Petersen, Hjerting, and Reche 1033*), ×1.

lower pairs petiolulate or sessile and decreasing in size towards the base, the apex obtuse or pointed, and the base cuneate to rounded or obliquely rounded. Pseudostipular leaves subfalcate, short and broad, 5-7(-9) mm long by 2.5-3 (-3.8) mm wide. Inflorescence cymose or cymose-paniculate, 3-4-flowered, borne on slender peduncles 2-4(-7) cm long by 1.8 mm in diameter at base, and pigmented light brown as in the pedicels and calyces. Pedicels articulated about the midpoint, the upper part 5-7 mm long and the lower 8-15 mm long. Calyx 6-9 mm long, the narrowly ovate-lanceolate lobes with shortly acuminate apices, and with abruptly narrowed and pointed acumens. Corolla rotate to rotate-pentagonal, usually small, 2-2.5(-3) cm in diameter, uniformly purple or violet with yellowish-green nectar guides, and occasionally with short, whitish acumens. Staminal cone shortly conical-truncate; anthers 4.5-5.5 mm long, borne on glabrous 0.5-1.5 mm-long filaments. Style 8-10 mm long, densely papillose along the lower two-thirds of its length; stigma small, capitate. Fruit globose to ovoid, 1.5 cm long, green with 1-2 vertical purple-colored stripes or occasionally glaucous and verrucose.

Chromosome number: 2n=2x=24.

Type: BOLIVIA. Department La Paz [possibly in the Province Murillo], La Paz, 3800 m alt., Andean region. April, 1913; O. *Buchtien 3982* (lectotype, here designated, US; isotype, GOET).

Figure 101. – Floral dissection of *Solanum leptophyes* (*Ochoa and Salas 15455*), ×1.

Affinities

This species is related to the Peruvian *S. bukasovii*, which has two ecotypic forms that have been previously given specific status by taxonomists. When recognized as distinct species, they are called '*S. multidissectum*' and '*S. canasense*.' It also resembles high altitude forms of *S. brevicaule*, these similar in habit and leaflet shape.

Although *S. leptophyes* shares some characteristics in common with the Bolivian *S. vidaurrei*, they are nevertheless sufficiently distinct from one another as to maintain them here as separate entities. However, the Argentine species *S. spegazzini*, *S. famatinae*, and *S. sleumeri* are exceedingly similar to *S. leptophyes*. With further study, these may need to be placed in synonomy with *S. leptophyes*.

Habitat and Distribution

This species occurs in arid or semiarid regions of the altiplano. In Bolivia, it occurs in poor, rocky soils along with acacia, cacti, and many other spiny plants. This species, however, has a wide distribution in the central Andes, extending from the Province of Antabamba, Department of Apurimac in southern Peru, to northwest Argentina. Its altitudinal limits in Bolivia extend from 2600 m near Sucre to over 4000 m in the vicinity of Oruro.

Figure 102. – Berries of *Solanum leptophyes* (*Ochoa 13586*), ×1.

Specimens Examined

Department Cochabamba

Province Arque: Aguas Calientes Station, 3500 m alt., about halfway between Oruro and Cochabamba. January 26, 1949; *Brooke 3009* (BM). Ventilla, 4000 m alt., among cultivated plants. February, 1955; *Brücher 562* (LL).

Department Chuquisaca

Province Oropeza: Near Sucre, 2625 m alt., in hard, dry soil, on steep stony slope. March 8, 1939; *Balls, Gourlay, and Hawkes 6177* (CPC, K, UC). Villa Maria, Sucre. February, 1948; *Cárdenas 4513* (HNF). Guerraloma, 2800 m alt., in rocky soil, plants of 30 cm, flowers blue. April, 1958; *Cardenas 6114* (HNF, US, consists of two plants mounted on one sheet, only one of which is *S. leptophyes*). Punilla, ca. 2600 m alt., 18°55′ S and 65°25′ W, near km 15 NW of Sucre (air distance) on the Sucre-Ravelo road. Margins of weedy wheat field, with *S. boliviense*, *S. pachytrichum*, nontuberous *Solanum* sp., *Setaria plumosa*, *Calceolaria*, and *Veronica*. April 12, 1963; *Ugent and Cardenas 4913, 4914* (WIS). Villa Maria, 2850 m alt., 19°1′ S, 65°10′ W, a small farm about 10 km NE of Sucre (reached by jeep but not serviced by road). Margins of weedy wheat field and surrounding thickets, with *S. boliviense*, *S. pachytrichum*, *Tagetes*, *Medicago*, *Rumex*, and *Brassica*. April 12, 1963; *Ugent and Cardenas 4932* (WIS). Cerro Ckuchutambo, 3075 m alt., near Guerraloma. On rocky soil, near cave beneath a large stone. Very scarce. Plants barely 18 cm tall, with *S. boliviense*. March 13, 1978; *Ochoa 11933-B* (OCH).

Province Sud Cinti: Some 50 km from Villa Abecia on the road to Las Carreras, among bushes of acacias and cactus. February 22, 1960; *Correll, Dodds, Brücher, and Paxman B645* (LL, UC, US).

Department La Paz

Province Manco Kapac: San Pedro de Tiquina, 3850 m alt., near Copacabana. Plants small, occasionally rosette-like at base. March 8, 1953; *Petersen and Hjerting 1054* (LL, OCH). One km before Tiquina, 3800 m alt., between Copacabana and Tiquina. Plant and flowers small, corolla rotate-pentagonal, purple. February 15, 1984; *Ochoa and Salas 15455* (CIP, NY, OCH, US).

Province Murillo: Challapampa, near La Paz, 3700 m alt. January 26, 1921; *Asplund 2044* (UPS). La Paz, Andean zone, 3800 m alt. April, 1913; *Buchtien 3982* (lectotype, here designated, US; isotype, GOET). La Paz, 3750 m alt.; 1919; *Buchtien 4696* (US). Near Calacoto, 3700 m alt., on dry slope under small shrubs. February, 1946; *Cardenas 3641* (CPC, LL).

Department Oruro

Province Abaroa: Challapata, 3800 m alt. March 31, 1921; *Asplund 3240* (UPS, US). Challapata, 3800 m alt. March 31, 1921; *Asplund 3303* (UPS).

Province Cercado: Aguas de Castilla, suburb of Oruro, dry slope, 4000 m alt. February 1, 1949; *Brooke 3015, 3020 p.p.* (BM). Hacienda Huancaroma, 3800 m alt., near Eucaliptus. February 19-24, 1934; *Hammarlund 126* (S). In mountains above Oruro, 3950 m alt.; 1913; *Wight 362* (NA).

Department Potosi

Province Frias: Along the road from Potosi to Oruro, 3900 m alt., between La Puesta and Chulchucani. February, 1953; *Petersen, Hjerting, and Reche 1033* (OCH). Km 56 from Potosi on road to Oruro, 3900 m alt. Puna vegetation but with shrubs and herbs. Growing among rocks and walls in cultivated fields. Flowers medium red-purple (deep tone). February 3, 1971; *Hawkes, Hjerting, Cribb, and Huamán 4339* (CIP). Km 4 from Casa Cayara on the road to Potosi, weed, occasional, growing in the corner of a cultivated field in the shadows of *Salix* sp. with *093*. April 7, 1974; *Astley 092* (CIP, from plants grown at Sturgeon Bay, Wisconsin, USA, under No. PI-473495).

Figure 103. – Habitat of *Solanum leptophyes*, in the vicinity of La Paz, 3700 m altitude.

Department Tarija
Province Mendez: On the Iscayachi-Tarija road, upper slope of Mt. Sama, 3650 m alt., among large stones, associated with native grasses. Very scarce. March 21, 1978; *Ochoa 11992* (OCH).

Crossability
Brücher reports that *S. leptophyes* is resistant to root nematode (Brücher, 1985). According to recent studies made by Javier Franco of the Department of Nematology and Entomology at CIP, *S. leptophyes* incorporates a high degree of resistance to the root nematode, *Globodera pallida*. Moreover, it is also highly resistant to drought. Farmers in the vicinities of Guerraloma (near Sucre) and Achacachi (near Pico de Illampu) report that this species is capable of growth even under very low rainfall conditions. The above qualities of *S. leptophyes* make it of potential value to breeders interested in improvement of the cultivated potato. In addition, this species also has resistance to potato leafroll virus (Geransenkova, 1974) and tolerance to *Rhizoctonia solani* (Vlasova, 1974).

Crosses of *S. leptophyes* with other wild Bolivian potato species gave different results depending upon the ploidy level and EBN value of the species involved. Some of the results, obtained at CIP using this wild diploid species, with an EBN value of 2, are given below.

Diploid crosses involving *S. leptophyes* (14919) and *S. boliviense* (11935) or *S. leptophyes* (13822) and *S. candolleanum* (11896) (all EBN=2) were easily made. In several cases, more than 200 fertile seeds per fruit were obtained.

In contrast, crosses between *S. leptophyes* (13720) and *S. alandiae* (12014), *S. leptophyes* (13621 and 13720) and *S. brevicaule* (11934) or *S. leptophyes* (13720 and 13621) and *S. sparsipilum* (13705 and 13774) were very difficult, this despite the fact that the above species are all diploid and have an EBN value of 2. The average number of seeds per fruit obtained in the above crosses was 60, 50, and 40, respectively.

Similarly, crosses between *S. leptophyes* (13720) and *S. neovavilovii* (14994) gave only 50 fertile seeds per fruit. However, an average of only 10 seeds per fruit was obtained in crosses between *S. tarijense* (11994) and *S. leptophyes* (13720), and only parthenocarpic berries were obtained in crosses between *S. megistacrolobum* (14272) and *S. leptophyes* (13720).

Descriptions of some of the more interesting and successful crosses made at CIP follow:

Solanum leptophyes (OCH-13720) × *S. neovavilovii* (OCH-14994).– The morphological characteristics of the hybrids of this parentage combine the features of both species. The F_1 plants developed branched, light green, erect or decumbent stems, 40-50 cm tall and some 4-5 mm in diameter at base. The leaves of these hybrids are similar to *S. leptophyes* in being highly dissected, that is, in having up to 5 pairs of leaflets and as many as 12 interjected leaflets. These hybrids are also similar to *S. leptophyes* in having a reflexed, light green calyx, and a light green, white-flecked, globose fruit, up to 1.5 cm in diameter. However, with respect to leaflet shape, these plants are more similar to *S. neovavilovii*. Some intermediate traits of these hybrids include their sky blue, white-tipped flowers that range to 3 cm in diameter, and their 3-5-flowered, cymose inflorescences, upheld by slender, light green peduncles. The pedicels of these plants are articulated, some 5-6 mm below the calyx base, or in the upper one-third of their stalks.

Solanum leptophyes × *S. phureja*.– Crosses between *S. leptophyes* (*OCH-13731*) and *S. phureja* (*OCH-15133*) recombine, in general, the characteristics of both parents. However, they differ from either in being more branched and compact, and in having much smaller, light green stems, ranging between 25-30 cm tall and 2-3 mm in diameter. All have a small, short-peduncled inflorescence that bears, at most, 2-4 rotate-pentagonal flowers, rarely more than 1.5 cm in diameter. The pedicels of these hybrids are articulated in the upper one-third of their stalks. In the characteristics of their foliage, these hybrids are similar to *S. phureja*. For example, all have small, broad, poorly dissected leaves, with 2-4 pairs of lateral leaflets and up to a few interjected leaflets. In the characteristics of their calyces, however, these hybrids are more strongly like *S. leptophyes*. Thus, the calyx lobes of these plants are narrowly lanceolate, with 1.5 mm-long attenuate or acuminate apices. Their fruits are unknown. Hybrids of *S. leptophyes* (OCH-13733) and *S. phureja*, called *Chaucha de Chacanaqui* (OCH-15072), differ from the above cross in developing more vigorous green or light green branches, leaves, and stems, the latter ranging to 70 cm in height and 4-5 mm in diameter. Moreover, these have 6-7 pairs of lateral leaflets and numerous interjected leaflets. All have cymose-paniculate, 7-12-flowered inflorescences, borne on slightly pigmented 5-7 mm-long peduncles. The small, purple, rotate-pentagonal flowers of these hybrids range to 2 cm in diameter. These are borne on pedicels articulated in the lower one-third of their stalks. The globose, light green fruits are lighter in coloration at the tip.

24. Solanum microdontum Bitter, Repert. Spec. Nov. 10:535-536. 1912.

FIGS. 104, 105.

S. bijugum Bitt., Repert. Sp. Nov. 10:533. 1912. TYPE: *Fiebrig 2253*, Argentina. Toldos, near Bermejo, 1800 m alt. (B, destroyed: F-photo, US-photo).

S. simplicifolium var. *trimerophyllum* Bitt., Repert. Sp. Nov. 12:446. 1913. TYPE: *Spegazzini*, Argentina. Salta, Pampa Grande (B, destroyed).

S. simplicifolium Bitt. subsp. *microdontum* (Bitt.) Hawkes, Scottish Pl. Breed.

Sta., Ann. Rept. 92. 1956. TYPE: *Fiebrig 2498*, Argentina. Toldos, near Bermejo (lectotype, W; isotypes, F-fragment, M, SI, W).

S. cevallos-tovari Card., Bol. Soc. Peruana Bot. 5:13-15, Plate IV-A, phototype. 1956. TYPE: *Cardenas 5087*, Bolivia. La Fortaleza, Dept. Santa Cruz, Prov. Valle Grande, March, 1955 (CA).

S. higueranum Card., Bol. Soc. Peruana Bot. 5:20-21, Plate I-F, Figs. 1-8. 1956. TYPE: *Cardenas 5067*, Bolivia. Between Pucara and Higuera, Dept. Santa Cruz, Prov. Valle Grande, Dept. March, 1955 (HNF, LL).

S. microdontum subsp. *microdontum* (Bitt.) Hawkes and Hjerting, Phyton, Graz 9:144-146. 1960. TYPE: *Fiebrig 2498*, Argentina. Toldos, near Bermejo (lectotype, W; isotypes, F, M, SI, W).

Plants erect or decumbent, usually delicate. Stems 25 to 100 cm or more tall and to 4-7(-10) mm in diameter at base, simple or branched, triangular, green or green-flecked with purple to uniformly dark purple, subglabrous to pubescent, with 0.5-1.5 mm-wide entire margined wings, the latter glabrous to little pilose and either straight or undulate. Stolons 1 or more m long; tubers round, small, 1-1.5 cm in diameter, the skin light pinkish-purple to dark brown, strongly lenticulate and verrucose. Leaves irregularly dissected or rarely simple, usually with 1-2(-3) pairs of leaflets, the latter with entire or occasionally very finely denticulate and pilose margins, pubescent with white, short, and fine multicellular hairs above, and less pubescent and frequently glandular below, borne on pubescent, narrowly winged rachises. Interjected leaflets absent or rarely present. Terminal leaflet larger than the laterals, (2-)5-11(-16) cm long by (1-)2-7(-11) cm wide, elliptic or ovate-elliptic to ovate-lanceolate and lanceolate, less frequently ovate-suborbicular to suborbicular, the apex obtuse or acuminate, and the rounded or cuneate base clearly decurrent upon the petiolule. Lateral leaflets abruptly diminishing in size toward leaf base, the upper pair not much larger than the terminal leaflet, elliptic-lanceolate to ovate-lanceolate, the apex obtuse to acuminate, and the base obliquely rounded and shortly petiolulate or decurrent upon rachis. Pseudostipular leaves semilunate, frequently auriculate, 0.5-0.7(-2) cm long by 0.3-0.4 (-1) cm wide. Inflorescence cymose, 4-7(-15)-flowered. Peduncles to 10 cm long, occasionally provided with narrow wings, sparsely to densely pubescent, and bearing only few glandular hairs, as in the case of the pedicels. Pedicels articulated slightly below, at or above center, 1.5-2 cm long. Calyx 5-9 mm long, pubescent, the lobes narrowly ovate-subquadrate to lanceolate-acuminate, and provided with pointed to subspatulate acumens of unequal size. Corolla rotate to pentagonal, 2-3.5 cm in diameter, white or occasionally light purple with bluish-lavender stripes and with grayish-white to greenish-yellow nectar guides, the acumens to 6 mm long. Anthers 5-8 mm long by 2 mm broad, borne on glabrous, 1-2.5 mm-long filaments. Style 10-12 mm long, glabrous

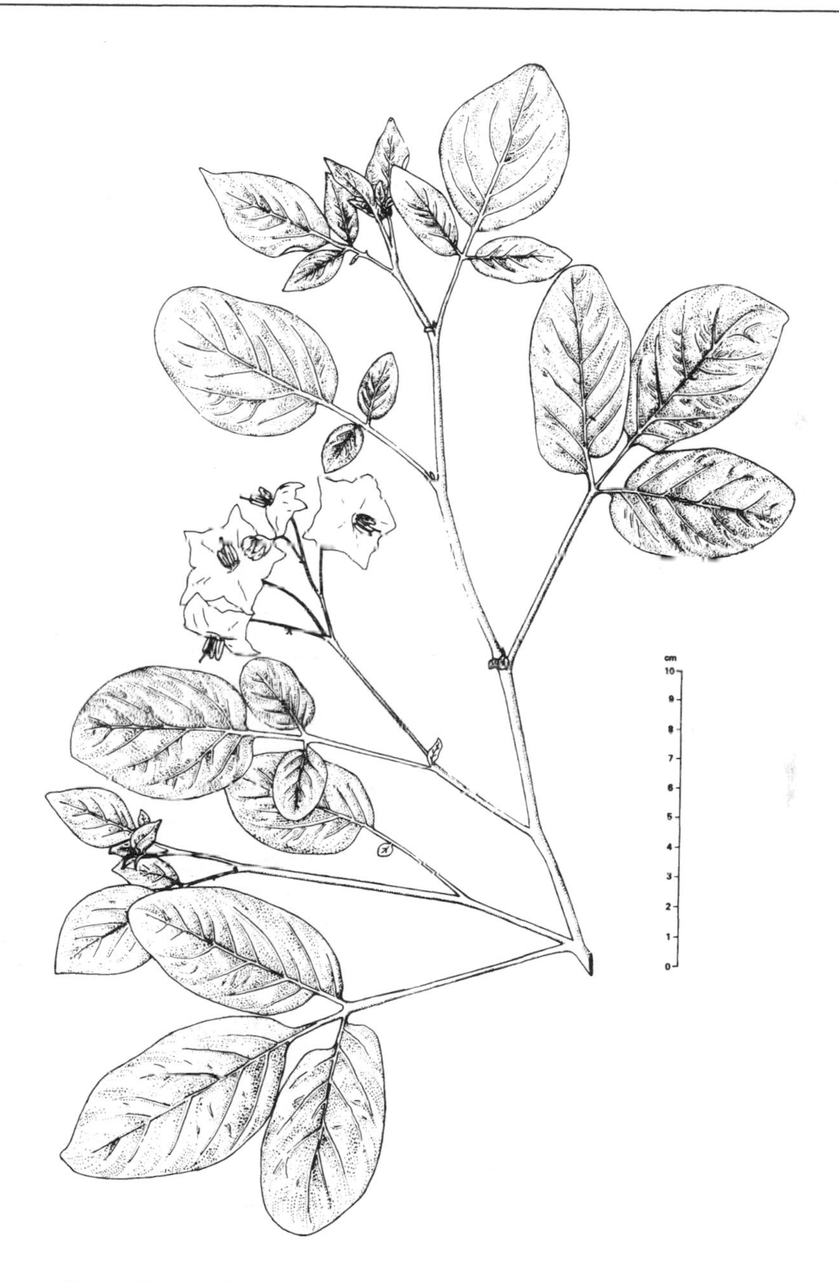

Figure 104. – *Solanum microdontum* (*Ochoa and Salas 15552*); upper part of flowering plant, ×1.

or sparsely papillose along the lower half or third of its length; stigma capitate, small. Fruit globose to ovoid, light green, to 2 cm in diameter.

Chromosome number: 2n=2x=24.

Type: ARGENTINA. Department Santa Victoria, Province Salta, Toldos, 1800 m alt., near Bermejo. December, 1903; *K. Fiebrig 2498* (lectotype, designated by Hawkes and Hjerting, W; isotypes, F-fragment, GOET, M, SI, U).

Bitter (1912) described *S. bijugum* and *S. microdontum* from material collected by Fiebrig at Toldos, Argentina (formerly Bolivian territory). Both are considered here as conspecific. The diagnosis of the first species was based on a unicate collected by Fiebrig in November, 1903. The latter specimen was destroyed during World War II, when the Dahlen Museum in Berlin was hit during a bombing raid. Although the second species was also destroyed in the same raid, isotypes of this plant exist in various herbaria (GOET, M, SI, U, and W). In 1960, Hawkes and Hjerting designated Fiebrig No. 2498 as the neotype of *S. microdontum*, and in so doing, gave validity to this taxon. A reproduction of a phototype of *S. bijugum* from the Gray Herbarium of Harvard is also presented by these same authors in a later publication (1969).

A specimen of *S. microdontum*, housed in the herbarium of the early twentieth century Argentine botanist, Miguel Lillo, who was aware of the enormous variation in this species, is labeled *S. tafiense*. The latter name, however, was never published.

As noted by many collectors, *S. microdontum* shows much variation in nature (Brücher and Ross, 1953; Brücher, 1957b, 1966; Hawkes and Hjerting, 1969). This fact is reflected in the many diverse collections of this species which have accumulated in worldwide herbaria.

The species, *S. simplicifolium*, which is here regarded as conspecific with *S. microdontum*, was described by Bitter (1912) from materials collected in March, 1873, by Lorentz and Hieronymus at Quebrada de San Lorenzo near Salta, northern Argentina. One year later, Bitter described the following three varieties of *S. simplicifolium*: vars. *mollifrons*, *metriophyllum*, and *trimerophyllum*.

In describing *S. simplicifolium*, as well as the variant of *S. microdontum* which he named *S. gigantophyllum*, Bitter (1912) expressed some reservations. Thus, he later realized that natural populations of *S. gigantophyllum* are highly polymorphic. This is perhaps why he reduced *S. gigantophyllum* in 1913 to a subspecies of *S. simplicifolium*.

Polymorphic variations in *S. simplicifolium* were investigated in northern Germany by Brücher and Ross (1953), who cultivated different forms of this plant under uniform environmental conditions. They concluded that the variations to be observed in *S. simplicifolium* are genetic, rather than environmentally induced. However, it is interesting to note that they found considerable segregation in leaf shape and dissection among the various introductions of this plant that were grown in their experimental field. Some of the segregate types observed by these investigators produced leaves with one or two pairs of leaflets, whereas other kinds were merely simple-leaved.

Affinities

Solanum microdontum is related to *S. venturii*, a plant endemic to northern Argentina. This species, however, is taller, more vigorous, larger-flowered, and more pubescent than *S. microdontum*. It also differs in its general habit, the shape of its leaves, and the decumbency of its stem.

Habitat and Distribution

This species is resistant to potato virus Y and *Heterodera rostochiensis* (Ross, 1958) and is also tolerant to *Rhizoctonia solani* (Vlasova, 1974). It is widely distributed in the central Andes, occurring from the Sierra de Famatina in the Province of La Rioja, Department of Chilecito, Argentina (29°7′ S and 67°38′ W) to the town of Tranca Jahuira in the Province of Inquisivi, Department of La Paz, Bolivia (17°2′ S and 68°47′ W). In various parts of its distributional range, *S. microdontum* occurs at elevations ranging between 1200 and 3200 m; however, it is most common between 1600 and 2700 m.

This species grows preferentially in the Selva Tucumano-Boliviana, the low-lying rain forest region of Bolivia and Argentina, a district that has been

Figure 105. – Floral dissection of *Solanum microdontum* (*Ochoa and Salas 15545*), ×1.

designated by Cabrera and Willink (1973) as 'Montano.' In the Argentine provinces of this region, according to Hawkes and Hjerting (1969), *S. microdontum* subsp. *microdontum* (=*S. microdontum* var. *microdontum*) grows at elevations between 1600 and 2000 m under shade of *Podocarpus*, *Cedrela*, and *Juglans*, while at elevations from 2600 to 2850 m, it occurs with *Alisos* (*Alnus jorullensis*), grasses and shrubs.

In the Province of Valle Grande, Department of Santa Cruz, var. *microdontum* occurs between 2000 and 2800 m. Here, the author has observed this plant growing in poor, dry soil, and rarely in association with a few herbs or shrubs. However, this same variety also grows in moist rain forests, as for example in the Paso de Siberia, between Cochabamba and Santa Cruz. At its highest altitudinal limits, between 2600 and 3200 m, var. *microdontum* grows in grasslands with shrubby composites, *Baccharis*, and plants of the heath family, or occasionally with wild potato species such as *S. boliviense* and others.

Hawkes and Hjerting (1969) also detail the Argentine habitat preferences of *S. microdontum* var. *metriophyllum* (Bitt.) Ochoa, a plant which they treat, however, as a subspecies of *S. microdontum*. According to these authors, subsp. *gigantophyllum* grows along the eastern slopes of the Andes in the Selva Tucumano-Boliviana. Here, it grows in midelevation zones (2500 m) with *Alisos* (*Alnus jorullensis*), and *Polylepis*. It is particularly abundant between Salta, Tucuman, and Jujuy, where it occurs, as in the case of Bolivia, at elevations between 3200 and 3350 m. This variety also grows beyond the limits of the rain forest in the western and southern reaches of Catamarca and La Rioja, where it occurs in open places, generally with a few other plant species.

Specimens Examined

Department Cochabamba

Province Carrasco: Incallajta, about 20 km from the Monte Punco junction along the Cochabamba-Santa Cruz road, 2900 m alt., in a wood 50 m below ruins of Incallajta. Under trees and glades by path, very wet. Plants to 1.5 m; white flowers, young fruits. February 27, 1971; *Hjerting, Cribb, and Huaman 4526* (CIP).

Department Santa Cruz

Province Valle Grande: On the way from Pucara to Higuera, 2000 m alt., among bushes in sandy soil. March, 1955; *Cárdenas 5067* (HNF, LL). La Fortaleza (Siberia) between Cochabamba and Santa Cruz, 2700 m alt. Among bushes in wet forest edges. March, 1955; *Cardenas 5087* (CA, holotype of *S. cevallos-tovari*). Fortaleza (Siberia) between Cochabamba and Santa Cruz, in very humid forest, 2800 m alt. February, 1960; *Cardenas 5517* (HNF, K, LL, topotypes of *S. cevallos-tovari*). Siberia, La Fortaleza, 2550 m alt., turning on south side of road, km 221 from Cochabamba. Ceja de Monte but scrub and secondary growth. Very frequent. Also seen across the road on descent to Aserradero La Fortaleza. 2n=24. February 7, 1971; *Hawkes, Hjerting, Cribb, and Huamán 4359* (CIP). Los Barriales, 2700 m alt., some 20 km from Valle Grande on the road to Pucara. On sandy-clay, stony talus slopes, some 10 m below the highway, among trees and bushes in forest, moist, foggy region. Chromosome number: 2n=2x=24. February 28, 1984; *Ochoa and Salas 15534* (CIP, NY, OCH, US). Near Cerro Merma, 2700 m alt., among shrubs, associated with nearby plants (100-200 m distant) of *S. candelarianum*, near stream banks. Plants with pilose, very light green foliage and dark purple stems; corolla white, pentagonal. Chromosome number: 2n=2x=24. February 28, 1984; *Ochoa and Salas 15540* (CIP, GH, NY, OCH, US). Cincho, 2800 m alt., on talus slope above roadside between Valle Grande and Pucara, in nearly barren, hard, dry clay soil. Chromosome number: 2n=2x=24. February 28, 1984; *Ochoa and Salas 15545* (CIP, OCH). Las Cañas, 2500 m alt., some 2 km from Pucara on the road to Valle Grande. Under forest trees and shrubs, in clay soil. February 28, 1984; *Ochoa and Salas 15547* (CIP, F, GH, NY, OCH, US). Rodeo, 2750 m alt., about 1 km from Agua de Oro and some 25 km from Valle Grande on road to Pucara. Arid Prepuna region, with many pasture grasses. Chromosome number: 2n=2x=24. February 28, 1984; *Ochoa and Salas 15552* (CIP, NY, OCH, US). Alto Grande, some 25-26 km from Valle Grande toward Pucara, 2400 m alt. Near cultivated fields of potatoes, ullucos, and other crops. Common name: 'papa de zorro' or 'apharu.' February 28, 1984; *Ochoa and Salas 15554* (CIP, OCH, US).

Indefinite: *Cárdenas 6117* (HNF) [because two other specimens of Cárdenas, Nos. 6116 and 6118, are labeled as 'collected along the Chapare-Cochabamba road in March of 1961,' it is possible that No. 6117 was collected at the same time and place].

24a. Solanum microdontum var. **metriophyllum** (Bitter) Ochoa comb. nov. Based on *Spegazzini* Hb. *14217*, Argentina, Tucuman, Arcas-Trancas, Jan 1897 (LP, labeled var. *metriophyllum* by Bitter)

FIGS. 106, 107.

S. gigantophyllum Bitt., Repert. Sp. Nov. 11:368-369. 1912. TYPE: *Lorentz and Hieronymus 802*, Argentina. Cuesta del Garabatal near Siambón, Sierra de Tucuman (GOET, not seen).

S. simplicifolium Bitt., Repert. Sp. Nov. 11:369-370. 1912. TYPE: *Lorentz and Hieronymus s.n.*, Argentina. Quebrada de San Lorenzo near Salta (B, destroyed; F-photo, GH-photo, US-photo).

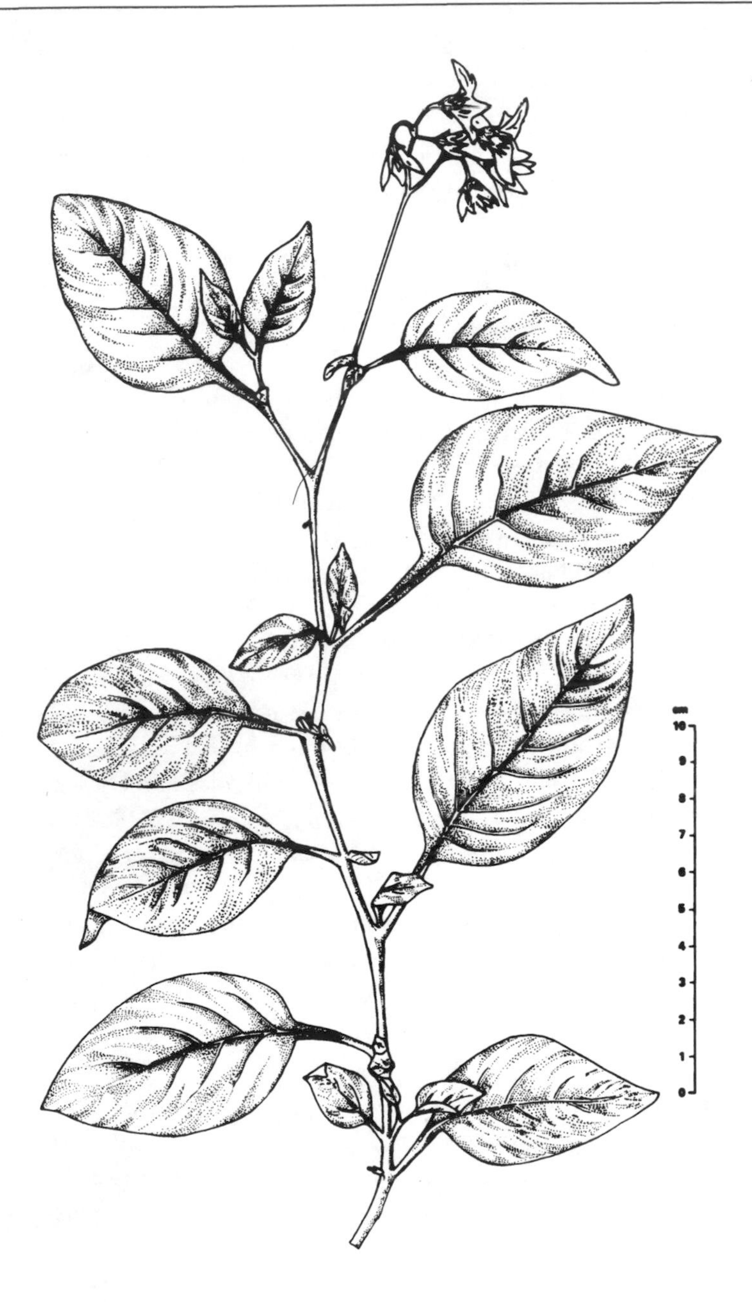

Figure 106. – *Solanum microdontum* var. *metriophyllum* (*Ochoa and Salas 15509*); upper part of flowering plant, ×1.

S. simplicifolium subsp. *gigantophyllum* (Bitt.) Bitt., Repert. Sp. Nov. 12:445. 1913. TYPE: *Lorentz and Hieronymus 802*, Argentina. Cuesta del Garabatal near Siambón, Sierra de Tucumán (GOET, not seen).

S. simplicifolium var. *metriophyllum* Bitt., Repert. Sp. 12:445. 1913. TYPE: *Spegazzini* Jan. 1897, Argentina. Prov. Salta, Pampa Grande; *Spegazzini* Feb. 1897, Tucumán, Arcas-Trancas; *Stuckert 22009*, Jan. 1911, Tucumán, Burruyacu (B, destroyed). Also ex Brücher & Ross, Lilloa 26:464, 1953.

S. simplicifolium var. *mollifrons* Bitt., Repert. Sp. Nov. 12:445-446. 1913. TYPE: *Spegazzini*, Argentina. Salta, Pampa Grande, (B, destroyed). Also ex Brücher & Ross, Lilloa 26:464, 1953.

S. simplicifolium var. *variabile* Brücher and Ross, Lilloa 26:465, Plate II, Fig. A. 1953 (as var. *variabilis*). TYPE: *Brücher 184, 185, 188, 264*, Argentina. Catamarca (GOET, not seen).

S. microdontum subsp. *gigantophyllum* (Bitt.) Hawkes and Hjerting, Phyton, Graz 9:140-146. 1960. TYPE: *Lorentz and Hieronymus 802*, Argentina. Cuesta del Garabatal near Siambón, Sierra de Tucumán (GOET, not seen; designated as lectotype of subsp. *gigantophyllum*).

S. microdontum var. *gigantophyllum* (Bitt.) Ochoa. Phytologia 57:321, 1985.

Plants robust. Stem erect or decumbent, more or less pubescent, 40-80(-200) cm tall by 7-20 mm in diameter at base, usually much branched with well-developed and conspicuous 4-5 mm-broad, undulate or strongly crisped wings, the latter with irregularly crenulate and denticulate margins, the teeth

Figure 107. – Floral dissection of *Solanum microdontum* var. *metriophyllum* (*Ochoa and Salas 15509*).

having short apical hairs. Leaves usually simple, consisting of a large blade with a rounded base that is strongly decurrent on a long petiole. The upper pair of leaflets, when present, normally half the size of the terminal. Terminal leaflet (3-)6-10(-16) cm long by (1.5-)3-5(-8) cm wide. Inflorescence 8-12(-25)-flowered. Corolla pentagonal.

Chromosome number: 2n=2x=24.

Type: ARGENTINA. Province Tucuman, Cuesta del Garabatal, near Siambon, Sierra de Tucuman. January 27, 1874; *P. G. Lorentz* and *G. Hieronymus 802* (GOET).

Specimens Examined

Department Cochabamba

Province Carrasco: 107.2 km from Cochabamba on road to Pojo, 3025 m alt., just before Uchungani, amongst bushes at roadside, one colony only, and very abundant there. Puna formation. Plants to 1 m; flowers white. January 29, 1971; *Hawkes, Hjerting, Cribb, and Huamán 4292* (CIP).

Department La Paz

Province Inquisivi: Tranca Jahuira, 2800 m alt., some 10 km NE of Quime, on the trail between Quime and Chichipaya and the barrens of Condorjonani, across the Irupaya River up to Tranca Jahuira, between moist thickets. February 24, 1984; *Ochoa and Salas 15509* (CIP, OCH, US).

Department Tarija

Province Tarija: Slopes of Sama, 2800 m alt., 32 km NW of Tarija, on road to Villazon; gulch in steep, boulder-strewn slope, among bushes and tall grass. Bush to 50 cm; flowers white. February 12, 1937; *West 8338* (UC). La Aguada, 3345 m alt., 33 km from from Tarija to Villazon, moist shady situations among rank grass and vegetation along small streams, on steep slope. February 27, 1939; *Balls, Gourlay, and Hawkes 6125* (BM, K, NA, US).

24b. **Solanum microdontum** var. **montepuncoense** Ochoa. *Phytologie* 46: 224, 1980. FIGS. 108, 109.

Plants very robust. Stems to 200 cm tall and 10-15 mm in diameter at base, densely short-pilose, intensely pigmented with anthocyanins (including the peduncles, pedicels, petioles, rachis, and nerves of the leaflets), broadly straight or sinuously winged; the wings 4-5 mm wide, entire (not denticulate), light green and glabrescent. Leaves large, to 30 cm long by 20 cm wide, with 1-2 pairs of conspicuously acuminate leaflets, the upper pair smaller than the terminal. Terminal leaflets (3-)6-9(-14) cm long by (1-)3-4(-8) cm wide.

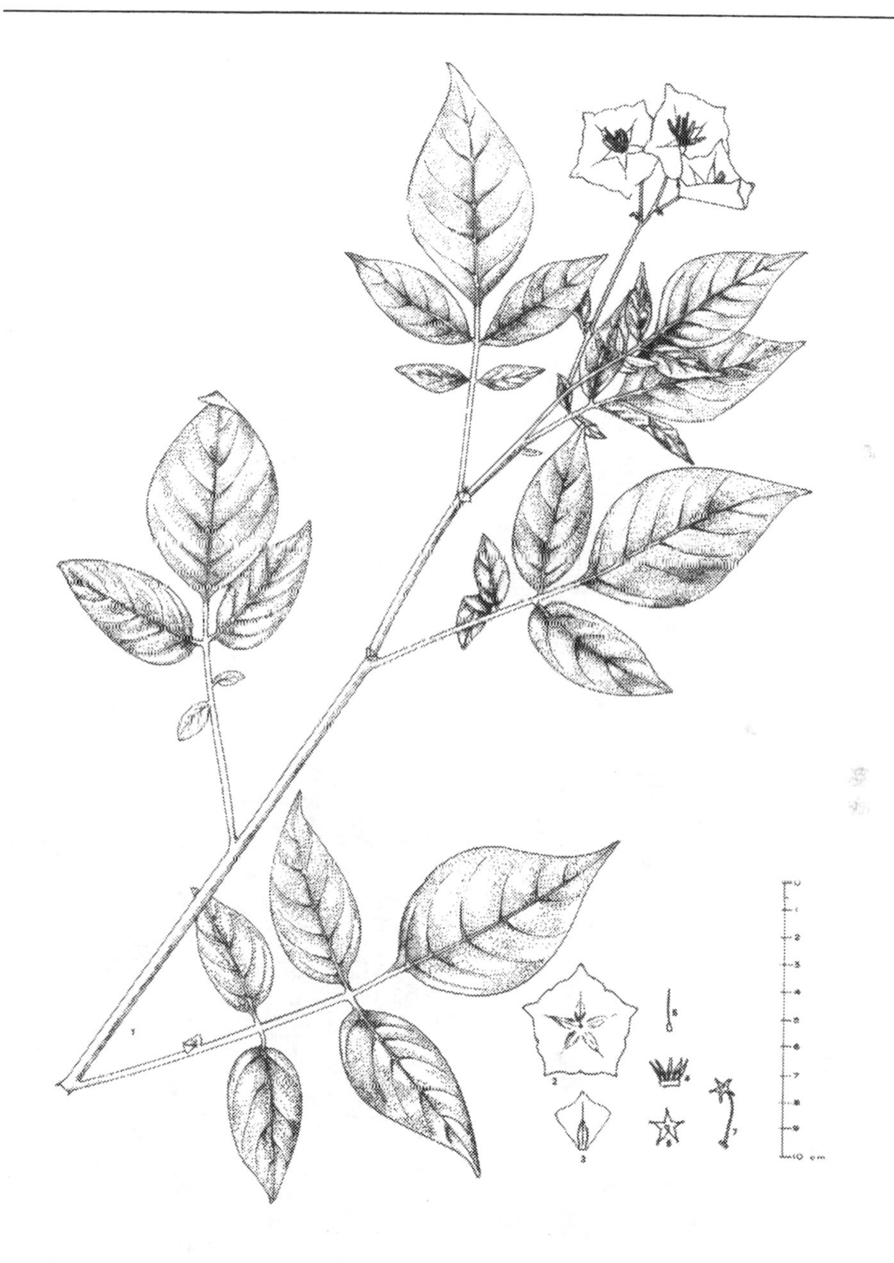

Figure 108. – *Solanum microdontum* var. *montepuncoense* (*Ochoa 12025*, holotype). 1. Upper part of flowering plant. 2. Corolla. 3. Petal. 4. Stamens, dorsal view. 5. Pistil. 6. Calyx. 7. Pedicel and calyx. All ×½.

Pedicels articulated in upper one-third of the stalk. Corolla rotate, to 4 cm in diameter, light purple with whitish acumens. Fruit long-conical, 2.5 cm long.
Chromosome number: 2n=2x=24.

Type: BOLIVIA. Department Cochabamba, Province Carrasco, between Montepunco and Schuencas, 2500 m alt. March 26, 1978; *C. Ochoa 12025* (OCH, holotype; CIP, isotype).

Specimens Examined

Department Cochabamba

Province Carrasco: Between Montepunco and Schuencas, 2500 m alt. Common name *Apharuma*. Tubers up to 5 cm long, ovate, pink-lilac. Leaves very pubescent, with two pairs of leaflets. Pseudostipular leaves very small. Flowers purple with whitish acumens. May 26, 1978; *Ochoa 12025* (OCH, holotype of *S. microdontum* var. *montepuncoense*; CIP, isotype).

Crossability

Natural crosses.– During our collecting trips in Valle Grande (Department Santa Cruz), we found two accessions which undoubtedly are natural hybrids

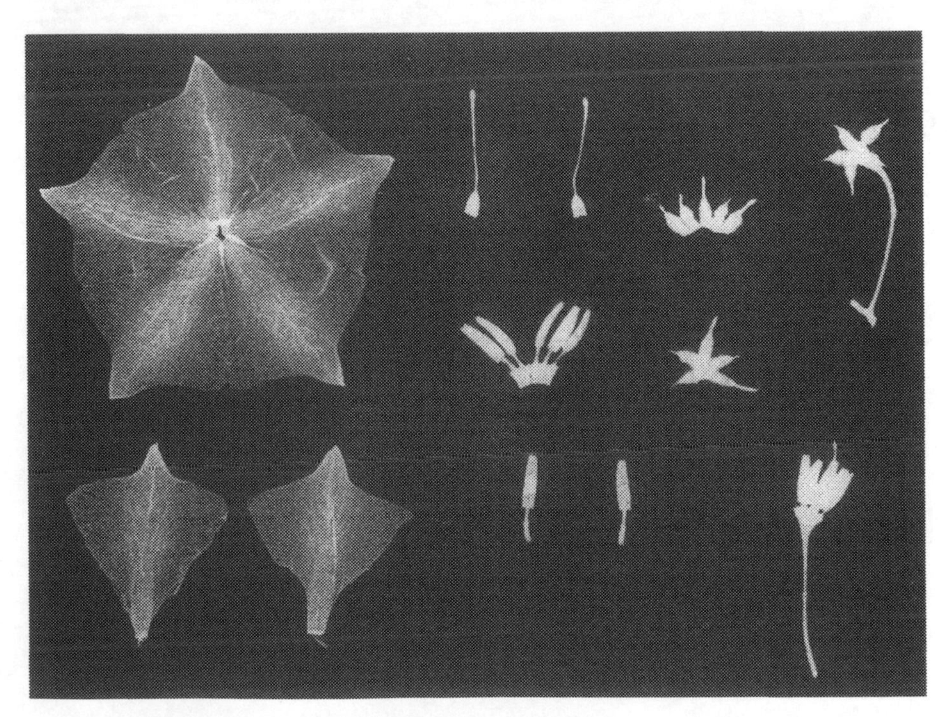

Figure 109. – Floral dissection of *Solanum microdontum* var. *montepuncoense* (*Ochoa 12025*), ×1.

involving *S. microdontum*. One of them, *Ochoa and Salas 15533* (CIP, OCH, collected at Cerro Guadalupe, 2000 m alt., near Guadalupe), is a triploid (2n=36). Similar to *S. microdontum*, the plants are vigorous, densely pubescent, and tall with broadly winged stems and well-dissected leaves. However, these plants are similar to *S. yungasense* in having a deeply stellate, creamy-white corolla. It is possible that this triploid arises as a result of fusion of a non-reduced gamete (2n) of *S. microdontum* with a normally reduced gamete (n) of *S. yungasense*.

The other accession, *Ochoa and Salas 15541* (CIP, OCH, collected near Mt. Merma, 2700 m alt., 25 km along the road from Valle Grande to Pucara), is a diploid hybrid (2n=24), which may represent a cross of *S. microdontum* with *S. candelarianum*. The hybrid differs from *S. microdontum* in having 2-3 jugate leaves and a rotate-pentagonal corolla with light blue or lavender acumens and shoulder lobes. In addition, the stems of these plants are smaller and more pubescent than *S. microdontum*, and the plants are less vigorous.

25. **Solanum neocardenasii** Hawkes and Hjerting, Bot. Jour. Linn. Soc. 86:411, 413, Fig. 5. 1983. FIGS. 110-114.

Plants graceful, usually small. Stems 15-20(-45) cm tall by 1.5-3.5 mm in diameter, pubescent with short and long glandular hairs, erect to decumbent, simple or branched, delicate, wingless, flexible, and bearing short internodes at base. Stolons delicate, to 30 cm or more long by 1-1.5 mm in diameter; tubers small, 1-1.5 cm long, round to ovoid, white-hyalinous. Leaves imparipinnate, long and narrow, delicate, (12-)15-17(-24) cm long by (5.5-)7-9 (-10.5) cm wide, borne on petioles 7 cm long, the leaves on the upper third of the plant bearing only 4 pairs of lateral leaflets, while the middle ones, which are larger and more dissected than the upper ones, have 5-6 pairs of lateral leaflets and a variable number of interjected leaflets. Leaflets triangular-lanceolate or obovate-lanceolate, nonglandular, borne on petiolules 20 mm or more long, sparsely pilose above and with only few hairs below, the apex obtuse and the base cordate. Terminal leaflets nearly the same size as laterals, 2.4-3.8(-4.5) cm long by 1.3-1.7(-2.3) cm wide. Upper lateral leaflets 2-3(-3.8) cm long by 1-1.4(-2) cm wide. Pseudostipular leaves extremely small, almost obsolete or absent. Inflorescence cymose, 5-7-flowered. Peduncles 3.5-6(-7.5) cm long, light green and glandular pubescent as in the case of the pedicels and calyx. Pedicel articulation high, just beneath calyx, the upper part approximately 2-3 mm long and the lower 10-25(-30) mm long. Calyx 4.5-5(-6) mm long, the lobes elliptic-lanceolate and with pointed, 1-1.5 mm-long acumens of unequal size. Corolla rotate, white with yellow nectar guides, small, 2.4-2.8 cm in diameter, the lobes rounded and clearly distinguishable, up to 8 mm

Figure 110. – *Solanum neocardenasii* (*Ochoa and Salas 15555*). 1. Flowering plant. 2. Corolla. 3. Petal. 4. Stamens, dorsal view. 5. Pistil. 6. Calyx. 7. Pedicel and calyx. 8. Fruit. All ×1.

long including the acumens, the latter short and distinctly delimited. Anthers short, 4-4.5 mm long, narrowly lanceolate, cordate at base, and borne on short, glabrous, greenish-white to white-hyalinous filaments, 0.5-0.8 mm long. Style 11.5-12 mm long, sparsely pilose, conspicuous, exerted to 5.5 mm; stigma very small, light green, slightly wider than style. Fruit globose to ovoid, 10-12 mm long, light green with 1-2 vertical, dark green stripes.

Chromosome number: 2n=2x=24.

Type: BOLIVIA. Department Santa Cruz, Province Valle Grande, 1.5 km from Mataral on road to Valle Grande, under shade of spiny bushes and between terrestrial bromeliads. February 29, 1980; *J. G. Hawkes, J. P. Hjerting, and Aviles 6496* (holotype, K, not seen; isotype, J.G.H.).

Affinities

This plant is not related to any other known potato species. Among the most important characteristics of this species are its graceful stems and dense

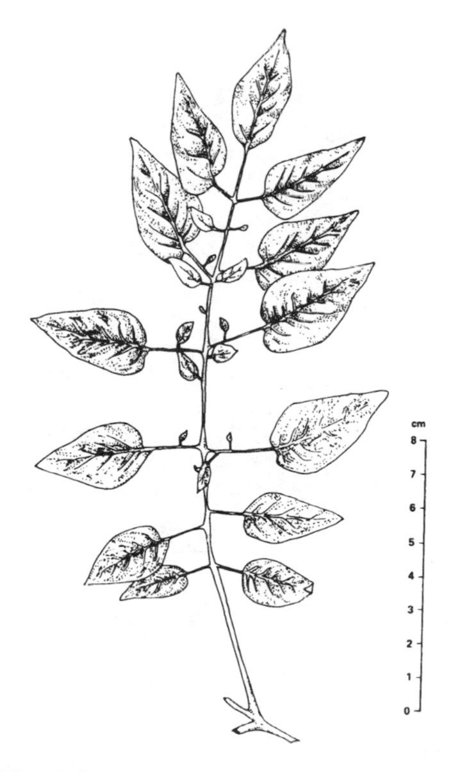

Figure 111. – *Solanum neocardenasii* (*Ochoa and Salas 15555*). Leaf, upper surface, ×1.

Figure 112. – Floral dissection of *Solanum neocardenasii* (*Ochoa and Salas 15555*).

Figure 113. – Type locality of *Solanum neocardenasii*. A hill, some 2 km from Mataral on the road to Valle Grande, Province Valle Grande, Department Santa Cruz.

pubescence, the latter having long, simple glandular hairs and short, glandular-tetralobular hairs. Also of importance here are the high articulation of the pedicels of this species, its very small pseudotipular leaves, and its long petiolules, often bearing interjected leaflets.

Habitat and Distribution

This species is found only in the arid, west-central Provinces of Florida and Valle Grande in the Department of Santa Cruz, Bolivia, where it occurs at elevations between 1400 and 1700 m. Here, it grows under spiny trees and shrubs, such as *Gymnosporia*, *Atamisquea*, and *Porlieria*, in hard, dry, gravelly soil. It is also associated with such other spiny plant genera as *Jatropha*, *Pitcairnia*, *Opuntia*, *Cleistocactus*, *Neoraimondia*, and *Cereus*. It is also found, however, within this same general area in more moist situations, for example, along river banks and walls of Vallegrandino River, between elevations of 1550 and 1700 m, where it grows with *Crotalaria*, *Verbena*, and various species of non-tuber-bearing *Solanum*.

Specimens Examined

Department Santa Cruz

Province Florida: Los Potreros, 1600 m alt., about 4 km NE of Pampa Grande, at the base of a wooden fence made out of wide logs, in moist soil rich in

Figure 114. – *Solanum neocardenasii* in its natural habitat at its type locality, protected by spiny plants and bromeliads.

organic matter, protected by *Jatropha*, some *Opuntias* and nettles. Common name 'papa de zorro.' Abundant. In full bloom and with immature fruits. February 28, 1984; *Ochoa and Salas 15559* (CIP, OCH, US). Los Negros, 1400 m alt., under the shade of spiny trees and under the shade and protection of extensive colonies of bromeliads. In hard and coarse gravelly soil. Arid region. February 28, 1984; *Ochoa and Salas 15560* (CIP, OCH).

Province Valle Grande: 1.5 km from Mataral on road to Valle Grande, under shade of spiny bushes and between terrestrial bromeliads. February 29, 1980; *Hawkes, Hjerting, and Aviles 6496* (holotype, K, not seen; isotype, J.G.H.). About 1.5-2 km from Mataral on the road to Valle Grande, 1600 m alt., near the top of a small hill situated along the right side of the Mataral-Valle Grande road. Among trees and spiny shrubs, abundant bromeliads and cacti associated with large colonies of *Jatropha* and bromeliads. February 28, 1984; *Ochoa and Salas 15555* (CIP, F, GH, NY, OCH, US, topotypes). Between Cerro La Tipa and Cerro El Estanque, 1700 m alt., some 7 km in a straight line SW of Mataral, under the shade and protection of spiny shrubs, *Opuntia* and *Pitcairnia*, and near plants of *Solanum gandarillasii*, the latter another tuber-bearing species abundant in this area. February 28, 1984; *Ochoa and Salas 15556* (CIP, NY, OCH, US).

26. Solanum okadae Hawkes and Hjerting, Bot. Jour. Linn. Soc. 86:414, 416-417, Fig. 7. 1983. FIGS. 115-118; PLATE XV.

S. venatoris Ochoa, Phytologia 55(5):297-298, illust. 1984. TYPE: *Ochoa 11917*, Bolivia. Between Quime and Inquisivi, Dept. La Paz, Prov. Inquisivi, March, 1978 (holotype, OCH; isotypes, CIP, US).

Plants small, light green. Stems 25-40(-60) cm tall by (2.5-)3.5-5 mm in diameter at base, branched, with lightly pigmented nodes and axils and 2-6 cm-long internodes, sparsely pilose with short, white hairs and straight, very narrow, barely distinguishable wings. Stolons 60-70 cm or more long; tubers small, 1-2.5 cm long, round, ovoid to oval-compressed to long subcylindrical, or occasionally moniliform, with white skin and flesh. Leaves imparipinnate, short and broad, (6-)11-14(-18) cm long by (1.5-)3-7.5(-13) cm wide, usually with (1-)2-3(-4) pairs of lateral leaflets and 0-4 interjected leaflets, borne on petiolules (1.5-)3.5-5 cm long. Leaflets dark green, sparsely pilose with short hairs above and much lighter green and pilose with very short hairs on the veins and veinlets below. Terminal leaflets much larger and broader than the laterals, (3-)4.5-5.5(-9.7) cm long by (2.2-)3.5-4(-6.5) cm wide, very broadly ovate or broadly elliptic-lanceolate to suborbicular, the apex acuminate and the base rounded to subcordate or cordate. Lateral leaflets elliptic-lanceolate or

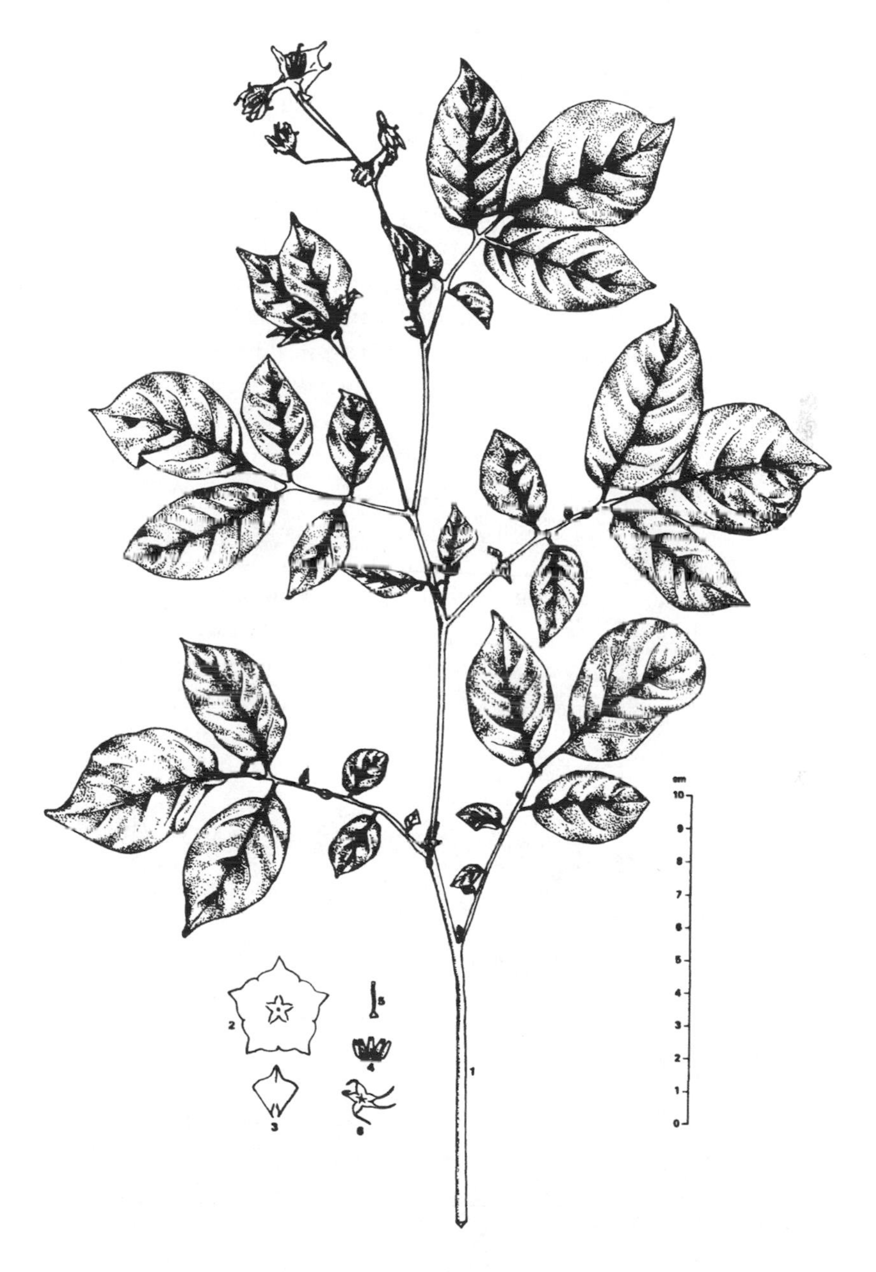

Figure 115. – *Solanum okadae* (*Ochoa 11917*). 1. Flowering plant. 2. Corolla. 3. Petal. 4. Stamens. 5. Pistil. 6. Calyx. All ×½.

broadly ovate, acuminately tipped. Pseudostipular leaves small, 6-8 mm long by 2.5-3 mm wide, narrow and obliquely elliptic-lanceolate or narrowly subfalcate. Inflorescence cymose-paniculate, 5-6-flowered. Peduncles slender, 5-7(-10) cm long, and light green and glabrescent as in the pedicels and calyx. Pedicel articulation high, in upper one-third of the stalk, the upper part 6-7 mm long and the lower 20-25 mm long. Calyx 7-9 mm long, the lobes elliptic-lanceolate and bearing narrowly subspatulate to almost linear, pointed acumens. Corolla rotate to rotate-pentagonal, white with yellowish-green or light-yellow nectar guides, 2.5-3.5 cm in diameter, including the 3.5-4 mm long, well-defined acumens. Anther column truncate-conical, well formed; anthers light-yellow, 5.5-6 mm long, cordate at base, borne on 0.5-1 mm-long glabrous, white-hyalinous filaments. Style 8.5-10 mm long, exerted 1-1.5 mm, densely papillose along the lower half of its length; stigma capitate, cleft. Fruit globose, dark green, 1.5 cm in diameter.

Chromosome number: 2n=2x=24.

Type: BOLIVIA. Department La Paz, Province Inquisivi, 0.5 km on road from Quime to Inquisivi on outskirts of Quime, 3060 m alt. March 14, 1981; *J. G. Hawkes, I. Aviles, and R. Hoopes 6727* (holotype, K, not seen; isotypes, BAL, C, J.G.H.).

Figure 116. – Floral dissection of *Solanum okadae* (*Ochoa and Salas 15506*), ×1.

Figure 117. – Roadside habitat of *Solanum okadae*, vicinity of Cancho, 2800 m altitude, between Quime and Inquisivi, Province Inquisivi, Department La Paz. Plants of this species grow in stony soil along roadsides bordered by introduced eucalyptus trees.

Affinities

Solanum okadae has been confused occasionally in the herbarium with *S. virgultorum*, perhaps because both have somewhat similar leaves. However, although this species retains a superficial resemblance to the latter plant, its true affinities lie with *S. microdontum*. Both have white flowers, sparsely pilose and poorly dissected leaves, and very large, terminal leaflets of similar shape.

Habitat and Distribution

This species prefers moist habitats between 2700 and 2800 m, where it grows between weeds in open fields or under shrubs, or along the margins of cultivated fields. It is distributed from Quime and Inquisivi in the Department of La Paz, to the Province of Valle Grande in the Department of Santa Cruz. According to Hawkes and Hjerting (1983), it is also found in the Argentine Provinces of Salta and Jujuy.

This species is very susceptible to potato late blight, *Phytophthora infestans*.

Specimens Examined

Department Cochabamba

Province Ayopaya: About 3 km from Independencia on footpath to Sivingani, 2450 m alt., lat. 17°06′, long. 66°55′. 2450 m alt. In fallow field near hedge.

Figure 118. – *Solanum okadae*, in its native habitat and type locality, outskirts of Quime.

Plant almost glabrous, flowers white, rotate-pentagonal, buds white, tipped with violet. March 11, 1980; *Hawkes, Hjerting, and Aviles 6585* (CIP, from plants grown at Sturgeon Bay, Wisconsin, USA under No. PI-498130).

Department La Paz

Province Inquisivi: Quime, 3110 m alt. [10,200 ft., according to the collector], some 160 km from Oruro on road to Eucaliptus, crossing the Serrania de Quimsacruz by the Pass of Tres Cruces. Beneath an over-ranging rock in a damp gorge. March 17, 1949; *Brooke 3030* (BM, F). Near Quime, 2500 m alt., among shrubs and along forest margins. March, 1978; *Ochoa 11917* (CIP, OCH, US). Km 0.5 on road between Quime and Inquisivi on outskirts of Quime, 3060 m alt. March 14, 1981; *Hawkes, Aviles, and Hoopes 6727* (K, BAL, C, J.G.H.). Cancho, 2800 m alt., some 3 km from Quime on road to Inquisivi, among maize and bean plants in cultivated fields bordered by eucalyptus trees. February 23, 1984; *Ochoa and Salas 15481, 15482* (CIP, OCH). Challani, 2750 m alt., near Cancho on the road between Quime and Inquisivi, in poor, stony soil, near forests of eucalyptus. February 23, 1984; *Ochoa and Salas 15483* (CIP, OCH). Some 300 m E of Challani, 2750 m alt. In poor, stony soil, near forests of eucalyptus. February 23, 1984; *Ochoa and Salas 15487* (CIP, OCH). Jañuccacca, 2700 m alt., about 6 km from Quime on road to Inquisivi, in a moist area among large stones and shrubs, the latter represented mainly by *Barnadesia* sp. Chromosome number: 2n=2x=24. February 23, 1984; *Ochoa and Salas 15488* (CIP, OCH). Canayapu, 2700 m alt., near Juñi, about 6 km from Quime on road to Inquisivi. Growing in coarse gravel as a weed in maize fields. Plants with white flowers, rotate corollas, and 10 mm-wide winged stem. February 23, 1984; *Ochoa and Salas 15498* (CIP, NY, OCH, US). Juñi, 2700 m alt., near Canayapu, about 16 km from Quime on road to Inquisivi. February 23, 1984; *Ochoa and Salas 15499* (CIP, OCH, US). Marquirivi, 2800 m alt., above and N of Juñi. February 23, 1984; *Ochoa and Salas 15500* (CIP, OCH). Near Quime, 2900 m alt., above right roadside along the Quime-Inquisivi road. Among stones and begonias, on moist, steep slope with abundant shrubs. Chromosome number: 2n=2x=24. February 23, 1984; *Ochoa and Salas 15501, 15506* (CIP, NY, OCH, topotypes of *S. okadae*). Rosanani, 2750 m alt., some 8 km from Quime on road to Inquisivi, about 300 m below roadside and in stony soil of maize field. February 23, 1984; *Ochoa and Salas 15507* (CIP, OCH). Cerro Impaya, 2800 m alt., about 3 km NW of Quime, in shrub forests, among stones in moist soil. February 24, 1984; *Ochoa and Salas 15508* (CIP, NY, OCH, US).

27. Solanum oplocense Hawkes, Bull. Imp. Bur. Pl. Breed. & Genet., Cambridge (June):39-40, 119, Figs. 22-23. 1944.

FIGS. 119-125; PLATE XVI.

S. subandigenum Hawkes var. *camarguense* Cárd., Bol. Soc. Peruana Bot. 5:25-26, Plate II-D, Figs. 2-3. 1956. TYPE: *Cárdenas 5078*, Bolivia. Near Camargo, 2200 m alt., Dept. Chuquisaca, Prov. Cinti, February, 1949 (HNF).

Plants pilose, erect, or occasionally with short internodes and rosette-like at base. Stems (10-)30-50(-70) cm tall by 3-4 mm in diameter at base, usually simple, slender, somewhat sinuous, and provided with very narrow, straight wings in the polyploid forms of this species, and strongly sinuous, upright wings in the diploid forms, the latter with light to dark purple stems. Stolons 1.5 m or more long, thick and fleshy; tubers small, ovoid to oval-compressed or rounded, white-skinned and lenticulate, 1-2(-3) cm long. Leaves imparipinnate, broad, (6-)15-16(-21) cm long by (3.8-)9-10(-14.5) cm wide, provided with 3-4 pairs of leaflets, these borne on a shortly pilose, narrowly winged rachis. Interjected leaflets, when present 1-8(-9), ovate or orbicular, membranous and subsessile, 3-6 mm long. Leaflets pilose, with thick, rough, white hairs alternating with glandular hairs above; nonglandular and more densely and finely pilose below, and with pilose and finely denticulate margins. Terminal leaflets somewhat larger or much larger than the upper pairs of lateral leaflets, (3-)6.5-7(-9) cm long by (2-)3-3.5(-4.5) cm wide, ovate-rhomboid to suborbicular or broadly elliptic-lanceolate to elliptic or ovate, the apex obtuse or abruptly pointed, and the base cuneate or broadly rounded. Lateral leaflets ovate to elliptic, the apex obtuse and the base attenuate to obliquely rounded, or narrowly rounded and narrowly to broadly decurrent along their basiscopic side to the rachis, the upper pair (2-)5-6(-8) cm long by (1-)2-3(-3.5) cm wide and borne on petiolules 3-6 mm long. Pseudostipular leaves broadly falcate to semiovate, 6-8 mm long in hexaploid forms, and nearly obsolete in diploids. Inflorescence cymose or cymose-paniculate, 2-7 (-14)-flowered. Peduncles 4-6(-10) cm long, pilose as in the pedicels. Pedicels articulated about midpoint or near base, (1-)1.5-2.5(-2.8) cm long. Calyx 5-8 mm long, pilose, the 2-4 mm-long lobes narrowly lanceolate and with narrow or linear acumens, the latter 1.5-2.5 mm long. Corolla pentagonal, 2-3 cm in diameter, purple to reddish-purple or light violet with gray or grayish-white nectar guides, the acumens 5-6 mm long and not well differentiated from the lobes. Anthers (4-)5-5.5(-7) mm long, lanceolate, cordate or distinctly lobulate at base, borne on glabrous, 0.2-0.4(-0.7) mm-long filaments. Style 10-11(-12) mm long, exerted 6 mm, densely papillose below center; stigma capitate, cleft, larger or equal in diameter to the style. Fruit large, globose to ovoid, to 25 mm

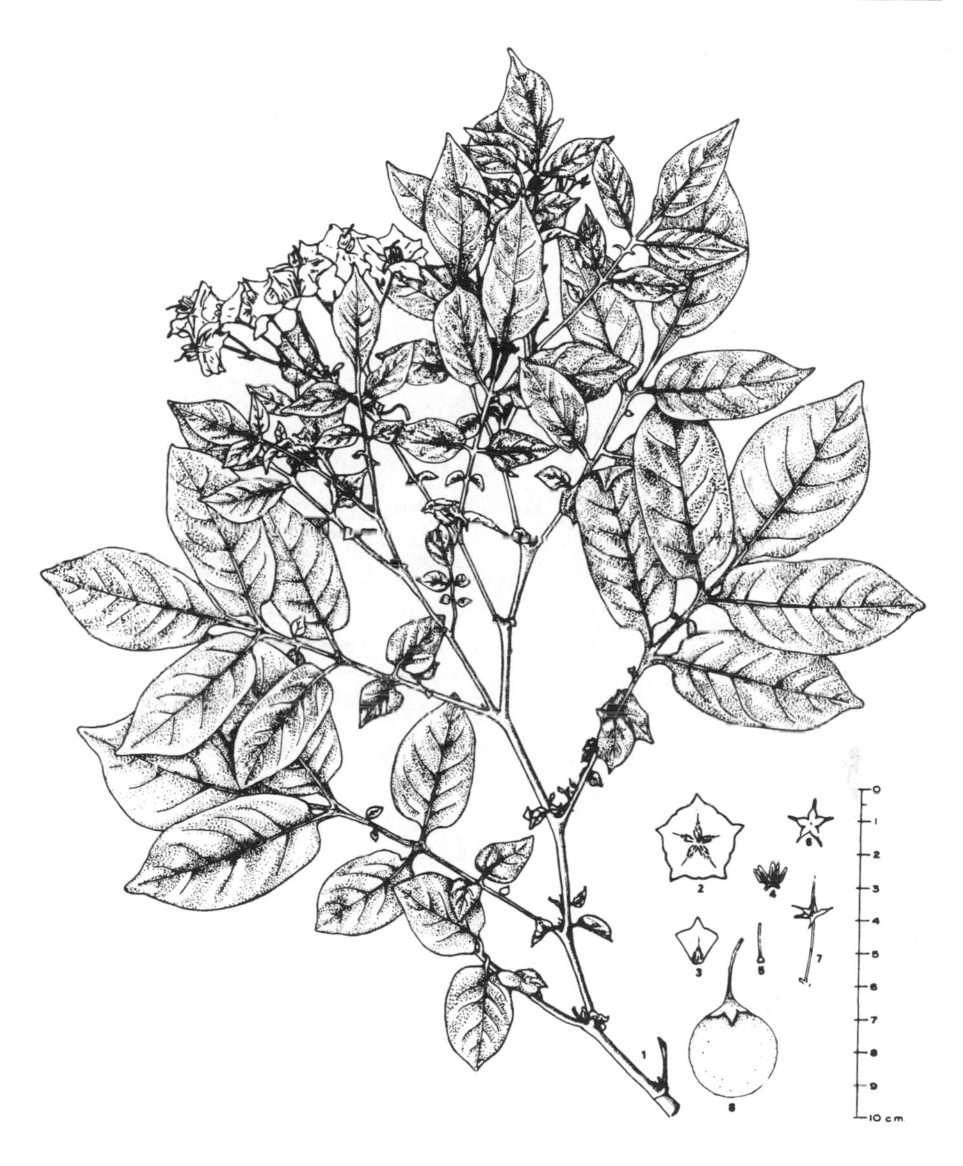

Figure 119. – *Solanum oplocense* (*Ochoa 11969*). 1. Upper part of flowering plant. 2. Corolla. 3. Petal. 4. Stamens, dorsal view. 5. Pistil. 6. Calyx. 7. Pedicel, calyx and pistil. 8. Fruit. All × ½.

long, green and white verrucose, or occasionally dark green and lilac-tinged at base.

Chromosome number: 2n=2x=24, 2n=4x=48, and 2n=6x=72.

Type: BOLIVIA. Department Potosi, Province Sud Chichas, between Oploca and Oro Ingenio, 3800 m alt. February 28, 1939; *E. K. Balls, W. B. Gourlay, and J. G. Hawkes 6127* (lectotype, CPC [= P.G.S.], designated by Hawkes and Hjerting; isotypes, BM, K, UC, US).

The stems of diploid plants of *S. oplocense* (*Ochoa 11927, 11928*) are either conspicuously straight or sinuously winged. These plants have more numerous interjected leaflets (5-9) than polyploid forms of this species, and their leaves tend to be less glandular and more pilose. The pedicel articulation occurs in the upper one-third of the stalk. Calyces average 10 mm in length and are provided with filiform acumens up to 5 mm in length.

Both diploid and tetraploid plants bear ovoid to ovoid-compressed tubers, whereas hexaploid forms bear only uniformly round to ovoid tubers. The tetraploid plants of this species can be further distinguished from diploids by the greater abundance of short glandular hairs. These cover the rachis, petiolules, axils, and lower surfaces of the leaflets.

In Pilquiza, a village located in the Province of Sud Chichas, Department of Potosi, this species is known by either its native name of *Lluttu-papa* or by its Spanish name of *papa de perdiz* ('partridge potato'). The berries of this plant are also known as *tulas*. Its bitter tubers, however, are rarely eaten.

Figure 120. – Floral dissection of *Solanum oplocense*, 2n=24 (*Ochoa 11927*), ×1.

According to Hawkes and Hjerting (1969, p. 379), a collection of *S. oplocense* made by Cárdenas (CPC 2503) near Oro Ingenio (its type locality) is diploid, 2n=2x=24, as is another collection made by Braun (CPC 2116). However, a collection of this species from Quebrada de La Quiaca, Department of Potosi, Bolivia, by Alandia and Ross (No. 106; MAX), as well as a germplasm accession from Yavi in northern Argentina by the same collectors (No. 123; MAX), is tetraploid, according to Hans Ross (*in* Hawkes and Hjerting, 1969).

All three polyploid levels occurred in living materials of this species collected in Bolivia by the author. Two samples of plants from Upituyo, near Sucre (Ochoa 11927 and 11928) were diploid; four samples from between Tacaquira and Aiquile (Ochoa 11945, 11947, 12010A, 12010B) were tetraploid; and four from the Department of Potosi (Ochoa 11969, 11970, 11971, 11972, including two topotypes) were hexaploid.

Affinities

This species appears to be closely related to *S. vidaurrei*. Diploid and tetraploid forms, in particular, resemble the latter species in having finely denticulate and pilose leaflet margins, narrowly decurrent upper leaflets, and pentagonal corollas. However, *S. oplocense* differs in leaf dissection, and in having broader and more ovate-elliptic leaflets that are borne on longer petiolules.

Habitat and Distribution

This species occurs from southern Bolivia to Humahuaca and Yavi in northern Argentina. In Bolivia, it grows preferentially in the Prepuna zones (Cabrera

Figure 121. – Floral dissection of *Solanum oplocense*, 2n=72 (*Ochoa 11969*), ×1.

and Willink, 1973), though it is also found to some extent in the higher reaches of the Puna as well. Its elevational limits extend from 2700 m in the vicinity of Camargo, Department of Chuquisaca, to 4000 m between Oploca and Oro Ingenio, Department of Potosi. It is often found in arid or semiarid places, where it grows among spiny plants or under the shade of *Prosopis* sp. and in cultivated fields.

Specimens Examined

Department Cochabamba

Province Campero: Near La Hoyada, 2300 m alt., between Aiquile and Totora, along the margins of maize fields, near *S. alandiae*. Chromosome number 2n=6x=72. March 25, 1978; *Ochoa 12014A* (OCH).

Province Mizque: Ichu Ttirana, 2200 m, above Huara Huara. Chromosome number 2n=4x=48. March 25, 1978; *Ochoa 12010A* (CIP, OCH). Vicinity of Huara-Huara, near Ichu Ttirana, 2200 m alt., E of the trunk road Aiquile-Totora. Chromosome number 2n=4x=48. March 25, 1978; *Ochoa 12010B* (OCH).

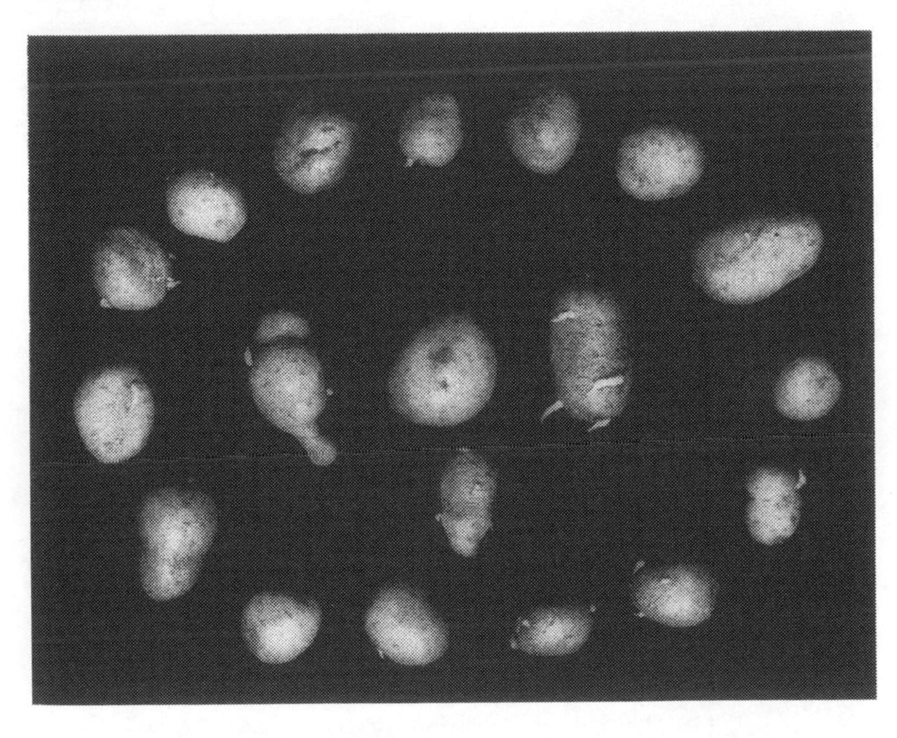

Figure 122. – Tubers of *Solanum oplocense*, 2n=48 (*Ochoa 11948*), ×1.

Department Chuquisaca

Province Nor Cinti: Near Camargo, 2200 m alt., growing at the borders of paths or even arising from the crevices of walls. February, 1949; *Cárdenas 5078* (HNF). Near hotel in Muyuquiri, km 161 of the Potosi-Camargo road, 3000 m alt. In barley fields and on rocky slopes, beneath *Dunalia*, possibly with a few cultivated potatoes. Flowers light blue with white nectar guides. Leaves similar to those of *S. sucrense*/*S. oplocense* that were collected at another locality. 2n=4x=48. March 9, 1971 *Hjerting, Cribb, and Huamán 4615* (CIP). Near Tacaquira, 3050 m alt., between Muyuquiri and Chaco, via Camargo-Tupiza. Fruits ovoid, dark green, densely white-speckled. Plants to 70 cm tall, graceful, stems somewhat simple, slender. In dry places in gravelly soil. Chromosome number: 2n=4x=48. March 14, 1978; *Ochoa 11945* (CIP, OCH). El Balcon, mountain W of San Pedro, 2900 m alt. In xerophytic or subxerophytic places, among small trees of *Acacia* sp. and abundant spiny colonies of *Cleistocactus*. Chromosome number: 2n=4x=48. March 15, 1978; *Ochoa 11947* (CIP, OCH). El Balcon, mountain W of San Pedro, 2900 m alt., among small trees of *Acacia* sp. and spiny colonies of *Cleistocactus*. Arid region.

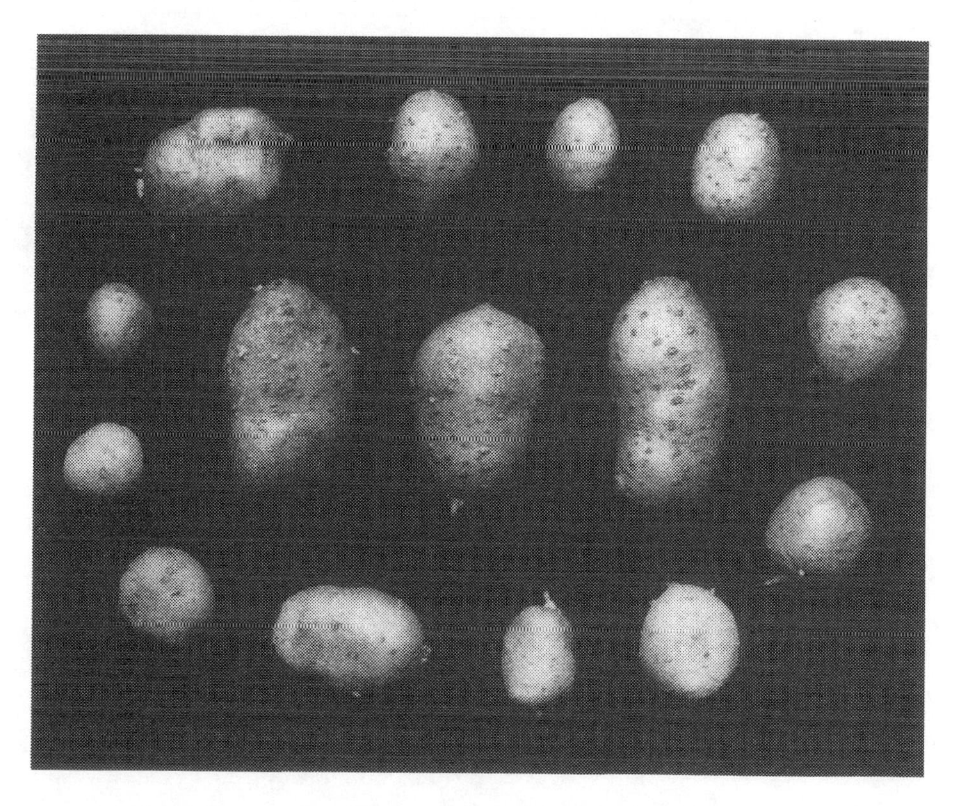

Figure 123. – Tubers of *Solanum oplocense*, 2n=72 (*Ochoa 11969*), ×1.

About 700 m from the above-cited collection (*Ochoa 11947*). Chromosome number: 2n=4x=48. March 15, 1978; *Ochoa 11948* (CIP, OCH).

Province Oropeza: Tacos, 2900 m alt., 12 km from Sucre on road to Zudañez. Growing as a weed of maize field. In shade, large plants. March 26, 1974; *Astley 021* (from plants growing at the Exp. St. of Sturgeon Bay, Wisconsin, USA). Vicinity of Azari, 2700 m alt., on adobe walls. Chromosome number: 2n=2x=24. March, 1978; *Ochoa 11927* (CIP, OCH). Upituyo, 2810 m alt., between Sucre and Azari, on rocky slopes and among shrubs of *Dodonea viscosa*. Fruits ovate conical, flowers subpentagonal, purple; articulation pigmented. Chromosome number: 2n=2x=24. March, 1978; *Ochoa 11928* (CIP, OCH). Upituyo, 2900 m alt., between Sucre and Azari, on rocky slopes in dry stony soil, under the shade of *Buddleja* sp. and among cacti, small algarobo trees and shrubs of *Dodonea viscosa*. Chromosome number: 2n=2x=24. March, 1984; *Ochoa 15572* (CIP, OCH, US). Campanario, 2430 m alt., between Sucre and Yotala, on rocky talus slopes, with abundant colonies of *Lupinus* sp. March, 1984; *Ochoa 15576* (CIP, OCH).

Province Sud Cinti: In acacia-cactus shrubs, about 70 km from Villa Abecia on road to Las Carreras. February 22, 1960; *Correll, Dodds, Brücher, and Paxman B649* (LL, MO, UC).

Figure 124. – Berries of *Solanum oplocense* (*Ochoa 11927*), ×1.

Figure 125. – Habitat of *Solanum oplocense*, subxerophytic region between Oploca and Tupiza, Province Sud Chichas, Department Potosi.

Department Potosi

Province Frias: Lecherias, below Potosi, about 3900 m alt. At the bases of rough stone walls; or under agricultural constructions, in well-fertilized soil. Corolla pentagonal, dark purple. January 31, 1971; *Hawkes, Hjerting, Cribb, and Huamán 4315* (CIP). Km 30 from Potosi on road to Oruro, 3600 m alt. Growing around borders of maize and potato fields (opposite *Salix babylonica* tree). February 3, 1971; *Hawkes, Hjerting, Cribb, and Huamán 4337* (CIP). Samasa, 3625 m alt., on road to Don Diego, about 155 km from Sucre on road to Potosi. March 6, 1971; *Hjerting, Cribb, and Huamán 4580* (CIP). About 27 km from Potosi on road to Oruro, 3550 m alt., above Tarapaya. In a moist hollow. March 8, 1971; Hjerting, Cribb, and Huamán 4596 (CIP).

Province Saavedra: Above Retiro, 3025 m alt., about 86 km from Sucre on road from Sucre to Potosi, on dry slopes among thorny shrubs. March 6, 1971; *Hjerting, Cribb, and Huamán 4578* (CIP). Above Retiro, 3025 m alt., about 86 km from Sucre on road from Sucre to Potosi, on dry slope among thorny shrubs, at end of barley field. March 6, 1971; *Hjerting, Cribb, and Huamán 4579* (CIP).

Province Sud Chichas: Between Oploca and Oro Ingenio (on railway from Villazon to Atocha), about 3800 m alt., growing amongst loose shaley stones on railway embankment. Plants rosette-like, small, 20 cm tall. Pedicel articulated almost at base, flowers purple-mauve. February 28, 1939; *Balls, Gourlay, and Hawkes 6127* (lectotype, CPC [= P.G.S.], designated by Hawkes and Hjerting; isotypes, BM, K, UC, US). Between Oploca and Oro Ingenio, 2700 m alt., under shrubs in dry sandy soil. March, 1952; *Cárdenas 5516* (CA, LL, erroneously considered as topotype of *S. oplocense* [note alt.]). About 87 km from Potosi on road to Sucre, at base of cliffs and in fields. February 18, 1960; *Correll, Dodds, Brücher, and Paxman B632* (LL, US). Morallas [*sic.* Morayas], growing in protection of acacias. February 23, 1960; *Correll, Dodds, Brücher, and Paxman B657* (LL, S, US). Near Mojo, about 3400 m alt., in acacia thicket, flowers blue. February, 1960; *Correll, Dodds, Brücher, and Paxman B659* (K, LL). About 39-40 km from the Villazon-Tupiza road, 3400 m alt., in thickets of acacia; without flowers. Plants wilting, but flowering plants obtained at km 40, flowers lilac. March 15, 1971; *Hjerting, Cribb, and Huamán 4713* (CIP; site of Correll's collection of *S. tacnaense*). Km 53 from Tupiza to Ramada, 3225 m alt. In hedges and under acacias in field by road. Shade forms with large leaves, light forms with smaller leaves and darker flowers, violet. March 16, 1971; *Hawkes, Hjerting, Cribb, and Huamán 4719* (CIP). Two km from Oploca on road to Tupiza, 3100 m alt. April 2, 1974; *Astley 058* (from plants grown at the Exp. St. at Sturgeon Bay, Wisconsin, USA). Between Oploca and Oro Ingenio [near type locality], 3750 m alt., along the rail route Atocha-Villazon-Uyuni, in stony soil, near railroad embankment. Chromosome number: 2n=6x=72. March 18, 1978; *Ochoa 11969* (CIP, OCH, US). Between Oploca and Tupiza, 3800 m alt., in stony soil, near embankment of the rail route Atocha-Villazon-Uyuni. Chromosome number: 2n=6x=72. March 18, 1978; *Ochoa 11970* (OCH, topotype). Above Pilquiza, 3400 m alt., near Ichupampa, under the shade of large trees of acacia, among grasses and plants of the genus *Bidens* and along margins of barley fields. Common name *Llutu papa*. Chromosome number: 2n=6x=72. March 18, 1978; *Ochoa 11971* (CIP, OCH, US). Some 5 km from Pilquiza, 3480 m alt., on road to Nazarenillo, in poor and stony soils. Chromosome number: 2n=6x=72. March 18, 1978; *Ochoa 11972* (CIP, OCH, US).

Department Tarija

Province Mendez: About 38 km from Las Carreras on the road to Tarija, 3350 m alt., under shrubs along the margins of maize and oca fields. Plants to 15 cm tall, these having decurrent upper lateral leaflets, some leaves with long, narrow leaflets, others with wide leaflets. Flowers blue-violet. March 11, 1971; *Hjerting, Cribb, and Huamán 4636* (CIP).

Province Indefinite: Samana, 3600 m alt., in rocky soil, under shrubs. Plants to 40 cm tall, flowers blue. April, 1963; *Cárdenas 6120* (US, WIS).

Putative Hybrids

Department Chuquisaca

Province Sud Cinti: Between Carreras and Villa Abecia, 2600 m alt. Chromosome number: 2n=4x=48. March 22, 1978; *Ochoa 12002* (CIP, OCH) [possibly *S. oplocense* × *S. chacoense* or *S. oplocense* × *S. litusinum*].

Crossability

Natural hybrids.– A collection of plants (OCH-12002) obtained by the author from a roadside locality between Carreras and Villa Abecia in southern Bolivia appears to be of hybrid origin. Inasmuch as these plants are tetraploid (2n=48), it is possible that they were derived from natural crosses between a hexaploid chromosomal form (2n=72) of *S. oplocense* and some locally abundant wild diploid (2n=24) species, such as *S. chacoense* or *S. litusinum*. A description of the salient morphological features of these hybrids is given below:

Plants more or less pilose. Stems slender, 0.3-0.5 m tall by 3-5 mm in diameter, very narrowly winged. Tubers ovoid to oval-compressed, white, 2.5 cm long. Leaves imparipinnate, with 4-5 pairs of leaflets and numerous interjected leaflets and with narrowly winged rachises. Leaflets oblong to ovate or elliptic-lanceolate, the apex shortly acuminate and the base rounded or cuneate to narrowly decurrent on the 20 mm or more long petiolule. Inflorescence cymose-paniculate. Pedicels 18-20 mm long, articulated near the midpoint. Corolla pentagonal, to 3 cm in diameter. Fruits globose, to 2 cm in diameter, dark green, white-flecked.

Chromosome number: 2n=4x=48.

The characteristics of the above hybrids correspond very closely to the description of *S. hondelmannii*, a plant recently described by Hawkes and Hjerting (1983, 1985a). Although authenticated material of this species was not available to the author, it would seem clear from its original diagnosis that this plant would probably be better classified as a hybrid variant of *S. oplocense*.

Artificial hybrids.– Because it shows resistance to *Heterodera rostochiensis* (Ross and Huijsman, 1969) and to *Globodera pallida* (Intl. Pot. Ctr., 1982), *S. oplocense* may be of value to plant breeders concerned with the improvement of cultivated potatoes. In this respect, it may be worthy to note here that crosses between diploid forms of *S. oplocense* and *S. goniocalyx*, a cultivated diploid species, are easily made.

Many crosses involving *S. oplocense* and various wild Bolivian and Peruvian potato species have been made at CIP. The results of these efforts are detailed in the following paragraphs.

All crosses made between diploid accessions of *S. candolleanum* (14459) and *S. oplocense* (11927) were successful. These yielded, on the average, 100 fertile seeds per fruit. However, crosses between diploid accessions of *S. candolleanum* (11814, 11835) and hexaploid *S. oplocense* (11972) were sterile. All pollinations between tetraploid *S. oplocense* (11945) and diploid *S. megistacrolobum* (11916) resulted in fruits; these hybrids, however, averaged only 30 fertile seeds per berry. The F_1 plants derived from still another introduction of *S. oplocense* (11947) and *S. megistacrolobum* were completely sterile.

Crosses between diploid *S. tarijense* (11994) and a tetraploid accession of *S. oplocense* (12010-A) yielded only parthenocarpic fruits containing 1-2 seeds. Similarly, crosses between tetraploid *S. acaule* (9835) and a hexaploid accession of *S. oplocense* (11972) yielded berries containing an average of 9 seeds per fruit. On the other hand, not even a single seed was obtained in crosses between tetraploid *S. acaule* (13214) and a tetraploid accession of *S. oplocense* (11947), nor were any obtained in crosses between *S. oplocense* (11947) and another tetraploid accession of *S. acaule* (11917-A).

The following crosses involving diploid and polyploid accessions of *S. oplocense* were also sterile: diploid *S. tarijense* (12001) × hexaploid *S. oplocense* (11972); diploid *S. berthaultii* (12029) × hexaploid *S. oplocense* (11972); diploid *S. megistacrolobum* (11941, 11963) × hexaploid *S. oplocense* (11972); diploid *S. brevicaule* (11934) × tetraploid *S. oplocense* (11948); diploid *S. bukasovii* (7619) × hexaploid *S. oplocense* (11972); and diploid *S. raphanifolium* (7610) × hexaploid *S. oplocense* (11972).

Lastly, crosses involving hexaploid *S. oplocense* (11969) and tetraploid *S.*

oplocense (11947) were only partially compatible, yielding, on the average, 4 seeds per fruit.

Descriptions of the more successful crosses involving *S. oplocense* follow:

Solanum acaule × *S. oplocense*.– The F_1 hybrids derived from crosses between tetraploid *S. acaule* and hexaploid *S. oplocense* have broadly decurrent leaflets and terminal leaflets that are 2-3 times the size of their laterals. These plants, which are only partially fertile, resemble *S. acaule* in leaf dissection, but in habit and in floral characteristics are more like *S. oplocense*. All developed 20-30 cm tall, subrosette-like stems. Significantly, both species are known from Oro Ingenio and Oploca, Potosi, where they occur in similar habitats.

Solanum oplocense × *S. megistacrolobum*.– The F_1 hybrids derived from this tetraploid × diploid cross are similar to *S. oplocense* in their leaf and floral characteristics. All have light violet to dark purple pentagonal flowers, borne on pedicels that range up to 30 cm long. Pedicel articulation occurs at the midpoint, or in the upper one-third of the stalk.

Solanum oplocense × *S. goniocalyx*.– The F_1 hybrids derived from crosses between tetraploid *S. oplocense* and the cultivated diploid species *S. goniocalyx* were similar to *S. goniocalyx* in habit. These developed many-flowered inflorescences that bore highly ornamental clusters of rotate flowers, these varying in color from purple to blue-violet.

28. Solanum sparsipilum (Bitter) Juzepczuk and Bukasov, in Vavilov, Theor. Bases Pl. Breed. 3:11, 1937. Plate XX. 1937.

Figs. 126-130; Map 9; Plate XVII.

S. tuberosum subsp. *sparsipilum* Bitt., Repert. Sp. Nov. 12:152-153. 1913. Type: *Buchtien 771*, Bolivia. Obrajes, near La Paz, Dept. La Paz [Prov. Murillo], January 26, 1907 (E, G, MO, US).

S. aracc-papa Juz., Bull. Acad. Sci. U.S.S.R., ser. Biol. 2:306-307. 1937. Type: *Juzepczuk 1455*, Peru. Cerro Huilcacalle, near the village of San Sebatian not far from Cusco, Dept. Cusco, tubers collected on August 1, 1927 (LE, type collection from plants grown near Leningrad, U.S.S.R).

S. catarthrum Juz., Bull. Acad. Sci. U.S.S.R., ser. Biol. 2:307-308. 1937. Type: *Kauffman s.n.*, Peru. Lucre near Cusco, November 18, 1932 (only tubers collected) (WIR, plant from tubers grown near Leningrad, U.S.S.R).

S. anomalocalyx Hawkes, Bull. Imp. Bur. Pl. Breed. & Genet., Cambridge 52:126-127, Fig. 42. 1944. Type: *Balls, Gourlay, and Hawkes 6222*, Bolivia. Near Cochabamba, Dept. Cochabamba, Prov. Cercado, March 14, 1939 (BM, CPC, K, US).

S. brevimucronatum Hawkes, Bull. Imp. Bur. Pl. Breed. & Genet., Cambridge

52:127, Fig. 43. 1944. TYPE: *Balls, Gourlay, and Hawkes 5890*, Bolivia. Obrajes, near La Paz, Dept. La Paz, January 28, 1939 (CPC); paratype: *Balls, Gourlay, and Hawkes 5895*, Bolivia, Calacoto, Obrajes Valley, near La Paz, Dept. La Paz, January 28, 1939 (CPC, UC, US).

S. lapazense Hawkes, Bull. Imp. Bur. Pl. Breed. & Genet., Cambridge 52:127-128, Fig. 44. 1944. TYPE: *Balls, Gourlay, and Hawkes 5903*, Bolivia. Below La Paz, Dept. La Paz, January 29, 1939 (CPC, K).

S. calcense Hawkes, Bull. Imp. Bur. Pl. Breed. & Genet., Cambridge 53:128-129, Fig. 46. 1944. TYPE: *Balls and Hawkes 6744*, Peru. Urco farm, 3350 m alt., near Calca, Dept. Cusco, Prov. Calca (CPC).

S. fragariaefructum Hawkes, Bull. Imp. Bur. Pl. Breed. & Genet., Cambridge 53:129, Figs. 47. 1944. TYPE: *Balls and Hawkes 6931*, Peru. Pallca, 3200 m alt., Dept. Ayacucho, Prov. Cangallo (CPC, K).

S. anomalocalyx var. *llallaguanianum* Cárdenas and Hawkes, Jour. Linn. Soc., Bot. 53:104–105. 1945. TYPE COLLECTIONS: *Gandarillas s.n., =Cárdenas 3508*, Bolivia. Llallaguani, Anzaldo, Dept. Cochabamba, Prov. Tarata, January, 1944 (CPC); and *Cárdenas 3518*, Bolivia. Near Angostura, 2550 m alt., Dept. Cochabamba, Prov. Tarata, February, 1944 (CPC).

S. anomalocalyx var. *brachystylum* Cárdenas and Hawkes, Jour. Linn. Soc., Bot. 53:105. 1945 (described as var. *brachystyla*). TYPE: *Cárdenas 3519*, Bolivia. La Maica, Dept. Cochabamba, Prov. Cercado, February, 1944 (CPC).

S. anomalocalyx var. *murale* Card. and Hawkes, Jour. Linn. Soc., Bot. 53:106. 1945 (described as var. *muralis*). TYPE: *Cárdenas 3507*, Bolivia. La Maica, Dept. Cochabamba, Prov. Cercado, March, 1944 (CPC).

S. calcense var. *urubambae* Vargas, Las Papas Sudperuanas, Part II (Publ. Univ. Nac. Cusco):57, Fig. 9. 1956 (as var. *urubambense*). TYPE: *Vargas C. 9824*, Peru. Phiri, Dept. Cusco (VAR, LL, MO).

S. membranaceum Vargas, Las Papas Sudperuanas, Part II (Publ. Univ. Nac. Cusco):62, Fig. 20. 1956. TYPE: *Vargas C. 7595*, Peru. Pumahuanca, Dept. Cusco, Prov. Urubamba (VAR, OCH).

S. sparsipilum var. *llallaguanianum* (Card. and Hawkes) Correll, in Correll, The Potato and Its Wild Relatives, Renner, Texas, p. 465. 1962. TYPE COLLECTIONS: *Gandarillas s.n., =Cárdenas 3508*, Bolivia. Llallaguani, Anzaldo, Dept. Cochabamba, Prov. Tarata, January, 1944 (CPC); and *Cárdenas 3518*, Bolivia. Near Angostura, Dept. Cochabamba, Prov. Tarata, February, 1944 (CPC).

S. ruiz-zeballosii Card., Rev. de Agricultura, Cochabamba 11:13–14, 1 Fig. s.n. 1968. TYPE: *Cárdenas 6229*, Bolivia. Las Cuadras, Dept. Cochabamba, Prov. Cercado, February, 1966 (CA).

Plants robust, sparsely pilose with white, thick hairs. Stems 60–70(–120) cm tall by 10–12 mm in diameter at base, often branched, erect to decumbent,

Figure 126. – Lectotype of *Solanum sparsipilum* (*Buchtien* 771). Photo: courtesy of the Smithsonian Institution.

Figure 127. – *Solanum sparsipilum* (*Ochoa 662*). 1. Flowering plant. 2. Corolla. 3. Petal. 4. Stamens, dorsal view. 5. Pistil. 6. Calyx. 7. Pedicel and pistil. 8. Fruit. All × ½.

slightly pigmented, with 1-2 mm-wide straight wings. Stolons to 2 m or more in length; tubers white, ovoid to somewhat rounded or compressed, 1.5-3(-5) cm long. Leaves imparipinnate, (7-)12-25(-30) cm long by (4-)7-16(-19) cm wide, somewhat rugose, usually dark green above and light green below, with 4-5(-6) pairs of leaflets and (2-)7-11(-20) interjected leaflets (the latter in two or three different sizes, 2, 5, or 10 mm), the rachis and petiolules pigmented or slightly pigmented, and borne on 1.5-3 cm-long petioles. Leaflets ovate-elliptic or broadly elliptic to elliptic-lanceolate, the apex obtuse to subobtuse or pointed and very shortly acuminate and the base obliquely rounded-cuneate to rounded or subcordate, and borne on petiolules 2-6(-10) mm long. Terminal leaflets occasionally slightly larger than the laterals, (3-)4.5-8(-11) cm long by (1.5-)3.5-5.5(-6.5) cm wide, broadly elliptic-lanceolate to orbicular-lanceolate, the apex obtuse and the base cuneate or narrowly decurrent on the rachis. Pseudostipular leaves broadly falcate, 7-12(-18) mm long by 4-7(-9) mm wide. Inflorescence cymose to cymose-paniculate, 8-12(-16)-flowered. Peduncles 4-8(-12) cm long, slightly pigmented and sparsely pilose as in the pedicels. Pedicel articulation above center or in upper one-third of the stalk, the lower part 15-30 mm long and the upper 6-7(-12) mm long. Calyx small, 5.5-6.5(-8)

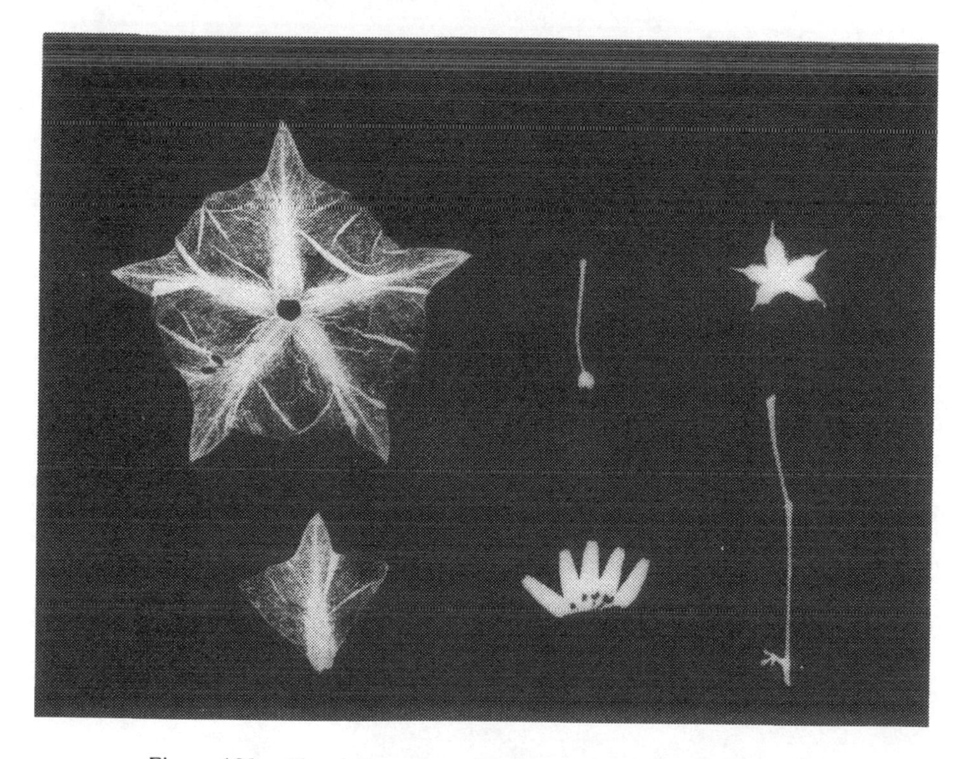

Figure 128. – Floral dissection of *Solanum sparsipilum* (*Ochoa 12028*), ×2.

mm long, invested with short, fine, white hairs, symmetric or asymmetric (bilabiate), the lobes broadly elliptic-lanceolate, shortly acuminate or occasionally apiculate. Corolla rotate to rotate pentagonal or pentagonal, 2.6-3 cm in diameter, dark purple to purple-violet, or very occasionally blue, sky blue, or whitish. Staminal column truncate-conical, asymmetric; anthers 5.5-6.5 mm long, lanceolate or narrowly elliptic-lanceolate, and cordate at base, borne on glabrous, 0.8-1.5 mm long filaments. Style 9 mm long, exerted 2.5-3 mm, shortly papillose about center or along the basal one-third of its length, or occasionally glabrous; stigma shortly capitate. Fruit globose to ovoid, 1.5-2 cm in diameter, green or green with widely scattered small, white dots, not verrucose.

Chromosome number: 2n=2x=24.

Type: BOLIVIA. Department La Paz, Province Murillo, Obrajes, near La Paz, 3350-3800 m alt. January 26, 1907; *O. Buchtien 771* (lectotype, here designated, US; isotypes, E, G, MO).

This species is highly polymorphic. A specimen collected by Balls (No. 5895) at Calacoto, Valle de Obrajes, La Paz, is the type of *S. brevimucronatum*

Figure 129. – Fruits of *Solanum sparsipilum* (*Ochoa 12028*), ×1.

Hawkes, while another specimen from the same locality by the same collector (No. 5903) is the type of *S. lapazense* Hawkes. These two plants, together with the species collected by Buchtien (No. 771) from the same locality and described by Bitter as *S. tuberosum* subsp. *sparsipilum*, are morphologically similar. All three are treated here as synonyms of *S. sparsipilum*.

The author has seen two specimens collected by Cárdenas in the Department of Cochabamba, Bolivia (Llallaguani near Anzaldo, No. 3508, and Angostura, No. 3518) that were labeled *S. anomalocalyx* var. *llallaguanianum*. The specimens in question have large, poorly dissected leaves and resemble the plants collected by Balls at Calacoto (see above). In the author's opinion, these should be considered as variants of *S. sparsipilum*.

The plant described as *S. calcense* var. *urubambae* Vargas is based on a series of vigorous, large-leafed plants collected by Cesar Vargas (No. 9824) from Phiri (Cañon del Urubamba, Department Cusco, Peru). This collection, which differs from Buchtien's type collection of *S. sparsipilum* in having somewhat more numerous interjected leaflets, is considered here to be another variant of that species.

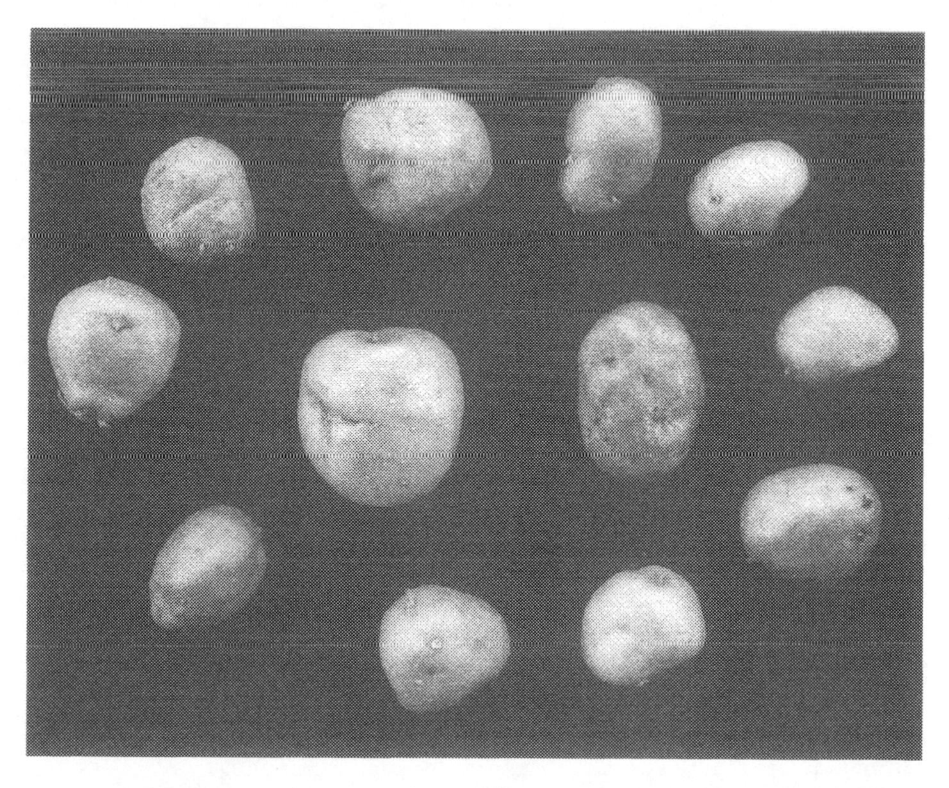

Figure 130. – Tubers of *Solanum sparsipilum*, 2n=24 (*Ochoa 12028*), ×½.

In Bolivia, *S. sparsilipum* is known locally as *apharuma* (wild), *jamppatu papa* (frog potato), *piscco papa* (bird potato), or *alcco papa* (dog potato). In Peru, where this species is much more common, it is regionally known by many other names. For example, between Tinta and Oropeza in the Vilcanota Valley of the Department of Cusco, it is known as *atoc papa* (fox potato), *c'kita papa* (wild), and *llutu arac* (partridge potato). On the other hand, between two other villages in the Department of Cusco, Andahuaylillas and Sailla, it is sometimes called *charca*, this in deference to its rough-textured tubers, or *ckascke*, because the tubers have a bitter taste.

Habitat and Distribution

Solanum sparsipilum is distributed from the southern Peruvian Departments of Apurimac, Cusco, and Puno to the Department of Chuquisaca in central Bolivia. Although this species grows under many different soil and climatic conditions throughout its broad range, it prefers the temperate to slightly cool climates of the inter-Andean valleys. Here it is found most frequently between 2400 and 3000 m altitude on barren, eroded, or cultivated soil. Within this zone, it occurs most frequently along rock or adobe walls in the vicinity of dwellings and maize fields. It is also found less frequently at elevations ranging between 3700 and 4200 m, where climatic conditions are often much more frigid.

Affinities

This wild diploid species shares some characteristics with the cultivated tetraploid *S. tuberosum* subsp. *andigena*. Plants of *S. sparsipilum* from the Cerveceria Taquina, near Cochabamba, Bolivia (Ochoa 12028), resemble *S. tuberosum* subsp. *andigena* in being very robust, floriferous, and highly branched. These also bear numerous small to medium size, ovoid and elongated tubers. However, unlike *S. tuberosum* subsp. *andigena*, *S. sparsipilum* bears bitter tubers, these always white and borne at the ends of long stolons. Moreover, the floral characteristics of these two taxa are quite different.

In southern Peru, where this plant is often called *atoc papa*, it grows spontaneously as a weed with other wild potato species in fields of cultivated potatoes and maize.

Solanum sparsipilum also has some resemblance to *S. sucrense*, another wild Bolivian potato species with inedible tubers. Both are similar in habit, leaf shape and dissection, corolla, and calyx. It is possible that both *S. sparsipilum* and *S. sucrense* have contributed genes at one time or another to the cultivated potato. It should be noted, however, that unlike *S. sparsipilum*, *S. sucrense* is tetraploid.

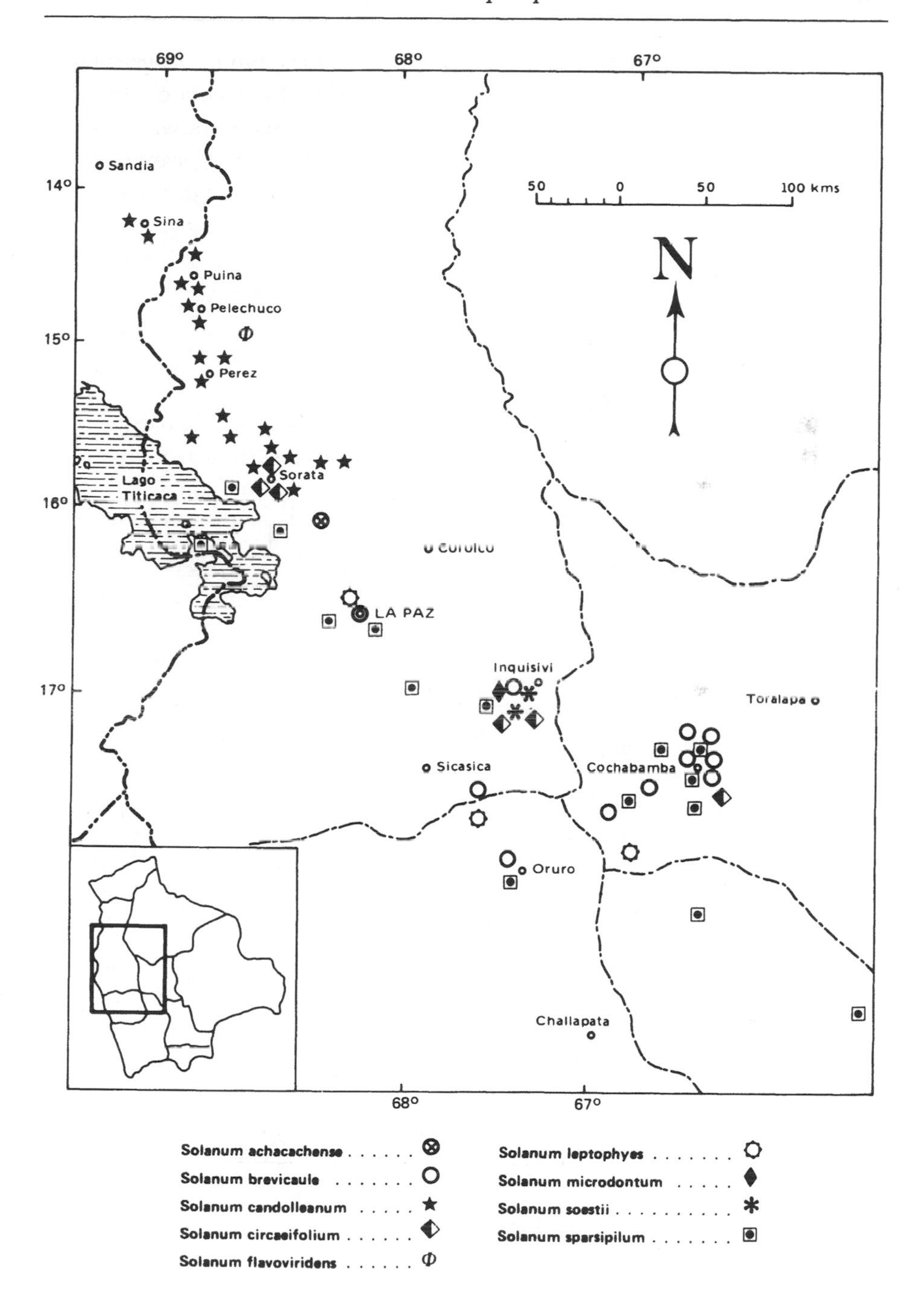

Map 9. – Distribution of wild potato species, showing *Solanum sparsipilum* (▣).

Specimens Examined

Department Cochabamba

Province Arani: Above Arani, 2700 m alt. March, 1944; *Cárdenas 3520* (CPC). One km past Arani on road to Mizque, 2700 m alt. In a maize field, typical. 2n=2x=24. February 16, 1971; *Hjerting, Cribb, and Huamán 4428* (from plants grown at Exp. St. at Sturgeon Bay, Wisconsin, USA).

Province Carrasco: Copachuncho, 2600 m alt., about 150 km from Cochabamba to Santa Cruz. On margins of maize fields and under bushes. Flowers pale blue-violet; calyx regular. Plant 20-30 cm tall. February 9, 1971; *Hawkes, Hjerting, Cribb, and Huamán 4399* (CIP). Copachunco, 2600 m alt., near km 150 of the Cochabamba-Santa Cruz highway, between Epizana and La Habana, among stones and at the base of an old, destroyed adobe wall. February 26, 1984; *Ochoa and Salas 15522* (CIP, OCH, US).

Province Capinota: Along road from Parotani to Capinota, 2400 m alt., passing 1 km from Itapaya. Along margins of maize fields. Some plants with very large terminal leaflets. Flowers blue-violet. Chromosome number: 2n=2x=24. February 15, 1971; *Hjerting, Cribb, and Huamán 4418* (CIP). On road from Parotani to Capinota, 1 km past Charamoco, 2400 m alt., by roadside. Small blue-violet flowers, 1.0-1.5 cm diam. February 15, 1971; *Hjerting, Cribb, and Huamán 4424* (CIP). One km past Itapaya on the road from Parotani to Capinota, 2400 m alt., on opposite side of road on a river cutting. Growing on perimeter of maize field, rachis winging and strong decurrency of primary lateral. March 22, 1974; *Astley 012* (from plants grown at the Exp. St. at Sturgeon Bay, Wisconsin, USA).

Province Cercado: Brewery Colon, 2745 m alt., near Cochabamba. Common name: 'Apharuma.' Growing as weed in potato fields and other cultivated fields. It is said that it is the result of spontaneous growth from cultivated varieties such as Runa. March 14, 1939; *Balls, Gourlay, and Hawkes 6222* (BM, CPC, K, US). Near the brewery Colon, 2745 m alt. [9,000 ft., according to the collector], a few km from Cochabamba, among stones. January 16, 1949; *Brooke 3006* (BM, F). La Maica, 2560 m alt., on dry sandy walls. March, 1944; *Cárdenas 3507* (CPC). La Maica, 2550 m alt., in open fields in dry sandy soil. February, 1944; *Cárdenas 3519* (CPC). Las Cuadras, Cochabamba, 2567 m alt. February, 1966; *Cárdenas 6229* (CA). Near the brewery Taquiña, above Cochabamba, on slope covered with thickets. February 13, 1960; *Correll, Dodds, Brücher, Cárdenas, and Paxman B604* (LL). About 1 km N of Sacaba, 2500 m alt., in land of Luciano Cornejo. Common name: 'Aparuma,' it is said to be harmful for the stomach. Flowers blue. November 1, 1942; *Cutler 7424* (MO, US). Near Cerro San Pedro on edge of Cochabamba, 2600 m alt., in grounds of Botanic Garden. Amongst grass and weeds. Flowers blue-purple.

January 25, 1971; *Hawkes, Hjerting, Cribb, and Huamán 4242* (CIP). Vicinity of the brewery Colon, 2700 m alt., near Cochabamba. January 27, 1949; *Ochoa 661* (GH, OCH, US). About 2 km N of Calacala, 2700 m alt., near Cochabamba, on stony soil among thickets. Abundant. Plants some 0.6-0.7 m tall, flowers dark violet, fruits globose to ovoid, up to 2 cm in diameter. January 27, 1949; *Ochoa 662* (OCH). Vicinity of the brewery Taquiña, near Cochabamba, 2700 m alt. March 30, 1978; *Ochoa 12028* (CIP, OCH). University of San Simon, growing as weed in the experimental gardens of the University. April 1, 1963; *Ugent and Cárdenas 4579-1, 4579-2, 4580* (WIS). Brewery Colon, about 2 km N of Cochabamba, 2700 m alt., in forests along the road, bordering small gardens, with *S. anomalocalyx, Cleome, Eryngium, Tagetes, Clematis, Euphorbia, Oenothera, Polygonum, Nicotiana glauca,* and nontuberous *Solanum* sp. April 3, 1963; *Ugent 4581* (F), *4582, 4583-1c, 4583-2c, 4584-1b, 4584-2b, 4586-1a, 4586-2a* (WIS), and *4585* (US).

Province Mizque: La Hoyada, 2450 m alt., on road from Aiquile to Totora. March 25, 1978; *Ochoa 12019* (OCH).

Province Quillacollo: On adobe walls along road about 5 km above Quillacollo. February 14, 1960; *Correll, Dodds, Brücher, and Paxman B612* (LL). Km 56.5 from Cochabamba on road to Oruro, 3350 m alt. Barley field and rocks below under bushes. Medium purple flowers, plants 20 cm in field but 70 cm under bushes. February 4, 1971; *Hawkes, Hjerting, Cribb, and Huamán 4344* (CIP). Km 46.5 from Cochabamba on road to Oruro, 2950 m alt., bushy slopes under tree. Flowers medium purple. February 4, 1971; *Hawkes, Hjerting, Cribb, and Huamán 4346* (CIP). Between Suticollo and Parotani, at junction between oil-pipe and highway, 2450 m alt., along roadside in dry place, growing with *S. berthaultii* [No. 4411]. February 15, 1971; *Hjerting, Cribb, and Huamán 4412.* About 20 km from Cochabamba, 3000 m alt. Chromosome number: 2n=2x=24. March 30, 1978; *Ochoa 12030, 12031* (CIP, OCH). Liriuni, above Escuela Bella Vista, 3500 m alt., about 17 km NW of Cochabamba (air distance), 17°20′ S and 66°20′ W. Roadside thicket with *S. brevicaule, Galium, Tagetes, Salvia,* and nontuberous *Solanum* sp. April 9, 1963; *Ugent 4656* (OCH).

Province Tarata: Llallaguani, 2900 m alt., near Anzaldo. In potato fields, under the shade of shrubs in moist soil. January, 1944; [collected by H. Gandarillas, but assigned under *Cárdenas 3508*, in Cárdenas' herbarium (CPC)]. Near Angostura, 2550 m alt., under shrubs and trees. February, 1944; *Cárdenas 3518* (CPC, cotype of *S. anomalocalyx* var. *llallaguanianum*). [As in the case of certain other Cárdenas' collections, his number *3518* does not follow chronologically with number *3510* (=*S. brevicaule*), collected in March, 1944.]

Department Chuquisaca

Province Nor Cinti: Vicinity of Tacaquira, 3000 m alt., between Muyuquiri-Chaco and Camargo. In maize fields. Chromosome number: 2n=2x=24. March, 1978; *Ochoa 11946* (OCH).

Province Oropeza: Above Sucre, 2592 m alt. [8500 ft., according to the collector]. Growing apparently as a weed in maize fields, in very dry soil, almost confined to the hills. It also grows among stones and margins of cultivated fields. Tubers round to ovoid, 4 cm long by 2 cm wide, skin brown, flesh somewhat transparent and aqueous. Leaves to 9 cm long, provided with 4 pairs of lateral leaflets and numerous interjected leaflets. March 6, 1939; *Balls 6146* (US). Near Sucre, 2592 m alt. [8500 ft., according to the collectors], steep dry sandy or shaley slope. March 6, 1939; *Balls, Gourlay, and Hawkes 6149* (BM, CPC, UC, US). Guerraloma, 3019 m alt. [9900 ft., according to the collectors], near Sucre. March 8, 1939; *Balls, Gourlay, and Hawkes 6189* (CPC). Villa Maria, about 10 km NE of Sucre. April 12, 1963; *Ugent 4930-39* (from plants grown at Exp. St. at Sturgeon Bay, Wisconsin, USA).

Department La Paz

Province Loayza: Viloco, 4500 m alt., large tin mine, approximately 160 km from Oruro on road to Eucaliptus. March 27, 1949; *Brooke 3034* (BM).

Province Manco Kapac: Copacabana, 4300 m alt. [13,000 ft., according to the collector], growing in stone walls and in sandy soil, among moist, organic matter along the margins of potato fields. Tubers oval-round or flattened, 1.5 x 1.3 x 1.0 cm, usually much smaller. April 18, 1939; *Balls 6537* (US). Vicinity of Casani, near Copacabana, 3850 m alt. Chromosome number: 2n-2x-24. December 24, 1977; *Ochoa 11820* (CIP, NY, OCH, US).

Province Murillo: Obrajes, 3660 m alt. [12,000 ft., according to the collector], near La Paz. Garden weed. Common name *quipa choque*. January 28, 1939; *Balls, Gourlay, and Hawkes 5890* (CPC). Calacoto [not 'Chalcoto'], 3500 m alt. [11,500 ft., according to the collectors], Valle de Obrajes, descending to La Paz. Growing as weed along trails and gardens. Plants to 15 cm tall, flowers purple, 2 cm in diameter, tubers very small, white. January 28, 1939; *Balls, Gourlay, and Hawkes 5895* (CPC (=EPC-72), UC, US). Below La Paz, 3800 m alt. [12,500 ft., according to the collectors], growing wild in dry, rocky steep slopes. January 29, 1939; *Balls, Gourlay, and Hawkes 5903* (CPC, K). La Paz, 3500 m alt., in gardens. February, 1955; *Brücher 563* (LL). Obrajes, 3350-3800 m alt. January 26, 1907; *Buchtien 771* (US, lectotype of *S. sparsipilum*, here designated; isotypes, E, G, MO; type collection of *S. tuberosum* subsp. *sparsipilum*). Calacoto, 3600 m alt. February, 1946; *Cárdenas 3640* (CPC). Near Obrajes, 3550 m alt., south of La Paz. February, 1953; *Ochoa 2074* (OCH, topotype of *S. tuberosum* subsp. *sparsipilum*).

Province Omasuyos: Huichi Huichi, 3810 m alt., near Achacachi. Chromosome number: 2n=2x=24. March 7, 1978; *Ochoa and Salas 11911* (CIP, F, NY, OCH, US).

Department Potosi

Province Tomas Frias: Finca Cayara, about 20 km NW of Potosi, along walls. February 19, 1960; *Correll, Dodds, Brücher, and Paxman B635* (LL, S).

Province Modesto Omiste: Moraya, growing under the protection of acacias in maize fields. February 23, 1960; *Correll, Dodds, Brücher, and Paxman B658* (LL).

Province Saavedra: On brushy slope about stone wall, about 50 km from Sucre on road to Potosi. Plants large, to 1 m tall; flowers lilac, rotate-pentagonal. February 17, 1960; *Correll, Dodds, Brücher, and Paxman B625* (K, LL); *B626* (LL, S, US); and *B627* (LL). Some 100 km from Sucre on road to Potosi, among acacias near stone walls. February 18, 1960; *Correll, Dodds, Brücher, and Paxman B629* (K, LL). Cancha Cancha, 3030 m alt., between Millares and Betanzos, fruits globose 12 mm in diameter, dark green along the basal third, slightly pigmented with purple-lilac. Chromosome number: 2n=2x=24. March 14, 1978; *Ochoa 11940* (OCH).

Crossability

Natural hybrids.– Few naturally occurring hybrids of this species are known from Bolivia. However, putative hybrids of *S. sparsipilum* × *S. leptophyes* and *S. sparsipilum* × *S. tapojense* have been located by the author at Casani (3840 m), a small Bolivian village situated near the border of Peru on the Lake Titicaca peninsula of Copacabana. These hybrids tend to be intermediate between their parental forms with respect to plant habit and/or vigor, corolla shape and color, pubescence, leaf dissection, and leaflet size and shape.

Solanum sparsipilum crosses naturally with *S. raphanifolium* in southern Peru. Hybrid plants from the Cusco Valleys of Ollantaytambo, Sagrado, and Vilcanota, tend to resemble *S. raphanifolium* in habit, leaf shape, and in the form and color of their flowers and fruit.

In other parts of southern Peru, this species hybridizes naturally with the above species and with *S. tapojense*, *S. bukasovii*, and *S. leptophyes*. Such crosses may be further complicated by backcrosses of the F_1 hybrids with one or more of their parental species, making exact identification of the taxa that are involved here very difficult.

Pure populations of *S. sparsipilum* in southern Peru are rarely seen. Thirteen plants collected at random by Iltis and Ugent (No. 933) from a maize field in the Urubamba Valley of Peru, for example, contained twelve typical plants of this species and one hybrid of *S. sparsipilum* and *S. raphanifolium*. Still other population samples that were obtained by these same collectors from Oropeza (*Ugent 3983*) and the Quebrada de Pumahuanca (*Iltis and Ugent 982*) contained even greater mixtures of pure and hybrid plants.

Artificial hybrids.– Although more studies need to be conducted with regard to the crossability of *S. sparsipilum* with other species, it is obvious from the results already obtained at CIP that this plant may have some potential for further economic development. In particular, it should be noted that *S. sparsipilum* shows resistance to several different plant pests and diseases. Thus, the author has been informed by his fellow co-worker at CIP, Parvis Jatala, that this is the only species known so far that shows some resistance to the root nematode, *Meloidogyne incognita acrita* (Intl. Pot. Ctr., 1975) and the cyst nematode, *Globodera pallida* (*ibid*, 1979, 1980, 1982). Similarly, Jatala and Martin (1978) and Schmiediche and Martin (1983) of CIP have reported that it is resistant to *Pseudomonas solanacearum*, and to the damaging potato tuber moth species, *Phthorimaea operculella*. Resistances to *M. incognita, M. javanica*, and *M. arenaria*, as well as to *P. solanacearum*, have been also found in *S. sparsipilum* by Gómez et al. (1983a, b). Moreover, according to S. K. Bhattacharya, who also conducted some studies at this institution, this species strongly resists the fungal attack of *Macrophomina phaseoli. Solanum sparsipilum* is also resistant to drought (Vargas, 1952).

To the taxonomist, much valuable information regarding the origins of naturally occurring hybrids can be obtained through the study of experimental crosses. The results of some of the more significant studies conducted along these lines at CIP are summarized in the following paragraphs.

In general, few genetic barriers to gene transfer between *S. sparsipilum* and the cultivated diploid species exist. Abundant berries, for example, are obtained in crosses of *S. sparsipilum* (7611, 11911, 13786) and *S. goniocalyx* f. *papa amarilla*. These contained an average of 120 fertile seeds per fruit. Berries were also easily obtained in crosses between *S. sparsipilum* (13757) and *S. phureja* (15133). However, the latter cross yielded an average of only 20 fertile seeds per fruit.

Much success was also had in crosses between *S. sparsipilum* and certain other wild Bolivian diploid species. Berries, for example, were easily obtained in crosses of *S. sparsipilum* (12030) with *S. berthaultii* (12029), *S. candolleanum* (11814) with *S. sparsipilum* (12028), and *S. sparsipilum* (12030) with *S. tarijense* (12001). These hybrid combinations yielded an average of 80 seeds per fruit.

In contrast to the above crosses, hybrids of *S. leptophyes* (13542) and *S. sparsipilum* (13564) were more difficult to obtain. Many pollinizations were required to obtain berries with hybrid seed, and these contained an average of only 40 seeds per fruit.

Even less fertility was observed in crosses involving two other introductions of the same two species. Thus, very few berries, containing an average of only

10 seeds per fruit, were obtained in crosses between *S. leptophyes* (13731) and *S. sparsipilum* (13774). The seeds from the above crosses all had well-formed embryos, but showed little development of the endosperm. A reduced level of fertility was also observed in the reciprocal cross, *S. sparsipilum* (13718) × *S. leptophyes* (13558). It is of interest to note that some F_1 individuals obtained in crosses between *S. leptophyes* and *S. sparsipilum* resembled natural hybrids of *S. sparsipilum* from Casani, the latter located on the Copacabana Peninsula of Lake Titicaca. However, the Casani hybrids, as pointed out above, may have their origin not only in crosses and backcrosses involving these two species, but also in crosses between *S. sparsipilum* and *S. tapojense*.

Experimental crosses between different introductions of *S. sparsipilum* and *S. megistacrolobum* gave different results. For example, crosses between *S. megistacrolobum* (11941) and *S. sparsipilum* (12028) or *S. megistacrolobum* (11987) and *S. sparsipilum* (12030) were unsuccessful. However, crosses between *S. megistacrolobum* (11964) and *S. sparsipilum* (12028, 12030), and *S. megistacrolobum* (11987) and *S. sparsipilum* (12028) yielded an average of 70 viable seeds per berry.

Unexpected results were also obtained in a cross of *S. sparsipilum* (12028) with another Bolivian diploid species, *S. alandiae* (12018). In this instance, although berries were produced, the number of seeds per fruit were unusually low, averaging less than 20.

Crosses between *S. sparsipilum* (13564) and the Bolivian tetraploid (2n=48) species, *S. acaule* (13865), yielded abundant fruit, containing an average of 100 viable seeds per berry. As might be expected from a diploid × tetraploid cross, the F_1 plants of this parentage were sterile.

Examples of the range of morphological variation encountered in some typical crosses involving this species are given in the following paragraphs.

Solanum alandiae (12018) × *S. sparsipilum* (12028).– Individuals of this parentage greatly resemble *S. alandiae* in leaf shape and in having long, many-flowered peduncles. They are also similar to this species in having long white-tipped, lilac-colored pentagonal corollas.

Solanum sparsipilum (12030) × *S. berthaultii* (12029).– Plants of this cross resemble *S. sparsipilum* in having a small, dark purple corolla. The pentagonal shape of the corolla, the long styles and the peculiar form and dissection of the leaves, however, are features derived from *S. berthaultii*. But, unlike either parent, these hybrids have very low pedicel articulation, the joint being located, in many cases, almost at the base of the stalk.

Solanum candolleanum × *S. sparsipilum*.– Hybrids of this parentage differ in appearance depending on the particular accession used and the direction of the cross. Plants derived from crosses of *S. candolleanum* (11835) with *S. sparsipilum* (12030), and *S. sparsipilum* (12030) with *S. candolleanum* (11913), for example, are similar to *S. candolleanum* in leaf shape and in having tall, vigorous stems, but with respect to their floral features, they are more like *S. sparsipilum*. Plants derived from crosses of *S. candolleanum* (11814) with *S. sparsipilum* (12028), and *S. candolleanum* (11897) with *S. sparsipilum* (12028), on the other hand, while similar to the first series of crosses in stem and leaf characteristics, they differ in having very long pedicels, these ranging from 40-50 mm in length, and an articulation that occurs at or below the center of the stalk. These also show segregation with respect to floral characteristics. In contrast to the above hybrids, plants derived from crosses of *S. candolleanum* (11835) with *S. sparsipilum* (12028) show much segregation with respect to both leaf and floral characteristics. These ranged from individuals that were unlike both parents in having large pubescent leaves that were dissected into 8 pairs of elliptic-lanceolate, sessile leaflets and numerous interjected leaflets, to plants that were very similar in appearance to *S. candolleanum*. All were like *S. sparsipilum*, however, in having light violet to dark purple-colored flowers. Moreover, all had pedicels that were articulated in the upper one-third of their stalks.

29. Solanum sucrense Hawkes, Bull. Imp. Bur. Pl. Breed. & Genet. Cambridge (June):51, 126, Fig. 41. 1944. FIGS. 131-133.

S. sucrense var. *brevifoliolum* Hawkes, Bull. Imp. Bur. Pl. Breed. & Genet., Cambridge (June):51. 1944, *nom. nud.* Based on *Balls, Gourlay, and Hawkes 6149*, Bolivia. Near Sucre, Dept. Chuquisaca, Prov. Oropeza, March 6, 1939 (CPC, not seen).

Plants light green, vigorous. Stems with 2-3(-5) cm-long internodes, to 60 cm tall by 15 mm or more in diameter at base, flexible, branched, glabrescent, and with sinuous, 1.5-2 mm-broad wings. Stolons 30-40 cm long; tubers abundant, small, white, 1.5-2.5 cm long, ovoid or rarely round. Leaves imparipinnate, short, and broad, (5.5-)7-16.8(-28) cm long by (3.5-)4-11.5 (-19.5) cm wide, with 3-4(-5) pairs of leaflets, borne on petioles 1.5-2.5(-5) cm long, sparsely pilose with short, stiff multicellular hairs above and nearly glabrous below except for veins and veinlets. Interjected leaflets 4-8, sessile or subsessile, of several different sizes, ranging from 2 to 6 mm long. Leaflets elliptic-lanceolate or broadly elliptic-lanceolate to ovoid-lanceolate, the apex obtuse or pointed to subacuminate, and the base slightly attenuate to rounded. Terminal leaflets somewhat larger and broader than the laterals, (2.6-)4.7-7 (-10.6) cm long by (1-)2-3(-6.5) cm wide. Lateral leaflets asymmetrically rounded to obliquely rounded at base and borne on petiolules to 10 mm long. Pseudostipular leaves conspicuous, large and broadly falcate, to 20 or more mm long and 10 mm or more wide. Inflorescence cymose to cymose-

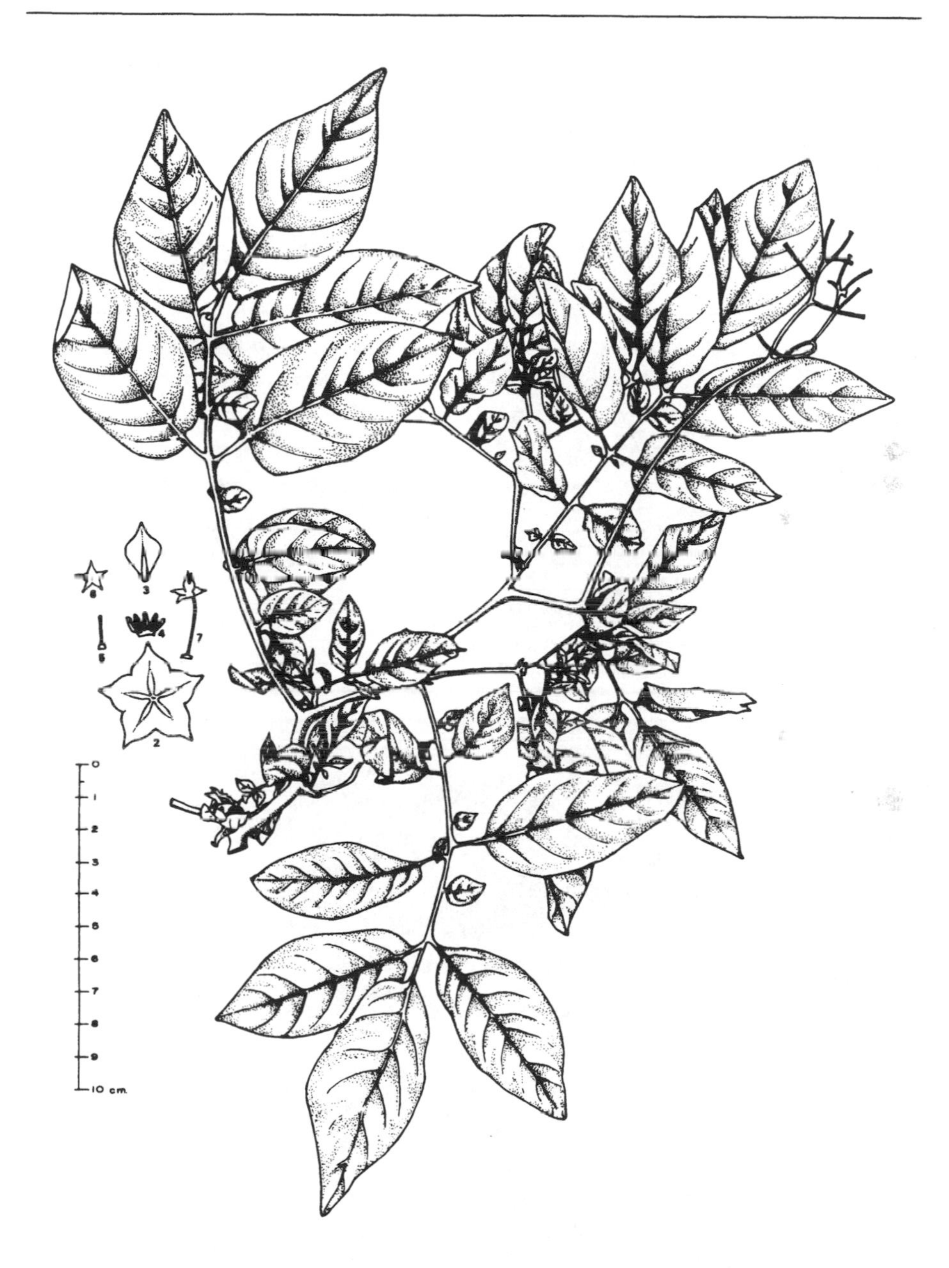

Figure 131. – *Solanum sucrense* (*Ochoa 11926*). 1. Upper part of plant. 2. Corolla. 3. Petal. 4. Stamens, dorsal view. 5. Pistil. 6. Calyx. 7. Pedicel and calyx. All × ½.

paniculate, 8-14-flowered, borne on light green, 10-15 mm-long peduncles, the latter sparsely pilose as are the slender pedicels. Pedicel articulation high, above center or in the upper one-third of the stalk, the upper part 6-8(-13) mm long and the lower 10-14(-17) mm long. Calyx usually small, 5.5-6.5(-8) mm long, slightly pigmented, sparsely and shortly pilose, symmetrical or occasionally bilabiate, the elliptic-lanceolate, shortly attenuate lobes bearing 1.5 mm-long acumens. Corolla pentagonal, light blue-violet or purple with light-gray nectar guides, 2.5-3 cm in diameter, and bearing 6-8 mm-long lobes, the latter with poorly delimited acumens. Staminal column truncate-conical, asymmetric, small; anthers 5.5 mm long, broadly lanceolate, cordate at base, and borne on short, glabrous, 0.5-0.8 mm-long filaments. Style 10.5-11.5 mm long, exerted 5 mm, sparsely papillose toward base; stigma capitate, cleft. Fruit deciduous, globose, 1.5-2 cm in diameter, dark green except for the brownish-purple base.

Chromosome number: 2n=4x=48.

Type: BOLIVIA. Department Chuquisaca, Province Oropeza, Azari, 2550 m alt., 7 km from Sucre. March 7, 1939; *E. K. Balls, W. B. Gourlay, and J. G. Hawkes 6169* (lectotype, here designated, K; isotypes, CPC, UC).

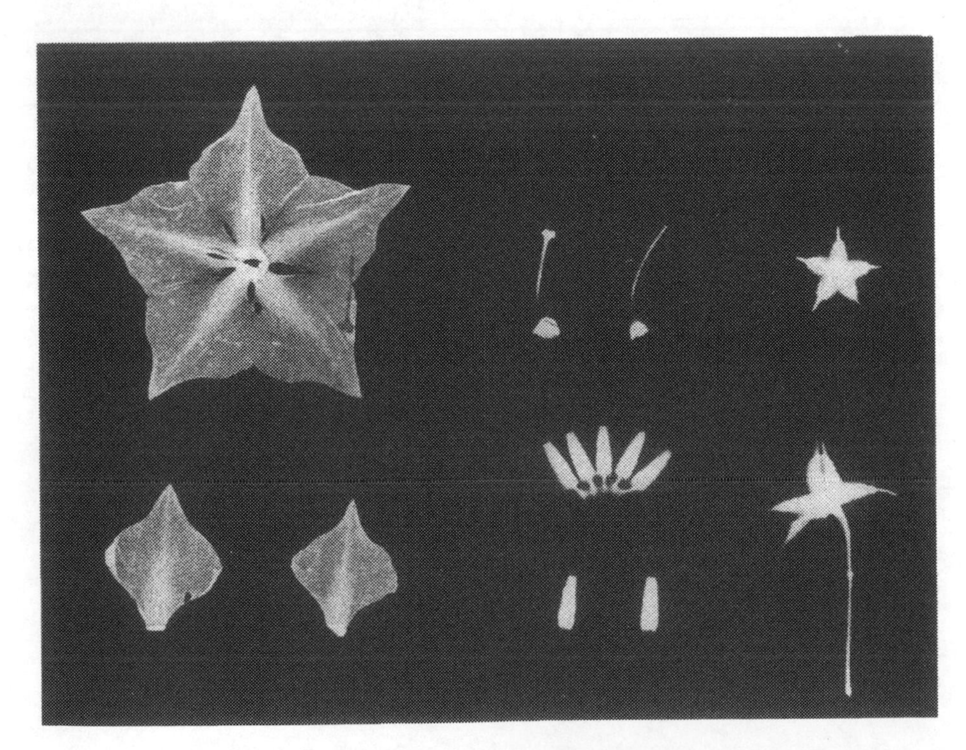

Figure 132. – Floral dissection of *Solanum sucrense* (*Ochoa 11926*), ×1.

Affinities

According to Astley and Hawkes (1979), this species, which is known in Bolivia as *Alcco papa*, or 'dog potato,' may be a hybrid of *S. tuberosum* subsp. *andigena* and a tetraploid chromosomal race of *S. oplocense*. This origin is supported in part by the following two facts: 1) segregate plants of *S. sucrense* and *S. oplocense* occur with *S. tuberosum* subsp. *andigena* in central Bolivia; and 2) experimental hybrids of *S. tuberosum* subsp. *andigena* (4x) and *S. oplocense* (4x) are very similar to naturally occurring plants of *S. sucrense*. Cytologically, however, *S. sucrense* appears to be an autotetraploid species; thus, Astley and Hawkes' hypothesis requires further verification.

It may be of interest to note here that *S. sucrense* grows as a 'weed potato' in Bolivia. Although tetraploid (2n=48), *S. sucrense* greatly resembles the diploid (2n=24) species *S. sparsipilum*, it also has some resemblance to the cultivated tetraploid potato, *S. tuberosum* subsp. *andigena*. In particular, it shares with the latter taxon a tendency to form vigorous, branched stems, and both have a common leaf shape. Thus, it is possible that *S. sucrense* and *S. sparsipilum* have played an important role in the evolution of the cultivated potato.

Figure 133. – *Solanum sucrense* in its type locality, Valle de Azari, 2700 m altitude, Province Oropeza, Department Chuquisaca, Bolivia. Growing on flat stones topping old adobe walls.

Habitat and Distribution

This species grows preferentially in the Subpuna or the Zona de los Valles de Piso Montano, as well as in the spiny steppe regions of the Departments of Potosi and Chuquisaca at altitudes ranging between 2400 and 3000 m. In these two regions, it grows as a weed in cultivated fields and farm gardens. It also occurs on eroded lands and in poor, dry, sandy-clay soils.

In the dry, narrow valley of Azari, the type locality of *S. sucrense*, it is found along old fence rows and in maize fields. In Azari, as well as between Sucre and Yotala, it also occurs under the shade of *Schinus molle*, *Prunus serotina*, *Juglans regia*, and *Prosopis ferox*. In addition, it occurs under the branches of shrubs, such as *Sonchus*, *Dodonaea*, *Physalis*, and *Solanum radicans*, between bunches of the invading grass species, *Pennisetum clandestinum*, and under the protective branches of columnar cacti, such as *Trichocereus* and *Cleistocactus*. Occasionally, it is also found with *S. oplocense* and *S. berthaultii*.

Specimens Examined

Department Chuquisaca

Province Nor Cinti: Sivingamayo. March, 1949; *Cárdenas s.n.* (HNF, No. 0071).

Province Oropeza: Azari, 2550 m alt., about 7 km from Sucre, 'growing as weed in cultivated field.' March 7, 1939; *Balls, Gourlay, and Hawkes 6169* (lectotype, K, here designated, includes a good floral dissection and an excellent specimen of a complete plant; isotypes: CPC, UC). Azari, 2600 m alt., near Sucre, 2600 m alt., 'growing as weed in cultivated field.' March 7, 1939; *Balls, Gourlay, and Hawkes 6170* (CPC, topotype). Azari, 2700 m alt., growing abundantly in maize fields. Stolons very long, 1 m or more long, tubers small, 3-4 cm in diameter, ovate, skin white or white-yellowish, buds creamy-white, slightly pigmented with light purple at base and apex, flesh white. Berries abundant, globose, deciduous or subdeciduous. Common name: 'Alcco papa' or 'dog potato.' Chromosome number: $2n=4x=48$. March 11, 1978; *Ochoa 11926* (CIP, OCH, US, topotypes). Azari, 2700 m alt., near old adobe walls. March, 1984; *Ochoa and Cagigao 15571* (CIP, OCH, topotypes). Between Sucre and La Glorieta, 2800 m alt. March, 1984; *Ochoa and Cagigao 15573* (CIP, OCH). Between La Glorieta and Campanario, 2550 m alt. March, 1984; *Ochoa and Cagigao 15574* (CIP, OCH, US). Campanario, 2450 m alt., along road from Sucre to Yotala. March, 1984; *Ochoa and Cagigao 15575* (CIP, OCH, US). Between Campanario and Cabezas, 2400 m alt., along road between Sucre and Yotala, on stony slope in cultivated field. March, 1984; *Ochoa and Cagigao 15577* (CIP, OCH, US).

Province H. Siles: Some 25 km from Sucre, 2650 m alt., along road from Sucre to Potosi. Among spiny shrubs. Leaflets small; growing about 20 m from *S. tarijense*, hybrids not present. March 6, 1971; *Hjerting, Cribb, and Huamán 4567* (CIP).

Province Yamparaez: Between Yotala and Calera, 2900 m alt., on road to Potosi. March, 1984; *Ochoa and Cagigao 15581* (CIP, OCH). Between Sucre and Yamparaez, 2900 m alt., along road between Sucre and Tarabuco. March, 1984; *Ochoa and Cagigao 15582* (CIP, OCH).

Department Potosi

Province Frias: Before Tarapaya, 3500 m alt., 23 km from Potosi on road to Oruro, in maize field and at edge of road. March 8, 1971; *Hjerting, Cribb, and Huamán 4594* (CIP).

Crossability

Solanum sucrense has been reported to be a potentially valuable source of germplasm for potato improvement programs. According to Rothacker et al. (1966), this species is resistant to some pathogenetic races of the golden nematode, *Heterodera rostochiensis* (=*Globodera pallida*), as are some clones of *S. sparsipilum*, *S. oplocense*, and *S. tuberosum* subsp. *andigena*.

The results of studies conducted at CIP on the reproductive behavior of *S. sucrense* suggest that this species is partially or completely self-incompatible. Fruits are not obtained when this species is artificially selfed, and only few fruits and seeds are obtained in sibling crosses. This behavior appears to be rare in polyploids, except among plants of autotetraploid origin.

Crosses between *S. sucrense* and *S. tuberosum* were made with difficulty. Generally, only few berries and few seeds were obtained.

According to Luis Salazar (CIP, 1983, personal communication), at least one accession of *S. sucrense* (*Ochoa 11926*) has a hypersensitive type of resistance to the potato virus complex, PVX-HB. No other wild potato species is known to have resistance to this virus. Moreover, Salazar reports that *S. sucrense* shows an immune type of resistance to complex PVX-CP. Both immunogenic reactions are governed by a tetrasomic, monogenic duplex condition, i.e., by a simple dominant gene (Intl. Pot. Ctr., 1983; C. R. Brown, 1983, personal communication). Because quadrivalent formation occurs in this species, it appears to be autotetraploid in origin.

Finally, Jaworski et al. (1984) have shown that at least one accession of this species (PI 458391) is resistant to race 1, biotype 1 of *Pseudomonas solanacearum*. The only other tuber-bearing species of *Solanum* that was determined by them to have a high level of resistance to this pathogen was *S. tuberosum* cv. Noordeling (PI 109760). It should be noted, however, that Jaworski et al. examined some 514 accessions representing 72 species of tuber-bearing *Solanum* and over 200 crosses before this information was obtained.

30. Solanum vidaurrei Cárdenas, Bol. Soc. Peruana Bot. 5:26-30, Plate II-E, Figs. 1-2, Plate IV-B. 1956. FIGS. 134-136; PLATE XVIII.

Plant delicate, usually small and shrubby, sparsely pilose throughout with white, short, thick hairs. Stems 15-40(-60) cm tall by 1.5-2.5 mm in diameter at base, erect, cylindrical, usually simple, spotted with purple along the lower two-thirds of its length, wingless, or with very narrow, barely distinguishable linear wings. Stolons 1-2 m long; tubers small, yellowish-white, 1-1.5(-3) cm in diameter, rounded to ovate. Leaves imparipinnate, 6.5-14(-22) cm long by 3.5-7(-15) cm wide, with 3-4(-5) pairs of leaflets. Interjected leaflets, several or absent. Leaflets linear-lanceolate to narrowly elliptic-lanceolate, with sparsely pilose upper and lower surfaces and slightly denticulate and pilose margins. Terminal leaflets slightly larger to about the same size as the laterals, narrowly lanceolate to oblanceolate. Lateral leaflets, 2-2.5(-5) cm long by 0.4-0.6(-1) cm wide, the upper pair slightly smaller than the second and third pairs, and frequently decurrent on the narrowly winged rachis, sessile or shortly petiolulate, the apex obtuse or pointed, the base narrowly cuneate, borne upon 1-3 mm-long petiolules. Pseudostipular leaves small, narrowly subelliptic or subfalcate, 7-8 mm long by 2.5-3 mm wide. Inflorescence cymose, 2-5(-9)-flowered, borne on 4-6 cm-long, light green peduncles, the latter sparsely pilose with short hairs and occasionally spotted with light purple. Pedicels articulated above the center or in the upper one-third of the stalk, the upper part 5-9(-11) mm long and the lower 10-15 mm long. Calyx 5-6(8) mm long, light green and slightly pigmented, pilose, the narrowly lanceolate lobes abruptly narrowed to pointed, nearly filiform acumens, 2-3 mm long. Corolla subpentagonal to rotate-pentagonal or rarely rotate, 2-2.5 cm in diameter with acumens up to 5 mm long, dark lilac to purple with greenish-yellow nectar guides. Staminal cones truncate-conical, compact, slightly asymmetrical; anthers narrowly lanceolate, 5.5 mm long, cordate at base, borne on 1 mm-long, glabrous filaments. Style narrow, 10-11 mm long, exerted 3 or more mm, densely papillose below center; stigma small, very dark green, capitate. Fruit globose to ovoid, to 2 cm long, dark green and white verrucose and irregularly tinged with violet-purple.

Chromosome number: 2n=2x=24 and 2n=2x=48.

Type: BOLIVIA. Department Chuquisaca, near km 150, between Potosi and Camargo, 2200 m alt. February, 1949; *M. Cárdenas 5075* (syntype, HNF-00092).

The type collection of *S. vidaurrei*, which is housed in the Herbario Nacional Forestal de Cochabamba, Bolivia (HNF), consists of two specimens mounted on one sheet. One of the two specimens has no flowers and its label includes

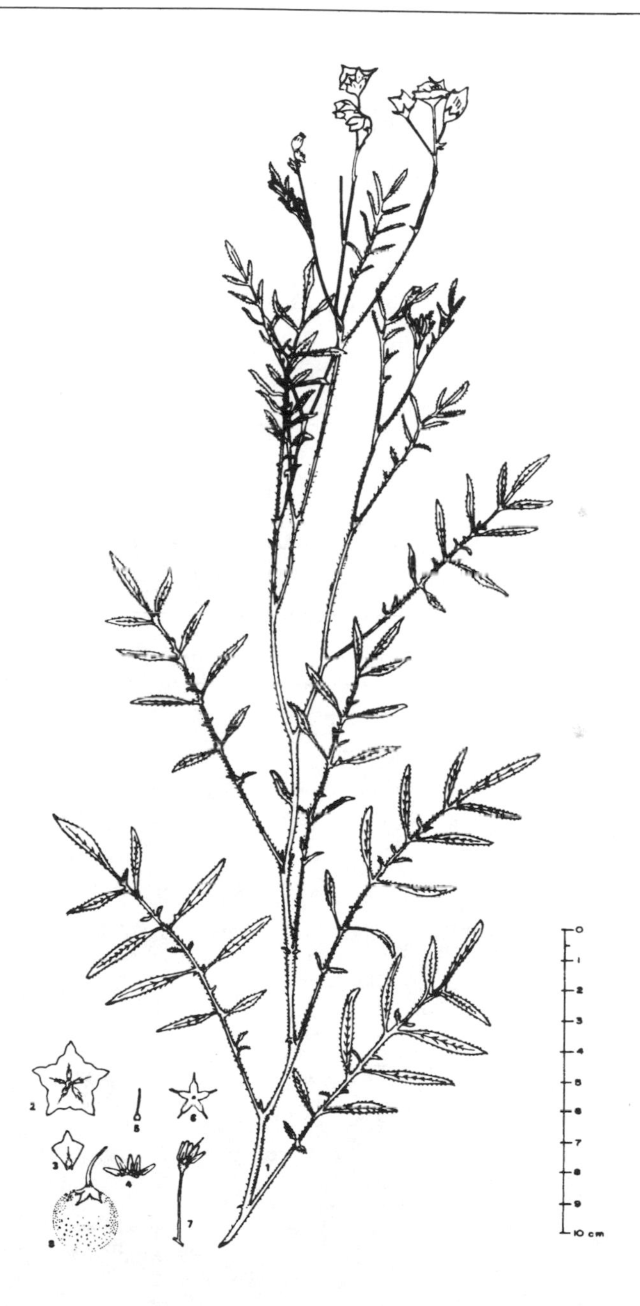

Figure 134. – *Solanum vidaurrei* (*Ochoa 11952*). 1. Flowering plant. 2. Corolla. 3. Petal. 4. Stamens, dorsal view. 5. Pistil. 6. Calyx. 7. Pedicel and stamens. 8. Fruit. All × ½.

only the collection number and field data. The other specimen, which is complete, well preserved, and retains its natural colors, has long, narrowly lanceolate light green leaflets and small, rotate-pentagonal purple flowers. Its label, aside from the usual collection data, has been annotated 'cotype' in the handwriting of the collector.

Few collections of *S. vidaurrei* are known. Only 4 Bolivian and 2 Argentine collections of this species are cited by Hawkes and Hjerting (1969). The author himself has collected this plant from only two localities, Sajlina (Sud Cinti) and km 150 between Potosi and Camargo (Nor Cinti), both in the Department of Chuquisaca, Bolivia. The latter locality, known as 'Quebrada Honda,' is situated at an elevation of 3400 m, and not at 2200 m, as reported originally by Cárdenas.

Affinities

This plant is more closely related to *S. oplocense* than to any other known wild potato species. Characteristics shared by *S. vidaurrei* and *S. oplocense* include a pentagonal to subpentagonal corolla and decurrent upper leaflets with denticulate and pilose margins. While the leaves and leaflets of hexaploid (2n=72) *S. oplocense* are, in general, much larger and more broadly ovate to elliptic than those of *S. vidaurrei*, tetraploid (2n=48) plants of *S. oplocense* have leaves and leaflets that compare more favorably in size and shape to those of *S. vidaurrei*. Moreover, while the berries of *S. oplocense* are larger than those of *S. vidaurrei*, they are of similar shape and color.

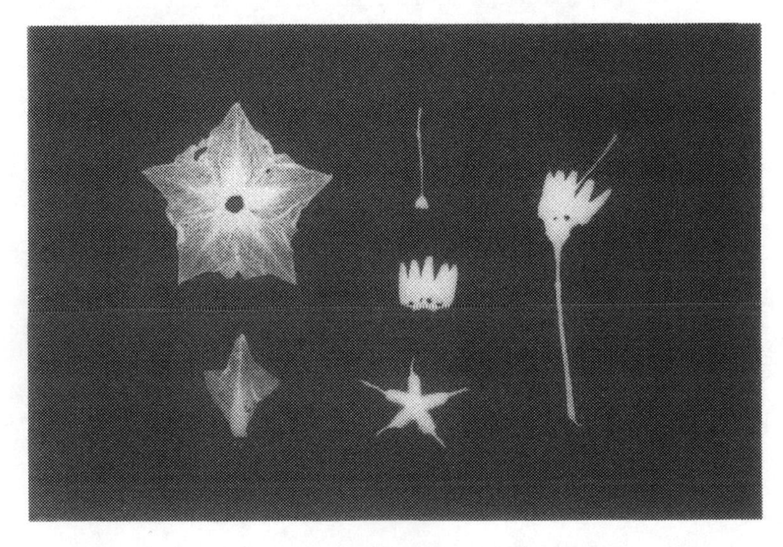

Figure 135. – Floral dissection of *Solanum vidaurrei*, 2n=48 (*Ochoa 11952*), ×1.

Solanum vidaurrei also shares some features in common with *S. leptophyes*; for example, both have delicate stems and narrow leaflets. They differ, however, in characteristics of the indumentum, leaflet margins, and floral details.

Correll (1962) believes this species is related to a complex, closely knit group of Argentine and Bolivian species. The various members of this group, which include *S. infundibuliforme*, *S. setulosistylum*, *S. leptophyes*, and *S. puberulofructum*, tend to hybridize with one another in nature; thus, it is often difficult to arbitrarily delimit species within this group. Nevertheless, Correll compares this species with *S. infundibuliforme*, to which he feels it is most closely related. It should be noted that small, immature, 15-20 cm-tall plants of *S. vidaurrei* are especially similar to *S. infundibuliforme*. However, the two are markedly different with respect to the shape, decurrency, and pilosity of their leaflets.

Habitat and Distribution

This species grows preferentially in the Prepuna formations of the Department of Chuquisaca in southern Bolivia. There it grows in dry, rocky soil in the shade of acacia and other xerophytic to subxerophytic plant species at elevations ranging between 2600 and 3400 m, though it is more commonly found between 2700 and 2800 m. It also occurs in the Provinces of Catamarca and Salta in northern Argentina.

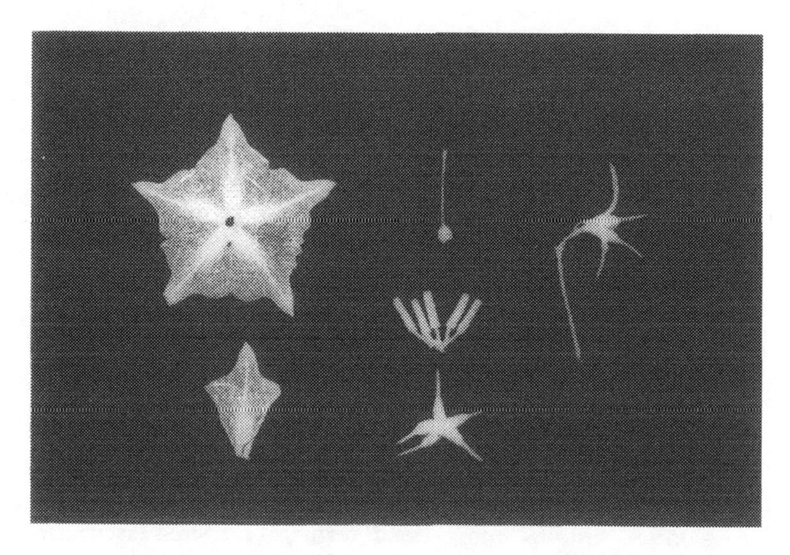

Figure 136. – Floral dissection of *Solanum vidaurrei*, 2n=24 (*Ochoa 12003*), ×1.

Specimens Examined

Department Chuquisaca

Province Nor Cinti: Km 150 along the road from Potosi to Camargo, 2200 m alt., on slate soil. February, 1949; *Cárdenas 5075* (HNF-00092). About 6 km above Quebrada Honda, 3400 m alt., near km 150 of the road from Potosi to Camargo. Possible type locality of this species. Chromosome number: 2n=2x=24. March 22, 1978; *Ochoa 12003* (CIP, OCH).

Province Sud Cinti: About 25 km from Culpina along the road to Camargo. On rocky slope, under the shade of acacias; flowers purple. February 21, 1960; *Correll, Dodds, Brücher, and Paxman B642* (LL). Near Sajlina, 2980 m alt., along the road from Camargo to Culpina. Between forests of *Chulque* or *Acacia* sp., dry region. Chromosome number: 2n=4x=48. March 16, 1978; *Ochoa 11952* (CIP, OCH).

Crossability

Tetraploid individuals of this species are self-fertile and can be crossed with diploid plants of *S. sparsipilum*, though with some difficulty. Although the resultant hybrids are triploid and tend to be highly sterile. In some cases, these hybrids yield empty fruits or fruits with only 1-3 seeds per berry, containing an embryo but no endosperm. Crosses between tetraploid *S. vidaurrei* and two wild Bolivian diploid species, *S. boliviense* (11935) and *S. tarijense* (12000 and 12001), are completely incompatible. On the other hand, crosses between tetraploid *S. vidaurrei* and tetraploid *S. oplocense* yield normal-sized fruits containing up to 40 seeds per berry, these with endosperm and fertile embryos.

Crosses involving diploid *S. vidaurrei* and other wild diploid or tetraploid species of potato were, in general, easily made. The results of reciprocal cross-pollinations between *S. vidaurrei* (12003) and *S. sparsipilum* (13644) were 100% successful. These yielded an average of 160 to 180 fertile seeds per berry. Similarly, much success was obtained in crosses involving the above introduction of *S. vidaurrei* and tetraploid *S. acaule* (14947), in which an average of 130 fertile seeds per berry was obtained. Somewhat less successful were the crosses between the above introduction of *S. vidaurrei* and *S. candolleanum* (15024), in which the plants yielded an average of 50 seeds per berry. And finally, the fewest seeds of all were obtained in crosses involving this species and diploid *S. megistacrolobum* (14273). In this latter case, an average of only 12 seeds per berry was obtained.

Solanum vidaurrei × *S. boliviense*.– Individuals derived from crosses between diploid introductions of these two species tend to be similar to *S. vidaurrei*. These have slender, sinuous stems, broad terminal leaflets that are larger than the laterals, 1-3-flowered cymose inflorescences and 20-25 mm-long pedicels that are articulated in the upper one-third of their stalks. All have small light violet to dark purple pentagonal corollas that vary from 2-2.5 cm in diameter.

Solanum vidaurrei × *S. sparsipilum*.– The F_1 populations derived from diploid crosses of these two species are highly variable. Some plants of these populations have leaves that are similar to *S. vidaurrei*, but are less dissected. Others have leaflets that are either entire or denticulate and more broadly elliptic-lanceolate than *S. vidaurrei*, while still others show intermediate leaflet characteristics. All have a small blue or purple pentagonal corolla that varies from 2 to 2.5 cm in diameter. Their pedicels are articulated centrally or in the upper one-third of their stalks.

31. Solanum virgultorum (Bitter) Cárdenas and Hawkes, Jour. Linn. Soc., Bot. 53:103, Fig. 8. 1945. FIGS. 137, 138.

S. boliviense subsp. *virgultorum* Bitt. Repert. Sp. Nov. 12:153. 1913. TYPE: *Mandon 399*, Bolivia. Munaypata, near and above Colomi, vicinity of Sorata, Dept. La Paz, Prov. Larecaja, February, 1860 (lectotype, here designated, K; isotypes, G, P).

Plants small, more or less pubescent. Stems slender, erect, 25-30 cm tall by 2-4(-5) mm in diameter at base, slightly decumbent and sinuous, sparsely pilose, simple or few-branched and narrowly winged, the internodes short, generally 8-15(-20) mm long at base. Tubers small, 1.5-2 cm in diameter. Leaves imparipinnate, small, 6-10 cm long, poorly dissected, with 1-2 pairs of leaflets. Interjected leaflets absent. Terminal leaflets much larger than the laterals, 3.4-5.5 cm long by 2.6-3.8 cm wide, broadly ovate to ovate lanceolate, the apex pointed or shortly acuminate and the base rounded to subcordate. Lateral leaflets ovate-elliptic, the apex usually pointed and the base obliquely rounded, shortly petiolulate, occasionally narrowly decurrent on the rachis, the first upper pair prominently larger than the second pair, 2.4-3.6 cm long by 1.6-1.8 cm wide. Pseudostipular leaves broadly falcate, small and inconspicuous. Inflorescence cymose-paniculate, 3-6(-9)-flowered, borne on slender 4-5 cm-long peduncles, the latter shortly and sparsely pilose as are the pedicels. Pedicels 15-20(-25) mm long, articulated usually at the midpoint. Calyx small, 6 mm long, sparsely pilose, the elliptic-lanceolate acuminate lobes with narrowly triangular-lanceolate, 1-1.5 mm-long acumens. Corolla rotate to rotate-pentagonal, 2.5-3 cm in diameter, light purple or bluish-lilac. Anthers broadly lanceolate, 5.5 mm long, not cordate at base, borne on thick, glabrous, 1-1.5 mm-long filaments. Style 9-10 mm long, papillose below

center; stigma oval to oval-capitate, much broader than the style apex. Fruit globose, green, 1.8 cm in diameter.
Chromosome number: 2n=2x=24.

Type: BOLIVIA. Department La Paz, Province Larecaja, Munaypata, 2650 m alt., near and above Colomi, vicinity of Sorata, among bushes. February, 1860; G. *Mandon 399* (lectotype, here designated, K; isotypes, G, P).

Some thirty-two years following Bitter's (1913) description of *S. boliviense* subsp. *virgultorum* (an entity based on a collection by Mandon, No. 399), Cárdenas and Hawkes (1945) proposed that this plant be given the rank of species. The author, in order to gain a better understanding of the classification of *S. virgultorum*, visited its type locality but was unable to collect it due to adverse weather conditions. Moreover, while the author examined one of Mandon's type collections housed in the Conservatory of Botany at Geneva, this visit did not prove to be especially fruitful due to the poor quality of the specimen involved. However, the author became convinced of the validity of this species after viewing the type collections that exist in the Paris Museum of Natural History and the Kew Herbarium at London, England. Both of the aforementioned specimens are in excellent condition.

The type collections of this species housed in the Herbaria of Kew and Geneva are annotated in Bitter's handwriting. The labels read '*Solanum* (Tuberarium) *boliviense* subsp. *virgultorum* Bitt.' Despite the fact that there are morphological differences between the above specimens, it has not been possible to determine which of the two was used by Bitter for drawing up his diagnosis of *S. boliviense* subsp. *virgultorum*. However, not even Bitter was sure of his determinations. On one of the two sheets, for example, he wrote, '...greater information may be needed.'

Mandon's type collection of this species at Kew Gardens, which is here chosen as the lectotype of *S. virgultorum*, is represented by two plants mounted on one sheet. The smaller of these was used by Cárdenas and Hawkes (1945, p. 103) to illustrate their concept of *S. virgultorum*, while the larger was used by Correll for the same purpose (1962, p. 482).

Affinities

Although this species has several characteristics in common with *S. candelarianum*, it differs in corolla color, pedicel articulation, and leaf dissection. Moreover, this species has smaller and less conspicuous pseudostipular leaves than *S. candelarianum*.

This plant also appears to be related to *S. microdontum* in that they both have similar leaves. However, they differ from one another with respect to many other characteristics.

Figure 137. – Isotype of *Solanum virgultorum* (*Mandon 399*), ca. ×½.
Photo: courtesy of the Natural History Museum of Paris.

Figure 138. – Type locality of *Solanum virgultorum*, Munaypata, 2650 m altitude, in the vicinity of Sorata, Province Larecaja, Department La Paz.

Habitat and Distribution

The type locality of *S. virgultorum*, Munaypata, a place today known as Villa Rosa, is situated on the northeast flanks of the mountain that hovers above the small town of Sorata (*see* Fig. 138). In the vicinity of this locality are found grand *barrancos* and very steep talus slopes, both partially covered with a lush mantle of vegetation.

The area of Villa Rosa is very much as the author remembers since his first visit there some three decades ago. Still to be seen are some palms, trees of *Erythrina*, colonies of *Opuntia tuna* and *Trichocereus* sp., bromeliads, *Solanum radicans*, and other nontuberous species of *Solanum*, composite genera such as *Eupatorium* and *Bidens*, and extensive colonies of *Verbena*. Here the soils are mostly underlaid by shale and are generally very poor. However, crops such as maize and potatoes are still grown in small plots along the rich valley bottoms.

It is in this moist and somewhat temperate environment where Mandon, some 125 years ago, first encountered *S. virgultorum* growing in thickets. While it is possible that this plant may still be lingering in certain isolated localities of this region, it is obvious from the author's own reconnaissance of this area that this species is today very close to extinction.

Solanum virgultorum is a species endemic to Bolivia. It occurs from Ilabaya, Sorata, and Tacacoma in northwest Bolivia (Province Larecaja, Department La Paz) to the serranias of Valle Grande in the southeastern region of the country (Department Santa Cruz). It is found between 2600 and 3600 m altitude.

Specimens Examined

Department La Paz

Province Larecaja: Munaypata, 2650 m alt., near and above Colomi, vicinity of Sorata, among bushes. February, 1860; *Mandon 399* (lectotype, here designated, K; isotypes, G, P). Between Sorata and Ilabaya, 3600 m alt., near maize fields. February, 1946; *Cardenas 3687* (HNF).

Department Santa Cruz

Province Valle Grande: Cincho, 2800 m alt., between Valle Grande and Pucara, on talus slope, some 100 m below highway. Chromosome number: 2n=2x=24. February 28, 1984; *Ochoa and Salas 15546* (OCH).

4

Systematic Treatment: Cultivated Species

Taxonomically, the cultivated potatoes of South America are much more complicated as a group than those grown elsewhere in the world. Several thousand native-named clones have been identified from the Central Andean Region alone. These are mostly grown in small household gardens and field-plots as a subsistence or local market crop, though commercial production occurs in some areas. Geographically, the full range of indigenous cultivation of this crop is enormous, extending from Venezuela and Colombia to Ecuador, Peru, Bolivia and northwestern Argentina. Within this vast area are grown local populations of this cultigen that have different chromosome numbers, or, in some cases, mixtures of chromosome numbers. Thus, in some areas of the Andes a single field may contain plants that range in chromosome number from diploid (2n=24) to triploid (3n=36), tetraploid (2n=48) and/or pentaploid (2n=60). However, the greatest concentration of both chromosomal types and morphological forms occurs in the Lake Titicaca region of Bolivia and Peru. Here are found more potato varieties that are useful to the geneticist and breeder for crop improvement purposes than anywhere else in the world.

Many potato collecting expeditions to the South American Andes have been sponsored during the past decade by the International Potato Center (CIP), in Lima, Peru (Intl. Pot. Ctr., 1973, 1975, 1976b; Huaman, 1979a, b; Lopez, 1979; Ochoa, 1979a, c; Okada, 1979; Huaman et al., 1982). One result of this intense collecting activity has been the assemblage of more than 16,000 samples of wild and cultivated potatoes at this center, which have now been reduced to approximately 6500 (Huaman, 1986). These collections constitute the total holdings of the world potato germplasm collection at CIP in Peru.

Insight into the evolution of the potato is of particular significance to those involved in the improvement of cultivated strains. In this regard, it is important to bear in mind that the cultivated potatoes with higher chromosome numbers have evolved from lower-numbered forms.

As has been pointed out in many other works, the first stage in the evolutionary analysis of any group of species is to work out their taxonomic relationships. It is often simple, in practice, to identify those elements of a cultivated population that correspond closely to a taxonomic type. Such is the case with cultivars *Phinu* and *Pitiquina* of *S. stenotomum* from Bolivia and southern Peru and *Papa Amarilla* or *Runtus* of the central Peruvian species *S. goniocalyx*. On the other hand, when a plant does not correspond closely to a previously published type, then, depending on the magnitude of the differences involved, a new species may be designated.

Modern experimental techniques are often used by taxonomists today to help in the identification of species. Of particular value are the application of colorimetry to problems involving the identification of specific chemical compounds of plants, the use of chromatography for the identification of phenolics, and the use of electrophoresis in the identification of plant proteins. These methods have been particularly valuable in the determination of duplicate living collections of potato that are currently being maintained at CIP. These accessions are being analyzed by Dr. H. Stegemann of the Biochemical Institute of Braunschweig, Germany. Up to the present, about half of the total living potato collections that are maintained clonally at CIP have been determined to be duplicates. These duplicate clones are converted to true seed and then discarded. The remaining collections are believed to be representative of the totality of variation to be found within each species.

The collections maintained at CIP have also been widely used in studies aimed at unraveling the phylogeny of the wild and cultivated species. Of particular importance in this regard have been the probing of the interfertility relationships of the various species and the synthesis in the field or greenhouse of naturally occurring hybrids.

Contrary to the reports of Bukasov (1933) and Lechnovitch (1971), there are several species of potato that are common to both Bolivia and southern Peru. According to the author's observations, all cultivated species of potato are grown along the altiplanic border region of these countries except *S. goniocalyx* and the Chilean, North American, and European varieties of *S. tuberosum*. Three clones of *S.* × *juzepczukii* (*rucki*, *ckaisalla* and *orcco-rucki*) and four clones of *S.* × *curtilobum* (*occocuri*, *yuracc-huaña*, *azul-huaña* or *laram-rucki*, and *choque-pitu*) are cultivated along the northern and southern shores of Lake Titicaca. Likewise, clones of *S.* × *ajanhuiri* (*laram-ajahuiri* and *janck'o-ajahuiri*) are also cultivated in Bolivia as well as in the Departments of Cusco and Puno in southern Peru. Certain forms of *S. stenotomum* (e.g., *p'pitiquina* [=*pitiquilla*] and *phinu*) and *S. phureja* (e.g., *chaucha* [=*phureja*]) are grown in the vicinity of

Puina and Sorata in the Department of La Paz, Bolivia, and also in the region of Cuyo Cuyo and Sandia in the Department of Puno, Peru. The forms *occocuri* (=*chojllu*), *ccoesullu*, and *lomo* (=*puca-suit'tu*) of *S.* × *chaucha* are also reported to be grown on both sides of the Peruvian-Bolivian border (Ochoa, 1958, 1975a).

Many forms of *S. tuberosum* subsp. *andigena* are also cultivated along both the northern and southern shores of Lake Titicaca. Examples include *Janck'o-Imilla* (='White Girl' or *Ccompis*), *Alccai-Huarmi*, *Yana-Imilla* (=*Chiar-Imilla* or 'Black Girl'), *Pucamama* (=*Huila-Imilla* or 'Red Girl'), *Sali-Imilla* (='Gray Girl'), *Laram-Imilla* (='Blue Girl'), *P'ako-Imilla* (='Blond Girl'), *Chupica-Imilla* (='Bright Red Girl'), *Alkka-Imilla* (='Blossom-colored Girl'), and *Sackam-paya*. However, some care must be given in the use of native names. The same vernacular name may be used to designate different botanical forms and varieties on either side of the border. Examples of names that are commonly confused in this way are *ckuchi-aca*, *puca-p'palta*, and *cuchillu-p'paqui*.

The botanical origins of the cultivated potatoes are much better known than those of the majority of the wild species. For this reason, the order in which they are presented in this chapter is by chromosome number and by relative stage of morphological advancement, rather than by strict alphabetical sequence, as used in Chapter 3. In the present chapter, the author also preferred to follow the traditional Linnean system, rather than use the classification of cultivated plants at the infra-specific level (International Code of Nomenclature for Cultivated Plants, 1980 and *Actae Horticulturae 182*, 42–53, 433–436, 1986).

Key to Cultivated Species
Based on Living Collections

1. Plants forming a semirosette or rosette; pedicel articulation high, within 3-4 mm of the calyx.
   2. Corolla typically rotate, the lobes very short, approximately 1-2 mm long excluding the acumen.
      3. Corolla small, 1.5-2.5 cm in diameter ...... 35. *Solanum* × *juzepczukii* (p. 358).
      3. Corolla large, 3-3.5 cm in diameter .... 36. *Solanum* × *curtilobum* (p. 369).
   2. Corolla subrotate, almost pentagonal, the lobes 4-5 mm long excluding the acumen .... 32. *Solanum* × *ajanhuiri* (p. 306).
1. Plants never rosette-like; pedicels normally articulated more than 4 mm below calyx.
   4. Leaflets elliptic-lanceolate or broadly elliptic-lanceolate; corolla 3-4 cm in diameter.
      5. Calyx symmetrical, usually with elliptic-lanceolate lobes and well-defined acumens; fructiferous; berries large, 20 mm or more in diameter, with abundant well-formed seed .... 38. *Solanum tuberosum* subsp. *andigena* (p. 382).

5. Calyx symmetrical or asymmetrical, usually with triangular-lanceolate lobes and poorly defined acumens; fruits very few or absent; berries small, 10 mm or less in diameter, without or with few poorly formed seeds .... 37. *Solanum* × *chaucha* (p. 373).

4. Leaflets more narrowly elliptic-lanceolate; corolla usually small, 2.5-3(-3.5) cm in diameter.
   6. Corolla lobes about half as broad as the length of petals, acumens long; leaves dull, densely pubescent .... 34. *Solanum stenotomum* (p. 335).
   6. Corolla lobes broader than half the length of the petals, acumens short; leaves somewhat shiny, less densely pubescent .... 33. *Solumn phureja* (p. 315).

32. Solanum × ajanhuiri Juzepczuk and Bukasov, Proc. U.S.S.R. Congr. Genet., Pl. & Anim. Breed. 3:605. 1929.

FIGS. 139-142; PLATES XIX-1, XXIII-4.

S. × *ajanhuiri* var. *azul* Lechn., in Bukasov (ed.), Flora of Cult. Pl., Vol. IX, Potato, Leningrad, pp. 70-71. 1971, *nom. nud.* Based on *Juzepczuk 1518*, Bolivia, La Paz (WIR, not seen); Tiahuanaco: *Juz. 1643, 1661, 1744* (WIR), *Schick K-480* and *Zhukowski K-3372* (WIR). Peru, Pomacauchi, *Juz. 1181* (WIR).

Plants rosette-like, more or less pilose with short hairs. Stems, when mature, to 40(-50) cm tall, narrowly straight-winged, subpigmented or pigmented at base and along the axils of the leaves. Tubers either fusiform or subfalcate with a dark bluish-violet periderm, deep eyes, dark bluish-violet sprouts. Leaves with 5-6(-7) pairs of lateral leaflets and up to 11 or more pairs of interjected leaflets, the rachis pigmented principally on or about the petiolules. Leaflets sparsely pilose, elliptic lanceolate, the upper pair almost always narrowly decurrent on the rachis, the apex pointed or subpointed and the base obliquely rounded, subsessile or shortly petiolulate. Peduncle 10-15 cm long by 1.5-1.8 mm in diameter at base, pigmented or subpigmented in the upper one-third of its length, as in the pedicels and calyx. Pedicel 25-30 mm long, the upper part 4-5 mm long. Calyx regular or asymmetric, 5-6 mm long, with narrowly elliptic lobes shortly acuminate, bearing 1 mm-long acumens. Corolla rotate-pentagonal, 2.5-2.8 cm in diameter, dark blue with greenish-yellow or dark-gray nectar guides, the lobes 6-7 mm long by 10-15 mm wide, and with 4 mm-long by 3 mm-wide acumens. Anthers 4.5-5 mm long, narrowly lanceolate with a well-defined dorsal groove and the base slightly cordate, borne on 0.6-0.7 mm-long filaments. Style 7.5-8 mm long, sparsely papillose toward the base; stigma capitate, about 1 mm in diameter, somewhat cleft. Fruit infrequently formed, globose to ovoid, 10-15 mm long.

Chromosome number: $2n=2x=24$.

Figure 139. – *Solanum* × *ajanhuiri* (*Ochoa 3112*).

Types: BOLIVIA. Department La Paz, Bolivian altiplano, above La Paz [Province Murillo], n.v. *Ahanhuiri*; *S. Juzepczuk 1518, 1661, 1699, 1800* (WIR), and *Juz. 1744* (WIR, lectotype here designated).

Solanum × *ajanhuiri* is a valuable source for frost resistance. According to Bukasov (1933) and Bavyko (1982), this cultigen can withstand temperature extremes from −4° to −5° C. It is also resistant to potato virus PVX, and Bukasov (1933) reports that its tubers have a high content of dry matter.

According to the evaluation work being carried out on *S.* × *ajanhuiri* at CIP, this cultigen is resistant to *Globodera*, a nematode of wide distribution in the Central Andes of South America.

Solanum × *ajanhuiri* also flowers and tuberizes much earlier than *S.* × *juzepczukii* and *S.* × *curtilobum*. Moreover, unlike the two previously mentioned cultigens, its tubers are not bitter and can be eaten without special processing. Nevertheless, not all clonal types of *S.* × *ajanhuiri* are equally palatable. The variety *yari*, for example, is much less popular than *laram-ajahuiri* of pigmented tubers, which represents the typical species *S.* × *ajanhuiri*.

Affinities

Solanum × *ajanhuiri* is related to *S. stenotomum*. Both have rather similar leaf and floral details. According to the investigations of Huaman (1975), Hawkes (1979), and Huamán et al. (1982, 1983), this hybridogenic cultigen has evolved as a result of natural crossing between *S. stenotomum* and *S. megistacrolobum*. Moreover, Huamán believes that form *yari* of *S. ajanhuiri* represents the F_1 cross of *S. stenotomum* × *S. megistacrolobum*, whereas form *ajawiri* develops as a backcross of the F_1 cross with *S. stenotomum*. Recently, Timothy Johns (1985), who has worked with the glycoalkaloids (TGA) of these cultigens, has confirmed these results. According to Johns, form *yari* is more closely related to *S. megistacrolobum* than to *S. stenotomum*, whereas in form *ajawiri*, the reverse is true.

Distribution

This cultigen is grown in the cooler potato-producing zones of the Peruvian-Bolivian altiplano, principally between 3800 and 4000 m altitude. According to Huaman et al. (1980), the area of greatest importance or genetic diversity of this species lies between 16°30′ and 18°45′ S and 67°30′ and 69° W. Moreover, he indicates that *S.* × *ajanhuiri* form *ajawiri* is cultivated extensively in the Provinces of Ingavi and Pacajes, Department of La Paz; whereas *S.* × *ajanhuiri* form *yari* is cultivated predominantly in the Provinces of Carangas, Cercado, and Litoral, Department of Oruro.

Figure 140. – *Solanum* × *ajanhuiri* f. *janck'o-ajanhuiri* (*Ochoa 3881*).
1. Leaf. 2. Corolla. 3. Petal. 4. Pistil. 5. Stamens. 6. Calyx.
7. Inflorescence. 8. Tuber. All ×½.

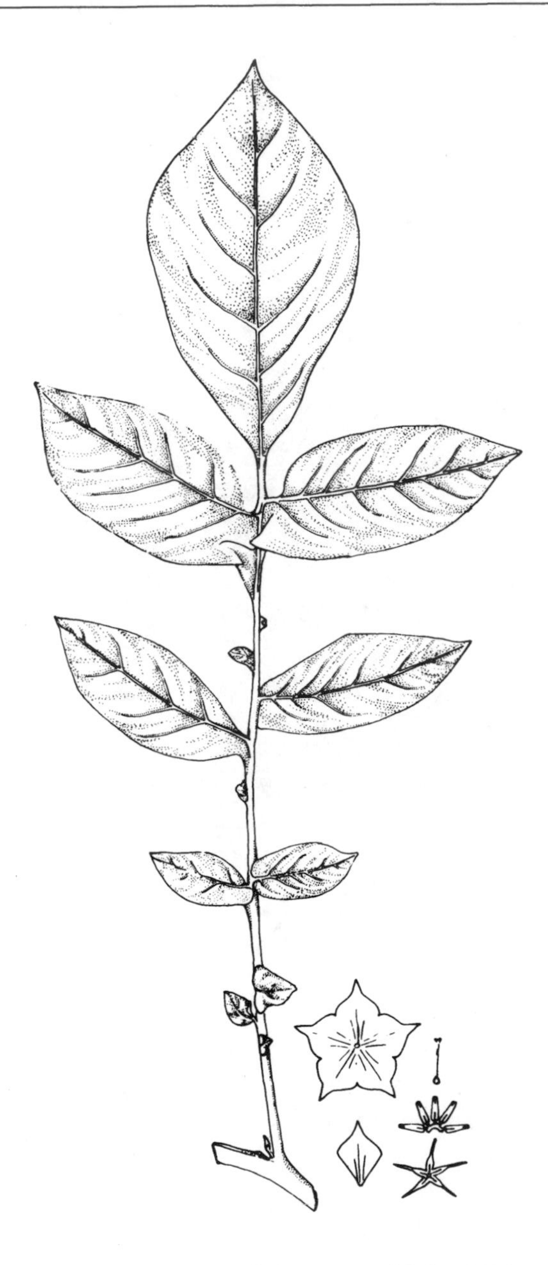

Figure 141. – *Solanum* × *ajanhuiri* var. *yari* (*Ochoa 3888*), × ½.

Specimens Examined

HERBARIUM COLLECTIONS

Department La Paz

Province Ingavi: Tiahuanaco; *Juzepczuk 1643, 1699, 1661* (WIR); *Juzepczuk 1744* (lectotype, here designated, WIR). Tiahuanaco, 3700 m alt., n.v. *Laram Ajahuiri* (=*Ajahuiri azul*); *Ochoa 3112, 3889* (OCH). Huancollo, near Tiahuanaco, 3650 m alt., n.v. *Sillco Ajahuiri*; *Ochoa 3890* (OCH).

Province Murillo: La Paz; *Juzepczuk 1518* (WIR). La Paz; *Schick K-480* (WIR). La Paz; *Zhukowski K-3372* (WIR).

Department Oruro

Province Poopo: Urmiri, 3750 m alt., n.v. *Ajahuiri*; *Ochoa 3885* (OCH).

Figure 142. – Floral dissections of *Solanum* × *ajanhuiri* var. *yari* (above, *Ochoa 3887*; below, *Ochoa 3888*), ×1.

GERMPLASM COLLECTIONS

Department La Paz

Province Aroma: Ayoayo, n.v. *Kara Ajanhuiri* (CIP-702658). Patacamaya, n.v. *Ajanhuiri* (CIP-702324); n.v. *Huilla* (CIP-702678). Sica Sica, 3800 m alt., n.v. *Laram Ajanhuiri* (HUA-865). Vilaque (CIP-702681). Vilaque, n.v. *Ajanhuiri* (HUA-876). Vilaque, n.v. *Ajanhuiri Negra* (CIP-702683, CIP-702684). Vilaque, n.v. *Kara Ajanhuiri* (HUA-875). Vilaque, n.v. *Laram Ajanhuiri* (HUA-877, HUA-897).

Province Ingavi: Tiahuanaco, 3880 m alt., n.v. *Ajahuiri* (EET-1171) (HUA-796). Guaqui, n.v. *Laram ajahuiri* (EET-1314). Titicaca, n.v. *Ajanhuiri* (CIP-702661).

Province Los Andes: Cura Pucara, 3860 m alt, 2n=24, n.v. *Ajanhuiri* (HAW-6187). Laja, n.v. *Laram Ajanhuiri* (EET-1209, EET-1321).

Province Murillo: El Alto, n.v. *Laram Ajanhuiri* (EET-1362). La Paz, 3800 m alt., 2n=24, n.v. *Kasca* (OCH-10527). Santa Ana, n.v. *Chiar Ajahuiri* (CIP-702664).

Province Omasuyos: Belen, 3800 m alt., 2n=24 (CIP-702285). Belen, n.v. *Ajahuiri* (CIP-702286, CIP-702803). Coromata, n.v. *Ajanhuiri* (CIP-702657).

Province Pacajes: Caquiaviri, 4000 m alt. (CIP-702666). Caquiaviri, n.v. *Ajanhuiri* (CIP-702323). Caquiaviri, n.v. *Ajanwiri* (CIP-702674). Caquiaviri, n.v. *Ajawiri* (CIP-702320); 2n=24, n.v. *Ajahuiri* (CO-2402). Caquiaviri, n.v. *Kara Ajanhuiri* (HUA-1034). Caquiaviri, n.v. *Laram Ajanhuiri* (HUA-753). Caquiaviri, n.v. *Luki Ajanhuiri* (HUA-782). Caquiaviri, n.v. *Pisla Ajahuiri* (HUA-1035). Santiago Machaca, n.v. *Ajawiri* (CIP-702668). Ticoniri, n.v. *Kara Ajanhuiri* (CIP-702673).

Department Oruro

Province Cercado: Condoriri (CIP-702654, CIP-702667, CIP-702671). Condoriri, n.v. *Ajanhuiri Negra* (CIP-702656).

Department Potosi

Province Frias: Tinguipaya, 2n=24, n.v. *Cichi Aca Papa* (CIP-702562). Tinguipaya, n.v. *uichu Aka* (CIP-702554).

Province Linares: Karachipampa, 4000 m alt., 2n=24, n.v. *Ayaviri* (OCH-10476). Palca, 3750 m alt., n.v. *ayaquiri* (OCH-10627). Locality unknown (CIP-702655, CIP-702800).

Province Unknown: Atacollo, n.v. *Koullu Papa* (CIP-702526).

32A(a). Solanum × ajanhuiri var. **ajanhuiri** f. **jancko-ajanhuiri** Ochoa f. nov., Phytologia 65(2): 103-113. 1988. FIG. 140; PLATE XIX-2.

S. × *ajanhuiri* var. *janck'o* Lechn., in Bukasov (ed.), Flora of Cult. Pl., Vol. IX, Potato, Leningrad, p. 71. 1971, *nom. nud.* Based on *Juzepczuk 1661-b, 1800,* Bolivia. La Paz (WIR); Vavilov K-97, Bolivia. La Paz (WIR); and *Erlanson and MacMillan 454-Beltsville, K-864, Bolivia.* La Paz (WIR).

This form differs from the typical species in lacking pigmentation of the tubers and stems. The tubers are yellowish-white or creamy-white, with sprouts tinged light purple at base, and creamy-white buds.

Specimens Examined

HERBARIUM COLLECTIONS

Department La Paz

Province Ingavi: Huanccollo, 3650 m alt., near Tiahuanaco, n.v. *Janck'o ajanhuiri*; *Ochoa 3886* (OCH).

Province Murillo: La Paz; *Juzepczuk 1661-b, 1800* (WIR). La Paz; *Vavilov K-97* (WIR). La Paz; *Erlanson and MacMillan 454*-Beltsville, *K-864* (WIR).

Department Oruro

Province Cercado: Near Eucaliptus, n.v. *kaisalla*; *Balls 6394* (CPC).

Province Poopo: Urmiri, 3750 m alt., 2n=24, n.v. *Janck'o Ajanhuiri* (=*Ajanhuiri blanco*); *Ochoa 3881* (OCH, type of f. *janck'o ajanhuiri*).

GERMPLASM COLLECTIONS

Department La Paz

Province Aroma: Chiar Umani, n.v. *Ajanhuiri* (CIP-702651). Conavi, n.v. *Jancko Ajanhuiri* (EET-1170). Patacamaya, n.v. *Ajanhuiri* (CIP-702325, CIP-702326, CIP-702802). Patacamaya, n.v. *Jancko Ajanhuiri* (EET-1353). Sica Sica, 3800 m alt., n.v. *Jancko Ajanhuiri* (HUA-866). Vilaque, n.v. *Ajanhuiri Blanca* (CIP-702682). Vilaque, n.v. *Jancko Ajanhuiri* (HUA-898).

Province Murillo: Calacoto, n.v. *Ajanhuiri Blanca* (CIP-702676).

Province Pacajes: Caquiaviri, 3900 m alt., n.v. *Janko Ajanhuiri* (HUA-1037). Caquiaviri, n.v. *Luki Ajanhuiri* (HUA-1036). Caquiaviri, n.v. *Luky Ajanhuiri* (HUA-1043).

Department Oruro

Province Cercado: Condoriri (CIP-702659, CIP-702672).

32B. Solanum × ajanhuiri var. **yari** Ochoa var. nov., Phytologia 65(2): 103-113. 1988. FIGS. 141, 142; PLATE XIX-3, 4.

Plants somewhat rosette-like and pubescent. Tubers with white flesh and obtuse apex, oblong or long-compressed, the periderm dark violet to dark pinkish-violet or white, occasionally variegated, with shallow or superficial eyes, purple buds, and creamy-white sprouts. Leaves 7-10 cm long by 3.5-5.8 cm wide, terminal leaflet broadly elliptic-lanceolate to rhombic-lanceolate, the apex pointed or abruptly acuminate and the base narrowly attenuate and decurrent on the petiolule. Lateral leaflets diminishing gradually in size toward the base, sessile or narrowly decurrent, much smaller than the terminal, the upper pair 4.5-7 cm long by 2-3.5 cm wide, and broadly decurrent on the rachis (*see* Fig. 141). Peduncles short, very slender, 4-6 cm long and about 1.5 mm in diameter at base. Calyx symmetrical or asymmetrical, 6-8 mm long, with narrowly elliptic-lanceolate lobes and abruptly narrowed apices terminating in 3.5 mm-long acumens. Corolla usually small, 2-2.5 cm in diameter, light bluish-purple or pale blue, and with whitish acumens.

Specimens Examined

HERBARIUM COLLECTIONS

Department Oruro

Province Poopo: Urmiri, 3750 m alt., n.v. *Yari*; *Ochoa 3887*, (OCH, type of var. *yari*), and *3888* (OCH) [both 2n=24].

Department Potosi

Province Linares: Palca, 3750 m alt., 2n=24, n.v. *Ayahuiri*; *Ochoa 10628* (OCH).

Province Quijarro: Condoriri, 3900 m alt., 2n=24, n.v. *Yari*; *Ochoa 10680* (OCH).

GERMPLASM COLLECTIONS

Department La Paz

Province Aroma: Patacamaya, n.v. *Laram Yari* (EET-1201). Patacamaya, n.v. *Yari* (CIP-702677).

Province Pacajes: Caquiaviri, n.v. *Yari* (HUA-754).

Department Oruro

Province Abaroa: Challapata, n.v. *Yari* (OCH-10687).

Province Carangas: Valle Bonito, 3800 m alt., n.v. *Laram Yari* (HUA-823).

Province Sajama: Kosapa, n.v. *Yari* (CIP-702650).

Department Potosi
Province Frias: Lecherias, 3800 m alt., n.v. *Yari* (HAW-5701).

32B(a). Solanum × ajanhuiri var. **yari** f. **janck'o-yari** Ochoa.

This form differs largely from the typical variety in having a white corolla and tuber periderm.

Specimens Examined

HERBARIUM COLLECTIONS

Department Oruro
Province Cercado: Local marketplace, 3700 m alt., 2n=24, n.v. *Yari*; *Ochoa 3944* (OCH).

GERMPLASM COLLECTIONS

Department La Paz
Province Aroma: Patacamaya, n.v. *Yari* (CIP-702679, 702680) [both 2n=24].

Department Oruro
Province Carangas: Valle Bonito, 3800 m alt., n.v. *Jango Yari (HUA-835).*

33. Solanum phureja Juzepczuk and Bukasov, Proc. U.S.S.R. Congr. Genet. 3:604-605. 1929. FIGS. 143-152; PLATES XX-1, 7, XXI-1-4.

S. erlansonii Buk., Bull. App. Bot. Leningrad, ser. A 19:83-85. 1936, ***nom. nud.*** Based on *Erlanson s.n.*, Bolivia. La Paz (WIR=K-451).
S. macmillanii Buk., Physis (Buenos Aires) 18:43. 1939, ***nom. nud.*** Based on *MacMillan s.n.*, Bolivia (WIR=K-447).
S. cardenasii Hawkes, Bull. Imp. Bur. Pl. Breed. & Genet., Cambridge (June):67, 129-130, Fig. 77. 1944. TYPE: *Balls 6299* (=EPC 284), Bolivia. Colomi, near Cochabamba, Dept. Cochabamba, Prov. Chapare (CPC [?], not seen).
S. phureja var. *pujeri* Hawkes, Bull. Imp. Bur. Pl. Breed. & Genet., Cambridge (June):68, 130, Fig. 68. 1944. TYPE: *Balls 6510* (=EPC), Bolivia. Huatajata farm, Dept. La Paz, Prov. Omasuyos (CPC[?], not seen).
S. macmillianii Buk. et Lechn., in Bukasov (ed.), Flora of Cult. Pl., Vol. IX, Potato; Lechnovitch, Cult. Potato Species, Leningrad, p.65. 1971, ***nom. nud.*** Based on *MacMillan s.n.* =PI-33411, Bolivia [La Paz?] (LE=K-447) and (K-857, LE [?], not seen).

S. phureja var. *flaviarticulatum* Lechn., in Bukasov (ed.), Flora of Cult. Pl., Vol. IX, Potato, Leningrad, p. 64. 1971, *nom. nud.* Based on *Juzepczuk 1815*, Bolivia. La Paz (WIR).
S. phureja var. *variegatum* Lechn., in Bukasov (ed.), Flora of Cult. Pl., Vol. IX, Potato, Leningrad, pp. 64-65. 1971, *nom. nud.* Based on *Juzepczuk 1815-b*, Bolivia. Sorata, Dept. La Paz, Prov. Larecaja (WIR, not seen).
S. phureja var. *albescens* Lechn., in Bukasov (ed.), Flora of Cult. Pl., Vol. IX, Potato, Leningrad, p. 64. 1971, *nom. nud.* Based on *Juzepczuk 1685*, Bolivia. Sorata, Dept. La Paz, Prov. Larecaja (WIR).

Stems simple or branched, 30-60(-80) cm tall by 6-8(-12) mm in diameter at base, upright to semidecumbent or decumbent, wingless or narrowly winged, usually pigmented with purple. Tubers broadly oblong or oval-elongate to long subcylindric, deep eyes, dark blue-violet sprouts, periderm variegated with red-violet and yellowish-white to yellow, flesh white or white-grayish. Leaves usually broad and abbreviated, 15-23(-30) cm long by 8-10(-14) cm wide, with 5-6(-7) pairs of leaflets and a variable number (18-30) of interjected leaflets of different sizes. Leaflets more or less pilose, somewhat shiny above, narrowly ovate to elliptic-lanceolate or narrowly elliptic, the apex pointed or subpointed and the base obliquely rounded, sessile or borne on short petiolules or occasionally somewhat decurrent on the rachis. Terminal leaflet, 6-7.5 cm long by 2.5-3.5 cm wide, slightly larger than the first and second pairs of lateral leaflets. Peduncles, 5-7(-12) cm long, 5-9-flowered, slender, 1.8-2.0 mm diameter at base, more or less pilose as in the pedicels and calyx, invested with short hairs and glandular trichomes. Pedicels slender, 15-20(-25) mm long, articulated about the center or in upper one-third of the stalk. Calyx, 7-8 mm long, usually strongly pigmented, asymmetric with 1 or 2 pairs of united lobes, or occasionally symmetric, the pointed lobes narrowly lanceolate and abruptly narrowed to 1.5-2 mm-long acumens, the latter reflexed in bud. Corolla rotate to rotate-pentagonal, usually pleated, 2.5-3(-3.5) cm in diameter, dark reddish-violet or deep violet, the lobes generally short and conspicuously broad and with short, broadly triangular acumens, whitish below. Staminal cone cylindric-conical to subdoliform; anthers 4.5-5.5 mm long and 1.8 mm broad, cordate at base, borne on 1.5-2 mm-long filaments, the latter 1 mm or less in diameter. Style slender, 9-10 mm long, exerted up to 3.5 mm, densely covered with very short papillae along the basal third of its length; stigma capitate, small, less than 1 mm in diameter, slightly cleft. Fruit globose, ovoid to oval-conical, 1.5-2.5 cm long, light green or green with light purple vertical stripes.

Chromosome number: 2n=2x=24 (rarely autotetraploid, or 2n=2x=48).

Figure 143. – *Solanum phureja* (*Ochoa 3536*), × ½.

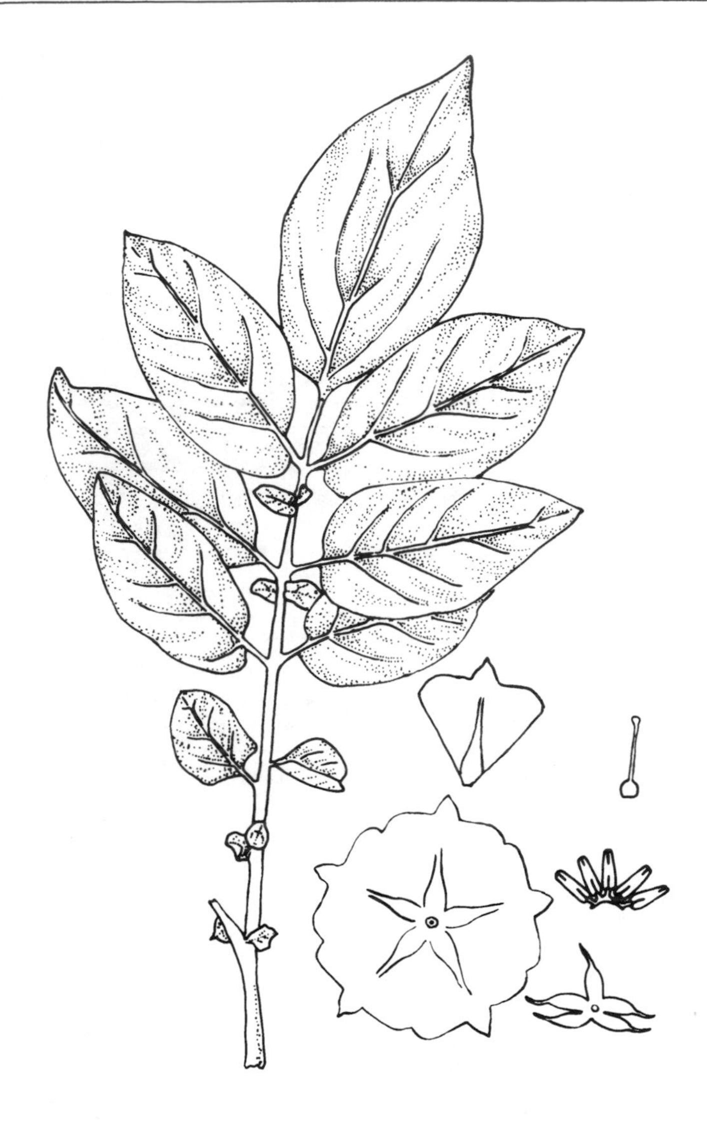

Figure 144. – *Solanum phureja* var. *caeruleus* (*Ochoa and Salas 14973*). Leaf, ×½. Floral dissection, ×1.

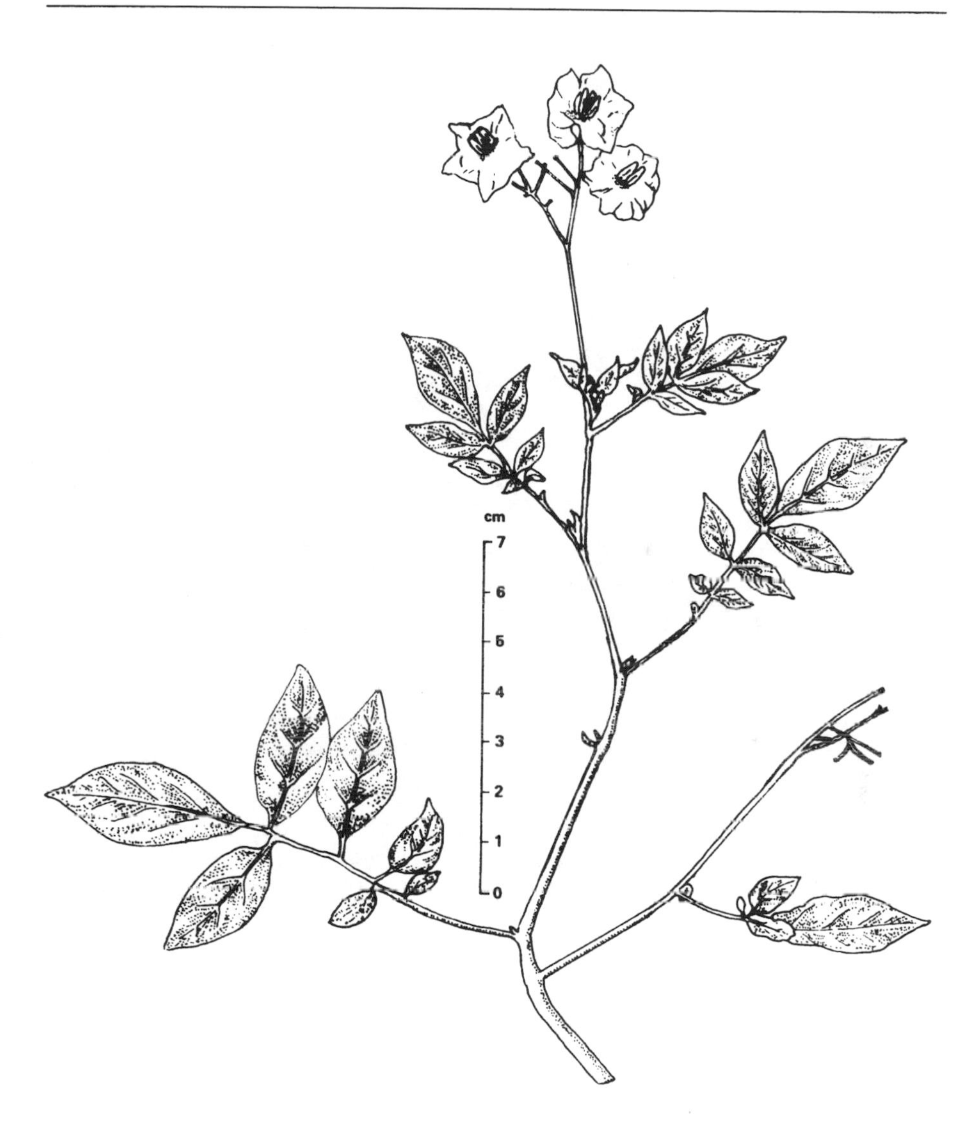

Figure 145. – *Solanum phureja* var. *erlansonii* (LE=K-451), ×1.

Figure 146. – *Solanum phureja* var. *macmillani* (LE=K-447), ×1.

Figure 147. – *Solanum phureja* var. *rubro-rosea* (*Ochoa and Salas 14971*). Leaf, ×½. Floral dissection, ×1.

Types: BOLIVIA. Department La Paz, Province Larecaja, Sorata [2700 m alt.], n.v. *phureja*; *S. Juzecpzuk 1654, 1655, 1685, 1801, 1810, 1813, 1815* (WIR).

Solanum phureja is a cultivated diploid species that has been much used in genetic studies and potato improvement programs. It is a valuable source of resistance to bacterial wilt caused by *Pseudomonas solanacearum* (Ciampi and Siqueira, 1980; Martin et al., 1980; French and De Lindo, 1982). It is also known to be resistant to (*Phytophthora infestans*) (Ramanna and Abdalla, 1970; Turkensteen, 1979; Martin and French, 1980; Sequeira, 1980; Bavyko, 1982) and to potato viruses PVX, PSV, PMV, and PAV (Bavyko, 1982). Moreover, hybrids of *S. phureja* × *S. tuberosum* subsp. *andigena* were found resistant to frost (Estrada, 1977). *S. phureja* is also a valuable source for heat tolerance (Gautney and Haynes, 1983) and adaptation to the temperate zones (Haynes, 1972).

Affinities

Relationships.– This species is related to *S. stenotomum*. Both have very similar leaves, however, they differ from one another substantially with regard to certain floral characteristics, such as in the form of the corolla lobes and acumens.

Solanum phureja is also related to the Colombian and Ecuadorian species, *S. rybinii*. Both species have similar flowers and have vines that mature very early, generally in about 3 or 4 months. Both also lack a tuber dormancy period. However, the two differ from one another with respect to habit, leaf shape, size of the terminal leaflet, characteristics of the calyx, and form and size of the stigma.

Origin.– *Solanum phureja* shares many characteristics in common with the Peruvian *S. limbaniense* and the Bolivian *S. neovavilovii* and *S. candolleanum*– three wild species which may have played an important role in the origin and evolution of this cultigen. The latter three species grow at an altitude of 2600–3400 m in the moist, shrub forests that skirt the eastern flanks of the Andean mountains, an area believed to comprise the center of genetic diversity for *S. phureja*. Both *S. limbaniense* and *S. phureja* lack a tuber dormancy period.

A tetraploid clone of *S. phureja*, discovered by the author at Queara (Department La Paz) is also of interest here. This new chromosomal form, described below, has been given the name *S. phureja* var. *quearanum*. This discovery would seem to support an origin for the tetraploid cultivated species, *S. tuberosum*, from *S. phureja*. As suggested by Mendiburo and Peloquin (1977) and several others, *S. tuberosum* may have arisen from a cultivated diploid species by the accidental production and chance fusion of two unreduced gametes.

Figure 148. – Floral dissection of *Solanum phureja* var. *caeruleus* (*Ochoa and Salas 14973*), ×1.

Figure 149. – Floral dissection of *Solanum phureja* var. *rubro-rosea* (*Ochoa and Salas 14971*), ×1.

The results of a recent numerical taxonomic study by C. S. Tay on accessions of *S. goniocalyx*, *S. phureja*, and *S. stenotomum* maintained at CIP suggest that the first two species are derivatives of *S. stenotomum* (Tay, 1979). This study included an analysis of morphological and chemical characters.

Distribution

Solanum phureja is cultivated largely in the eastern Andes, between 2000 and 3400 m altitude. It is grown widely in northwestern Bolivia and along the entire eastern range of the Peruvian Andes. Throughout this zone there is little danger of early frost, and rainfall is fairly abundant and uniform throughout the entire growing season, enabling many growers to irrigate their crops.

The Aymaran name of this species means 'early' or 'precocious.' This cultigen yields tubers in about 3 or 4 months, and it may be planted 3 or 4 times a year. The tubers of this species lack a dormancy period.

Although *S. phureja* is grown, to some extent, in the central region of Bolivia, it is cultivated largely in the northwestern part of the country along the eastern ranges of the Andes (Bravo, 1939). Below Apacheta and Paso Alto, for example, where the natural vegetation is very poor or even non-existent, a large tract of land has been given over almost exclusively to the cultivation of *S. phureja*. It is grown especially between Huarisata and Sorata in the vicinities of Paccoya and Umanata at elevations from 3200 to 3400 m. In this same area, but at altitudes of 2800 and 3000 m, are found a number of small farming communities famous for their cultivation of *S. phureja*. Especially notable in this regard are the villages of Curupampa, Chiripaca, Ilabaya, Chihuani, and Chullusirca. Much of the tuber crop of this species sold in the markets of La Paz originates in fields harvested in the Sorata vicinity, where this crop is grown at an altitude of 2600 to 2700 m. Another market source for La Paz potatoes, including a form of *S. phureja* called *zapallo*, is the famous farming region of Chojnacota (2900-3100 m alt.), which lies southwest of Achacachi. Traditionally, forms of *S. phureja* are also grown in the vicinity of Timusi (north of Combaya, Province Muñecas, Department La Paz) and Chojlla (region of the Yungas, near La Paz) (*see* Fig. 153).

Solanum phureja is also cultivated in the vicinities of Queara and Puina in the Province of Franz Tamayo, Department of La Paz, at an elevation of 3000-3200 m. In this little explored part of Bolivia, the species is called by its native name of *Chaucha*.

Figure 150. – Tubers of *Solanum phureja* var. *flavum* (*Ochoa and Salas 14972*), ×1.

Figure 151. – Tubers of *Solanum phureja* var. *flavum* (*Ochoa and Salas 15000*), ×1.

Specimens Examined

HERBARIUM COLLECTIONS

Department La Paz

Province Larecaja: Sorata, 2700 m alt.; *Juzepczuk 1654, 1655* (WIR); *Juzepczuk 1685* (WIR) [as *S. phureja* var. *albescens* Lechn.]. Sorata, 2700 m alt.; *Juzepczuk 1815* (lectotype, here designated, WIR) [as *S. phureja* var. *flaviarticulatum* Lechn.]. Sorata, 2650 m alt., n.v. *Phureja*; *Ochoa 3146* (OCH). Sorata, 2700 m alt., n.v. *Phureja Manzana*; *Ochoa 3155* (OCH). Sorata; *Ochoa and Salas 15473* (OCH, CIP).

Province Murillo: La Paz; *Vavilov K-99, K-100* (WIR). La Paz; *Juzepczuk 1815b* (WIR) [as *S. phureja* var. *variegatum* Lechn.].

Figure 152. – Tubers of *Solanum phureja* var. *janck'o-phureja* (*Ochoa and Salas 15472*), ×½.

Figure 153. – Los Yungas, a humid valley zone between La Paz and Chuspipata, Bolivia. *Solanum phureja* is cultivated on mountain tops near the hamlet of Chojlla (in background), and *Solanum yungasense* grows wild in the valley bottoms.

33A(a). Solanum phureja var. **phureja** f. **viuda** Ochoa f. nov., Phytologia 65(2): 103-113. 1988. PLATE XXI-4.

S. phureja var. *nigriarticulatum* Lechn., in Bukasov (ed.), Flora of Cult. Pl. Vol. IX, Potato, Leningrad, p. 64. 1971, *nom. nud.* Based on *Juzepczuk 1801, 1810, 1813*, Bolivia. Sorata, Dept. La Paz, Prov. Larecaja (WIR).

Pedicels strongly pigmented, articulated slightly above center. Corolla violet, 2.0-2.5(-3) cm in diameter. This form *viuda* is typically that of variety *phureja* except that the tuber periderm has very dark bluish-violet areas, almost black, from which its native name of *Viuda* originates. These darker areas are interrupted with yellow halos around the eyes. Tubers strongly irregular, varying in shape from almost round to broadly oblong, flesh white to pale grayish-white, the dark bluish-violet nondormant buds with whitish adventitious root nodes.

Specimens Examined

HERBARIUM COLLECTIONS

Department La Paz

Province Camacho: Catalone, 3000 m alt., n.v. *Alkka Phureja*; *Ochoa and Salas 11838* (OCH). Catalone, n.v. *Phureja*; *Ochoa and Salas 11840* (OCH).

Province Larecaja: Sorata, 2000-2700 m alt., n.v. *Phureja*; *Juzepczuk 1801, 1810, 1813* (WIR) [as *S. phureja* var. *nigriarticulatum* Lechn.]. Sorata, 2700 m alt., n.v. *Viuda*; *Ochoa 3536* (OCH, type of f. *viuda*). Sorata, n.v.*Phureja*; *Ochoa and Salas 15459* (OCH). Sorata, 2600 m alt., local market, n.v.*Phureja*; *Ochoa and Salas 15471* (OCH).

Province Murillo: Viacha, 3800 m alt., above La Paz city, local market, brought from Sorata, 2n=24, n.v. *Katari*; *Ochoa 10528* (OCH). La Paz, Rodriguez Market, brought from Sorata, n.v. *Viuda Phureja*; *Ochoa 11787* (CIP, OCH) and *Ochoa 11788* (F, OCH) [both 2n=24]. Rodriguez Market; *CIP-702287, CIP-702288* (CIP) [both 2n=24].

33B. Solanum phureja var. **caeruleus** Ochoa var. nov., Phytologia 65(2): 103-113. 1988. FIG. 144; PLATE XX-4.

Stems to 60 cm tall, very narrowly winged, pigmented or subpigmented along the basal one-third of its length and in the leaf axils. Tubers with dark bluish-violet vascular ring, and creamy-white flesh, nondormant, long subcylindrical but irregularly formed, 8-12 cm long by 3-3.5 cm in diameter, with obtuse apex and narrow base, the periderm dark bluish-violet with deep eyes and slender, very branched whitish sprouts, the latter colored dark blue-violet at base and apex. Leaves green to dark green, somewhat brilliant above, with 3-4(-5) pairs of leaflets and 0-2 pairs of sessile interjected leaflets. Leaflets 3.5-8 cm long by 2-4 cm wide, broadly elliptic lanceolate, the apex obtuse to subpointed and the base obliquely rounded, borne on petiolules to 10 mm long. Peduncles slender, 4-7(-10) cm long, 3-5-flowered. Pedicels 15-20 mm long, articulated in the upper one-quarter of the stalk, the upper part 4 to 5 mm long. Calyx 7-8 mm long, asymmetrical to symmetric, the narrowly elliptic-lanceolate lobes abruptly narrowing in nearly filiform, 1.5 mm-long acumens. Corolla small, light sky blue with 2 mm-long white acumens, 2-2.5 cm in diameter, the lobes 13-15 mm long by 18-20 mm wide. Anthers elliptic-lanceolate, 5 mm long, borne on 0.4-0.7 mm-long filaments. Style 7-9 mm long.

Specimens Examined

Department La Paz

Province Franz Tamayo: Qucara, 3350 m alt., 2n=24, n.v. *Chaucha Negra*; *Ochoa and Salas 14973* (CIP, F, OCH, type of var. *caerulus*).

33C. **Solanum phureja** var. **sanguineus** Ochoa var. nov. Phytologia 65(2): 103-113. 1988. PLATE XXI-3.

Stems 60 cm tall by 5-7 mm in diameter at base, slightly pigmented, slender, cylindrical, with or without straight, narrow wings. Tubers with pale grayish-white or cream-colored flesh, long and thick, almost accordion-like, strongly irregular with obtuse ends, the dark wine-red periderm dotted occasionally with deep, narrow eyes toward apex, the sprouts dark bluish-violet and nondormant. Leaves dark green, with 5 pairs of leaflets and 4 pairs of interjected leaflets. Leaflets elliptic lanceolate, the apex subpointed and the base obliquely rounded, borne on petiolules 2-3 mm long. Peduncles 4-6 cm long, very slender, bearing 2-4 flowers. Pedicels very thin, 1.7-2 cm long; the upper part 7-8 mm long. Corolla purple-violet, 2.5-2.8 cm in diameter.

This variety appears to be a natural cross between *S. phureja* and *S. stenotomum*.

Specimens Examined

Department La Paz

Province Inquisivi: Quime, 2900 m alt., 2n=24, n.v. *Wila Phureja*; *Ochoa and Salas 15515* (CIP, OCH, type of var. *sanguineus*).

33C(a). Solanum phureja var. **sanguineus** f. **puina** Ochoa. PLATE XX-1.

Plants and leaves very similar to var. *sanguineus*. Tubers with light beige-white or pale-yellow flesh, oblong to oval-compressed, of uniform shape, with dark reddish-violet periderm and pale yellow superficial or shallow eyes; sprouts bluish-violet with white adventitious root nodes.

Specimens Examined

Department La Paz

Province Franz Tamayo: On road from Pauchinta Cucho to Puina, 3600 m alt., 2n=24, n.v. *Chaucha Roja o Phureja; Ochoa and Salas 15008* (OCH).

33D. Solanum phureja var. **erlansonii** (Bukasov and Lechnovitch) Ochoa, comb. nov. FIG. 145.

S. erlansonii Buk. and Lechn., in Bukasov (ed.), Flora of Cult. Pl., Vol. IX, Potato; Lechnovitch, Cult. Potato Species, Leningrad, p. 65. 1971, *nom. nud.* Based on *Erlanson s.n.*, Bolivia. La Paz (LE= K-451) and K-848 (WIR [?], not seen).

Plants of small or medium stature. Stem 30-50 cm tall. Leaves poorly dissected, provided with 3-4 pairs of leaflets. Peduncles slender, 3.5-5 cm long, and bearing 6-8 small flowers. Pedicels short, the upper part 4-5 mm long and the lower 10-12 mm long, densely pubescent as in the calyx, with very short hairs. Calyx 8.5-9.5 mm long with broadly lanceolate lobes. Corolla 18-20 mm in diameter, the acumens poorly developed or barely distinguishable. Staminal column truncate-conical; the anthers 5 mm long by 1.8 mm wide at base. Style 6.5-7.5 mm long, barely exerted; the stigma capitate.

Specimens Examined

Department La Paz

Province Murillo: La Paz, n.v. *Pulo Colorado*; *Erlanson s.n.* (LE=K-451, WIR).

33E. Solanum phureja var. **flavum** Ochoa var. nov., Phytologia 65(2): 103-113. 1988. FIG. 150; PLATE XX-6, 7.

Plants light green. Stems less than 50 cm tall, very narrowly winged. Tubers without dormancy period, with bright yellow flesh, round to somewhat irregular in shape, with yellow well-defined periderm or occasionally with dark pink and yellow stripes or a pink tinge along the upper one-third of the tuber at the point of its attachment to the stolon, with deep eyes, pink buds, and green or greenish-yellow apical sprouts. Leaves somewhat brilliant above, with 4-5 pairs of lateral leaflets and 4-6(-11) interjected leaflets. Leaflets 3.5-7 cm long by 1.8-3 cm wide, broadly elliptic or elliptic-lanceolate, the apex obtuse and the base obliquely rounded and borne on 2-4 mm-long petiolules. Peduncles slender, 4-7(-9) cm long, and 4-6(-10)-flowered. Pedicels 12-15 mm long, the upper part 3-4 mm long. Calyx small, 6-7 mm long, symmetric or asymmetric, the lobes abruptly narrowed to 1.5 mm-long pointed acumens, sparsely pilose as in the pedicels and peduncles. Corolla small, light lilac, 2.5 cm in diameter. Anthers short, 4 mm long, borne on very short filaments, the latter aproximately 0.5-0.7 mm long. Style short, 6.5-7.5 mm long, glabrous; the stigma broad and capitate.

Specimens Examined

Department La Paz

Province Franz Tamayo: Queara, 3350 m alt., n.v. *Chaucha Amarilla*; *Ochoa and Salas 14970* (CIP, OCH, type of var. *flavum*). Queara, n.v. *Chaucha*; *Ochoa and Salas 14972* (CIP, F, OCH) [both 2n=24]. Between Ayamachay and Sayhuanimayo, near Puina, 3300 m alt., n.v. *Chaucha Redonda*; *Ochoa and Salas 15000* (F, OCH).

Province Murillo: La Paz market, 3650 m alt., brought from Chojnaccota, 2n=24, n.v. *Zapallo*; *Ochoa 3913* (OCH). La Paz, 3650 m alt., Lanza Market, 2n=24, n.v. *Yema de Huevo*; *Ochoa 3915* (OCH).

33E(a). Solanum phureja var. **flavum** f. **sayhuanimayo** Ochoa f. nov., Phytologia 65(2): 103-113. 1988.

Tubers with yellow flesh, long subcylindric, narrowing toward base, the apex somewhat thick and obtuse, with light smoky-pink skin and deep to semideep eyes, the white or creamy-white buds pale pink at base and apex. Leaves somewhat more dissected than in variety *flavum*. Leaflets more narrowly lanceolate. Corolla similar in size and color to variety *flavum*. Calyx asymmetric, very reflexed. Pedicels articulated 2 mm below base of calyx.

Specimens Examined

Department La Paz

Province Franz Tamayo: Vicinity of Sayhuanimayo, 3200 m alt., 2n=24, n.v. *Chaucha Larga*; *Ochoa and Salas 15001* (F, OCH, type of f. *sayhuanimayo*).

33F. Solanum phureja var. **janck'o-phureja** Ochoa var. nov., Phytologia 65(2): 103-113. 1988. FIG. 152.

Plants vigorous, light green. Stems 40-50 cm tall, very branched. Tubers with yellow flesh, long cylindrical or subcylindric, slender, somewhat irregular in shape, with obtuse apices and bases and uniformly yellowish-white skin, semideep to superficial eyes, and creamy-yellow nondormant eyes with pale lilac bases and greenish sprouts (*see* Fig. 152). Leaves short and broad, poorly dissected, with 2-3(-4) pairs of leaflets and usually one pair of interjected leaflets. Terminal leaflet somewhat larger than the laterals, 8-10.5 cm long by 5-6 cm wide, the apex obtuse to subpointed and the base slightly asymmetrical, borne on short petiolules. Peduncles longer than usual in *S. phureja*, to 9 cm in length, 8-12-flowered, light green and glabrescent as in the pedicels and calyx. Pedicels 20-30 mm long, articulated in the upper one-third of the stalk. Calyx symmetrical or asymmetric, 7-8 mm long, the narrow lobes with 1-2 mm-long acumens. Corolla white with greenish-yellow nectar guides, 2.5-3 cm in diameter, with bulbiform lobes, 7-8 mm long by 12-15 mm wide, and very short acumens. Anthers 5-5.5 mm long. Style short and slender, 6.5-7 mm long, densely papillose along the lower two-thirds of its length. Fruit ovoid.

Specimens Examined

Department La Paz

Province Larecaja: Tacacoma, 3500 m alt., n.v. *Phureja*; *Ochoa and Salas 11815, 11816* (CIP, F, OCH) [both 2n=24]. Sorata, 2600 m alt., 2n=24, n.v. *Janck'o Phureja* (=*Blanca Precoz*); *Ochoa and Salas 15458* (CIP, OCH). Sorata, 2n=24, n.v. *Zapallo*; *Ochoa and Salas 15472* (CIP, OCH, type of var. *janck'o phureja*).

Department Cochabamba

Province Tarata: Near Anzaldo, 3600 m alt., white flowers, tubers of good flavor. January 9, 1943; *Cutler, Cardenas, and Gandarillas 7630* (F, US).

33F(a). Solanum phureja var. **janck'o-phureja** f. **timusi** Ochoa f. nov., Phytologia 65(2): 103-113. 1988. PLATE XXI-2.

Vegetatively similar to var. *janck'o phureja* but varying slightly with respect to flower color. This form differs from the typical variety of this species by its tubers, which are subcylindrical, narrowed at base and obtuse and enlarged at apex. These have yellowish periderm, semideep eyes, and nondormant, light violet or lilac buds.

Specimens Examined

Department La Paz

Province Larecaja: La Paz market, 3600 m alt., brought from Timusi, 2n=24, n.v. *Janck'o Phureja*; *Ochoa 3910* (OCH, type of f. *timusi*).

33G. **Solanum phureja** var. **macmillanii** (Bukasov and Lechnovitch) Ochoa comb. nov. FIG. 146.

S. macmillanii Buk. and Lechn., in Bukasov (ed.), Flora of Cult. Pl., Vol. IX, Potato, Leningrad, p. 65. 1971, *nom. nud.* Based on *MacMillan s.n.* (=PI-33411), Bolivia (LE=K-447).

Plants usually very branched. Stems slender, cylindrical. Leaves light green and somewhat lax or drooping, with 4-5 pairs of shortly petiolate or subsessile leaflets, and 1-2(-3) pairs of interjected leaflets. Peduncles 3-4 cm long, 3-5-flowered. Calyx 6.5-7.5 mm long, with conspicuously nerved lanceolate lobes. Corolla 25-30 mm long, dark red-violet with whitish acumens, the shortly acuminate lobes 10 mm long by 15 mm wide. Staminal column cylindrical conical; anthers to 5.5 mm long. Style 8-9 mm long, exerted 3.5 mm; stigma capitate, slightly cleft.

According to Bukasov and Lechnovitch, this plant may be a natural hybrid of *S. ajanhuiri* and *S. phureja*. The author, however, feels otherwise. In particular, he wishes to point out the distinct morphology and distribution of the plants involved.

Specimens Examined

Department La Paz

Province Indefinite: La Paz[?], n.v. *Pina* or *Yunka Tujru*; *MacMillan s.n.* (LE=K-447, WIR).

33H. **Solanum phureja** var. **quearanum** Ochoa. PLATE XX-2.

Plants light green, of medium stature. Stems slender, to 50 cm tall and 5-7 mm in diameter at base, very narrowly winged, sparsely pubescent with 1-2 mm-long, silvery white hairs. Tubers with yellow flesh, very long and slender, somewhat irregular (or bulging), cylindrical to subcylindric, obtuse and enlarging toward the narrow apex and pointed at base, the periderm yellow with deep eyes and nondormant buds, the latter with pinkish violet sprouts and creamy-green apices. Leaves broad and short, poorly dissected, with 3-4 pairs of leaflets and 0-5 interjected lealets, the latter borne on pubescent petiolules and rachis. Leaflets sparsely pubescent on both sides. Terminal leaflets much larger than the laterals, the apex pointed or subpointed and the base rounded to cordate, borne on 1-2 mm-long petiolules. Pedicels and calyx densely pubescent. Pedicels articulated aproximately 1 mm below base of calyx. Corolla 2-2.5 cm in diameter, light lilac with very short whitish acumens.

Variety *quearanum* is distinguished from all other varieties and forms of *S. phureja* principally by its silvery-white pubescence, its poorly dissected leaves, and its extremely high pedicel articulation. Moreover, it is a tetraploid (2n=48). Possibly it originated from a cross of two *S. phureja* clones of unreduced gametes (Mock et al., 1975). A remote possibility could also be that variety *quearanum* originated from a bilateral sexual tetraploidization by intercrossing a diploid *S. phureja* × other diploid cultivated species (Mendiburu and Peloquin, 1977).

Specimens Examined

HERBARIUM COLLECTIONS

Department La Paz

Province Franz Tamayo: Queara, 3000 m alt., n.v. *Chaucha*; *Ochoa and Salas 14976* (CIP, OCH).

33I. **Solanum phureja** var. **rubro-rosea** Ochoa, var. nov., Phytologia 65(2): 103-113. 1988. FIGS. 147-149; PLATE XX-3.

Stems erect, 40-50 cm tall, simple or few-branched, dark brown except for the light green tip, narrowly winged, very sparsely pilose. Tubers with white flesh, nondormant, long subcylindric to broadly ovoid, very irregular (or bulging), the apex obtuse and the base somewhat narrow, with a dark pink-reddish periderm and numerous deep eyes, the latter provided with dark pink or reddish sprouts. Leaves with 6-7 pairs of lateral leaflets and 25-30 interjected leaflets. Leaflets slightly rugose, very sparsely pilose with short hairs, narrowly elliptic-lanceolate, apex obtuse and base asymmetrically rounded, borne on 1-1.5 mm-long petiolules or the upper sessile, and narrowly decurrent on the rachis. Peduncle light green, short, 1-4 cm long, 6-12-flowered and puberulent as in the pedicels and calyx. Pedicels dark brown, the upper part 4-5 mm long and the lower 10-12 mm long. Calyx dark brown, 8-9 mm long, strongly reflexed, the lobes narrowly lanceolate. Corolla white, 3 cm in diameter. Anthers short, 4.5 mm long, broadly lanceolate, the base cordate, borne on 0.4

mm-long filaments. Style 7.5-8 mm long, thick, exerted 1.5-2 mm, sparsely papillose at base. Fruit ovoid to oval conical.

Chromosome number: 2n=24.

Specimens Examined

HERBARIUM COLLECTIONS

Department La Paz

Province Franz Tamayo: Quecara, 3400 m alt., n.v. *Chaucha Roja*; *Ochoa and Salas 14971* (CIP, OCH, type of f. *rubro-rosea*), *14974* (F, OCH), *14975* (CIP, F, OCH). Between Ayamachay and Sayhuanimayo, near Puina, n.v. *Chaucha Roja*; *Ochoa and Salas 14999* (OCH). Capilla Pampa, 3600 m alt., n.v. *Chaucha*; *Ochoa and Salas 15007* (CIP, OCH).

33I(a). Solanum phureja var. **rubro-rosea** f. **orbiculata** Ochoa f. nova, Phytologia 65(2): 103-113. 1988.

Plants light green, usually of medium stature. Stems 40-50 cm tall by 6-8 mm in diameter at base, very narrowly winged. Tubers with yellow flesh, round, irregular (or bulging), the dark wine-red skin with deep eyes and white sprouts, and with dark pink and red buds at base. Leaves similar in form, although less dissected than var. *rubro-rosea*. Corolla white, usually 2-2.5 cm in diameter.

Specimens Examined

HERBARIUM COLLECTIONS

Department La Paz

Province Franz Tamayo: Between Ayamachay and Sayhuanimayo, 3200 m alt., near Puina, n.v. *Chaucha Roja*; *Ochoa and Salas 15009* (CIP, F, OCH, type of f. *orbiculata*), *15010* (CIP, F, OCH) [both 2n=24].

34. Solanum stenotomum Juzepczuk and Bukasov, Proc. U.S.S.R. Congr. Genet. 3:604. 1929. FIG. 154; PLATE XXII-2.

S. churuspi Hawkes, Bull. Imp. Bur. Pl. Breed. & Genet., Cambridge (June):64-65, 129. Fig. 73, 129. 1944. TYPE COLLECTIONS: *Yabar 25 and 27*, Peru. Paucartambo, Cusco (CPC, not seen).

S. yabari Hawkes, Bull. Imp. Bur. Pl. Breed. & Genet., Cambridge (June):65-66, 129. 1944. TYPE: No collector cited, Peru. Probably Paucartambo, Cusco (CPC, not seen).

S. yabari var. *pepino* Hawkes, Bull. Imp. Bur. Pl. Breed. & Genet., Cambridge (June):65, Fig. 74. 1944, *nom. nud.* TYPE COLLECTIONS: *Yabar 698 and 1077*, Peru. Paucartambo, Cusco (CPC, not seen).

S. yabari var. *cuscoense* Hawkes, Bull. Imp. Bur. Pl. Breed. & Genet., Cambridge (June):65, Fig. 75. 1944, *nom. nud.* TYPE: *Yabar 29*, Peru. Paucartambo, Cusco (CPC, not seen).

S. stenotomum var. *phiñu* Lechn., in Bukasov (ed.), Flora of Cult. Pl., Vol. IX, Potato, Leningrad, p. 68. 1971. Based on Clone 1, 2n=24, *Juzepczuk 1556-b*, Bolivia. Tiahuanaco (WIR, not seen); *Juzepczuk 1681, 1710-b*, Bolivia. La Paz (WIR); and Clone 2, 2n=36, *Juzepczuk 1523, 1662, 1700*, Bolivia. La Paz (WIR, not seen).

Plants erect, light green. Stems 50-70 cm tall, branched or simple, very slightly pigmented at base and leaf axils. Tubers long subcylindrical to pyriform, falcate, or U-shaped with an enlarged, obtuse apex and narrow, almost pointed base, uniformly pinkish-violet periderm, semideep eyes, and pink-violet sprouts, white to light beige flesh. Leaves minutely pilose, long and narrow, not overlapping, very narrowly decurrent on the stem, strongly dissected, with 5-6(-8) pairs of leaflets and up to 20 or more interjected leaflets of different sizes. Leaflets usually very narrow, 2 to 3 times longer than broad, narrowly elliptic-lanceolate or narrrowly obovate, the apex subpointed to obtuse. Terminal leaflets usually same size or somewhat smaller than the second and third pairs of laterals, symmetrical to obliquely rounded at base or slightly attenuate, borne on 3-7 mm-long petiolules. Petiolules of the first upper pair, occasionally sessile or slightly decurrent. Peduncles 4-5(-8) cm long, 4-9(-12)-flowered. Pedicels 17-28 mm long, the upper part articulated near the upper one-third of the stalk, the upper part 7-8 mm long and the lower up to 20 mm long. Calyx conical, slightly aristate, 8-10 mm long, the attenuate, narrowly elliptic-lanceolate lobes usually asymmetrical and tipped with 2.5-3.5 mm-long acumens, the latter reflexed especially in bud stage. Corolla light lilac-violet with white acumens, rotate-pentagonal, 2-2.5 cm in diameter, the lobes strongly divided, 7 mm long. Anthers small, 4.5-5 mm long by 1.8 mm wide, borne on slender 1-1.5 mm-long, glabrous filaments. Style short, 7-8 mm long, papillose at base. Stigma capitate, slightly thicker than the apex of the style, more or less narrowly cleft. Fruit globose to ovoid.

Chromosome number: 2n=2x=24 or, very rarely, autotriploid (2n=3x =36).

Types: BOLIVIA. [Department La Paz, Province Murillo], La Paz, altiplano of Bolivia; *S. Juzepczuk 1681* (WIR) and 1743 (WIR, not seen).

The full potential of this species for plant breeding has yet to be realized.

Figure 154. – 1. Plant of *Solanum stenotomum* (*Ochoa 2799*).
2. Inflorescence and upper part of plant. 3. Tubers of *S. stenotomum* var. *pitiquiña* (*Ochoa 3919*).

Gautney and Haynes (1983) report that it seems to have some value for breeding for heat tolerance. Some forms are rich in starch and dry matter, and others can withstand temperatures of −3° C (Bavyko, 1982). In Peru, according to the collection made by Juzepczuk *1743* (No. 1743), this species is called *Phiñu*.

Affinities

In the author's opinion, this species is closely related to *S.* × *ajanhuiri*. The leaves of both are of similar shape, and the corollas are of similar size and form. Other authors, however, have compared this plant with *S. leptophyes* (Hawkes, 1956a, 1979; Ugent, 1970; Bukasov 1971b, 1978).

Evidently, the most primitive variety of *S. stenotomum* is *poccoya* from southern Peru. This variety not only has very narrow leaves, but resembles *S. leptophyes* quite closely in habit, or life form. Other varieties and botanical forms of *S. stenotomum* appear to have some features in common with both *S. bukasovii* and *S. candolleanum*.

As pointed out above, Hawkes (1956a, 1979) believes that this species may have originated from the wild potato *S. leptophyes*. However, *S. stenotomum* is highly polymorphic. The leaflets of this species range from extremely narrow in some varieties and forms to very broadly elliptic or elliptic-lanceolate in others. This suggests that the origin of this species may have been polyphyletic. Possibly, the plant comprising this cultigen may have arisen in more than

Figure 155. – Floral dissection of *Solanum stenotomum* var. *pitiquiña* f. *laram* (*Ochoa 3926*), ×1.

one place and at more than one time from such commonly distributed wild potato species as *S. bukasovii*, *S. soukupii*, and *S. brevicaule*.

Habitat and Distribution

This species is cultivated only very occasionally along the northeastern ranges of the Bolivian Andes. In this long, narrow area, which includes the important potato-producing districts bordering the shores of Lake Titicaca, *S. stenotomum* has been replaced largely by *S. phureja* and some forms of ssp. *andigena*. The major center of cultivation of *S. stenotomum* occurs in the altiplano or Puna region, and in the higher inter-Andean valleys between 3000 and 3600 m altitude (Ochoa, 1958) from central Bolivia to central Peru.

Specimens Examined

HERBARIUM COLLECTIONS

Department La Paz

Province Aroma: Patacamaya n.v. *Pituhuayaca*; *CIP-702601* (CIP). Market-place Lahuachaca, n.v. *Pitu Huayaca*; *Hawkes et al. 5593* (CIP).

Province Camacho: Puerto Acosta, 3835 m alt., 2n=24, n.v. *Areckero*; *Ochoa 3534* (OCH).

Province Ingavi: Tiahuanaco; *S. Juzepczuk 1556-b* (WIR). Tiahuanaco, 3750 m alt., 2n=24, n.v. *Phinu* (='sweet'); *Ochoa 3142* (OCH). Tiahuanaco, 3700 m alt., 2n=24; *Ochoa 3537* (OCH). Tiahuanaco, 3680 m alt., 2n=24, n.v. *Huila*

Figure 156. – Floral dissection of *Solanum stenotomum* var. *pitiquiña* f. *phiti-kalla* (*Ochoa 3928*), ×1.

Phinu (='sweet red'); *Ochoa 3882* (OCH). Tiahuanaco, 3720 m alt., 2n=24, n.v. *Huahua Chara*; *Ochoa 3883* (OCH).

Province Los Andes: Pucarani, 3840 m alt., n.v. *Pinola*; *Ochoa 2799* (OCH).

Province Murillo: La Paz; *Juzepczuk 1681* (WIR, type of *S. stenotomum*). La Paz; *Juzepczuk 1710-b* (WIR, as *S. stenotomum* var. *phinu* Lechn.). La Paz, 3650 m alt., market, 2n=24, n.v. *Phinu*; *Ochoa 10525* (CIP).

Province Pacajes: Corocoro, 3800 m alt., 2n=24, n.v. *Huila Phinu*; *Ochoa 2807* (OCH). Caquiaviri, 3800 m alt., n.v. *Wila Phinu*; *Huaman 781* (CIP).

Department Oruro

Province Poopó: Urmiri, 3700 m alt., 2n=24, n.v. *Pepino Rosado*; *Ochoa 3884* (OCH).

GERMPLASM COLLECTIONS

Department Cochabamba

Province Cochabamba: Cochabamba, 2800 m alt., market, 2n=24 (CO-2528).

Department La Paz

Province Aroma: Calamarca, 3600 m alt., n.v. *Wila Pinu* (HUA-870). Lahuachaca, n.v. *Huila Phinu* (HAW-5587). Occopampa, 3650 m alt., 2n=24, n.v. *Pitu Huayaca* (CO-2461). Sica Sica, 3800 m alt., n.v. *Wilapituwaya* (HUA-861). Sica Sica, 3800 m alt., n.v. *Pituwaya* (HUA-864). Vilaque, 3800 m alt., n.v. *Wila Pinu* (HUA-895).

Province Ingavi: Aipa, 4000 m alt., 2n=24, n.v. *Huila Phinu* (CO-2362). Tiahuanacu, 3880 m alt., 2n=24, n.v. *Phinu* (CO-2165).

Province Murillo: Huaillara, 3800 m alt., 2n=24, n.v. *Huila Surimana* (CO-2191). La Paz market, 3600 m alt., 2n=24, n.v. *Huila Phinu* (CO-2206). La Paz market, 3600 m alt., 2n=24, n.v. *Khati* (HAW-5568). La Paz market, 3600 m alt., n.v. *Phinu* (HAW-5563-5564). La Paz market, 3600 m alt., n.v. *Phiñu Rojo* (HAW-5570).

Province Pacajes: Caquiaviri, 3960 m alt., 2n=24, n.v. *Ckellu Phiñu* (CO-2364). Caquiaviri, 3960 m alt., 2n=24, n.v. *Huila Phinu* (CO-2374). San Andres de Machaca, 3980 m alt., 2n=24, n.v. *Phinu* (CO-2420).

Department Oruro

Province Carangas: Toledo-Corque, 3700 m alt., n.v. *Pinu* (HUA-820).

Province Cercado: Condoriri, n.v. *Pitu Huayaca* (CIP-702579).

Province Oruro: Oruro market, 3700 m alt., 2n=24, n.v. *Pitu Huayaca* (CO-2485). Oruro, 3706 m alt., 2n=24, n.v. *Pitu Huayaca* (OCH-10490, 10491).

34A(a). Solanum stenotomum var. **stenotomum** f. **alkka-phiñu** Ochoa f. nov., Phytologia 65(2): 103-113. 1988. PLATE XXII-5.

Tubers with light beige flesh, slender, subcylindrical or cylindric, slightly irregular (or bulging), with red skin, pink buds, and narrow or usually narrow yellow halo areas about the eyes.

Specimens Examined

HERBARIUM COLLECTIONS

Department La Paz

Province Camacho: Puerto Acosta, 3840 m alt., 2n=24, n.v. *Wila Puya*; *Ochoa 3154* (OCH).

Province Murillo: La Paz market, 3650 m alt., 2n=24, n.v. *Alkka Phinu*; *Ochoa 3911* (OCH, type of f. *alkka-phinu*).

34A(b). Solanum stenotomum var. **stenotomum** f. **añahuaya-culli** Ochoa.

Plants similar to *S. stenotomum* var. *phinu* except with irregularly pigmented stems and dark green leaves. Tubers long and slender, with yellow flesh, fusiform, subcylindric or somewhat irregular or bulging with deep eyes, dark blue-violet buds and periderm, violet-pink stripes, and yellowish-white, narrow halo areas about the eyes.

The native name of this form means 'persistent purple' or 'morada persistente' in Spanish.

Specimens Examined

HERBARIUM COLLECTIONS

Department Potosi

Province Frias: Tinguipaya, 3900 m alt., n.v. *Anahuaya*; *Ochoa 10537* (CIP).

GERMPLASM COLLECTIONS

Department La Paz

Province Ingavi: Chama, 4100 m alt., 2n=24, n.v. *Alkka Phinu* (CO-2352).

Province Pacajes: Caquiaviri, 3960 m alt., 2n=24, n.v. *Anahuaya Culli* (CO-2391). Nazacara, 3860 m alt., 2n=24, n.v. *Alkka Phiñu* (CO-2404). San Andres Machaca, 3980 m alt., 2n=24, n.v. *Alpaca Chuchuli* (CO-2430).

Department Oruro

Province Cercado: Local market, 3700 m alt., 2n=24, n.v. *Ana Huaya* (CO-2503). Local market, 3700 m alt., 2n=24, n.v. *Anahuaya Culli* (CO-2488). Local market, 3700 m alt., 2n=24, n.v. *Alpaca Chuchuli* or *Canilla de Alpaca* (CO-2486).

34A(c). Solanum stenotomum var. **stenotomum** f. **chiar-ckati** Ochoa f. nov., Phytologia 65(2): 103-113. 1988. FIG. 158; PLATE XXIV-3.

This plant is generally similar to *S. stenotomum* var. *phinu* f. *chiar-phinu*. However, its folioles are much more narrowly lanceolate. Moreover, this form has long-compressed and subcylindrical tubers with superficial eyes and dark purple buds, light beige flesh, and a dark blue-violet periderm with light brown stripes.

Chromosome number: 2n=2x=24.

Specimens Examined

HERBARIUM COLLECTIONS

Department La Paz

Province Murillo: La Paz, Rodriguez Market, n.v. *Chiar-ckati*; *Ochoa 3925* (OCH, type of f. *chiar-ckati*).

34A(d). Solanum stenotomum var. **stenotomun** f. **chiar-phiñu** Ochoa f. nov., Phytologia 65(2): 103-113. 1988. FIG. 157.

Stems lightly pigmented. Tubers similar in shape to variety *phiñu* but with very dark blue-violet skin, deep eyes, uniformly blue-violet buds and light yellow flesh. Leaves dark green, with 6-7 pairs of leaflets with slightly undulate margins and many interjected leaflets of different sizes. Pedicel articulation slightly above center. Calyx about 10 mm long, asymmetrical, in two pairs of two plus one. Corolla small, 2 cm in diameter, light sky blue with white acumens. Anthers 4 mm long. Style 7 mm long.

Chromosome number: 2n=2x=24 (except CO-2397 which has 2n=3x=36).

Figure 157. – Tubers of *Solanum stenotomum* f. *chiar-phiñu*, ×1.

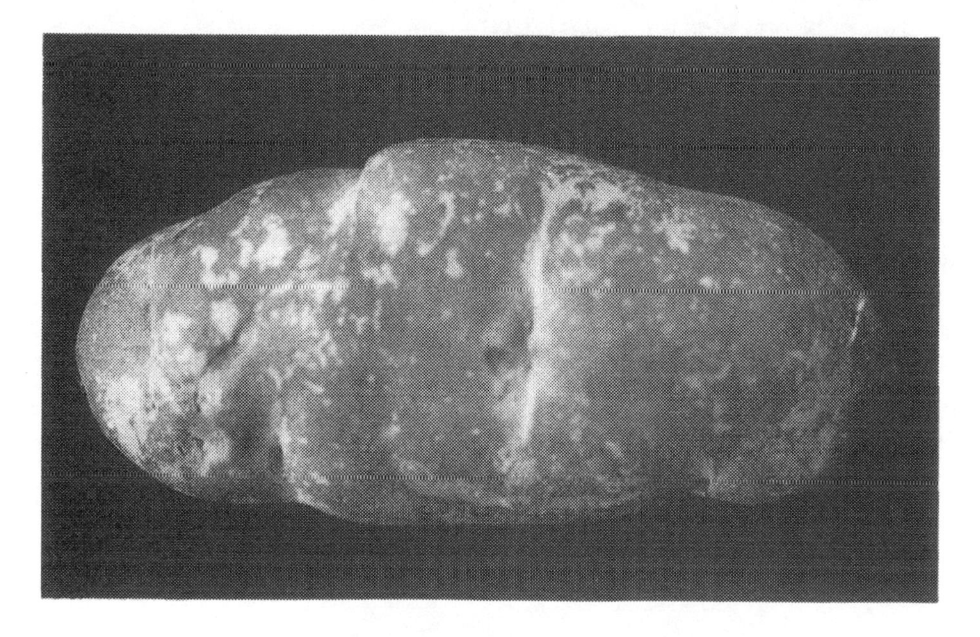

Figure 158. – Tuber of *Solanum stenotomun* f. *chiar-ckati*, ×1.

Specimens Examined

HERBARIUM COLLECTIONS

Department La Paz

Province Ingavi: Tiahuanaco, 3850 m alt., 2n=24, n.v. *Ckati*; *Ochoa 3918* (OCH).

Province Los Andes: Pucarani, 3850 m alt., 2n=24, n.v. *Chiara Pinola*; *Ochoa 3149* (OCH).

Province Murillo: La Paz, Buenos Aires Market, 3650 m alt., brought from the altiplano, 2n=24, n.v. *Chiar-phiñu* (=*Dulce Negra*); *Ochoa 3914* (OCH, type of *chiar-phiñu*). La Paz, Rodriguez Market, 2n=24, n.v. *Ckati*; *Ochoa 3924* (OCH).

Province Pacajes: Caquiaviri, 3960 m alt., n.v. *Chiar Phinu*; Ochoa 2397 (OCH).

Department Oruro

Province Carangas: Valle Bonito, n.v. *Pinta Cayllo*; *Huamán 833* (CIP).

Province Cercado: Indefinite, 3706 m alt., 2n=24, n.v. *Huaycku Pituhuaycca* (=*Bolsa de Harina* or 'flour bag'), for boiling; *Ochoa 10483* (OCH).

GERMPLASM COLLECTIONS

Department La Paz

Province Aroma: Sicasica, 3800 m alt., n.v. *Yana Pituhuaya* (HUA-863). Vilaque, 3800 m alt., n.v. *Iquishko* (HUA-885).

Province Murillo: Near Viacha, 3800 m alt., n.v. *Sarama* (OCH-10518).

Province Pacajes: Caquiaviri, 3960 m alt., 2n=36, n.v. *Chiar Phiñu* (CO-2397).

34A(e). Solanum stenotomum var. **stenotomum** f. **pacajes** Ochoa f. nov., Phytologia 65(2): 103-113. 1988.

Plants delicate. Stems slender. Tubers long and slender, with yellow flesh, cylindrical or subcylindric with obtuse and subpointed apex and red periderm, small deep eyes and pink buds. Leaves with 5 pairs of leaflets and 6-12 interjected leaflets, the latter attached in some cases to the petiolules of the first and second upper pairs of leaflets. Leaflets narrowly lanceolate, the apex pointed or subpointed, and the base slightly attenuate or rounded, borne on 10 mm-long petiolules. Corolla small, white, 2 cm in diameter.

This form is closely related to typical *S. stenotomum*.

Specimens Examined

HERBARIUM COLLECTIONS

Department La Paz

Province Pacajes: Yuriwaya, 3850 m alt., near Corocoro, 2n=24, n.v. *Orureña*; *Ochoa 2804* (OCH). Caquiaviri, 3960 m alt., 2n=24, n.v. *Piña*; *Ochoa 3551* (OCH, type of f. *pacajes*).

GERMPLASM COLLECTIONS

Department La Paz

Province Aroma: Aroma, n.v. *Pina* (HAW-5583).

Province Murillo: La Paz, n.v. *Pina* (HAW-5575).

Province Pacajes: San Andres de Machaca, 3980 m alt., 2n=24, n.v. *Piña* (CO-2423).

Department Oruro

Province Cercado: Indefinite, local market, 3700 m alt., n.v. *Quisca Pali* (CO-2475).

34A(f). Solanum stenotomum var. **stenotomum** f. **pulu-wayk'u** Ochoa f. nov., Phytologia 65(2): 103-113. 1988. PLATE XXII-3.

Plants graceful. Stems slender, 40-50 cm tall, irregularly pigmented. Tubers with white flesh, long subcylindric, somewhat irregular (or bulging), the obtuse apex narrowing toward base, with semideep to deep eyes, very dark violet periderm or occasionally with yellowish, very narrow halo areas, and uniformly dark blue-violet buds. Leaves long and narrow, 17-18 cm long by 7-8 cm wide. Peduncles 5-7 cm long. Pedicels 29-35 mm long, the upper part 4-5 mm long and the lower 25-30 mm long. Calyx asymmetrical, 10-12 mm long with subspatulate acumens. Corolla dark violet.

Chromosome number: 2n=2x=24.

This form has tubers very similar to form *chiar-phinu* of *S. stenotomum*.

Specimens Examined

HERBARIUM COLLECTIONS

Department La Paz

Province Murillo: Near Viacha, 3820 m alt., n.v. *Sura Kayana*; *Ochoa 10520* (CIP, OCH).

Department Oruro
Province Cercado: Indefinite, local market, 3650 m alt., n.v. *Pulu Wayk'u*; *Ochoa 10493* (CIP, OCH, type of f. *pulu-wayk'u*).

34B. Solanum stenotomum var. **ari-chuwa** Ochoa var. nov., Phytologia 65(2): 103-113. 1988.

Stems light green, spotted irregularly with light brown, and narrowly winged. Tubers with pink buds and white flesh, rounded, strongly irregular (or bulging), the yellowish-white skin with pinkish-purple stripes about the deep eyes or irregularly patterned. Leaves light green, with 6 pairs of leaflets and numerous interjected leaflets of several sizes. Leaflets narrowly lanceolate, the apex obtuse or subpointed and the base rounded, borne on 3-5 mm-long petiolules, the latter brown-spotted as in the rachis and petioles. Peduncles short, about 2-4 cm long and 1.5 mm in diameter at base. Pedicels articulated in the upper one-third of the stalk, the upper 6-7 mm long and the lower 12-14 mm long. Calyx asymmetrical, the sepals in two pairs of two plus one, or one fused pair plus three singles, 7-8 mm long, the lobes attenuate or abruptly narrowed in 1 mm-long pointed acumens. Corolla deep purple, somewhat pleated, large 3-3.5 cm in diameter. Style 8.5-9 mm long, exerted about 1 mm, papillose at base; stigma capitate, about 1 mm in diameter. Anthers 6 mm long, noncordate, with a poorly defined dorsal groove. Fruit small, globose and light green, up to 15 mm in diameter.

Chromosome number: $2n=2x=24$.

The common name of this variety is *Ari-Chuwa*, or 'Plato fino' in Spanish which means 'delicious meal.' This is in reference to its fine culinary qualities.

Specimens Examined

HERBARIUM COLLECTIONS

Department La Paz
Province Ingavi: Huatajata, 3850 m alt., 2n=24, n.v. *Ari Chuwa*; *Ochoa 3937* (F, OCH, type of var. *ari-chuwa*).

Province Pacajes: Corocoro, 3880 m alt., n.v. *Durasnillo*; *Ochoa 3553* (OCH).

Department Potosi
Province Quijarro: Condoriri, 3900 m alt., n.v. *Puca Ñahui*, or *Ojos rojos* (='red eyes'); *Ochoa 10677* (CIP).

GERMPLASM COLLECTIONS

Department La Paz

Province Aroma: Occopampa, 3850 m alt., n.v. *Durasnillo* (CO-2463).

Province Ingavi: Aipa, 4000 m alt., n.v. *Durazno* (CO-2363).

Province Omasuyos: Sisasani, 4160 m alt., n.v. *Huamán P'pecke* (='falcon head') (CO-2287).

Province Pacajes: Ballivian, 4300 m alt., n.v. *Durasnillo* (CO-2433). Caquiaviri, 3960 m alt., n.v. *Durazno* (CO-2388). Omaccollo, 3870 m alt., n.v. *Durazno* (CO-2410). San Andres de Machaca, 3980 m alt., n.v. *Chiquiña* (CO-2421).

Department Oruro

Province Cercado: Oruro, local market, 3700 m alt., n.v. *Ari-chuwa* (CO-2478, CO-2497).

34C. Solanum stenotomum var. **ccami** Bukasov in Bukasov, The Potatoes of South America and their Breeding Possibilities, p.33. 1933, *nom. nud.* Based on *Juzepczuk 1645* (WIR).

S. stenotomum Juz. et Buk. f. *ccami* Lechn., in Bukasov(ed.), Flora of Cult. Pl., Vol IX, Potato, Leningrad, p.69. 1971, *nom. nud.* Based on *Juzepczuk 1645* (WIR).

Stems vigorous, to 8 mm in diameter at base. Tubers round to ovoid, irregular (or bulging), with red or dark wine skin, red buds, and small deep eyes. Leaflets lanceolate or narrowly oval. Terminal leaflets of the same size, or slightly smaller than the first and second upper pairs of laterals. Pedicels articulated in the upper one-third of the stalk, the upper part 8-9 mm long and the lower 15-18 mm long. Calyx asymmetrical, 8 mm long, the lobes broadly elliptic-lanceolate, narrowing towards the apex in 1-1.5 mm-long acumens. Corolla violet, 2.6 cm in diameter. Anthers 5 mm long, borne on short, slender filaments. Style short and thick, 8 mm long by 0.5 mm in diameter; stigma large, broadly capitate, cleft.

Chromosome number: 2n=2x=24.

Specimens Examined

HERBARIUM COLLECTIONS

Department La Paz

Province Ingavi: Tiahuanaco, n.v. *Ccami*; *Juzepczuk 1645* (WIR, type of var. *ccami*).

GERMPLASM COLLECTIONS

Department La Paz

Province Ingavi: Cachuma, 4000 m alt., n.v. *Ccami* (CO-2335). Chama, 4100 m alt., n.v. *Ckami* (CO-2350).

Province Pacajes: Ballivian, 4030 m alt., n.v. *Chaucha* (CO-2431). Caquiaviri, 3960 m alt., n.v. *Ckunurana* (CO-2379). Omaccollo, 3870 m alt., n.v. *Ccami* (CO-2409).

Department Oruro

Province Cercado: Indefinite, 3700 m alt., n.v. *Ph'anama* (CO-2480).

34D. **Solanum stenotomum** var. **chaquina** Ochoa.

Stems slender, slightly pigmented. Tubers round, with dark purple flesh and blue-violet periderm. Leaves 6-jugate, dark green, with pigmented veins and veinlets. Leaflets usually small, ovate to elliptic, 2.5-3 cm long by 1.5-2 cm wide. Pedicels articulated about 4-5 mm below base of calyx. Calyx symmetrical or asymmetrical, 7 mm long. Corolla white, 2-2.5 cm in diameter. Anthers about 4.5 mm long, borne on 0.5-0.7 mm-long filaments. Style short and thick, 7 mm long; stigma broadly capitate.

Chromosome number: $2n=2x=24$.

The dye extracted from the dark purplish tubers of this species is used by the Aymara Indians for the coloring of yarn.

Specimens Examined

Department La Paz

Province Larecaja: Yamuco, 3200 m alt., n.v. *Chaquiña*; *Ochoa and Salas 11797, 11801* (OCH).

34E. **Solanum stenotomum** var. **chojllu** Ochoa var. nov., Phytologia 65(2): 103-113. 1988. PLATE XXII-4.

Tubers large, subcylindrical or occasionally ovoid, strongly irregular (or bulging), slightly compressed, apex and base obtuse, yellow periderm, deep eyes, violet and creamy-white buds, light yellow flesh. Leaflets narrowly elliptic-lanceolate, apex obtuse and base obliquely rounded, the upper pair occasionally narrowly decurrent on the rachis, or shortly petiolate or almost sessile. Calyx 9-10 mm long, the lobes narrowly elliptic-lanceolate and narrowing to 1.5 mm-long pointed acumens. Corolla violet, pentagonal, or occasionally hexagonal.

Chromosome number: 2n=2x=24.

This variety is frequently called *chokkllo* in the Quechuan language or *chhojjllu* (=*Puya*) in Aymara, this in reference to the peculiar corncob-like shape which frequently appears among the tubers of this species. These names should not be confused with the Aymara word for another variety of potato, *chhojjlla* (meaning 'grass' or 'fine herb'), which has differently colored leaves and more graceful stems. As noted below, the adjectives *jancko* ('white') and *wila* ('red') are frequently combined with the nouns *Chhojjllu* or *Puya* to indicate different clonal forms of varieties *chojllu* and *pitiquina*.

Specimens Examined

HERBARIUM COLLECTIONS

Department La Paz

Province Murillo: La Paz marketplace, brought from the altiplano, 3800 m alt., n.v. *Chojllu*; *Ochoa 3912* (OCH, type of var. *chojllu*).

34E(a). Solanum stenotomum var. chojllu f. janck'o-chojllu Ochoa f. nov., Phytologia 65(2): 103-113. 1988. Plate XXI-7.

Plants delicate. Tubers long, subcylindrical or slightly compressed, strongly irregular (or bulging) but not concertina-like, somewhat narrowed at apex and enlarged at basal end, light yellow flesh. Leaves light green, with 5-6(-7) pairs of leaflets and 7-11(-18) interjected leaflets. Peduncles, 3-5(-10) cm long by 1.5-2.0 mm in diameter at base. Pedicels, 15-18 mm long, articulated slightly above center. Calyx asymmetrical. Corolla, 2-2.8 cm in diameter. Anthers, 4-5 mm long borne on 1-1.5 mm-long filaments. Style very slender, 7 mm long, stigma small, capitate.

Chromosome number: 2n=2x=24 (except CO-2295, which is 2n=3x=36).

Specimens Examined

HERBARIUM COLLECTIONS

Department La Paz

Province Murillo: La Paz, 3800 m alt., marketplace, brought from the altiplano, n.v. *Zapallo*; *Ochoa 3938* (OCH, type of f. *janck'o-chojllu*).

GERMPLASM COLLECTIONS

Department La Paz

Province Camacho: Cariquina, 3940 m alt., n.v. *Chojlla* (CO-2324). Puerto Acosta, 3880 m alt., n.v. *Chinchilla* (CO-2302). Puerto Acosta, 3880 m alt., n.v. *Janck'o-Puya* (CO-2295). Soracko, 4150 m alt., n.v. *Janck'o-Ckolla* (CO-2327).

Province Larecaja: Millipaya, 3550 m alt., n.v. *Chojllu* (CO-2262). Millipaya, 3550 m alt., n.v. *Orcco Puya* (CO-2263). Paccolla, 4150 m alt., n.v. *Chojllu* (CO-2257).

Province Los Andes: Tambillo, 3880 m alt., n.v. *Chojlla* (CO-2158).

Province Pacajes: Caquiaviri, 3960 m alt., n.v. *Janck'o Chojllu* (CO-2385). Caquiaviri, 3960 m alt., n.v. *Ckellu Chojllu* (CO-2395).

34E(b). **Solanum stenotomum** var. **chojllu** f. **wila-chojllu** Ochoa f. nov., Phytologia 65(2): 103-113. 1988.

Tubers broadly elongate-compressed, slightly wider at base, strongly irregular (or bulging), with deep eyes, pink to dark wine skin, light pink buds and yellowish-white flesh. Leaflets elliptic-lanceolate, the apex subpointed or subacuminate and the base rounded, borne on 2-3(-6) mm-long petiolulates. Calyx, 7-8 mm long, the lobes abruptly narrowed to 1-1.5 mm-long acumens. Corolla lilac.

Chromosome number: 2n=2x=24.

Specimens Examined

HERBARIUM COLLECTIONS

Department La Paz

Province Camacho: Puerto Acosta, 3840 m alt., n.v. *Wila Puya*; *Ochoa 3552* (OCH).

Province Murillo: La Paz, local market, brought from the altiplano, 3650-3850 m alt., n.v. *Wila Chojllu; Ochoa 3942* (OCH, type of f. *wila-chojllu*).

GERMPLASM COLLECTIONS

Department La Paz

Province Camacho: Italaque, 3590 m alt., n.v. *Wila Puya* (CO-2328).

Province Larecaja: Paccolla, 4150 m alt., n.v. *Huailli* (CO-2258).

Province Pacajes: Caquiaviri, 3960 m alt., n.v. *Phureja* (CO-2377). Caquiaviri, 3960 m alt., n.v. *Wila Chojllu* (CO-2383).

Department Oruro
Province Oruro: Oruro Market, 3700 m alt., n.v. *Wila Chojlluraya* (CO-2473).

34F. Solanum stenotomum var. **kkamara** Ochoa var. nov., Phytologia 65(2): 103-113. 1988. PLATE XXII-7.

Plants vigorous. Stems thick, to 70 cm tall, branched, and pigmented dark blue-violet along the lower two-thirds of its length. Tubers with light beige flesh, long and thick, irregular (or bulging), cylindrical, subcylindrical or long pyriform with obtuse apex and narrow base, and dark blue-violet or almost black periderm, deep eyes and blue-violet buds. Leaves dark green, 16.5-22.5 cm long by 10-13 cm wide, with 5(-6) pairs of leaflets and 9-11 interjected leaflets. Leaflets elliptic-lanceolate. Terminal leaflets larger than the laterals, about 7.2 cm long by 4.2 cm wide. Lateral leaflets 5.8-5.5 cm long by 3-2.8 cm wide, the first and second upper pair approximately of equal size, borne on 3-5 mm-long petiolules. Corolla dark blue-violet.

Chromosome number: 2n=2x=24.

Specimens Examined

Department La Paz
Province Murillo: La Paz, Camacho Market, 3650 m alt., brought from the altiplano, n.v. *Kkamara*; *Ochoa 3920* (OCH, type of var. *kkamara*).

34G. Solanum stenotomum var. **luru** Ochoa var. nov., Phytologia 65(2): 103–113. 1988.

Plants delicate, light green. Stems to 70 cm tall and up to 6 mm in diameter at base. Tubers with white flesh, rounded or slightly compressed, irregular (or bulging), with yellowish-white periderm, deep eyes and creamy-white buds tinged with pink at base and apex. Leaves long and narrow, 18-22.5 cm long by 8.5-10.5 cm wide, with 6-7 pairs of leaflets and 8-12 interjected leaflets. Terminal leaflets about 5.6 cm long by 2.5 cm wide, rhombic-lanceolate. Lateral leaflets narrowly elliptic-lanceolate, the three lower ones almost the same size as the terminal. Peduncles, 8-10 cm long, 5-7-flowered.

Chromosome number: 2n=2x=24.

Variety *luru* shares some features in common with *S. stenotomum* var. *phinu* f. *pacajes*. Both are small and white-flowered, and both have similar pedicel and calyx characteristics. However, the two differ from one another substantially in their leaf and tuber characteristics.

This variety is known as *luru* (='dwarf') or *huerfana*.

Specimens Examined

HERBARIUM COLLECTIONS

Department Potosi

Province Frias: Near Tinguipaya, 3900 m alt., n.v. *Luru*; *Ochoa 3949* (OCH, type of var. *luru*).

GERMPLASM COLLECTIONS

Department Potosi

Province Frias: Lecherias, 3800 m alt., n.v. *Lurun* (HAW-5695). Miraflores, 3800 m alt., 2n=24, n.v. *Lurun* (OCH-10664). Tinguipaya, 3900 m alt., 2n=24, n.v. *Luru Papa* (OCH-10647, 10648A).

34H. Solanum stenotomum var. **pitiquiña** Ochoa var. nov., Phytologia 65(2): 103-113. 1988. FIGS. 154-3, 159; PLATE XXI-6.

Plants erect to decumbent. Stems 60-70 cm tall. Tubers long, broad and concertina-like, subcylindric with obtuse ends and with numerous deep eyes and red violaceous or pink skin, occasionally with yellowish-white, narrow halo areas, pink or pink-violet buds, and white or light beige flesh, occasionally with a pigmented vascular ring. Leaves light green, long and narrow, strongly dissected, 6-7 pairs of leaflets and numerous interjected leaflets. Leaflets narrowly elliptic-lanceolate to obovate, borne on 4-7 mm-long petiolules. Peduncles 3-4(-6) mm long, 5-9-flowered, sparsely pilose as in the pedicels. Corolla small, 2-2.5 cm in diameter, occasionally hexagonal, usually lilac with whitish acumens. Calyx 8-9(-11) mm long.

Chromosome number: 2n=2x=24.

The many forms of this variety are distinguished principally by the pigmentation of the tuber and buds.

It is possible that *S. stenotomum* var. *pitiquiña* may be related to *S. stenotomum* var. *pitiquilla*, or to one of the many forms of *S. stenotomum* var. *pitiquilla-ccosilinle* (Hawkes, 1944; Lechnovitch, 1971). However, material of the latter two taxa were not available for examination.

The variety *pitiquina* is the most widely distributed variety of *S. stenotomun* in Bolivia. It is also commonly cultivated in southern Peru.

Specimens Examined

HERBARIUM COLLECTIONS

Department La Paz

Province Ingavi: Tiahuanacu, 3650 m alt., 2n=24, n.v. *Yanarico; Ochoa 3140, 3755* (OCH). Tiahuanacu, 3650 m alt., 2n=24, n.v. *Pitiquilla*; *Ochoa 3535* (OCH, type of var. *pitiquina*). Tiahuanacu, 3650 m alt., 2n=24, n.v. *Pitiquina*; *Ochoa 3919* (OCH). Potomanta, 3800 m alt., 2n=24, n.v. *Pitiquilla*; *Ochoa 3564* (OCH).

Province Los Andes: Huancane, 3830 m alt., n.v. *Pitiquilla*; *CPP-1815* (OCH). Pucarani, 3840 m alt., n.v. *Piticalla*; *Ochoa 2797, 3151* (OCH).

Province Murillo: La Paz Market, 3650 m alt., 2n=24, n.v. *Puca Pitiquiña*; *CIP-700891* (CIP).

Department Oruro

Province Poopo: Cercado, 3750 m alt., 2n=24, n.v. *Huaycku Papera*; *Ochoa 10484* (CIP).

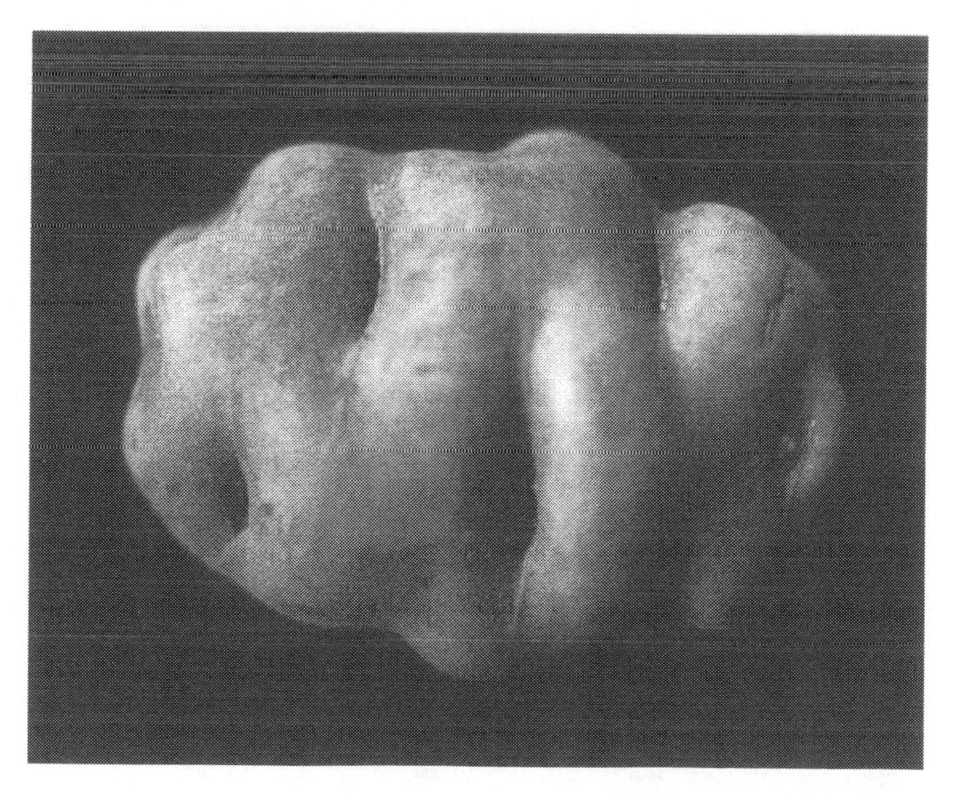

Figure 159. – Tuber of *Solanum stenotomum* var. *pitiquiña*, ×2.

GERMPLASM COLLECTIONS

Department La Paz

Province Aroma: Vilaque, 3800 m alt., n.v. *Chulo* (HUA-886).

Province Ingavi: Cachuma, 4000 m alt., n.v. *Cholo* (CO-2336). Patarani, 3850 m alt., n.v. *Chiquina* (CO-2174). Patarani, 3850 m alt., n.v. *Pitiquilla* (CO-2183).

Province Los Andes: Keallani, 3880 m alt., n.v. *Piticalla* (CO-2147). Pucarani, 3900 m alt., n.v. *Piticalla* (CO-2222). Tambillo, 3880 m alt., n.v. *Piticalla* (CO-2156).

Province Murillo: Huaillara, 3800 m alt., n.v. *Piticana* (CO-2192). La Paz Market, 3800 m alt., n.v. *Pitiquina* (CIP-701072).

Province Omasuyos: Sisasani, 4160 m alt., n.v. *Pitiquina* (CO-2288).

Province Pacajes: Caquiaviri, 3960 m alt., n.v. *Misi Cayu* (CO-2381). Omaccollo, 3870 m alt., n.v. *Chojllu (CO-2412)*. San Andres de Machaca, 3980 m alt., n.v. *Chiquina* (CO-2415).

34H(a). Solanum stenotomum var. **pitiquina** f. **alkka-pitiquiña** Ochoa f. nov., Phytologia 65(2): 103-113. 1988.

Stems slightly pigmented. Tubers with bicolored periderm, white flesh, white halo areas about the eyes, concertina-like, with black skin and purple buds, or with red skin and pink buds.

The common name for this form is *Puca alkka pitiquiña*.

Specimens Examined

HERBARIUM COLLECTIONS

Department La Paz

Province Los Andes: Palcolo, 3800 m alt., 2n=24, n.v. *Yana Alcca Piticalla*; *Ochoa 3148* (OCH, type of f. *alkka-pitiquina*). Pucarani, 3840 m alt., 2n=24, n.v. *Puca Alcca Pitiquina*; *Ochoa 2793* (OCH).

34H(b). Solanum stenotomum var. **pitiquiña** f. **azureo-ckati** Ochoa f. nov., Phytologia 65(2): 103-113. 1988.

Stems, peduncles, pedicels, and leaf rachises strongly pigmented. Tubers with deep eyes and white flesh, broadly ovoid, strongly irregular (bulging) or subconcertina-like, with dark bluish-violet skin, narrow yellowish-white halo areas about the eyes, and dark bluish-violet buds. Leaflets elliptic-lanceolate, with pointed apices and rounded to obliquely rounded bases. Corolla dark violet, to 3 cm in diameter.

This form is related to *S. stenotomum* var. *pitiquiña* f. *laram.*

Specimens Examined

HERBARIUM COLLECTION

Department La Paz

Province Inquisivi: Quime, 3200 m alt., 2n=24, n.v. *Ckati* or *Azul Ckati*; *Ochoa and Salas 15514* (CIP, OCH, type of f. *azureo-ckati*).

34H(c). Solanum stenotomum var. pitiquiña f. ch'asca Ochoa.

Plants graceful. Tubers broadly oblong with light beige flesh, strongly irregular (bulging) to subconcertina-like, with yellowish-white skin, deep eyes and with creamy-white buds, the latter painted pink at base and apex. Leaves light green, very dissected. Leaflets narrowly lanceolate.

Specimens Examined

HERBARIUM COLLECTION

Department Potosi

Province Chayanta: Maragua, 3700 m alt., 2n=24, n.v. *Ch'aska*; *Ochoa 10608* (OCH).

34H(d). Solanum stenotomum var. pitiquiña f. laram Ochoa f. nov., Phytologia 65(2): 103-113. 1988. Fig. 155.

Stems, petioles, rachises, peduncles, and pedicels colored dark wine-red. Tubers oblong, with light beige flesh, strongly irregular (bulging), with dark bluish-pink or blue-violet periderm, deep eyes and dark blue-violet buds. Leaves dark green. Corolla violet.

The form *laram* is also distributed in southern Peru.

Specimens Examined

HERBARIUM COLLECTIONS

Department La Paz

Province Murillo: La Paz Market, brought from the altiplano, 3800 m alt.; *Ochoa 3927 and 3926* (OCH, type of f. *laram*) [both 2n=24].

34H(e). Solanum stenotomum var. **pitiquiña** f. **phiti-kalla** Ochoa f. nov., Phytologia 65(2): 103-113. 1988. FIG. 156; PLATE XXI-5.

Tubers long and thick, with light beige flesh, subcylindric to subconcertina-like, with numerous deep eyes and white buds, the latter with dark bluish-violet bases and apices. Corolla light violet.

This form, which is also known locally as *Loro escarbador*, has similar tuber characteristics to *S. stenotomum* var. *pitiquiña* f. *alkka-pitiquiña*, but differs in having a light violet corolla.

Specimens Examined

HERBARIUM COLLECTIONS

Department La Paz

Province Murillo: La Paz Market, 3650 m alt., n.v. *Phiti Kalla* (='diger parrot'); *Ochoa 3928* (OCH, type of f. *phiti-kalla*) *and 3930* (OCH) [both 2n=24.]

34H(f). Solanum stenotomum var. **pitiquiña** f. **quime** Ochoa f. nov., Phytologia 65(2): 103-113. 1988.

Tubers with light beige flesh, broadly concertina-like, with very dark red-violet periderm, and narrow, yellow halo areas around the eyes, and red-violet buds.

Specimens Examined

HERBARIUM COLLECTION

Department La Paz

Province Inquisivi: Quime, 2900 m alt., 2n=24; *Ochoa and Salas 15513* (CIP, OCH, type of f. *quime*).

34I. Solanum stenotomum var. **zapallo** Ochoa var. nov., Phytologia 65(2): 103-113. 1988. PLATE XXII-6.

Plants uniformly light green. Stems 30-40 cm tall, much branched. Tubers round, with bright or pale yellow flesh and dark yellow skin tinged smoky-pink at apex, deep eyes and creamy-white buds, the latter tinged pale pink at base and apex. Leaves light green, short and broad, with 4-5(-6) pairs of leaflets and 7-8(-12) interjected leaflets of different sizes. Terminal leaflets slightly larger, or same size as laterals. Lateral leaflets elliptic-lanceolate or broadly elliptic-lanceolate, with pointed to obtuse apices and rounded bases, borne on 4-6(-12) mm-long petiolules, the latter occasionally with 1-2 interjected leaflets. Peduncles 5-8(-12) cm long, 7-10-flowered. Pedicels articulated 3-4 mm below base of calyx, the lower part 10-15 mm long. Calyx 5 to 6 parted and 10-11 mm long, symmetrical, or asymmetrical with two fused pairs of two plus two single sepals or one fused pair and three single sepals, the lobes broadly elliptic-lanceolate and narrowing in broadly spatulate acumens or in an obtuse apex. Corolla white to creamy-white, with short lobes and acumens. Anthers about 5 mm-long, borne on 0.5 mm-long filaments. Style more or less slender, 9-10 mm-long, exerted 4-5 mm, sparsely papillose about center; stigma light green, somewhat broader than style, capitate. Fruit ovoid, light green.

Chromosome number: 2n=2x=24.

The native farmers of the Bolivian altiplano frequently call this variety *Zapallo* (or 'squash'). However, this name is also indiscriminately applied to potato varieties having round tubers and white or sky blue flowers, or to varieties with long tubers and violet flowers. In all cases, it is important to remember that the most important feature to the native farmer is the bright yellow flesh color of these varieties, which is similar in appearance to the squash, *Cucurbita maxima* Duch.

Specimens Examined

HERBARIUM COLLECTIONS

Department La Paz

Province Ingavi: Yanarico, 3800 m alt., n.v. *Zapallo*; *Ochoa 2790* (OCH). Tiahuanaco, 3680 m alt.; *Ochoa 3145* (OCH, type of var. *zapallo*).

Province Los Andes: Pucarani, 3900 m alt.; *Ochoa 2795, 3150* (OCH).

Province Pacajes: Yuriwaya, 3850 m alt., n.v. *Zapallo*; *Ochoa 3529* (OCH).

34I(a). **Solanum stenotomum** var. **zapallo** f. **churi-puya** Ochoa f. nov., Phytologia 65(2): 103-113. 1988.

Stems pigmented or slightly pigmented toward base and in the upper leaf axils. Leaves darker green than the typical variety, pigmented at base of petioles. Tubers similar to *S. stenotomum* var. *zapallo* f. *zapallo* but with the cream colored buds tinted dark blue violet at base and apex. Corolla sky blue.

Although this form has been classified under *S. stenotomum*, it has the dark green leaves, sky blue flowers and tuber characteristics of the south Peruvian cultigen, *S. goniocalyx* var. *coeruleum* Vargas. It is here kept as a form of *S. stenotomum* as it has the typical leaflet shape and leaf dissection of this species.

Specimens Examined

HERBARIUM COLLECTION

Department La Paz

Province Murillo: Walata Chico, 3850 m alt., n.v. *Churipuya*; *Ochoa 3152* (OCH, type of f. *churi-paya*).

34I(b). Solanum stenotomum var. **zapallo** f. **kkarachi-pampa** Ochoa f. nov., Phytologia 65(2): 103-113. 1988.

Plants somewhat more robust and branched than in the preceding form. Tubers similar to those of the typical variety but with more dissected leaves. Leaflets ovoid to elliptic, with obtuse apices and rounded bases, and borne on long petiolules, the latter sustaining one or more acroscopic and basiscopic interjected leaflets. Calyx much longer than the typical variety, with broader lobes. Corolla white or whitish, with pale lilac-tinged acumens. Style shorter and thicker than in the typical variety; stigma more broadly capitate, bifid or trilobed. Fruit globose.

Specimens Examined

HERBARIUM COLLECTION

Department Potosi

Province Linares: Kkarachipampa, 4000 m alt., n.v. *Zapallo*; *Ochoa 10481* (CIP, OCH, type of f. *kkarachi-pampa*).

35. Solanum × juzepczukii Bukasov, Proc. U.S.S.R. Congr. Genet. Pl. & Anim. Breed. 3:603-604. 1929. FIGS. 160, 161; PLATE XIX-6.

S. juzepczukii f. *orcco-malko* Lechn., in Bukasov (ed.), Flora of Cult. Pl., Vol. IX, Potato, Leningrad, pp. 71-73. 1971, *nom. nud.* Based on *Juzepczuk 1166, 1178-b*, Peru. Pomacanchi, Dept. Cusco, Prov. Acomayo (WIR); and *Juzepczuk 1641, 1647, 1687-b*, Bolivia. Tiahuanaco, Dept. La Paz, Prov. Ingavi (WIR).

Plants small, rosette-like when young. Stems 30-40 cm tall, light green or irregularly mottled or slightly pigmented at base; stoloniferous, frequently forming sucker sprouts. Tubers oblong to elongated-compressed, to long

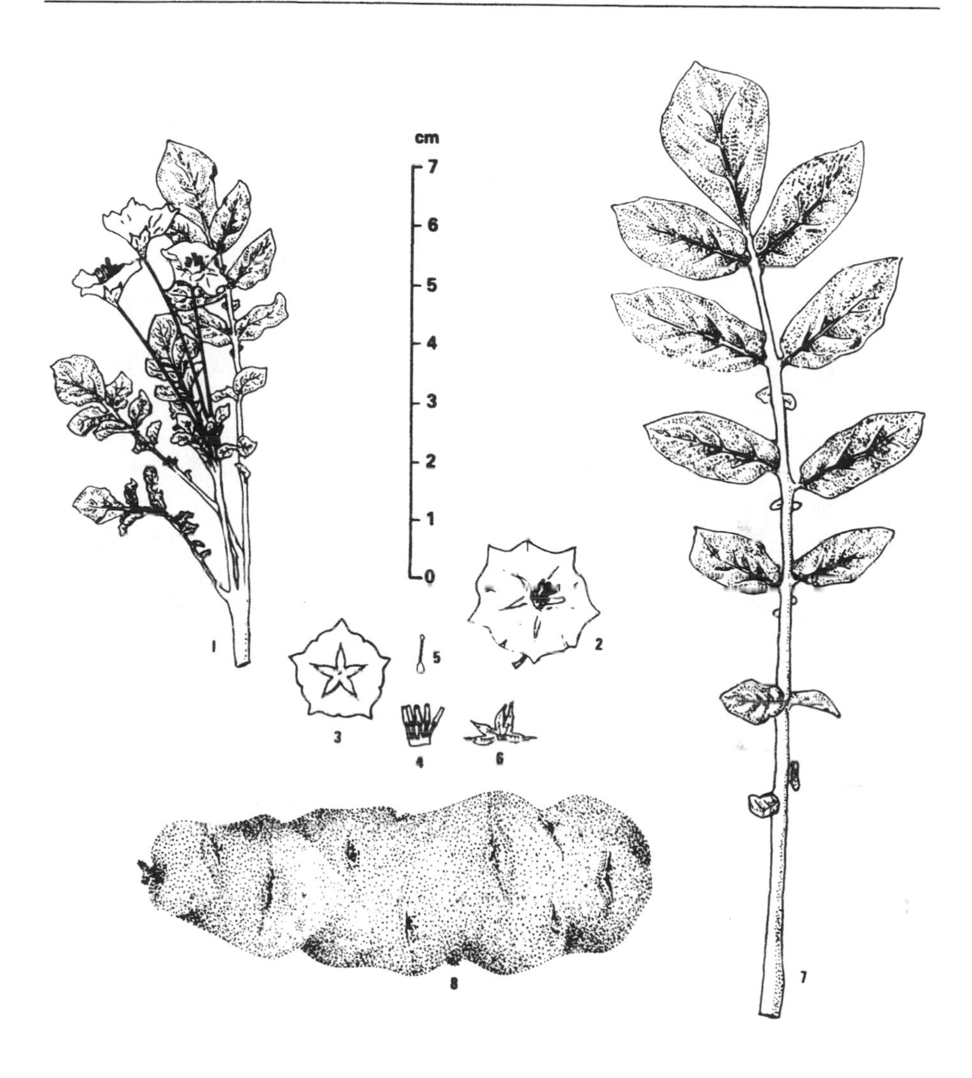

Figure 160. – *Solanum* × *juzepczukii* (*Ochoa-2806*). 1. Upper part of flowering plant. 2. Flower. 3. Corolla. 4. Stamens. 5. Pistil. 6. Calyx. 7. Leaf. 8. Tuber.

subcylindrical, with obtuse apex with white, or occasionally with pale smoky-purple periderm and superficial or semideep eyes, creamy-white buds, long brows, white flesh. Leaves green, usually long and narrow, 14–28 cm long by 6–10 cm wide, with 5–7 pairs of leaflets and (3-)7-9(-12) interjected leaflets. Leaflets ovate or elliptic to elliptic-lanceolate, shortly petiolulate, rugose, 2.5-5(-6.5) cm long by 1-2.5(3.5) cm wide, with rounded and sessile or subsessile bases, sometimes the upper first pair narrowly decurrent on rachis, the second pair occasionally slightly larger than the first pair. Terminal leaflets broadly elliptic to rhombic-lanceolate, 3-7 cm long by 2-4 cm wide. Inflorescence cymose-panniculate, 5-10-flowered. Peduncles short, 2-3 cm long. Pedicels slender, light green or very slightly pigmented, 4-6 cm long, the upper part 4-7 mm long. Calyx small, 5-6 mm long, occasionally asymmetrical, the triangular-lanceolate or narrowly elliptic-lanceolate lobes terminating in very short, pointed acumens. Corolla rotate, small, 1.5-2 cm in diameter, blue, more intensely pigmented along the outer borders of the grayish-white nectar guides, with short lobes, these approximately 3.5 mm long including the acumens. Staminal column cylindrical; anthers small, 3.5-4 mm long with a well-defined dorsal groove, cordate at base, borne on 1-2 mm-long filaments. Style short, 5.5-6 mm long and less than 0.5 mm in diameter, barely exerted, very papillose along the lower one-third of its length; stigma capitate, cleft.

Chromosome number: 2n=3x=36.

Types: PERU. Department Cusco, Province Acomayo, Pomacanchi, n.v. *Orcco-malcco*; *S. Juzepczuk 1166* (WIR). Department La Paz, Province Ingavi, Tiahuanaco, n.v. *luqui*; *S. Juzepczuk 1641* (WIR).

In Bolivia this species is known by many names, including *Lucki*, *Orcco-malcco*, *Chocke pito*, *Pinkula*, *Ck'aisa*, *Mullunku*, *Pocco tturu*, *Nasari*, *Sisu*, *Janck'o rai*, and *Pingo*.

Affinities

Solanum × *juzepczukii* resembles *S. acaule* in having a rosette-like habit and small rotate corollas, small anthers, and long, slender filaments. Moreover, it is similar to *S. acaule* in having a pedicel that is not clearly articulated and in having similarly shaped leaves. These characteristics prompted Bukasov (1939, 1941) to suggest *S. acaule* as one hybrid parent of *S.* × *juzepczukii*. Lechnovitch (1971), who agrees with Bukasov on this point, places this cultigen within subseries Acaulia of series Acaulia. The other ancestral species of this plant, according to Hawkes (1962, 1979), is *S. stenotomum*. However, the morphological features of *S. stenotomum* are completely masked in *S.* × *juzepczukii*.

Bukasov (1939, 1941), who first proposed the hybrid origins of *S.* × *juzepczukii* and *S.* × *curtilobum*, suggested the first plant may be a natural cross of a diploid wild or cultivated species, and the second, a hybrid of *S.* × *juzepczukii* with *S. andigenum*. Hawkes (1956a, 1962) later reported the artificial synthesis of these two hybridogenic cultivars, *S.* × *juzepczukii* from *S. acaule* and *S. stenotomum*, and *S.* × *curtilobum* from *S.* × *juzepczukii* and *S. tuberosum* cv. Teton. According to Schmiediche et al. (1982), the artificial synthesis of *S. juzepczukii* has also been achieved at CIP.

As in the case of *S.* × *curtilobum*, which also has *S. acaule* in its parentage, this hybridogenic species is frost-resistant. It reportedly survives temperatures as low as −5° C (Bukasov, 1933; Schmiediche et al., 1980; Bavyko, 1982).

A series of crossing experiments at CIP, utilizing *S. acaule* and the diploid species *S.* × *ajanhuiri*, *S. stenotomum*, *S. phureja*, and *S. goniocalyx*, yielded some frost-resistant types. Various segregates from this crossing experiment were able to survive temperatures as low as −5° C. The strongest and most tolerant plants were obtained when *S. goniocalyx* was used as one of the parents. The tetraploid cultivated potato, *S. tuberosum* subsp. *andigena*, was also used experimentally in this crossing program. However, only tetraploid plants were obtained when *S.* × *juzepczukii* was hybridized artificially with *S. tuberosum* subsp. *andigena*. Although the experiment failed to recreate *S.* × *curtilobum*, some of the obtained tetraploid plants survived temperatures down to −2° C (Schmiediche et al., 1982).

Figure 161. – Habitat of *S. acaule*, *S.* × *juzepczukii* and *S.* × *curtilobum*, 4000 m altitude, near La Paz, with the Cordillera Real in the background.

Aside from frost tolerance, some morphotypes of *S.* × *juzepczukii* and *S.* × *curtilobum* show resistance to the wart-causing fungus, *Synchytrium endobioticum* (Bukasov, 1940b; Kameraz 1957; Rothacker and Müller, 1960). Other clonal forms are resistant to *Alternaria solani*, the causal agent of early blight (Bukasov and Kameraz, 1959). Resistance to potato virus PVX has also been reported for *S.* × *juzepczukii* (Cockerham, 1943; Bavyko, 1982) as well as *S.* × *curtilobum* (Bavyko, 1982). Moreover, Stelter and Rothacker (1965) report that *S.* × *juzepczukii* is resistant to *Globodera*, a genus of cyst nematodes. The latter results have also been obtained at CIP (Intl. Pot. Ctr., 1979a).

The bitter tubers of *S.* × *juzepczukii* and *S.* × *curtilobum* are high in glycoalkaloids. Thus, both are normally only used for the making of *chuño* or *moraya*, two dehydrated products of the potato which can be kept almost indefinitely (Cevallos Tovar, 1914; Bukasov, 1933; Vargas, 1949; Schmiediche, 1977a; Schmiediche et al., 1980).

Distribution

Solanum × *juzepczukii* is cultivated from the Department of Ancash in northern Peru to the Province of Jujuy in northern Argentina. However, the major center of production for this cultigen lies in the Peruvian-Bolivian altiplano, where it is cultivated extensively along the margins of the Lake Titicaca basin. The altitudinal limits of this species lie between 3400 and 4100 m.

Present day cultivation of *S.* × *juzepczukii* takes place in areas once largely dominated by Puna, a high grassland formation. In general, no other cultivated species are grown in these very cold areas except *S.* × *curtilobum*. The only wild species of this high Puna formation (*see* Fig. 161) is *S. acaule*.

The following varieties and forms of *S.* × *juzepczukii* were all tested electrophoretically with the exception of variety *sisu* and forms *suitu-lucki* and *lucki-pinkula*. These analyses, conducted by Schmiediche et al. (1980) at CIP, were aimed at the elimination of duplicate collections.

Specimens Examined

HERBARIUM COLLECTIONS

Department La Paz

Province Camacho: Puerto Acosta, 3850 m alt., 2n=36; *Ochoa 2806* (OCH).

Province Ingavi: La Paz; *Juzepczuk 1641, 1647, 1678B* (WIR).

Province Pacajes: Near Corocoro, 4000 m alt., n.v. *Lucki*; *Ochoa 2808* (OCH).

Department Oruro

Province Poopó: Urmiri, 3750 m alt., n.v. *Lucki*; *Ochoa 3945* (OCH).

35A. Solanum × juzepczukii var. **ckaisalla** (Lechnovitch) Ochoa comb. nov.

S. juzepczukii f. *ckaisalla* Lechn., in Bukasov (ed.), Flora of Cult. Pl., Vol. IX, Potato, Leningrad, p. 72. 1971, *nom. nud.* Based on *Juzepczuk 1637*, Bolivia. La Paz (WIR), and *Erlanson and MacMillan 429* (WIR=K-852).

Plants compact, subrosette-like. Stem irregularly pigmented. Stolons 25-35 cm long, occasionally forming adventitious plants. Tubers with white flesh, oval-compressed to long-compressed, or almost flattened subspatullate, broad at apex and narrower and somewhat pointed at base, with superficial and semideep eyes and blue-violet periderm, the latter striped yellowish-white. Leaves dark green, with 4-5(-6) pairs of leaflets and 0-5 interjected leaflets. Leaflets ovate to elliptic, with very obtuse apices and sessile bases.

Chromosome number: 2n=3x=36.

Specimens Examined

HERBARIUM COLLECTIONS

Department La Paz

Province Aroma: Patacamaya, 3780 m alt.; *CIP-701283* (CIP).

Province Murillo: La Paz, n.v. *Ccaisalla*; *Juzepczuk 1637* (WIR). La Paz; *Erlanson and MacMillan 429* (WIR, K-852).

GERMPLASM COLLECTIONS

Department La Paz

Province Aroma: Patacamaya n.v. *Lucky* (CIP-701248). Patacamaya, n.v. *Kaisalla* (CIP-702637). San Antonio, 3700 m alt., n.v. *Laram Luque* (HUA-777).

Department Oruro

Province Cercado: Belen, 3700 m alt., n.v. *Sultuma* (HUA-802).

Department Potosi

Province Sud Lipez: Lipez (CIP-702620).

35A(a). Solanum × juzepczukii var. **ckaisalla** f. **ck'oyu-ckaisalla** Ochoa f. nov. Phytologia 65(2): 103-113. 1988.

This form differs from the typical variety in having only a light purple-violet, occasionally striped yellowish-white tuber periderm.

Specimens Examined

HERBARIUM COLLECTION

Department Oruro

Province Carangas: Jango Cala, n.v. *Kaisalla*; *Huamán 821* (OCH, type of f. *ck'oyu-ckaisalla)*.

GERMPLASM COLLECTION

Department Oruro

Province Carangas: Valle Bonito, 3800 m alt., n.v. *Kaisalla* (HUA-832).

35A(b). Solanum × juzepczukii var. ckaisalla f. janck'o-ckaisalla Ochoa f. nov., Phytologia 65(2): 103-113. 1988.

This form differs from the last two variants in having light green leaves and stems, with white tuber periderm.

Chromosome number: 2n=3x=36.

Specimens Examined

HERBARIUM COLLECTION

Department Oruro

Province Poopó: Toledo, 3700 m alt., n.v. *Ckaisalla*; *Huamán 809* (CIP, type of f. *janck'o-ckaisalla*).

GERMPLASM COLLECTIONS

Department Oruro

Province Poopó: Saucari-Toledo, 3700 m alt., n.v. *Janck'o Ckaisalla* (HUA-816). Toledo, 3700 m alt., n.v. *Kaisalla* (HUA-811). Saucari-Toledo, 3700 m alt., n.v. *Jhanco Sullthuma* (HUA-836).

35A(c). Solanum × juzepczukii var. ckaisalla f. wila-ckaisalla Ochoa f. nov., Phytologia 65(2): 103-113. 1988.

This cultigen differs from the typical variety in having a pink-violet tuber periderm.

Specimens Examined

HERBARIUM COLLECTION

Department Oruro

Province Poopó: Saucari-Toledo, 2n=36; *Huamán 815* (OCH, type of f. *wila-ckaisalla*).

35B. Solanum × juzepczukii var. lucki Ochoa var. nova, Phytologia 65(2): 103-113. 1988.

Plants forming a rosette or subrosette. Stoloniferous, and forming many sucker sprouts. Tubers round to ovate, elongated or slightly obliquely compressed, with superficial to semideep eyes and white periderm or occasionally white with light blue-violet apical stripes. Leaves relatively short, 8-12 cm long.

This plant, which very closely resembles *S. acaule*, is probably the most widespread of the various forms of this species in the altiplanic region of Bolivia and southern Peru. Out of nearly 140 collections of this species at CIP, more than 60 correspond to this variety. Some appear to set tubers earlier than others.

Chromosome number: 2n=3x=36.

Specimens Examined

HERBARIUM COLLECTIONS

Department La Paz

Province Ingavi: Indefinite; *Huamán 788* (WIR) and *789* (WIR, type of var. *lucki*).

GERMPLASM COLLECTIONS

Department La Paz

Province Ingavi: Guaqui, 3800 m alt. (HUA-786, 787, 790, 791).

Province Pacajes: Caquiaviri, 3900 m alt. (HUA-785, 1041).

Department Oruro

Province Saucari: Belen, 3700 m alt., n.v. *Luque* (HUA-803). Toledo, 3700 m alt., n.v. *Laram Chockepito* (HUA-839). Toledo, 3700 m alt., n.v. *Laram Luque* (HUA-841).

35B(a). Solanum × juzepczukii var. **lucki** f. **lucki-pechuma** Ochoa f. nov., Phytologia 65(2): 103-113. 1988.

Plants very similar to the typical variety but differing in having tubers with very deep eyes. Tubers round, periderm and flesh white.

Chromosome number: 2n=3x=36.

Specimens Examined

HERBARIUM COLLECTIONS

Department La Paz

Province Aroma: Patacamaya, n.v. *Luqui*; *CIP-702635.* (CIP, type of f. *lucki-pechuma*).

Department Oruro

Province Indefinite: Challa; *CIP-702631* (OCH).

GERMPLASM COLLECTIONS

Department La Paz

Province Aroma: Pomani, n.v. *Pincu* (CIP-702626). Vilaque (CIP-702630).

Department Oruro

Province Carangas: Totora, n.v. *Pechuma* (CIP-702699).

35B(b). Solanum × juzepczukii var. **lucki** f. **lucki-pinkula** Ochoa f. nov., Phytologia 65(2): 103-113. 1988. PLATE XIX-5.

Plants small, resembling *S. acaule*. Stems 20-25 cm tall, thick and with very short internodes, light green or irregularly pigmented. Stoloniferous, frequently developing sucker sprouts. Tubers with white flesh, round pyriform to subfalcate, with a thick, obtuse apex and a narrow, pointed base, superficial eyes, and white buds and periderm. Leaves dark green, with 6-7 pairs of leaflets and 5-6 interjected leaflets.

Specimens Examined

HERBARIUM COLLECTION

Department Potosi

Province Frias: Challactiri, 3900 m alt., 2n=36, n.v. *Sack'ampaya Lucki*; *Ochoa 10571* (CIP, type of f. *lucki-pinkula*).

35C. Solanum × juzepczukii var. **parco** Hawkes in Hawkes, Potato Collecting Expeditions in Mexico and South America. II. Systematic Classification of the Collections, pp. 73, 131. 1944. Based on EPC-747, 1106, 1107 (= CPC [?], not seen).

Plants not strongly rosette-forming. Stolons long, not forming adventitious plantlets. Tubers with yellowish flesh and periderm, long pyriform-compressed. Leaves broadly ovate, conspicuous, more than 20 cm long, with 6-7 pairs of leaflets and 10 or more interjected leaflets.

This variety was described originally by Hawkes from collections made in Peru: *747* (Cusco) and *1106*, *1107* (Puno) (Hawkes, 1944, pp. 73, 131).

Specimens Examined

GERMPLASM COLLECTIONS

Department La Paz

Province Pacajes: Caquiaviri (HUA-1042).

Department Oruro

Province Atahuallpa Sillja: Indefinite, n.v. *Chullpa Choque* (CIP-702581).

35D. Solanum × juzepczukii var. **rosea** Vargas, in Vargas, Las Papas Sudperuanas, II Parte, p.20, Fig.2. 1956. Type: *Vargas 1145*, Peru. Macusani, Puno (CUZ).

Plants similar in habit and leaf morphology to variety *lucki*. Stoloniferous, but not forming sucker sprouts. Tubers with white flesh and red periderm. Corolla light pink-violet.

This variety was originally described by Vargas (1956) from a collection (*Vargas 1145*) made in Province of Carabaya, Department of Puno, Peru.

Chromosome number: 2n=3x=36.

Specimens Examined

GERMPLASM COLLECTIONS

Department La Paz

Province Aroma: San Antonio, 3700 m alt., n.v. *Wila Luque* (HUA 778). Vilaque, 3800 m alt., n.v. *Wila Pinku* (HUA-896).

35E. Solanum × juzepczukii var. **sisu** Ochoa.

Plants somewhat rosette-like at flowering time. Stems 40-50 cm tall, lightly pigmented. Tubers with white flesh, long, slender to subcylindrical or fusiform, with dark violet periderm, superficial eyes and white buds, the latter tinted purple-violet at base.

Chromosome number: 2n=3x=36.

Although the tubers of this triploid plant are sweet and not bitter, as in the case of all other varieties of *S.* × *juzepczukii*, there would appear to be little question that it belongs to this species. Johns (1985), who reports that the tubers of this variety are eaten directly without dehydration or further processing, cites other clones of this cultigen that have not been examined by the author. These include *Chaña Sisu, Huaycha Sisu, Parco Sisu, Huilla Sisu*, and *Ankanchi*. According to Johns, *Ankanchi* and *Sisu* are hybrids of *S. acaule* and a cultivated diploid species, possibly *S.* × *ajanhuiri*.

Specimens Examined

HERBARIUM COLLECTIONS

Department La Paz

Province Pacajes: Vicinity of Nazacara, 3860 m alt., n.v. *Laram sisu* or *Sisu*; *Ochoa 3921, 3922* (OCH).

GERMPLASM COLLECTIONS

Department La Paz

Province Pacajes: Nazacara, 3860 m alt., 2n=36, n.v. *Laram Sisu* (CO-2401). Nazara, 3860 m alt., n.v. *Sisu* (CO-2400). Omacollo, 3870 m alt., 2n=36, n.v. *Sisu* (CO-2408). San Andres de Machaca, 3980 m alt., 2n=36, n.v. *Sisu* (CO-2426).

35E(a). Solanum × juzepczukii var. **sisu** f. **janck'o-sisu** Ochoa.

PLATE XXIII-5.

This form is very similar in its vegetative characteristics to form *sisu*, but differs in having light green stems and a white tuber periderm with creamy-white buds.

Specimens Examined

Department La Paz

Province Pacajes: San Andres de Machaca, 3980 m alt., 2n=36, n.v. *Sisu* or *Jancko Sisu*; *Ochoa 3923* (OCH).

36. Solanum × curtilobum Juzepczuk and Bukasov, Proc. U.S.S.R. Congr. Genet. Pl. & Anim. Breed. 3:609. 1929. FIG. 162; PLATE XXIII-1.

S. × *curtilobum* f. *choque-pitu* Lechn., in Bukasov (ed.), Flora of Cult. Pl., Vol. IX, Potato; Lechnovitch, Cult. Potato Species, Leningrad, p. 74. 1971, *nom. nud.*. Based on *Juzepczuk 1642*, Bolivia. Tiahuanaco [Prov. Ingavi] (WIR).

Plants vigorous, subrosette-like, or forming a true rosette when young. Stems 30-40 cm tall, thick, pigmented mainly at base, with short internodes, conspicuously straight or sinuously winged. Tubers oval-compressed, with bluish-violet to dark purple periderm, superficial eyes, long brows, sprouts with dark blue-violet apices and bases, white flesh or white flesh with a vascular ring, or mottled occasionally with dark blue-violet. Leaf axils and rachises densely pigmented. Leaves 7-18 cm long by 3.5-9 cm wide, dark green, rigid, sparsely and shortly pilose, with 5-6 pairs of leaflets and 5-7(-12) interjected leaflets of different sizes. Leaflets 2.5-4.5 cm long by 1.5 to 2.7 cm wide, ovate to elliptic, the terminal slightly longer than the laterals and occasionally suborbicular, the apex shortly acuminate and the base symmetrical or asymmetrically rounded, borne on 2-3 mm-long petiolules. Pseudostipular leaves obsolete, or absent. Inflorescence cymose, 8-12-flowered. Peduncle thick, 3-8 cm long by 3 mm in diameter at base, pubescent with very short hairs as in the pedicels and calyx. Pedicels articulated near calyx, the upper part 4-5 mm long and the lower 12-15 mm long. Calyx 5.5-6.5 mm long, symmetrical, the elliptic-lanceolate lobes abruptly narrowed at apex to very short pointed acumens. Corolla rotate, typically with very short lobes, dark blue-violet, 3-3.5 cm in diameter, with short, pubescent acumens. Staminal column asymmetrically truncate-conical; anthers short, 4.5-5 mm long by 1.8 mm wide at base, with a well-defined apical dorsal groove, borne on glabrous 1.5-2 mm-long filaments. Style 8-9.5 mm long, exerted 2 mm, densely papillose below midpoint; stigma broadly capitate, cleft. Fruit globose, 15-20 mm in diameter, dark green.

Chromosome number: 2n=5x=60.

Types: BOLIVIA. Department La Paz [Province Murillo], La Paz., n.v. *Choque-pitu*; *S. Juzepczuk 1707* (WIR), Tiahuanaco *1642* (WIR). PERU. Department Cusco [Province Urubamba], Ollantaitambo, n.v. *China-malcco; S. Juzepczuk 1078* (WIR), and [Province Acomayo], Pomacanchi 1167 (WIR).

Affinities

This hybrid of *S.* × *juzepczukii* (*see* under Affinities of *S.* × *juzepczukii*) resembles its parental form in having a rotate corolla and a rosette-like or subrosette-like habit. However, it differs from *S.* × *juzepczukii* in being more vigorous and in having leaflets of different size and shape and a larger corolla. The leaflets of *S.* × *curtilobum* are also more shortly acuminate than those of *S.* × *juzepczukii.*

Distribution

Similar to *S.* × *juzepczukii*, this hybridogenic species is cultivated on the high Puna from the Provinces of Huaraz and Bolognesi in the northern Department of Ancash, Peru, to the Departments of Cochabamba and Potosi in central Bolivia. Its indigenous cultivation extends principally between 3400 and 4100 m.

This highly frost-tolerant species is grown principally in the altiplano region of southern Peru and northern Bolivia. Here it enjoys great popularity among the Aymara and Quechua Indians.

Electrophoretic studies were conducted on the two known morphotypes of *S.* × *curtilobum* (*china-malko* and *chocke-pito*) at CIP. These studies suggest there are no fundamental differences in the protein content of these two cultigens (Ochoa, 1979a; Schmiediche et al., 1980). Nevertheless, as the tubers of these differ notably in color, they are maintained here as separate entities, *S.* × *curtilobum* and *S.* × *curtilobum* f. *china-malko*. This treatment agrees with a classification proposed earlier by Lechnovitch (1971).

Specimens Examined

HERBARIUM COLLECTIONS

Department Cochabamba

Province Arani: Toralapa, n.v. *Lucki*; *CIP-702436* (CIP).

Department La Paz

Province Los Andes: Pucarani, 3900 m alt., n.v. *Chocke Pito*; *Ochoa 3939* (OCH). Huancané, 3900 m alt., n.v. *Chocke-Pito; Ochoa 3940* (OCH).

Province Larecaja: Paccolla, 4150 m alt., n.v. *Laram-Lucki*; *Ochoa 3931* (OCH).

Province Murillo: La Paz, n.v. *Choque-Pitu*; *Juzepczuk 1642, 1707* (WIR).

GERMPLASM COLLECTIONS

Department La Paz

Province Omasuyos: Belen, n.v. *Moro Luqui* (CIP-702281) and *Choque-Pitu morado* (CIP-702282) [both 2n=60].

Figure 162. – *Solanum* × *curtilobum*. 1. Flowering plant (*Juzepczuk 1707*). 2-7. *S.* × *curtilobum*, leaf and floral details (*Ochoa 3939*). 2. Corolla. 3. Petal. 4. Stamens. 5. Pistil. 6. Calyx. 7. Leaf. All ×1.

Department Oruro
Province Cercado: Condoriri, 2n=60, n.v. *Luque Morada* (CIP-702615).
Province Poopó: Cipatiti, 2n=60, n.v. *Choque Pitu Morada* (CIP-702614). Toledo, 3700 m alt., 2n=60, n.v. *Jancko Luke* [?] (HUA-840).

Department Potosi
Province Frias: Tinguipaya, 2n=60, n.v. *Huaca-Corota* (CIP-702543).

36A(a). Solanum × curtilobum var. **curtilobum** f. **china-malko** Lechnovitch, in Bukasov (ed.), Flora of Cult. Pl., Vol. IX, Potato, Leningrad, p. 74. 1971, *nom. nud.* Based on *Juzepczuk 1167*, South Peru (WIR).
PLATE XXIII-2,3.

Stems light green, irregularly and lightly subpigmented along the lower one third of its length. Tuber shape and sprouts similar to that of the species, but differing in having pure white flesh and periderm.

Specimens Examined

HERBARIUM COLLECTIONS

Department Cochabamba
Province Cercado: One km north of Sacaba, 2500 m alt., in fields of Luciano Cornejo, n.v. *Luki; Cutler 7425* (F).

Department La Paz
Province Ingavi: Tiahuanaco, 3690 m alt., n.v. *Lucki*; *Ochoa 3936* (OCH). Patarani, 3850 m alt., n.v. *Lucki*; *Ochoa 3934* (OCH).

Department Potosi
Province Quijarro: Condoriri, 3900 m alt., n.v. *Lucki*; *Ochoa 10675* (OCH).

GERMPLASM COLLECTIONS

Department Oruro
Province Poopó: Saucari-Toledo, 3700 m alt., 2n=60, n.v. *Jancko Chocko Pitu* (HUA-838).

Department Potosi
Province Frias: Yacalla, 3450 m alt. (HAW-6145). Tinguipaya, 2n=60, n.v. *Jatun Lucki Toro* (CIP-702566).

37. Solanum × chaucha Juzepczuk and Bukasov, Proc. U.S.S.R. Congr. Genet. Pl. & Anim. Breed. 3:609. 1929.

S. tenuifilamentum Juz. & Buk., Proc. U.S.S.R. Congr. Genet. Pl. & Anim. Breed. 3:603. 1929. TYPE COLLECTIONS: *Juzepczuk 1710,* Bolivia. La Paz (WIR); *Juzepczuk 1321, 1355* (WIR) and *1417* (WIR, not seen) Peru, Dept. Cusco, Chinchero.

S. mamilliferum Juz. & Buk., Proc. U.S.S.R. Congr. Genet. Pl. & Anim. Breed. 3:609. 1929. TYPE COLLECTIONS: *Juzepczuk 1001, 1051, 1169* (WIR) and *1338, 1339, 1944* (WIR, not seen), Peru, Dept Cusco.

S. coeruleiflorum Hawkes, Bull. Imp. Bur. Pl. Breed. & Genet. 76:131, Fig. 91. 1944. TYPE COLLECTIONS: *Balls 6941, 6955,* Peru. Ayacucho market, (CPC [?], not seen); and *Balls 7000,* Peru. Dept. Junin, Huancayo market (CPC [?], not seen).

S. chaucha f. *pigmentatum* Lechn., in Bukasov (ed.), Flora of Cult. Pl., Vol. IX, Potato, Leningrad, p. 62. 1971, *nom. nud.* Based on *Juzepczuk 1556, 1644,* Bolivia, Tiahuanaco (WIR), *1658, 1783,* La Paz (WIR not seen); *Juzepczuk 990,* Peru, Cusco, Ppisac (WIR, not seen), *Juzepczuk: 1010* Cusco, St Jeronimo, *1179* Cusco, Pomacanchi (WIR).

Stems usually vigorous, erect, branched, partly or entirely pigmented, with 1-2.5 mm-wide wings. Tubers mostly irregular (bulging), long subcylindric or cylindric, the periderm blue-reddish to violet-reddish, sprouts blue-violet. Leaves sparsely pubescent with very short hairs, more or less dissected (*see* Fig. 163), with 4-5(-6) pairs of leaflets and 5-8(-12) interjected leaflets. Leaflets ovate or broadly elliptic to elliptic-lanceolate, the apex obtuse and the base asymmetrically rounded, borne on 2-5 mm-long petiolules. Terminal leaflet slightly larger than the laterals. Peduncles 5-8(-12) cm long and 2.5-3 mm in diameter at base, 6-8(-12)-flowered. Calyx well delimited from the pedicel, usually asymmetrical and pigmented, the lobes broadly and deeply triangular-lanceolate. Corolla 3 cm in diameter, rotate pentagonal or occasionally pleated, the lobes frequently broadly conspicuous, commonly dark violet. Anthers 5-6.5 mm long by 1.5 mm broad at base, with a well-defined dorsal groove, borne on white, 2-3 mm-long filaments. Style 9 mm long, exerted 2-2.5 mm; stigma capitate with small, obtuse sinus, about 1 mm in diameter. Fruits rarely formed, or when present globose to ovoid, small, 10-15 mm long, without seeds, or with few poorly developed seeds.

Chromosome number: 2n=3x=36.

The tubers of this cultigen usually have excellent flavor and culinary qualities. The starch and dry matter content varies between 27% and 30%.

Types: BOLIVIA. [Department La Paz, Province Murillo], La Paz; *S. Juzepczuk 1658* (WIR, not seen). PERU. [Department and Province Cusco], San Jeronimo, [near] Cusco, n.v. *Chaucha; S. Juzepczuk 1010* (WIR).

Affinities

Undoubtedly, this hybridogenic cultigen is formed, not only by natural crosses of *S. tuberosum* subsp. *andigena* and *S. stenotomum* (Hawkes, 1963) but also in crosses between either of the cultivated diploid species including *S. goniocalyx* or *S. phureja* and the tetraploid *S. tuberosum* subsp. *andigena*. Usually, it resembles the latter species closely in habit, leaf and leaflet shape, and in the size and color of its flowers.

The artificial synthesis of *S.* × *chaucha* was attempted by Jackson et al. (1977). In crosses between *S. tuberosum* subsp. *andigena* and diploid *S. stenotomum*, and *S. goniocalyx* and *S. phureja*, almost 100% of the pollinations resulted in fruit. However, very few triploid seeds were obtained.

Jackson et al. (1978) also report the obtaining of a fairly high frequency of tetraploid plants (66%) in crosses between *S. tuberosum* subsp. *andigena* and diploid potato cultivars. However, in the latter case these resulted from the production of nonreduced gametes by the diploid parent. Moreover, in a series of crosses made between 20 morphotypes of *S.* × *chaucha* and various accessions of *S. tuberosum* subsp. *andigena* and cultivated diploid species, Jackson et al. recovered some hybrid forms with greater seed fertility than that normally found in diploid/diploid crosses. In general, however, it is probably preferable to attempt the transfer of diploid germplasm to *S. tuberosum* subsp. *andigena* directly through the intermediacy of unreduced gametes than to try to use the method of hybrid bridges, as in the case of *S.* × *chaucha*.

Distribution

This cultigen is grown throughout the northern and central Andes from the Paramos of Colombia and Ecuador to the Jalca formations of the inter-Andean valleys of northern Peru and the Puna or Prepuna formations of southern Peru, Bolivia, and the northwest of Argentina. It is grown principally between 2500 and 3800 m.

The most widely distributed morphotypes of this cultigen in southern Peru and Bolivia are *piña*, *lomo*, and *ccoe-sullu*. In total, there are about 30 different morphotypes of *S.* × *chaucha* in the central Andean region (Ochoa, 1975a). The range of genetic variation to be found within this cultigen is greater than that encountered in *S.* × *ajanhuiri*, *S.* × *juzepczukii*, and *S.* × *curtilobum*.

Specimens Examined

HERBARIUM COLLECTIONS

Department La Paz

Province Ingavi: Tiahuanaco; *Juzepczuk 1556, 1644* (WIR).

Province Murillo: La Paz; *Juzepczuk 1710* (WIR, same as *S. tenuifilamentum*).

37A. Solanum × chaucha var. ckati Ochoa var. nov., Phytologia 65(2): 103-113. 1988.

Stems subpigmented, shortly and sparsely pilose, winged. Tubers long and thick, irregular (bulging), with light beige flesh, the apex somewhat broad and obtuse and the base narrowed and subpointed, with dark-blue violet, almost black skin, deep eyes and creamy-white buds, the latter dark-blue violet at base and apex. Leaves 15-17 cm long by 7-8 cm wide, dark green above and light purple below, with 4-5 pairs of leaflets and 4-8 interjected laeflets. Leaflets 4-4.5 cm long by 2.4-2.7 cm wide, elliptic to elliptic-lanceolate, the apex pointed to subpointed and the base rounded to obliquely rounded, borne on short petiolules. Peduncles 10-15 cm long. Corolla 3-3.5 cm in diameter, dark violet with white acumens. Pedicels articulated near base of calyx, the upper part 6-7 mm long and the lower part 25-30 mm long. Calyx 7-8 mm long, symmetrical or asymmetrical and divided into either 5 or 6 lobes.

Because this variety has some of the tuber characteristics of *S. stenotomum* var. *stenotomum* f. *chiar phiñu*, and the leaf, leaflet, and calyx characteristics of some forms of *S. tuberosum* subsp. *andigena*, it may have arisen as a triploid (2n=36) hybrid of the two species.

Specimens Examined

HERBARIUM COLLECTIONS

Department La Paz

Province Murillo: La Paz, 3650 m alt., Rodriguez Market, brought from the altiplano, 2n=36, n.v. *Ckati*; *Ochoa 3909* (OCH). Buenos Aires Market, brought from the altiplano, n.v. *Chiar phiñu*, 2n=3x=36; *Ochoa 3929* (OCH, type of var. *ckati*).

GERMPLASM COLLECTION

Department La Paz

Province Murillo: La Paz, Buenos Aires Market, 2n=36, n.v. *Ari Chua*, brought from Chojñaccota (OCH-3907).

37B. Solanum × chaucha var. **ccoe-sullu** Ochoa var. nov., Phytologia 65 (2): 103-113. 1988.

Tubers with yellow flesh, long-compressed to oval compressed, with superficial to semideep eyes, the periderm bicolored, with dark pink-violet and light yellow areas (especially toward the apex), or occasionally dark pink violet with light yellow halos about the eyes, buds creamy-white with blue-violet apices and bases. Leaves 20-25 cm long by 11-13 cm wide, with 5 pairs of leaflets and 10-15 interjected leaflets of different sizes, including some basiscopic. Peduncle 3-5 cm long. Pedicel 19-22 mm long, the upper part 4 mm long and the lower 15-18 mm long. Calyx symmetric or asymmetrical. Corolla violet. Anthers 5-5.5 mm long. Style 8 mm long.

Chromosome number: 2n=3x=36.

All germplasm collections of *S.* × *chaucha* var. *ccoe-sullu* were determined to be identical electrophoretically. This variety, *ccoe-sullu* ('rabbit fetus' in Quechua), is widely distributed from southern Peru (Ochoa, 1975a) to Bolivia and northern Argentina.

Specimens Examined

HERBARIUM COLLECTIONS

Department Potosi

Province Frias: Tinguipaya, n.v. *Alcca Coillo* (CIP-702522, CIP). Tinguipaya, 3900 m alt., n.v. *Dominguillo*; *Ochoa 10541* (OCH). Tinguipaya, 3900 m alt., n.v. *Kkoyllu*; *Ochoa 10646* (OCH).

Province Saavedra: Lequezana, 3300 m alt., n.v. *Kkoyllur*; *Ochoa 3950* (OCH, paratype of var. *ccoe-sullu*).

GERMPLASM COLLECTIONS

Department Potosi

Province Chayanta: Cajõn Mayu, n.v. *Alcca Chuchu* (HAW-6175).

Province Frias: Azangaro, 3800 m alt., n.v. *Alka Koillu* (OCH-10501). Tinguipaya, 3900 m alt., n.v. *Alka Kkoyllu* (OCH-10536). Tinguipaya, 3900 m alt., n.v. *Kkoyllu* (OCH-10554). Tinguipaya, 3900 m alt., n.v. *Puka Kkoyllu* (OCH-10543).

Province Linares: Palca, 3750 m alt., n.v. *Yutu Runtu* (OCH-10633).

Province Quijarro: Condoriri, 3900 m alt., n.v. *Condor Coillu* (OCH-10673).

37C. Solanum × chaucha var. **kkoyllu** Ochoa var. nov., Phytologia 65(2): 103-113. 1988.

Plants vigorous, light green. Stems to 70 cm tall, winged. Tubers with light yellow flesh, round, with dark red-wine periderm dotted irregularly with violet or almost black areas, deep eyes and dark pink buds (*see* Plate XXV-5). Leaves 17.5-19 cm long by 8-9 cm wide, with 5-6 pairs of leaflets and many interjected leaflets of 2-3 sizes. Leaflets 4-4.5 cm long by 2-2.3 cm wide, elliptic-lanceolate, the apex shortly acuminate and the base attenuate, borne on 2-5 mm-long petiolules. Peduncles slender, 8-12 cm long. Pedicels articulated 3-4 mm below base of calyx. Calyx asymmetrical. Corolla usually small, 2.6 cm in diameter, violet-lilac. Anthers 6-7 mm long. Style short, 8 mm long.

The Quechua name of this variety, *Kkoyllu*, means 'resplendent' or 'lustrous.'

Specimens Examined

HERBARIUM COLLECTION

Department Cochabamba

Province Cercado: La Cancha, 3400 m alt., marketplace of Cochabamba, brought from Colomi, 2n=36, n.v. *Khoillu*; *Ochoa 3948* (OCH, type of var. *khoyllu*).

GERMPLASM COLLECTIONS

Department Cochabamba

Province Arani: Tiraque, 2n=36, n.v. *Koyllu* (HAW-5645). Tiraque, n.v. *Coyllu* (EET-1042). Tiraque, n.v. *Ccoyllu Imilla* (EET-1040).

Province Ayopaya: Independencia, n.v. *Ccoyllu* (EET-1043).

37D. Solanum × chaucha var. **piña** f. **chulluco** Ochoa. PLATE XIX-7.

Solanum × *chaucha* var. *piña* is native to Peru. The variety *piña* f. *chulluco* is the only variant so far found in Bolivia. Its tubers are oblong to shortly elongate-cylindrical or ananiform (i.e., pineapple-shaped), strongly irregular (or bulging) and obtusely pointed, with dark pink to pink-violet periderm, deep eyes, and dark blue-violet buds.

Specimens Examined

HERBARIUM COLLECTION

Department La Paz
Province Murillo: Camacho, 3650 m alt., local marketplace of La Paz, 2n=36, brought from the altiplano, n.v. *Chulluco*; *Ochoa 12130* (OCH).

37E. Solanum × chaucha var. **puca-suitu** Ochoa var. nov., Phytologia 65 (2): 103-113. 1988.

Stems to 80 cm tall, very narrowly winged. Tubers with white to light beige flesh, long cylindrical to subcylindric, slender, the apex obtuse and the base narrowing or almost pointed, periderm pink with yellowish-white stripes, with deep to semideep eyes, sprouts with pink apices and bases the rest white. Leaves with 4-5 pairs of leaflets and 5-7 interjected leaflets. Corolla dark violet.

Chromosome number: 2n=3x=36.

This variety, while occurring in Bolivia, is distributed mainly in central and southern Peru.

Specimens Examined

HERBARIUM COLLECTIONS

Department Potosi
Province Chayanta: Maragua, 3700 m alt., n.v. *Sallimani*; *Ochoa 10603* (OCH, paratype of var. *puca-suitu*).

Province Linares: Puna, 3850 m alt., n.v. *Solimana*; *Ochoa 10622* (OCH).

GERMPLASM COLLECTIONS

Department Potosi
Province Chayanta: Maragua, 3700 m alt., 2n=36, n.v. *Sulku* (OCH-10604).

Province Frias: Tinguipaya, 2n=36, n.v. *Curu Papa* ('worm potato') (CIP-702570).

Department La Paz
Province Aroma: Vilaque, 3800 m alt., n.v. *Wila Solimana* (HUA-883). Vilaque, 3800 m alt., n.v. *Wila Phiñu* (HUA-889).

37E(a). Solanum × chaucha var. puca-suitu f. solimana Ochoa.

A bud mutation of *S.* × *chaucha* var. *puca-suitu* probably gives rise to f. *solimana*, a form which has whitish-yellow tubers instead of pink or reddish periderm. Plants are much less pigmented and the leaves and flowers are similar to variety *puca-suitu*. However, these two clonal types are indistinguishable by electrophoresis. Thus, our samples of variety *puca-suitu* (HUA-889, HUA-883, and CIP-702570) and form *solimana* (OCH-10551) revealed no specific differences in proteins.

Specimens Examined

GERMPLASM COLLECTION

Department Potosi
Province Frias: Tinguipaya, 3900 m alt., n.v. *Solimana* (OCH-10551).

37F. Solanum × chaucha var. surimana Ochoa var. nov., Phytologia 65(2): 103-113. 1988. Fig. 163.

Plants similar to some morphotypes of *S. tuberosum* subsp. *andigena* var. *chiar-imilla*. Stems 40-50 cm tall. Tubers with white flesh, long subcylindrical to slightly compressed, somewhat irregular (or bulging), the skin dark pinkish-purple with irregularly distributed whitish stripes, semideep eyes, and dark blue-violet buds. Leaves 15-17 cm long by 8.5-10 cm wide, poorly dissected, with 4-5 pairs of leaflets and 4-7 interjected leaflets. Leaflets 4-5.5 cm long by 2.8-3 cm wide, elliptic, the apex obtuse to subpointed and the base obliquely rounded, borne on 2-3 mm long petiolules. Peduncles slender, 6-8 cm long by 2 mm in diameter at base. Pedicels 25-32 mm long, the upper part 10-12 mm long and the lower 15-20 mm long. Calyx 7-8 mm long, symmetrical or occasionally bilabiate. Corolla 3-3.5 cm in diameter, light violet. Anthers 6-7 mm long. Style 8-9 mm long; stigma thick, capitate.

Specimens Examined

HERBARIUM COLLECTIONS

Department La Paz
Province Murillo: La Paz, n.v. *Surimana*; *Juzepczuk 1658* (WIR, as *S.* × *chaucha* f. *pigmentatum* Lechn.).

Department Oruro
Province Poopó: Urmiri, 3750 m alt., n.v. *Surimana*; *Ochoa 3946* (OCH, type of var. *surimana*).

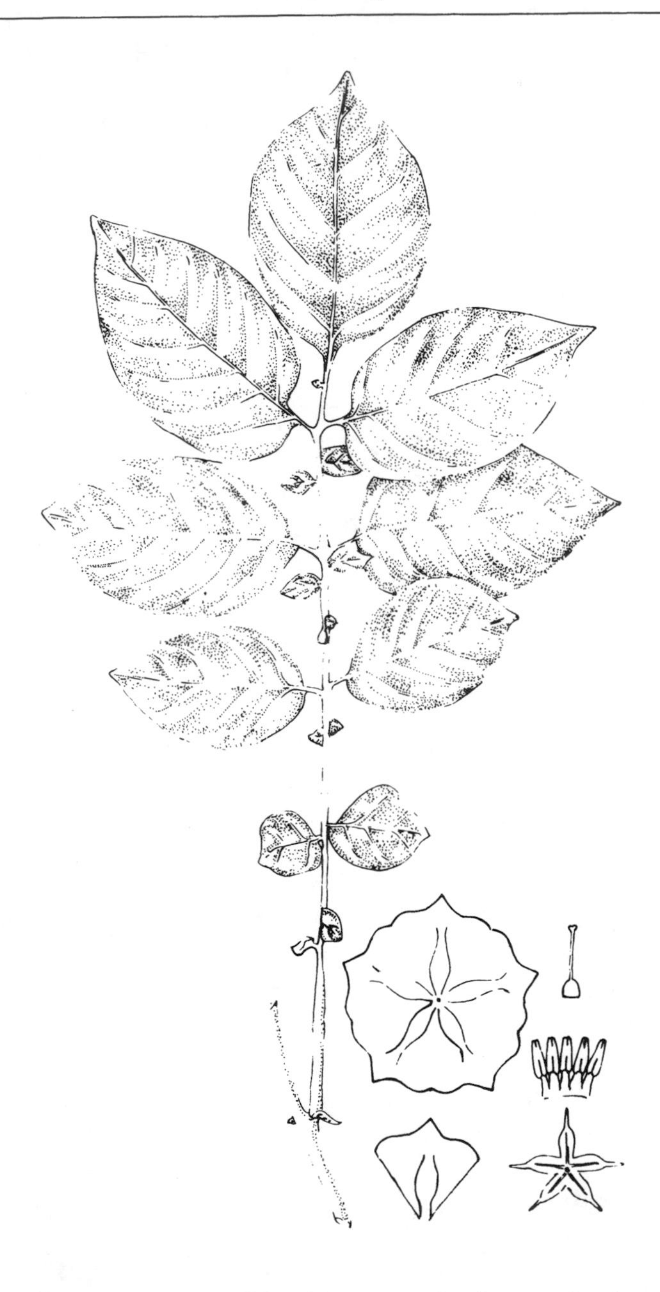

Figure 163. – Leaf and floral dissection of *Solanum* × *chaucha* var. *surimana* (*Ochoa 3946*), ×1.

GERMPLASM COLLECTIONS

Department La Paz

Province Ingavi: Patarani, 3850 m alt., 2n=36, n.v. *Surimana* (CO-2184).

Province Murillo: La Paz, 3650 m alt., local market, brought from the altiplano, 2n=36, n.v. *Surimana* (CO-2202).

37F(a). Solanum × chaucha var. surimana f. ccoyo Ochoa.

Tubers with light beige flesh, long subcylindrical, apex obtuse and base pointed, periderm dark blue-violet with irregularly distributed whitish-yellow stripes, superficial to semideep eyes and dark blue-violet buds.

Chromosome number: 2n=3x=36.

This form, known locally as *Ccoyo*, is cultivated throughout the Bolivian altiplano.

This and the following two forms of *S.* × *chaucha* are based on living collections.

Specimens Examined

GERMPLASM COLLECTIONS

Department La Paz

Province Ingavi: Patarani, 3850 m alt. (CO-2175). Jackara, 3960 m alt. (CO-2445). Chama, 4100 m alt. (CO-2347).

Province Pacajes: Caquiaviri, 3960 m alt. (CO-2375).

37F(b). Solanum × chaucha var. surimana f. tamatta Ochoa.

Tubers with white flesh and blue-violet medula, rounded to oblong, irregular (or bulging), with dark blue-violet periderm and deep eyes, narrow yellowish halo areas about the eyes, and blue-violet buds.

Chromosome number: 2n=3x=36.

This form, known locally as *Tamatta*, *Tamitta*, or *Damiatta*, is cultivated principally in the Province of Camacho, Department of La Paz.

Specimens Examined

GERMPLASM COLLECTIONS

Department La Paz

Province Camacho: Cariquiña, 3940 m alt. (CO-2322). Puerto Acosta, 3880 m alt. (CO-2296). Unahuaya, 3960 m alt. (CO-2304).

37F(c). Solanum × chaucha var. **surimana** f. **chiar-surimana** Ochoa.

Plants of medium stature. Stems 30-50 cm tall, strongly pigmented. Tubers long and thick, with obtuse apex, the periderm very dark blue-violet almost black with a few whitish-yellow stripes. Leaves dark green. Corolla dark violet. Ovary pigmented. Style pigmented in the upper two-thirds of its length.

All CIP collections of this form are identical morphologically and electrophoretically.

Specimens Examined

GERMPLASM COLLECTIONS

Department La Paz

Province Aroma: Vilaque, 3800 m alt., 2n=36, n.v. *Chiar Solimana* (HUA-881). Vilaque, 3800 m alt., n.v. *Higos Solimana* (HUA-882).

Province Ingavi: Ingavi, n.v. *Phiñu Negro* (HAW-5571).

Province Los Andes: Cura Pucara, 3860 m alt., 2n=36, n.v. *Surimana* (HAW-6186).

Province Murillo: La Paz, 3650 m alt., local market, n.v. *Surimana* (HAW-5562).

38. Solanum tuberosum L. subsp. **andigena** (Juzepczuk and Bukasov) Hawkes, Proc. Linn. Soc. Bot. 166:130-137, Pl.3-4. 1956. Figs. 164–180.

S. andigenum Juz. et Buk., Proc. U.S.S.R. Congr. Genet. Pl. & Anim. Breed. 3:609-610. 1929. Type: *Juzepczuk 598*, Peru. Vicinities of Titlicaen, Cerro de Pasco, Dept. Junin [Dept. and Prov. Pasco, Ticlacayan, near Cerro de Pasco] (LE, OCH; from plants grown near Leningrad).

S. herrerae Juz., Bull. Acad. Sci. U.S.S.R. 2: 310-311. 1937. Type: *Kaufman s.n.*, Peru. Kcaira, Dept. and Prov. Cusco (WIR=K-24).

S. subandigena Hawkes, Bull. Imp. Bur. Plant Breed. & Genet. Cambridge (June):128, Fig. 45. 1944. Type: *Balls 6146, 6171*, Bolivia. Sucre, Dept. Chuquisaca, Prov. Oropeza, March 7, 1939 (CPC [?], not seen).

S. apurimacense Vargas, Las Papas Sudperuanas, II Parte (Univ. Nac. Cusco) 2:58-59, Fig. 14. 1956 (as f. *apurimacensis*). Type: *Vargas C. 811*, Peru (VAR). Pamparackay, 3500 m alt., Apurimac.

S. estradae L. Lopez J., Mutisia 55:5-10, Figs. 3-4. 1983. Type: *Lopez J. CCC-4702*, Colombia. Municipio de Pijao, 1850 m alt., Vereda La Quiebra, Dept. Quindio, Nov. 15, 1973 (COL, holotype; Herb. Instituto Colombiano Agropecuario, Bogota, isotype).

S. andigenum subsp. *ecuatorianun* Lechn., in Lechnovitch, New Taxa within the Species *S. andigenum*. Trudy Prikl. Bot. Genet. Selek. 82: 46. 1983. TYPE: *Vavilov s.n.*, Ecuador. Riobamba (WIR=K-50, not seen).

S. andigenum subsp. *tarmense* Buk. et Lechn., in Lechnovitch, New Taxa within the Species *S. andigenum*. Trudy Prikl. Bot. Genet. Selek. 82: 46. 1983. TYPE: *Juzepczuk 587-b*, Peru. Chinchao, Cerro de Pasco, 1929 (WIR, not seen; from plants grown near Leningrad).

S. andigenum subsp. *centraliperuvianum* Lechn., in Lechnovitch, New Taxa within the Species *S. andigenum*. Trudy Prikl. Bot. Genet. Selek. 82: 46. 1983. TYPE: *Juzepczuk 191-c*, Peru. Tarma, 1929 (WIR, not seen; from plants grown near Leningrad).

S. andigenum subsp. *australiperuvianum* Lechn., in Lechnovitch, New Taxa within the Species *S. andigenum*. Trudy Prikl. Bot. Genet. Selek. 82: 46–47. 1983. TYPE: *Juzepczuk 1341-b*, Peru. Huaccoto, Cusco, 1928 (WIR, not seen; from plants grown near Leningrad, 1929).

S. andigenum subsp. *bolivianum* Lechn., in Lechnovitch, New Taxa within the Species *S. andigenum*. Trudy Prikl. Bot. Genet. Selek. 82: 48. 1983. TYPE: *Juzepczuk 1740*, Bolivia. La Paz (WIR, not seen; from plants grown near Leningrad.

S. andigenum subsp. *argentinicum* Lechn., in Lechnovitch, New Taxa within the Species *S. andigenum*. Trudy Prikl. Bot. Genet. Selek. 82: 48. 1983. TYPE: *R.D. Schechai 2186-3*, Argentina. Lillo, Tucumán, 1951 (WIR, not seen).

S. andigenum subsp. *mediamericanum* (Buk.) Lechn., in Bukasov (ed.), Flora of Cult. Pl., Vol IX, Potato, Leningrad, pp. 85-86. 1971, *nom. nud.* Based on Guatemala collections of *Bukasov 9-10* (Quetzaltenango), *12* (San Felipe), *13-15* Amatitlan, and *16* (Escuintla) [WIR [?], not seen].

S. andigenum subsp. *runa* Buk. et Lechn., in Bukasov (ed.), Flora of Cult. Pl., Vol. IX, Potato, Leningrad, pp. 224-225. 1971, *nom. nud.* Based on *Knappe 0004 and K-5590* (WIR [?], not seen).

Plants vigorous, light green, slightly or strongly pigmented. Stems thick and fleshy, 40 to 120 cm or more tall and 8-20 mm in diameter at base, branched, erect, straggling or occasionally decumbent, with narrow or broad wings, the latter straight or sinuous. Stolons 10 to 150 cm or more long, fleshy, white or pigmented. Tubers mostly long-dormant, with pure white, grayish white, light beige, or yellow flesh, or occasionally with a purple, blue-violet or pink vascular ring or medula (or both), varying enormously in shape from perfectly round, somewhat compressed, oblong, long cylindrical, falcate with obtuse ends, to long-fusiform, long-flattened, oval-compressed, pyriform, ofidiform, vermiform or palmate, the periderm varying from pure white, yellow, brown, red, pink, purple, violet, blue-violet or black to bicolored, or with halo areas of contrasting color about the eyes, or striped, with superficial, semideep or deep eyes, the latter rounded, oval or very elongated in shape, and with buds cylindrical at base and pleated at apex and varying greatly in color as in the periderm. Leaves strongly ascending or erect, imparipinnate, dull green or dark green above and light green and shining below, sparsely invested with multicellular hairs and short unicellular hairs, usually very dissected with 4-6(-8) pairs of leaflets and 5-12(-45) interjected leaflets, the latter sessile or shortly petiolulate and of varying size (but always smaller than the upper lateral pairs of leaflets) and situated principally on the rachis or occasionally on the petiolules where, depending upon their point of attachment, they may be either acroscopic, semiacroscopic, basiscopic or semibasiscopic. Terminal leaflet generally somewhat larger than the laterals, broadly elliptic, ovate or obovate to suborbicular or subcordate. Lateral leaflets usually decreasing in size from the first or second pair of leaflets toward base, oval-lanceolate, ovate or occasionally obovate to elliptic, or elliptic-lanceolate, and twice as long as wide, the apex usually acuminate and the base symmetrical or obliquely rounded to subcordate or occasionally cuneate, borne on 2-5 or occasionally 30 or more mm-long petiolules. Pseudostipular leaves 8-25(-35) cm long by 5-12(-20) cm wide, sublunate or reniform, clasping to subclasping or falcate. Inflorescence lateral or terminal, generally very floriferous, with 25 or more blooms. Peduncle vigorous, 5-20 cm long by 2.5-4 mm in diameter at base, forked or biforked one or more times and more or less shortly pilose as in the pedicels and calyx. Pedicels 15-40 mm long and of uniform diameter, articulated in the upper one-third or one-quarter of the stalk. Calyx 5-9 mm long, generally symmetrical, with broadly ovate, ovate-elliptic, elliptic, rectangular to subquadrate or acuminate lobes, the latter bearing 1-3(-5) mm-long acumens. Corolla 3-4 cm in diameter, rotate to rotate-pentagonal, varying from purple to dark violet, light violet, pale pink or lilac to white, intensely pigmented about the greenish-yellow, grayish-white or dark-violet to almost black nectar guides. Staminal column symmetrical or asymmetrical, compact, cylindric or subcylindric to truncate-conical; anthers highly fertile, well formed, broadly to narrowly lanceolate, (5.5-)6.5-7.5(-8.5) mm long by 2-2.8 mm wide, yellow or orange and occasionally with a red or dark brown apex, with a cordate or rounded base and a well-defined (or occasionally barely perceptible) dorsal groove, borne on slender, glabrous, 1.5-3 mm long by 0.5-1 mm broad, white (or occasionally slightly colored) hyalinous filaments. Style slender, straight, (8.5-)9-11(-13) mm long, densely papillose along the lower two-thirds of its length; stigma large, broader than the style apex, (0.8-)1-1.3 mm in diameter, capitate, ovoid, subconical or claviform, entire or cleft. Fruits abundant, 5-20 or more per raceme, containing 300 or more fertile seeds each, 2-3.5(-5) cm in diameter, usually globose or shortly pyriform, less frequently oblong or more rarely elongate conical, light green, dark green, or occasionally with a dark brown or purple-violet basal tinge, or speckled with white or painted with two or three dark violet stripes.

Chromosome number: 2n=4x=48.

Type: PERU. Pasco, Province Daniel Carrión [Prov. Pasco], heights of Huariaca, 3600 m alt.: *C. Ochoa 36* (JGH, OCH).

Although classified here as a single, highly variable subspecies, this cultigen has been treated in the past as a number of lesser species, as well as varieties and subspecies of *S. tuberosum* (Bukasov, 1933, 1939, 1955a, 1971a; Hawkes 1956a, b, 1963; Dodds, 1962; Ochoa 1965, 1979a; Hawkes and Hjerting, 1969; Lechnovitch, 1971).

Solanum tuberosum subsp. *andigena*, more than any other cultivated species of potato, represents the largest reserves of germplasm for modern potato improvement programs. Various clones of this tetraploid species, for example, are high in starch, dry matter, and protein (Bukasov, 1940a; Hawkes, 1945; Virsoo, 1956; Rothacker, 1961). Moreover, some clones are resistant to potato late blight (Ross, 1958; Ochoa, 1965), potato wart (Hawkes, 1945; Bukasov, 1955b; Kameraz, 1957; Ross, 1958; Rothacker and Müller, 1960), scab (Calderoni and Induni, 1952; Ross, 1958), powdery scab (Ross, 1958; Zadina, 1958), early blight (Ross, 1958; Bukasov and Kameraz, 1959), wilt *Verticillium alboatrum* (Bazan de Segura and Ochoa, 1959), bacterial wilt *Erwinia chrysanthemi* (Thung, 1947; French and De Lindo, 1980), or viruses such as PVY (Cárdenas, 1966; Muñoz et al., 1975; Fernandez-Northcote, 1983) and PLRV (Ross and Baerecke, 1950; Geransenkova, 1974).

Clones of *S. tuberosum* subsp. *andigena* resistant to the cyst nematode, *Heterodera rostochiensis*, have been recovered from the Peruvian-Bolivian potato growing districts of Lake Titicaca (Ellenby, 1952, 1954; Toxopeus and Huijsman, 1952, 1953; Howard, 1955; Stelter and Rothacker, 1965; Cárdenas, 1966; Howard et al. 1970). Similarly, Scurrah and Franco (1985) have identified, among the more than 3000 samples of *S. tuberosum* subsp. *andigena* maintained by CIP, clones that are fully resistant to pathotype R1A of *Globodera pallida* and partially resistant to pathotype P4A. Among the Bolivian clones that show partial resistance to this nematode are *Sipancachi* and *Atacama*.

Affinities

Solanum tuberosum subsp. *andigena* shares many characteristics in common with *S. tuberosum*. However, the latter species differs in having a smaller stem and more compact habit, shorter internodes, less dissected and weakly ascending not erect leaves, broader and more crowded leaflets, and pedicels that enlarge gradually toward the base of the calyx. Moreover, the anthers of *S. tuberosum* are frequently not as well developed as those of *S. tuberosum* subsp. *andigena*, and are often not as fertile.

According to Cadman (1942), *S. tuberosum* has four identical sets of chromosomes. Thus, it would appear to be an autotetraploid. If this were true, according to Hawkes, the immediate diploid ancestor of *S. tuberosum* subsp. *andigena* would be *S. stenotomum*. However, the calyces of both species are very distinct, and as this floral difference persists in artificially obtained parthenogens of *S. tuberosum*, Hawkes proposes that this species could be an amphidiploid.

In his article dealing with the origin of the tetraploid cultivated potato, Hawkes (1956b) suggests that *S. tuberosum* could have arisen as a result of the spontaneous doubling of a hybrid derived from crosses between *S. stenotomum* and the wild species *S. sparsipilum*. However, in more recent publications, Hawkes (1963) and Cribb (1972) have opted for an autopolyploid origin for this species. Bukasov (1978), however, remains convinced that *S. andigenum* (*S. tuberosum* subsp. *andigena*) is of polyphyletic origin.

Although it is possible that *S. tuberosum* subsp. *andigena* has arisen from one or more diploid species by a process of polyploidization, it could also be that the species has arisen directly at the 4x level as a result of crossing between tetraploid species. Possible candidates for such a cross might include *S. nubicola* and *S. sucrense*. Another possibility is that *S. tuberosum* subsp. *andigena* represents a doubled diploid hybrid of a wild species with a cultivated diploid species.

According to Dodds (1965), and Dodds and Paxman (1962), the cultivated potatoes all fall under a single species, *S. tuberosum*, and two hybrids, *S.* × *juzepczukii* and *S.* × *curtilobum*. The cultivated potato, *S. tuberosum*, is divided into five major horticultural groups based on chromosome number.

Of the above five groups, two are diploid. The first, Group Stenotomum ($2n=2x=24$), is distributed principally in central Peru and Bolivia, while the second, Group Phureja ($2n=2x=24$), is distributed from Venezuela to northern Perú. These two groups have originated from unknown ancestors. The intermediate triploid level, Group Chaucha ($2n=3x=36$), is distributed largely in central Peru and Bolivia. This hybrid originated from natural crosses between Groups Stenotomum or Phureja with Group Andigena ($2n=4x=48$). The last mentioned tetraploid group has a wide distribution from Venezuela to northern Argentina. It arose by the doubling of the chromosome numbers of Groups Stenotomum and Phureja. The remaining tetraploid group, Tuberosum ($2n=4x=48$), represented by the cultivated potatoes of Europe and North America, arose from Group Andigena potatoes through the process of selection.

Of the two hybrid potatoes previously mentioned, *S.* × *juzepczukii* ($2n=3x=36$) originated from crosses of Group Stenotomum with the wild species *S. acaule* (2n=4x=48), while *S.* × *curtilobum* (2n=5x=60) originated from crosses between *S.* × *juzepczukii* and Group Andigena. These cultigens are distributed from central Peru to Bolivia. Although this horticultural system of classification has received much support among potato breeders of Europe and the United States, for taxonomic purposes, at least, the author prefers to treat the various chromosomal groups of cultivated potatoes as separate species.

Distribution

Solanum tuberosum subsp. *andigena* is cultivated at elevations of 2500-4000 m in the Andean region of South America from the Serranias of northwest Argentina, the Punas and Prepunas of central Bolivia and southern Peru, and the Jalcas of northern Peru to the Paramos and Subparamos of Ecuador, Colombia, and Venezuela. It has the widest geographic distribution of any known cultivated potato. Its varieties and forms grown north of the equator have taller stems and larger leaflets than those grown southward.

In Peru and Bolivia, this species is frequently grown intermixed in small field plots with plants of *S. stenotomum* or *S. goniocalyx*, or with natural hybrids formed in crosses between the three. It is also cultivated near sea level in large commercial plantations along the central and southern coasts of Peru. This species was introduced into Guatemala and Mexico during post-conquest times. Because it is a short-day plant, requiring 9-12 hours of light to tuberize, *S. tuberosum* subsp. *andigena* does not do well in northern latitudes.

38A. Solanum tuberosum subsp. **andigena** var. **chiar-imilla** (Lechnovitch) Ochoa. FIG. 164.

S. andigenum Juz. et Buk. var. *imilla* Juz. et Buk., in Bukasov, The Potatoes of South America and their Breeding Possibilities, pp. 66-67, Fig. 16. 1933, *nom. nud.* Based on *Juzepczuk 1582, 1660*, Bolivia, La Paz (WIR); and *Juzepczuk 1680, 1683, 1714*, La Paz (WIR, not seen).

S. andigenum var. *chiar imilla* Lechn. f. *imilla* Lechn., in Bukasov (ed.), Flora of Cult. Pl., Vol. IX, Potato, Leningrad, pp. 186-187, Fig. 74. 1971, *nom. nud.* Based on *Juzepczuk 1520, 1660a, 1660c, 1680, 1683, 1714*, Bolivia, La Paz (WIR, not seen); and *Juzepczuk 1582*, La Paz (WIR).

Stems ascending to erect, decumbent at flowering time, slightly pigmented. Tubers round, with dark blue-violet or almost black periderm, irregularly spotted light violet or with light brown stripes, dark blue-purple buds, and light beige flesh. Leaves 9-15(-22) cm long by 5-9.5(-13) cm wide, with 4-5(-6) pairs of leaflets and 5-9 interjected leaflets, the latter occasionally acroscopic or basiscopic. Leaflets 3.3-5(-7) cm long by 1.8-3(-5.3) cm wide, elliptic or broadly elliptic-lanceolate, provided with pointed or shortly acuminate apices and rounded, shortly oblique bases. Peduncles many-flowered, 8-14 cm long. Pedicel articulation high, in the upper one-third of stalk or near base of calyx. Calyx short pilose, 6-10 mm long, symmetrical, the lobes attenuate or abruptly narrowed to short acumens. Corolla 3-3.5 cm in diameter, blue to sky blue except in hybrid forms or in plants propagated from seed (in which case the petals may be white or various tones of violet) and greenish or grayish-white nectar guides, the 8 mm-long lobes narrowed into 3-3.5 mm-long pubescent acumens. Staminal column truncate conical; anthers 6-7 mm long, borne on 1.5 to 2 mm-long filaments. Style 10-12 mm long; stigma large, compressed-conical, somewhat cleft. Fruits abundant, globose to ovoid.

This plant is the most widely grown of the various Bolivian forms of *Solanum tuberosum* subsp. *andigena* var. *chiar-imilla*. Its native name, '*chiar imilla*', means 'black girl' in Aymara.

Plants grown from the true seed of this form differ from the above description in having blue, sky-blue-violet or white corollas and light pink or reddish tubers. The tubers of the latter are similar in color and shape to those of *wila imilla* and *ccompis*, two other forms of var. *chiar-imilla* that are also cultivated on the Bolivian altiplano.

The *imilla* varieties of potato are among the most important in Bolivia and southern Peru. All produce round to oblong, yellowish-white or bicolored tubers. Because of the occurrence of hybrid forms of this variety, not all types are easily classified.

Specimens Examined

HERBARIUM COLLECTIONS

Department La Paz

Province Camacho: Puerto Acosta, 3850 m alt., n.v. *Killu chomani* or *Chiar imilla*; *CPP-1898* (OCH). Puerto Acosta, 3850 m alt., n.v. *Chiara imilla*; *CPP-1901* (OCH).

Province Ingavi: Tiahuanaco, n.v. *Chiar Imilla*; *Juzepczuk 1640* (WIR). Taraco, 3820 m alt., n.v. *Chiar imilla*; *CPP-2117* (OCH). Huanccollo, 3650 m alt., n.v. *Ccoyo*; *Ochoa 3935.*

Province Murillo: La Paz, n.v. *Chiar imilla*; *Juzepczuk 1582, 1660* (WIR). La Paz, 3650 m alt., local market, brought from the altiplano, n.v. *Chiar imilla*; *Ochoa 3908* (OCH). La Paz, 3650 m alt., local market, n.v. *Yana imilla*; *Ochoa 3933* (OCH).

Province Pacajes: Andres de Machaca, 3900 m alt., n.v. *Chiar imilla*; *CPP-2075* (OCH). Caquiaviri, 3960 m alt., n.v. *Monda negra*; *Ochoa 3962* (OCH).

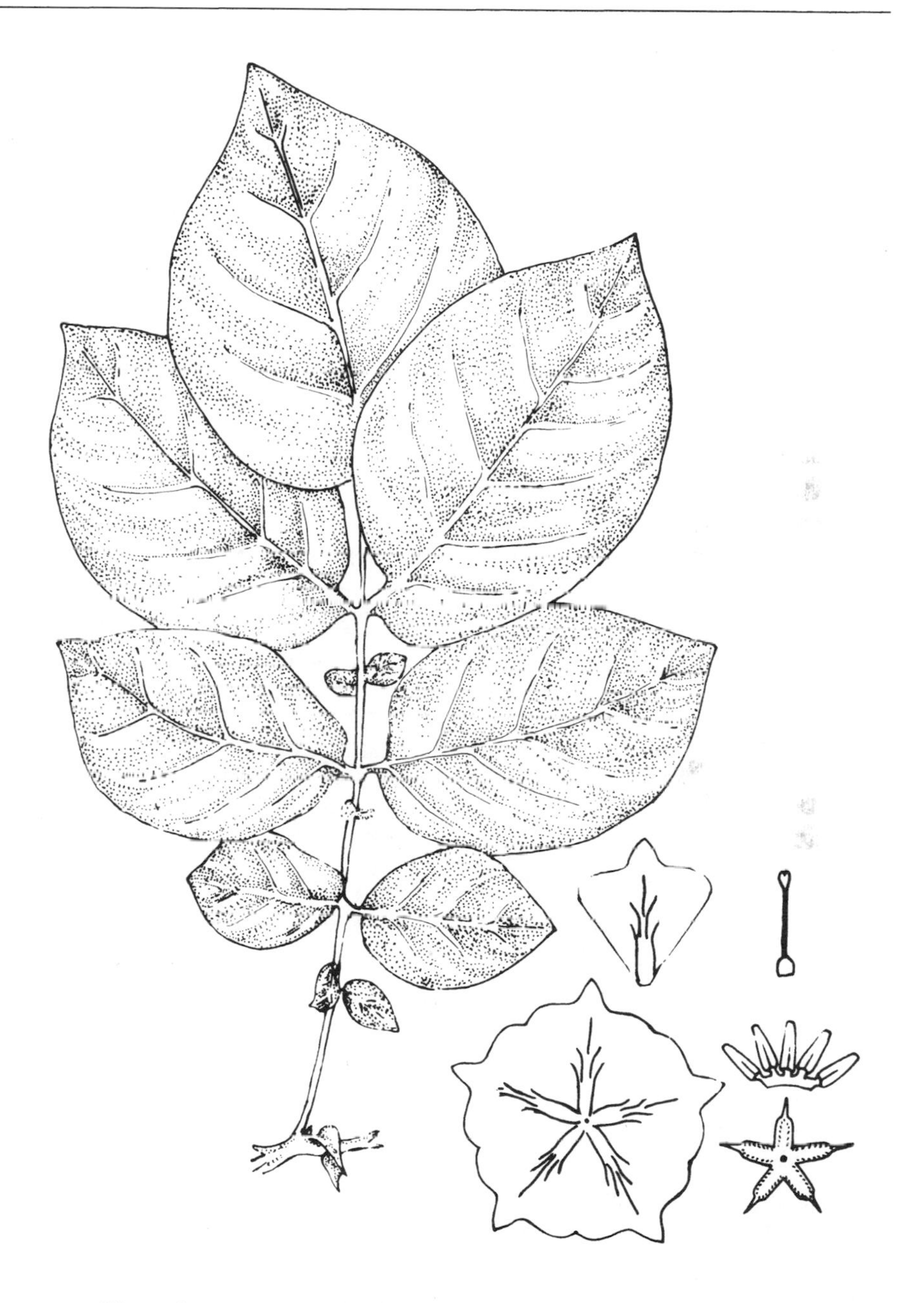

Figure 164. – Leaf and floral dissection of *Solanum tuberosum* subsp. *andigena* var. *chiar-imilla* (*Ochoa 3908*), ×1.

Department Potosi
Province Chayanta: Cajón Mayu, n.v. *Imilla negra*; *Hawkes et al. 6180* (CIP).

GERMPLASM COLLECTIONS
(examined electrophoretically)

Department La Paz
Province Aroma: Lahuachaca, 2n=48, n.v. *Alka Imilla* (HAW-5595).
Province Ingavi: Tiahuanaco, 3800 m alt., n.v. *Chiar Imilla* (HUA 798).
Province Murillo: La Paz, 2n=48, n.v. *Chiri Imilla* (OCH-10533). La Paz, 2n=48, n.v. *Imilla Negra* (OCH-10514).
Province Omasuyos: Hualata, 3850 m alt., 2n=48, n.v. *Llusku Imilla* (CIP-702294).

Department Oruro
Province Carangas: Saucari, Toledo, n.v. *Yana Imilla* (HUA-856).

Department Potosi
Province Chayanta: Maragua, 3700 m alt. (OCH-10593A). Maragua, 3700 m alt., n.v. *Yana Imilla* (OCH-10611, 10613A). Both collections 2n=48.
Province Frias: Lecherias, 3800 m alt., n.v. *Chia Papa* (HAW-5699). Tacabamba, 3500 m alt., 2n=48, n.v. *Yana Imilla* (OCH-10695). Yocalla, 3450 m alt. (HAW-6140).
Province Linares: Puna, 3850 m alt., 2n=48, n.v. *Alkka Imilla* (OCH-10624). Utacalla, 3750 m alt., 2n=48, n.v. *Yana Imilla* (OCH-10639).

GERMPLASM COLLECTIONS
(examined morphologically only)

Department Cochabamba
Province Chapare: Colomi (?), n.v. *Yana Imilla* (ETT-1135).
Province Punata: Punata, n.v. *Papa Collo* (CIP-702274).
Province Tapacari: Challa, n.v. *Yana imilla* (EET-1317).

Department Chuquisaca
Province Oropeza: Potolo, n.v. *Imilla Negra* (HAW-5659).

Department La Paz
Province Aroma: Calamarca, 3600 m alt., n.v. *Chiar Munta* (HUA-869). Calamarca, 3800 m alt., n.v. *Yana Imilla* (HUA-773). Colchani, n.v. *Chiar Imilla* (HAW-5592). Lahuachaca, n.v. *Chiar Monda* (HAW-5599). Panduro, n.v. *Chiar Imilla* (EET-1426). Patacamaya, n.v. *Chiar Imilla* (EET-1182, EET-1529). Quelkata, n.v. *Chiar Imilla* (EET-1396, EET-1408).

Province Camacho: Jupani, 3960 m alt., n.v. *Chiar Imilla* or *Imilla* (CO-2313, CO-2314). Both 2n=48. Puerto Acosta, 3800 m alt., n.v. *Alka Imilla* (CO-2298). Tajani, 3960 m alt., n.v. *Chiar Imilla* (CO-2318). Ullacsantía, 3870 m alt., n.v. *Chiar Imilla* (CO-2279). Unahuaya, 3960 m alt., n.v. *Chiar Imilla* (CO-2305). [These last four collections also 2n=48.]

Province Ingavi: Cachuma, 4000 m alt., 2n=48, n.v. *Chiar Imilla* (CO-2332). Chama, 4100 m alt., 2n=48, n.v. *Chiara Imilla* (CO-2345). Patarani, 3850 m alt., 2n=48, n.v. *Chiar Imilla* (CO-2186). Tiahuanaco, n.v. *Chiar Imilla* (EET-1447).

Province Larecaja: Ccochapata, 3600 m alt., n.v. *Papa Negra* (O/S-11808). Yamuco, 3600 m alt., n.v. *Negra* (O/S-11800). [Both 2n=48].

Province Los Andes: Cura Pucara, 3860 m alt., n.v. *Chiar Imilla* (HAW-6188). Huancane, 3900 m alt., n.v. *Chiar Imilla* (CO-2246) and *Wila Imilla* (CO-2245). Both 2n=48. Keallani, 3880 m alt., 2n=48, n.v. *Chiar Imilla* (CO-2146). Laja, n.v. *Chiar Imilla* (EET-1331). Tambillo, 3880 m alt., 2n=48, n.v. *Chiar Imilla* (CO-2162). Yana-Cachi, 3900 m alt., 2n=48, n.v. *Alkka Imilla* (CO-2231).

Province Murillo: El Alto, n.v. *Chiar Imilla* (EET-1526). Huaillara, 3800 m alt., 2n=48, n.v. *Imilla* or *Chiar Imilla* (CO-2198, CO-2199). La Paz, near Viacha, 3800 m alt., 2n=48, n.v. *Imilla* (OCH-10515, 10516). La Paz, 2n=48, n.v. *Jovera* (OCH-10534).

Province Omasuyos: Belen, n.v. *Imilla Negra* (CIP-702597). Near Tiquina, 3960 m alt., 2n=48, n.v. *Chiar Imilla* (CO-2248). Huatajata, n.v. *Chiar Imilla* (EET-1493). Warisata, n.v. *Chiar Imilla* (EET-1245, 1469, 1489).

Province Pacajes: Caquiaviri, 3960 m alt., 2n=48, n.v. *Chiar Imilla* (CO-2372).

Department Oruro

Province Poopó: Cala Cala, n.v. *Yana Imilla* (EET-1627). Pasto Pampa, n.v. *Imilla Negra* (EET-1636).

Province Carangas: Toledo-Corque, 3700 m alt., n.v. *Chiar Imilla* (HUA-819). Valle Bonito, 3800 m alt., n.v. *Chiar Imilla* (HUA-826).

Province Cercado: Caracollo, n.v. *Chejchi Imilla* (EET-1335). Caracollo, n.v. *Chiar Imilla* or *Imilla Negra* (EET-1482, EET-1528). Oruro, 3700 m alt., local market, 2n=48, n.v. *Alkka Takka* (CO-2479).

Department Potosi

Province Chayanta: Maragua, 3700 m alt., 2n=48, n.v. *Imilla* (OCH-10607). Ravelo, n.v. *Imilla Negra* (EET-1056).

Province Frias: Azangaro, 3800 m alt., 2n=48, n.v. *Koillu* (OCH-10511). Cerdas, n.v. *Yana Imilla* (CIP-702567). Gran Peña, 3800 m alt., 2n=48, n.v. *Imilla* (OCH-10672). Lecherias, 3800 m alt., n.v. *Imilla* or *Yana Imilla* (HAW-

5694, HAW-5708). Santa Lucia, 3800 m alt., 2n=48, n.v. *Imilla* (OCH-10683). Totora, n.v. *Imilla Negra* (HAW-6130). Cavara, n.v. *Imilla Negra* (CIP-702584). Yocalla, 3450 m alt., n.v. *Imilla Negra* (HAW-6147).

Province Ibañes: Sacaca, n.v. *Chiar Imilla* (CIP-702705). Sacaca, n.v. *Yana Imilla* (EET-1611).

Province Linares: Turuchipa, 3200 m alt., 2n=48, n.v. *Yana Imilla* (OCH-10693).

Province Saavedra: Cerdas, 3900 m alt., 2n=48, n.v. *Imilla* (OCH-10562).

38A(a). **Solanum tuberosum** subsp. **andigena** var. **chiar-imilla** f. **nigrum** (Lechnovitch) Ochoa f. nov., Phytologia 65(2): 103-113. 1988.

PLATE XXIV-1.

S. andigenum var. *chiar imilla* Lechn. f. *nigrum* Lechn., in Bukasov (ed.), Flora of Cult. Pl. Vol. IX, Potato, Leningrad, p. 86, Fig. 74. 1971, *nom. nud.* Based on *Juzepczuk 1660–b*, Bolivia. La Paz (WIR, not seen).

Plants of moderate height. Stems, peduncles, pedicels, and rachices strongly pigmented. Tubers round, with white flesh and black or uniformly dark blue-violet periderm, deep eyes and blue-violet buds. Leaves somewhat larger and more dissected than variety *chiar-imilla*, with up to 12 interjected leaflets of various sizes. Leaflets narrow, elliptic-lanceolate. Corolla dark blue, the acumens white externally.

This form is similar in many respects to *Solanum tuberosum* subsp. *andigena* var. *chiar-imilla*, and it is often known by the same name. However, it can be distinguished from the latter by the absence of violet or brown striping on the periderm of the tuber. The tuber and leaf characteristics of this form also reveal its relationship to *S. tuberosum* subsp. *andigena* vars. *bolivianum*, *sicha*, and *chiar-imilla* f. *wila imilla*.

Specimens Examined

HERBARIUM COLLECTIONS

Department La Paz

Province Murillo: Hacienda Huatajata, 3850 m alt., n.v. *Imilla Negra*; *Ochoa 3963* (OCH). La Paz; *Juzepczuk 1660-b* (WIR, type of f. *nigrum*). La Paz, Buenos Aires Market, 3700 m alt., n.v. *Negra*; *Ochoa-3903* (OCH). La Paz, Rodriguez Market, 3800 m alt., n.v. *Chiar Imilla*; *Ochoa-3904* (OCH).

Province Pacajes: Caquiaviri, 3960 m alt., n.v. *Monda Negra*; *CPP-2021* (OCH). Caquiaviri, 3960 m alt., n.v. *Monda Negra*; *CPP-2021A* (OCH).

Department Potosi

Province Chayanta: Maragua, 3700 m alt., n.v. *Yana Imilla*; *Ochoa 10612* (OCH). Ravelo, 3100 m alt., n.v. *Muki Papa*; *Ochoa 10641* (OCH).

Province Frias: Azangaro, 3800 m alt., n.v. *Yana Imilla*; *Ochoa 10507* (OCH).

Province Saavedra: Betanzos, 3400 m alt., n.v. *Jalka Yurac*; *Ochoa 10657* (OCH).

GERMPLASM COLLECTIONS

Department Oruro

Province Cercado: Oruro, 3706 m alt., local market, 2n=48, n.v. *Imilla* (OCH-10485). Oruro, 2n=48, n.v. *Sonsa Imilla* (OCH-10497).

Department Potosi

Province Frias: Tinguipaya, 3900 m alt., 2n=48, n.v. *Yankka Papa* (OCH-10547).

Province Saavedra: Cerdas, 3900 m alt., 2n=48, n.v. *Yana Imilla* (OCH-10565).

38A(b). Solanum tuberosum subsp. **andigena** var. **chiar-imilla** f. **sani-imilla** Ochoa f. nov., Phytologia 65 (2): 103-113. 1988. Plate XXIV-4.

Plants vigorous. Stems to 1 m tall, branched, densely purple-spotted below. Tubers with deep eyes and light beige flesh, round to slightly oblong, with violet eyes, purple buds and grayish-white periderm except for the apical end, which is colored dark violet for one-third of its length. Leaves 10-16 cm long by 6-10 cm wide, longer and more dissected than in the case of variety *chiar-imilla*, provided with 5 pairs of leaflets and 12 or more interjected leaflets. Leaflets 2.6-5 cm long by 1.5-3.4 cm wide. Pedicels 15-20 mm long, articulated in the upper one-third of the stalk. Calyx small, 6 mm long, with lobes abruptly narrowed to short, pointed acumens. Corolla 3.5 cm in diameter, dark sky blue. Staminal column truncate-conical; anthers 6 mm long. Style 11 mm long.

The native name of this potato is *sani-imilla*, which means ‘gray girl’ in Aymara.

Specimens Examined

HERBARIUM COLLECTIONS

Department La Paz

Province Camacho: Puerto Acosta, 3800 m alt., n.v. *Sani Imilla*; *CPP-1910* (OCH).

Province Ingavi: Taraco, n.v. *Sani Imilla*; *CPP-2096* (OCH).

Province Los Andes: Pucarani, 3900 m alt., n.v. *Phureja*; *Ochoa 2796* (OCH). Pucarani, n.v. *Suni Imilla*; *CPP-1834* (OCH). Palcoco; *CIP-702267*(OCH).

Province Murillo: La Paz, 3650 m alt., Buenos Aires Market, brought from altiplano, n.v. *Sani Imilla*; *Ochoa 3966* (OCH). La Paz, market, n.v. *Sala Imilla*; *Ochoa 3965* (OCH). La Paz, Huatay Market, n.v. *Sani Imilla*; *CPP-1801* (OCH). Hacienda Huatajata, 3850 m alt., n.v. *Sani*; *Ochoa 3972* (OCH, type of f. *sani-imilla*). Hacienda Huatajata, n.v. *Imilla*; *Ochoa 3973* (OCH).

Department Oruro

Province Chayanta: Toledo-Saucari, 3700 m alt., n.v. *Sani Colorada*; *Huamán 854* (CIP).

Department Potosi

Province Cercado: Potosi, n.v. *Sani Imilla*; *Hawkes 5667* (CIP).

GERMPLASM COLLECTIONS

Department La Paz

Province Aroma: Colchani, n.v. *Sali Imilla* (EET-1036). Lahuachaca, n.v. *Sali Imilla* (HAW-5597).

Province Ingavi: Desaguadero, n.v. *Sani* (EET-1289). Tiahuanaco, n.v. *Sani Imilla* (HUA-800).

Province Larecaja: Ccochapata, 3600 m alt., 2n=48, n.v. *Sani* (O/S-11811). Yamuco, 3600 m alt., 2n=48, n.v. *Sani* (O/S-11802).

Province Los Andes: Cullacachi, 3850 m alt., 2n=48, n.v. *Sali Imilla* (CO-2232). Cura Pucara, 3860 m alt., n.v. *Sani Imilla* (HAW-6191).

Province Murillo: La Paz, n.v. *Monda* (HAW-5578).

Department Oruro

Province Chayanta: Saucari-Toledo, 3700 m alt., n.v. *Sani Imilla* (HUA-853). Saucari-Toledo, n.v. *Alka Imilla* (HUA-849).

Department Potosi

Province Frias: Lecherias, 3800 m alt., n.v. *Pichilla* (HAW-5703). Lecherias, n.v. *Sani Imilla* (HAW 5706).

Province Nor Chichas: Ramada, n.v. *Sani Imilla* (HAW-6109).

38A(c). Solanum tuberosum subsp. **andigena** var. **chiar-imilla** fa **isla-imilla** Ochoa f. nov., Phytologia 65(2): 103-113. 1988.

Stems light green, densely pigmented with anthocyanins along the basal one-third of its length. Tubers with light beige flesh, round, the bicolored periderm with well-defined dark violet pink and yellowish-white areas. Leaves 12-17.6 cm long by 8-9 cm wide, with 4(-5) pairs of leaflets and 4-12 interjected leaflets. Leaflets 3.2-5.3 cm long by 2-3.6 cm wide. Peduncles to 12 cm long. Pedicels 15-18 mm long, articulated approximately 2-3 mm below base of calyx. Calyx 9-10 mm long, the lobes abruptly narrowed to 1-2 mm-long acumens. Corolla 3-3.5 cm in diameter, white or colored, with dark violet nectar guides. Anthers 6 mm long, borne on very short filaments, the latter less than 1 mm long. Style 8-8.5 mm long, exerted about 1 mm, glabrous; stigma large, ovoid.

There are some variants of *Solanum tuberosum* subsp. *andigena* var. *chiar-imilla* f. *isla imilla* that have more intensely pigmented tuber skins and buds, and occasionally deeper colored corollas. These would suggest that this cultigen has been derived in natural crosses between the white-flowered *S. tuberosum* subsp. *andigena* f. *janck'o imilla* and some other forms of variety *chiar-imilla.* However, this form can be distinguished from all other cultivated potatoes of the Bolivian altiplano by its dark violet (almost black) nectar guides.

Specimens Examined

HERBARIUM COLLECTIONS

Department La Paz

Province Ingavi: Huancollo, 3850 m alt., n.v. *Yaco*; *Ochoa 3975* (OCH). Tiahuanaco, 3900 m alt., n.v. *Isla*; *Ochoa 3976* (OCH).

Province Los Andes: Huancané, 3900 m alt., n.v. *Isla*; *CPP-1819* (OCH). Pucarani, 3900 m alt., n.v. *Isla Imilla*; *Ochoa 2798* (OCH).

Province Murillo: La Paz, Buenos Aires Market, 3700 m alt., n.v. *Chilena*; *Ochoa 3906* (OCH). La Paz, Buenos Aires Market, n.v. *Isla Imilla*; *Ochoa 3899* (OCH). La Paz, Camacho Market, 3800 m alt., n.v. *Isla Imilla*; *Ochoa 3969* (OCH, type of *isla-imilla*).

GERMPLASM COLLECTIONS

Department Cochabamba

Province Arani: Tiraque, 2n=48, n.v. *Lunca Imilla* (HAW-5643).

Department La Paz

Province Aroma: Calamarca, n.v. *Papa Isla* (EET-1225). Lahuachaca, 2n=48, n.v. *Ishla* (HAW-5588). Quelkata, n.v. *Isla* (EET-1442, EET-1444). Vilaque, 3800 m alt., n.v. *Isla* (HUA-888). Wayhuasi, n.v. *Isla* (EET-1365).

Province Ingavi: Tiahuanaco, n.v. *Isla* (EET-1356).

Province Larecaja: Yamuco, 3600 m alt., 2n=48, n.v. *Misi Alkka* (O/S-11804).

Province Murillo: La Paz, Rodriguez Market, brought from altiplano, 2n=48, n.v. *Isla*; (O/S-11790). La Paz Market, 2n=48, n.v. *Alkka Imilla* (HAW-5567). Ichucerca, n.v. *Isla* (EET-1260).

Province Omasuyos: Belen, n.v. *Isla* (EET-1505). Hualata, 3850 m alt., n.v. *Isla* (CIP-702299). Huatajata, n.v. *Wila Isla* (EET-1581). Ichucerca, n.v. *Isla* (EET-1181).

Department Potosi

Province Frias: Lecherias, 3800 m alt., n.v. *Sacca Imilla* (HAW-5702).

38A(d). Solanum tuberosum subsp. **andigena** var. **chiar-imilla** f. **ccompis** Ochoa.

S. andigenum Juz. et Buk. var. *imilla* f. *ccompis* Juz. et Buk., in Bukasov, The Potatoes of South America and their Breeding Possibilities, p.67, 1933, *nom.nud.* Based on Juzepczuk *1663, 1666* (WIR) and *1807a* (WIR, not seen), Bolivia. La Paz. Also many other Juzepczuk collections from Cusco, Peru.

Plants vigorous. Stems light green, narrowly winged. Tubers with deep eyes and white to light beige flesh, round, irregular (or bulging), with pink to light violet-pink periderm and dark pink eyes and brows, or occasionally, in the case of morphotypes called *alkka imilla* or *alkka ccompis*, with whitish halos about the eyes, and light violet-pink buds. Leaves light green, little dissected. Peduncles to 10 cm long. Pedicels approximately 18-19 mm long, the upper part about 3-4 mm long and the lower about 15 mm long. Calyx 7.5-8 mm long, the lobes abruptly narrowed in 1.5 mm-long acumens. Corolla 3-3.5 cm in diameter, white, with greenish-yellow nectar guides. Staminal column truncate conical; anthers 5.5 mm long. Style 10-11 mm long.

This cultigen is widely distributed in Southern Peru and Bolivia. In Peru, it is called *Ccompis*. However, on the altiplano of Bolivia this white-flowered and pink-tubered potato is known variously as *Imilla Rosada*, *Yurac Imilla*, *Imilla Blanca*, or *Janck'o Imilla*.

Of the several herbarium collections that were examined electrophoretically, the following numbers proved to be identical clones: *HUA-893*, *CIP-702265*, *HAW-6170, 5688, 5697*, and *5693*.

Specimens Examined

HERBARIUM COLLECTIONS

Department La Paz

Province Aroma: Vilaque, 3800 m alt., n.v. *Kello Monda*; *Huamán 893* (OCH).
Province Los Andes: Palcoco market; *CIP-702265* (CIP).
Province Murillo: La Paz; *Juzepczuk 826, 1663, 1807* (WIR).

Department Potosi

Province Chayanta: Suruhana, n.v. *Imilla Rosada*; *Hawkes 6170* (CIP).
Province Frias: Lecherias, 3800 m alt., n.v. *Puca Imilla*; *Hawkes 5688* (OCH). Lecherias; *Hawkes 5697* (OCH). Lecherias, n.v. *Yurac Imilla*; *Hawkes 5693*.

GERMPLASM COLLECTIONS

Department Cochabamba

Province Arani: Tiraque, 2n=48, n.v. *Yurac Imilla* (HAW-5651). Potolo, n.v. *Alcca Imilla* (HAW-5660).

Department La Paz

Province Aroma: Patacamaya, n.v. *Imilla Blanca* (EET-1372). Quelkata, n.v. *Janko Imilla* (EET-1450). Waywasi, n.v. *Janko Imilla* (EET-1358).
Province Barrón: Huancaroma, n.v. *Janko Imilla* (EET-1370).
Province Ingavi: Desaguadero, n.v. *Imilla Rosada* (EET-1390).
Province Los Andes: Cura Pucara, 3860 m alt., 2n=48, n.v. *Janco Imilla* (HAW-6192). Laja, n.v. *Janko Papa* (EET-1452).
Province Omasuyos: Belen, n.v. *Janko Imilla* (EET-1475).

Department Oruro

Province Carangas: Pasto Grande, n.v. *Imilla Blanca* (EET-1456).
Province Abaroa: Huancané, n.v. *Imilla Blanca* (EET-1229).

Department Potosi

Province Chayanta: Cajun Mayu, 2n=48, n.v. *Imilla Blanca* (HAW-6176). Hichuna Pampa, n.v. *Imilla Blanca* (HAW-6167).
Province Frias: Lecherias, 3800 m alt., 2n=48, n.v. *Alcca Imilla* (HAW-5704). Tinguipaya, 2n=48, n.v. *Yurac Imilla* (CIP-702558). Yocalla, 3450 m alt., 2n=48 (HAW-6141).

38A(e). Solanum tuberosum subsp. **andigena** var. **chiar-imilla** f. **janck'o-imilla** Ochoa. FIG. 165.

Plants light green. Stems narrowly winged. Tubers with white to light beige flesh, round, with white or yellowish-white periderm, deep eyes, white buds, pink at base. Leaves short and broad, dark green, with 4(-5) pairs of leaflets and 4-5 interjected leaflets. Leaflets broadly elliptic, the apex obtuse and very shortly subacuminate and the base obliquely rounded, borne on very short petiolules. Corolla white, the lobes short and broad. Fruit globose.

The native name of this potato, *jank'o-imilla*, means 'white girl' in Aymara.

This form may represent a somatic mutation of *Solanum tuberosum* subsp. *andigena* var. *chiar-imilla* f. *ccompis*.

Specimens Examined

HERBARIUM COLLECTIONS

Department La Paz

Province Ingavi: Tiahuanaco, 3900 m alt., n.v. *Jancko Imilla*; *Ochoa 3941* (OCH).

Province Murillo: La Paz, 3650 m alt., Buenos Aires Market, brought from altiplano, n.v. *Jancko Imilla*; *Ochoa 3900* (OCH). La Paz Market, n.v. *Imilla Blanca*; *Ochoa 3964* (OCH).

Province Omasuyos: Huarisata, 3900 m alt., n.v. *Janko Imilla*; *CIP-702313* (CIP).

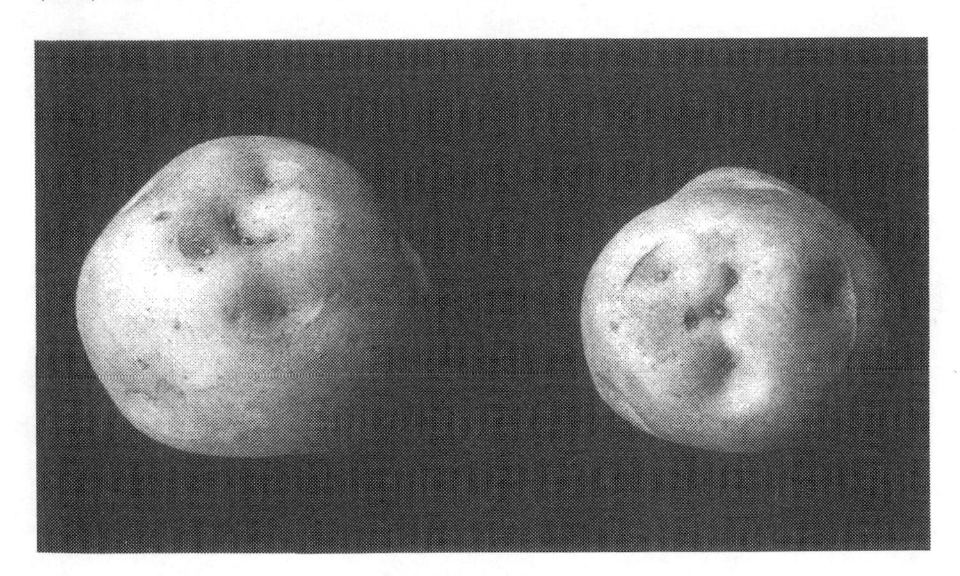

Figure 165. – Tubers of *Solanum tuberosum* subsp. *andigena* var. *chiar-imilla* f. *janck'o-imilla*, ×1.

GERMPLASM COLLECTIONS

Department Cochabamba

Province Chapare: Colomi, n.v. *Imilla Blanca* (EET-1015).

Department La Paz

Province Camacho: Puerto Acosta, 3880 m alt., 2n=48, n.v. *Janko Imilla* (CO-2292). Tajani, 2n=48, n.v. *Janko Imilla* (CO-2317). Unahuaya, 2n=48, n.v. *Janko Imilla* (CO-2308).

Province Ingavi: Patamanta, n.v. *Imilla Blanca* (EET-1216). Viacha, n.v. *Imilla Blanca* (EET-1026). Cachuma, 4000 m alt., 2n=48, n.v. *Janko Monda* (CO-2343).

Province Los Andes: Cullacachi, 3850 m alt., 2n=48, n.v. *Jancko Imilla* (CO-2233). Huancané, 3900 m alt., 2n=48, n.v. *Choke Jincho* (CO-2243). Laja, n.v. *Imilla Blanca* (EET-1219). Keallani, 3880 m alt., 2n=48, n.v. *Janko Imilla* (CO-2145). Pucarani, 3900 m alt., 2n=48, n.v. *Jancko Imilla* (CO-2220).

Province Murillo: El Alto, n.v. *Janko Imilla* (EET-1306). La Paz Market, 2n=48, n.v. *Jancko Imilla* (CO-2205).

Province Omasuyos: Ichucerca, n.v. *Imilla Blanca* (EET-1246). Tiquina, 3960 m alt., 2n=48, n.v. *Jancko Imilla* (CO-2249).

Province Pacajes: Caquiaviri, 3960 m alt., 2n=48, n.v. *Alto T'tanta* (CO-2384).

Department Oruro

Province Cercado: Caracollo, n.v. *Janko Imilla* (EET-1435). Oruro, 3700 m alt., local market, 2n=48, n.v. *Yurac Imilla* (CO-2470).

Province Indefinite: Cala Cala, n.v. *Yurac Imilla* (EET-1615).

Department Potosi

Province Chayanta: Ravelo, n.v. *Imilla Blanca* (EET-1055).

Province Frias: Frias, n.v. *Yurac Imilla* (EET-1084).

38A(f). Solanum tuberosum subsp. **andigena** var. **chiar-imilla** f. **janck'o-chockella** Ochoa f. nov., Phytologia 65(2): 103-113. 1988.

Tubers with light beige flesh, round, the yellowish-white skin with pink buds and pink stripes about the eyes and brows. Peduncle light green, 10-12 cm long, bearing 6-10 flowers. Corolla white.

Specimens Examined

HERBARIUM COLLECTIONS

Department La Paz

Province Ingavi: Sullcatiti, n.v. *Jancku Choquella*; *CIP-702616* (CIP, type of var. *janck'o-chockella*).

GERMPLASM COLLECTIONS

Department La Paz

Province Aroma: Vilaque, 3800 m alt., n.v. *Monta Blanca* (HUA-900).
Province Ingavi: Guaqui, 3800 m alt., (HUA-794).

Department Oruro

Province Carangas: Valle Bonito, 3800 m alt., n.v. *Paca Taca* (HUA-834).

38A(g). Solanum tuberosum subsp. **andigena** var. **chiar-imilla** f. **alkka-imilla** Ochoa f. nov., Phytologia 65(2): 103-113. 1988. PLATE XXIV-5.

Stems light green, pigmented light brown below center. Tubers with white to light beige flesh, round, somewhat irregular (bulging), bicolored, with red or dark pink skin and large yellowish halo areas about the eyes (or yellowish apical areas), and yellow buds, the latter pink at base and apex. Leaves with 4(-5) pairs of leaflets and generally few interjected leaflets. Peduncles 7-10 cm long. Pedicels articulated very near the base of calyx, the upper part 2-4 mm long and the lower about 20 mm long. Corolla 3 cm in diameter, lilac or light lilac-violet with white acumens.

Hybridization between different clones and segregates of the *imilla* complex has probably been important in the origins of the *alkka imilla* and *isla imilla* potatoes of Bolivia. Both are extremely variable in tuber skin and bud, and flower color.

Specimens Examined

HERBARIUM COLLECTIONS

Department La Paz

Province Aroma: Lahuachaca, n.v. *Alka Imilla*; *Hawkes 5600* (CIP).
Province Ingavi: Pillapi, n.v. *Alkka Imilla*; *CPP-1932* (OCH).
Province Murillo: La Paz Market, n.v. *Alkka Imilla*; *Ochoa 3901, 3902* (OCH, type of *alkka-imilla*).

Department Potosi

Province Frias: Lecherias, 3800 m alt., n.v. *Alkka Imilla*; *Hawkes 5696* (OCH). Tinguipaya, 3900 m alt., n.v. *Alkka Imilla*; *CIP-702525* (CIP). Tinguipaya, n.v. *Elena*; *Ochoa-10558* (OCH).

Province Linares: Puna, n.v. *Yurac Imilla*; *Ochoa 10626* (OCH).

Province Saavedra: Betanzos, 3407 m alt., n.v. *Melka Yurac*; *Ochoa-10661* (OCH).

GERMPLASM COLLECTIONS

Department Chuquisaca

Province Oropeza: Potolo, n.v. *Alka Imilla* (EET-1057).

Department La Paz

Province Aroma: Panduro, n.v. *Alka Imilla* (EET-1375). Quelkata, n.v. *Alka Imilla* (EET-1448).

Province Los Andes. Laja, n.v. *Alka Imilla* (EET-1552).

Province Murillo: El Alto, n.v. *Alka Monda* (EET-1300)

Department Oruro

Province Indefinite: Cala Cala, n.v. *Alka Imilla* (EET-1221.). Pasto Grande, n.v. *Alka Imilla* (EET-1339). Pasto Pampa, n.v. *Alka Imilla* (EET-1597).

Department Potosi

Province Frias: Frias, n.v. *Alka Imilla* (EET-1086).

38A(h). Solanum tuberosum subsp. **andigena** var. **chiar-imilla** f. **wila-imilla** Ochoa f. nov., Phytologia 65(2): 103-113. 1988. FIG. 166.

Stems light green, pigmented light brown along the basal one-third of its length. Tubers with white to light beige flesh, round to oblong, with red or dark pink periderm, deep eyes and pink to dark reddish-violet buds. Leaves 13-15 cm long by 7-8 cm wide, more dissected than in variety *chiar-imilla*, with 5(-6) pairs of leaflets and 6-8 interjected leaflets. Leaflets 3.5-3.9 cm long by 2.3-2.8 cm wide, elliptic to ovate-elliptic, the apex obtuse and the base rounded, borne on 2-3 mm-long petiolules. Peduncles to 18 cm long, light green, fructiferous. Pedicels 25 mm long, the upper part 5-7 mm long. Calyx 7-9 mm long, with elliptic-lanceolate, reflexed lobes, the apex abruptly narrowed in short, pointed acumens, the latter about 1.5 mm in length. Corolla 3.5 cm in diameter, white. Anthers 7(-7.5) mm long, borne on 2 mm-long filaments. Style 11-12 mm long; stigma thick, somewhat cleft. Fruit ovoid.

The native name of this potato, *wila imilla*, means ‘red girl’ in Aymara.

Specimens Examined

HERBARIUM COLLECTIONS

Department La Paz

Province Ingavi: Tiahuanaco, 3850 m alt., n.v. *Wila Imilla*; *CPP-1720* (OCH).

Province Murillo: Chilaya Grande, 3840 m alt., n.v. *Wila Imilla*; *CPP-1810* (OCH, type of f. *wila-imilla*). Llosa, n.v. *Wila Imilla*; *CIP-702696* (CIP).

Province Omasuyos: Achacachi, 3800 m alt., n.v. *Wila Imilla*; *CPP-1888* (OCH). Huarisata, 3800 m alt., n.v. *Wila Imilla*; *CPP-1877* (OCH).

Province Pacajes: Caquiaviri, 3960 m alt., n.v. *Monda Roja*; *CPP-2009* (OCH).

GERMPLASM COLLECTIONS

Department La Paz

Province Murillo: La Paz, 2n=48, n.v. *Monda* (HAW-5577).

Province Omasuyos: Hualata, n.v. *Morales Negra* (CIP-702296).

Province Pacajes: San Andres de Machaca, 3890 m alt., 2n=48, n.v. *Wila Imilla* (CO-2418).

Department Oruro

Province Cercado: Azafranal, n.v. *Wila Imilla* (EET-1397).

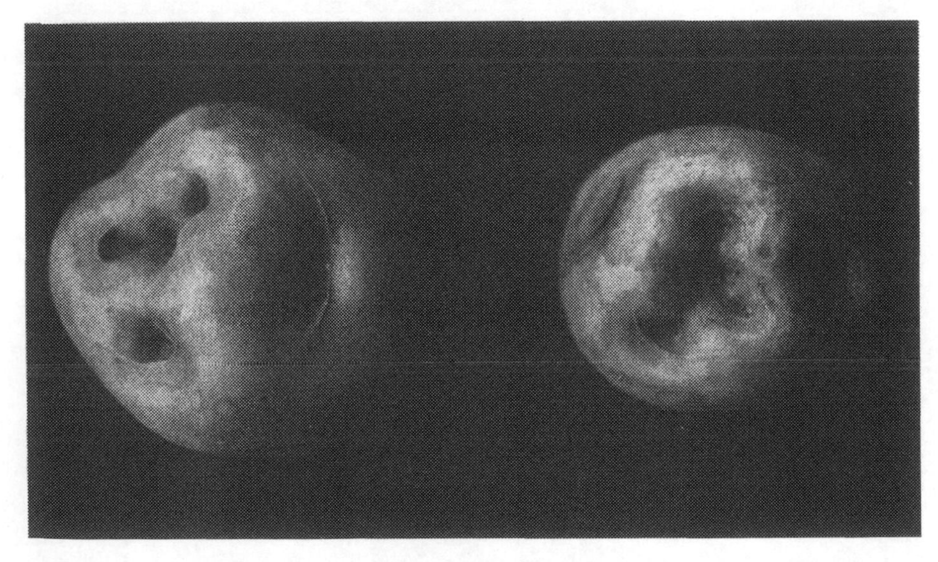

Figure 166. – Tubers of *Solanum tuberosum* subsp. *andigena* var. *chiar-imilla* f. *wila-imilla*, ×1.

38A(i). Solanum tuberosum subsp. **andigena** var. **chiar-imilla** f. **wila-monda** Ochoa f. nov., Phytologia 65(2): 103-113. 1988. PLATE XXIV-2.

Tubers round to oblong, with pink or red periderm similar to form *wila-imilla*. Leaves light green, similar in dissection to form *sani-imilla* with 5 pairs of leaflets and 8-12 interjected leaflets. Leaflets larger than form *wila-imilla*, elliptic to ovate-elliptic, the apex obtuse and the base rounded, borne on 5 mm-long petiolules. Peduncles to 15 cm long, light green. Pedicels 14-18 mm long, articulated in the upper one-third of stalk, the upper part light green and 4 mm long, and the lower part lightly pigmented and 10-14 mm long. Corolla violet to light lilac-violet with whitish acumens. Staminal column truncate-conical; the anthers 7 mm long and borne on 1.5-2 mm-long filaments. Style 11-12 mm long, densely papillose below center; stigma capitate, cleft.

Specimens Examined

HERBARIUM COLLECTIONS

Department Cochabamba

Province Yamparaez: Tarabuco, n.v. *Sonsa Imilla*; *Hawkes-5663* (CIP).

Department La Paz

Province Murillo: La Paz, 3650 m alt., Buenos Aires Market, brought from altiplano, n.v. *Puca Imilla*; *Ochoa 3967* (OCH, type of f. *wila-monda*). La Paz, Camacho Market, brought from altiplano, n.v. *Wila Monda* or *Puca Monda*; *Ochoa 3968* (OCH).

Province Ingavi: Huanccollo, 3850 m alt., n.v. *Wila Imilla*; *Ochoa 3974* (OCH).

Department Potosi

Province Chayanta: Ravelo, n.v. *Puca Runa*; *CIP-702538* (CIP).

GERMPLASM COLLECTIONS

Department Cochabamba

Province Chapare: Melga, 2n=48, n.v. *Puca Imilla* (HAW-5641).

Province Oropeza: Santa Catalina, 2950 m alt., 2n=48, n.v. *Sonsa Imilla* (HAW-5674).

Department Oruro

Province Cercado: Belen, 3700 m alt., 2n=48, n.v. *Puca Imilla* (HUA-804). Oruro, 3700 m alt., local market, 2n=48, n.v. *Puca Imilla* (OCH-10492).

Department Potosi

Province Chayanta: Cajon Mayu, n.v. *Sonsa Imilla* (HAW-6183). Ravelo, n.v. *Sonsa Imilla* (HAW-5666).

Province Frias: Cerdas, n.v. *Puca Imilla* (CIP-702560). Challactiri, 2n=48, n.v. *Puca Imilla* (OCH-10573).

Province Saavedra: Betanzos, 3700 m alt., 2n=48, n.v. *Puca Imilla* (OCH-10658).

Province Indefinite: Atacolla, n.v. *Puca Imilla* (CIP-702573).

38B. Solanum tuberosum subsp. **andigena** var. **bolivianum** (Juzepczuk and Bukasov) Ochoa.

S. andigenum Juz. et Buk. var. *bolivianum* Juz. et Buk., in Bukasov, The Potatoes of South America and their Breeding Possibilities, pp. 70-71, Fig. 19. 1933, *nom.nud.* Based on *Juzepczuk 1656* (WIR) and *1807-b* (WIR, not seen), Bolivia. La Paz, Sorata.

Stems lightly to moderately pigmented. Tubers oblong to oval compressed. Leaves with 5(-6) pairs of leaflets and 5-12 interjected leaflets, the latter occasionally with basiscopic or semibasiscopic segments, or less frequently with acroscopic segments. Pedicels articulated in the upper one-third of the stalk. Calyx 9-10 mm long. Corolla lilac, violet, blue or blue-violet.

Specimens Examined

HERBARIUM COLLECTIONS

Department La Paz

Province Larecaja: Indefinite, n.v. *Reina* or *Monda*; *Juzepczuk 1656* (WIR).

38B(a). Solanum tuberosum subsp. **andigena** var. **bolivianum** f. **chiarpala** Ochoa f. nov., Phytologia 65(2): 103-113. 1988. PLATE XXV-2.

Plants vigorous. Stems to 1 m tall, strongly pigmented. Tubers with light beige flesh, oval-compressed with very dark, bluish-violet (almost black) periderm, superficial eyes and dark pink-violet buds. Leaves 10-22.5 cm long by 5.5-9 cm wide. Leaflets 3-5 cm long by 1.6-2.5 cm wide, elliptic, borne on 5 mm-long petiolules, the latter pigmented as in the rachis and nerves of the leaflets. Peduncles 5-8 cm long. Pedicels 17-20 mm long, the upper part about 5 mm long and the lower 12-15 mm long. Calyx 7-8 mm long, the rectangular lobes abruptly narrowed to 1.5 mm-long pointed acumens. Corolla 3-3.5 cm in diameter, dark-blue violet with white acumens. Fruit 3 cm in diameter, globose.

Related to *Solanum tuberosum* subsp. *andigena* var. *bolivianum* f. *dilatatum* Juz. et Buk.

Specimens Examined

HERBARIUM COLLECTIONS

Department La Paz

Province Ingavi: Huanccollo, 3680 m alt., n.v. *Pala*; *Ochoa 3977* (OCH, type of f. *chiar-pala*).

Province Murillo: La Paz Market, 3650 m alt., n.v. *Chiar Pali*; *Ochoa 3978* (OCH). La Paz, Buenos Aires Market, brought from altiplano, n.v. *Chiar Pala*; *Ochoa 3981* (OCH).

Province Omasuyos: Achacachi, 3800 m alt., n.v. *Chiara Pala*; *CPP-1889* (OCH).

GERMPLASM COLLECTION

Department La Paz

Province Camacho: Carabuco, n.v. *Pali Morada* (EET-1484).

38B(b). **Solanum tuberosum** subsp. **andigena** var. **bolivianum** f. **janck'o-pala** Ochoa f. nov., Phytologia 65(2): 103-113. 1988. Plate XXV-3.

Plants similar to form *chiar-pala* except for the light green stems, the latter pigmented only in the lower one-third of the stalk. Tubers with light beige flesh, oval compressed, with yellowish-white skin or yellowish-white skin with pale smoky-pink apices and eyebrows, superficial eyes and white buds, the latter with light pink apices and bases.

Specimens Examined

HERBARIUM COLLECTIONS

Department La Paz

Province Murillo: La Paz, 3650 m alt., Camacho Market, brought from altiplano, n.v. *Janck'o-Pala*; *Ochoa s.n.* (OCH).

Department Oruro

Province Poopo: Urmiri, 3750 m alt., n.v. *Janck'o Pali*; *Ochoa 3996* (OCH, type of f. *janck'o-pala*).

Department Potosi
Province Frias: Tinguipaya, 3900 m alt., n.v. *Yurac Pali*; *Ochoa 3951* (OCH).

GERMPLASM COLLECTIONS

Department La Paz
Province Aroma: Calamarca, 2n=48, n.v. *Hanko Pala* (HUA-774). Panduro, n.v. *Janko Pali* (EET-1454). Patacamaya, n.v. *Janko Pala* (EET-1222, 1464). Sica Sica, n.v. *Hanko Pala* (HUA-859). Vilaque, n.v. *Janko Pala* (EET-1524).
Province Camacho: Carabuco, n.v. *Pala Blanca* (EET-1563).
Province Lareeaja: Ccochapata, 3600 m alt., 2n=48, n.v. *Pala*; O/S-11812.

Department Oruro
Province Barron: Huancaroma, n.v. *Pala Blanca* (EET-1323).
Province Cercado: Caracollo, n.v. *Janck'o Pali* (EET-1459).

Department Potosi
Province Chayanta: Ravelo, n.v. *Yumac Pali* (EET-1590).

38B(c). Solanum tuberosum subsp. **andigena** var. **bolivianum** f. **wila-pala** Ochoa f. nov., Phytologia 65(2): 103-113. 1988.
Fig. 167; PLATE XXV-1.

Stems irregularly pigmented light pink. Tubers with light beige or yellowish flesh, oval-compressed, with red periderm, or occasionally red with a few yellowish-white stripes, superficial to semideep eyes and pink or pink-violet buds. Leaves similar to variety *bolivianum*. Leaflets 7.5-8.5 cm long by 4-4.5 cm wide, the apex shortly acuminate and the base symmetrical or asymmetrically rounded, borne on 10 mm-long petiolules. Corolla dark lilac. Calyx 7 mm long, the lobes narrowing to short, pointed acumens, the latter 1-2 mm long. Anthers 6.5-7 mm long. Style 10 mm long; stigma large, capitate.

Specimens Examined

HERBARIUM COLLECTIONS

Department La Paz
Province Ingavi: Huanccollo, 3650 m alt., n.v. *Wila Pala*; *Ochoa 3986* (OCH, type of f. *wila-pala*).
Province Murillo: La Paz, 3650 m alt., Buenos Aires Market, brought from altiplano, n.v. *Pala*; *Ochoa 3979* (OCH). La Paz, 3650 m alt., Camacho Market, n.v. *Pala*; *Ochoa 3980* (OCH). La Paz, Murillo Market, n.v. *Pali*; *Ochoa 3983* (OCH).

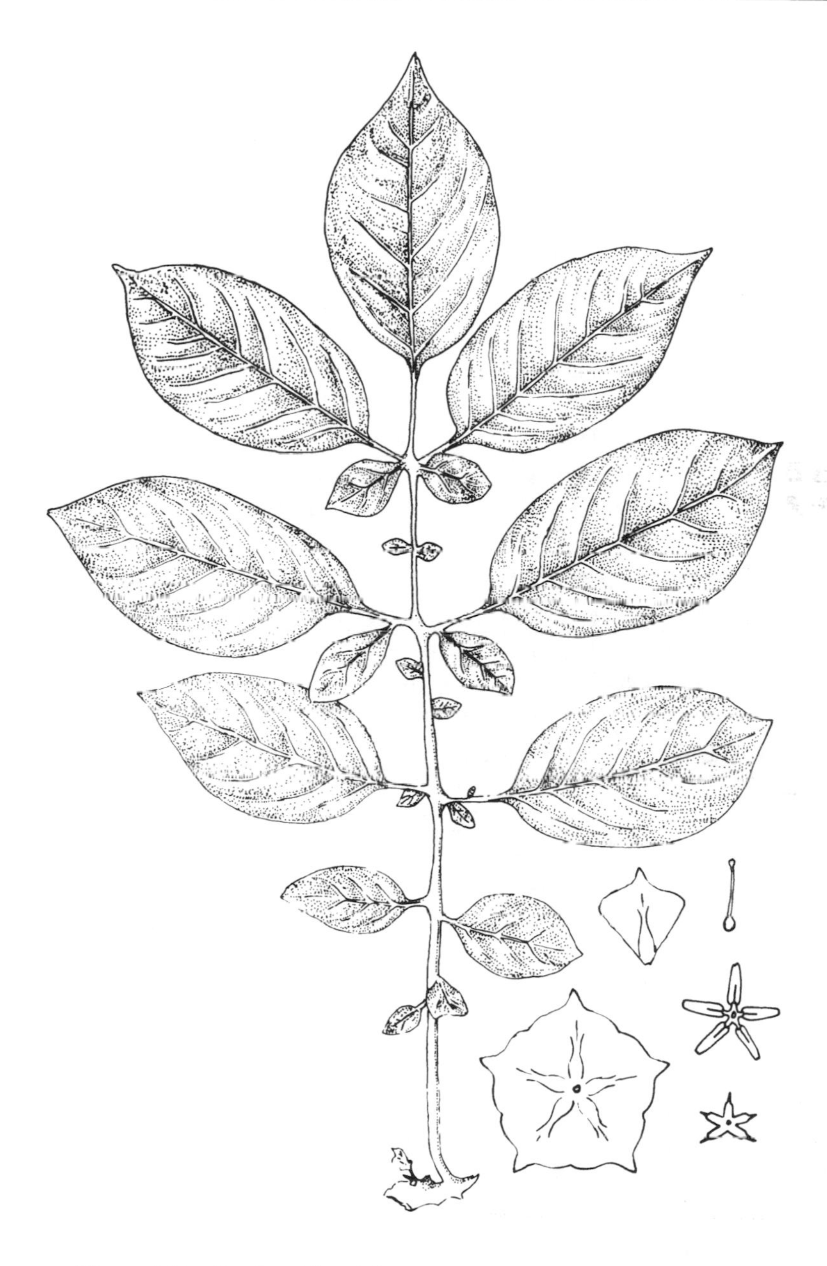

Figure 167. – Leaf and floral dissection of *Solanum tuberosum* subsp. *andigena* var. *bolivianum* f. *wila-pala* (*Ochoa 3983*), ×1.

Department Oruro

Province Poopo: Urmiri, 3750 m alt., n.v. *Wila Pali*; *Ochoa 3997* (OCH).

Department Potosi

Province Linares: Puna, 3400 m alt., n.v. *Phali*; *Ochoa 3961* (OCH).

Province Indefinite: Lequezana, 3300 m alt.; *Ochoa 3952* (OCH).

GERMPLASM COLLECTIONS

Department La Paz

Province Aroma: Lahuachaca, 2n=48, n.v. *Pala Pala* (HAW-5589). Lahuachaca, n.v. *Wila Pala* (EET-1233). Occopampa, 3850 m alt., 2n=48, n.v. *Wila Pala* (CO-2464). Panduro, n.v. *Pali* or *Wila Pali* (EET-1436, EET-1412, EET-1419). Patacamaya, n.v. *Paloma* (CIP-702577). Quelkata, n.v. *Wila Pala* or *Wila Pali* (EET-1401, EET-1399). Wayhuasi, n.v. *Wila Pala* (EET-1298).

Province Ingavi: Ingavi, n.v. *Huila Pala* (HAW-5554). Tiahuanaco, 2n=48, n.v. *Huila Pala* (CO-2166).

Province Larecaja: Ccochapata, 3600 m alt., 2n=48, n.v. *Ccallu Huahua* O/S-11809. Yamuco, 3600 m alt., 2n=48, n.v. *Ccallu Huahua* O/S-11798.

Province Los Andes: Cura Pucara, 3860 m alt., n.v. *Pala Pala* (HAW-6184).

Province Murillo: El Alto, n.v. *Wila Pala* (EET-1540). Viacha, near La Paz, 3800 m alt., 2n=48, n.v. *Paloma* (OCH-10522).

Province Omasuyos: Belen, 2n=48, n.v. *Wila Pala* (CIP-702277). Huatajata, n.v. *Wila Pala* (EET-1530). Warisata, n.v. *Wila Pala* (EET-1520).

Department Oruro

Province Abaroa: Challapata, 3770 m alt., 2n=48, n.v. *Pali* or *Puca Pala* (CO-2520, CO-2522). Huancane, n.v. *Wila Pala* (EET-1214). Tacagua, n.v. *Rosada Pali* (EET-1196).

Province Barron: Huancaroma, n.v. *Wila Pala* (EET-1329, EET-1437).

Province Cercado: Caracollo, n.v. *Wila Pala* (EET-1407). Jankoyu, n.v. *Puca Pala* (EET-1322). Oruro, 3706 m alt., local market, 2n=48, n.v. *Pali* (OCH-10488, OCH-10489). Oruro Market, 3700 m alt., 2n=48, n.v. *Puca Pali* (CO-2482).

Department Potosi

Province Alvaro Ibanez: Sacaca, n.v. *Pali* (EET-1634).

Province Chayanta: Macha, n.v. *Wila Pala* (EET-1257, EET-1625). Maragua, 3700 m alt., 2n=48, n.v. *Puca Pali* (OCH-10596).

Province Frias: Azangaro, 3800 m alt., n.v. *Paloma Janeka* (OCH-10502). Azangaro, n.v. *Puca Paloma* (OCH-10503) [both 2n=48]. Challactiri, 3900 m alt., 2n=48, n.v. *Puca Pali* (OCH-10572). Santa Lucia, 3800 m alt., n.v. *Puca*

Pali (OCH-10686). Tinguipaya, 3900 m alt., 2n=48, n.v. *Pali* (OCH-10649). Totora, n.v. *Pali Papa* (HAW-6129). Yoccalla, n.v. *Pala Roja* (EET-1230). [The last four collections 2n=48].

Province Linares: Hornos, 3400 m alt., n.v. *Puca Paloma* (OCH-10665). Pacasi, 3900 m alt., n.v. *Paloma* (OCH-10579). Utacalla, 3750 m alt., n.v. *Pali* (OCH-10636, OCH-10637).

38C. **Solanum tuberosum** subsp. **andigena** var. **longibaccatum** Juzepczuk and Bukasov.

S. andigenum Juz. et Buk. var. *longibaccatum* Juz. et Buk., in Bukasov, The Potatoes of South America and their Breeding Possibilities, pp. 73-74, Fig. 21. 1933, *nom.nud.* Based on *Juzepczuk 1526, 1678*, Bolivia (WIR).

Plants vigorous. Stems decumbent, branched, straight-winged. Leaves well dissected, with (4-)5-6 pairs of leaflets and usually many interjected leaflets or multiple interjected leaflets of the second or third order. Leaflets ovate to elliptic, shortly pubescent, the apex pointed or shortly acuminate and the base obliquely rounded. Berries long-ovate to long-conical.

Specimens Examined

HERBARIUM COLLECTIONS

Department La Paz

Province Murillo: La Paz, n.v. *Coyo*; *Juzepczuk 1526, 1678* (WIR).

38C(a). **Solanum tuberosum** subsp. **andigena** var. **longibaccatum** f. **cevallosii** (Juzepczuk and Bukasov) Ochoa. Fig. 168; Plate XXV-4.

S. andigenum Juz. et Buk. var. *longibaccatum* Juz. et Buk. f. *cevallosii* Juz. et Buk., in Bukasov, The Potatoes of South America and their Breeding Possibilities, p. 74. 1933, *nom.nud.* Based on *Juzepczuk 1659, 1739, 1707-b, 1814*, Bolivia, La Paz (WIR, not seen).

Plants light green. Tubers long straight, subfalcate to strongly falcate or narrowly subcylindrical to narrowly compressed, with yellow periderm, superficial eyes and creamy-white buds, the latter light purple at base, yellow flesh. Leaves with (4-)5-6 pairs of leaflets of the second and third order. Leaflets elliptic-lanceolate to elliptic, borne on long petiolules. Peduncles 8-12 cm long, light green as in the pedicels and calyx. Pedicels 25-28 mm long, the upper part 5-8 mm long and the lower part about 20 mm long. Corolla white, with grayish-white nectar guides, rotate-pentagonal, with narrow well-defined lobes and long acumens. Calyx 7-8 mm long, with narrow lobes and short, pointed acumens. Staminal column cylindrical-conical; anthers 7.5 mm long. Style 12 mm long. Fruit light green, 2-2.5 cm long by 1.6-1.8 cm in diameter at base, long-ovate to long-conical.

Specimens Examined

HERBARIUM COLLECTIONS

Department La Paz

Province Murillo: La Paz Market, 3800 m alt., n.v. *Sackampaya*; *Ochoa 3982* (OCH). La Paz, 3800 m alt., Rodriguez Market, n.v. *Sackampaya*; *Ochoa 3984* (OCH) and *Ochoa 3985* (OCH).

Department Potosi

Province Chayanta: Ravelo, 3500 m alt., n.v. *Sackampaya*; *Ochoa 3953* (OCH).

GERMPLASM COLLECTIONS

Department Cochabamba

Province Arani: Tiraque, n.v. *Sakampaya* (EET-1672).

Department La Paz

Province Aroma: Rio Seco, 2n=48, n.v. *Sacampaya* (HAW-5586). Panduro, n.v. *Sakampaya* (EET-1385). Patacamaya, n.v. *Sakampaya* (EET-1206). Quelkata, n.v. *Sakampaya* (EET-1418).

Province Ingavi: Patarani, 3850 m alt., *Sakampaya* (CO-2172).

Province Los Andes: Cura Pucara, 3860 m alt., near Tambillo, 2n=48, n.v. *Sacampaya* (HAW-6193). Laja, n.v. *Sakampaya* (EET-1238). Pucarani, 3900 m alt., n.v. *Sacampaya* (CO-2224).

Province Murillo: La Paz (?), n.v. *Khati* (HAW-5580). La Paz Market, n.v. *Malcachu* (OCH-10530). Indefinite, n.v. *Sakampaya* (EET-1032).

Province Saavedra: Koata, n.v. *Ruqqui* (CIP-702611).

Department Oruro

Province Abaroa: Challapata, 3770 m alt., 2n=48, n.v. *Sackampaya* (CO-2517). Challapata, 3720 m alt., 2n=48, n.v. *Sacampaya* (OCH-10688).

Province Barrón: Huancaroma, n.v. *Sakampaya* (EET-1327).

Province Carangas: Cala Cala, n.v. *Sacampaya* or *Sakampaya Candelero* (HUA-851, HUA-845) [both collections 2n=48]. Valle Bonito, 3000 m alt., n.v. *Sacampaya* (HUA-831). Calacala, n.v. *Sakampaya* (EET=1616).

Province Cercado: Belen, 3700 m alt., 2n=48, n.v. *Sacampaya* (HUA-805).

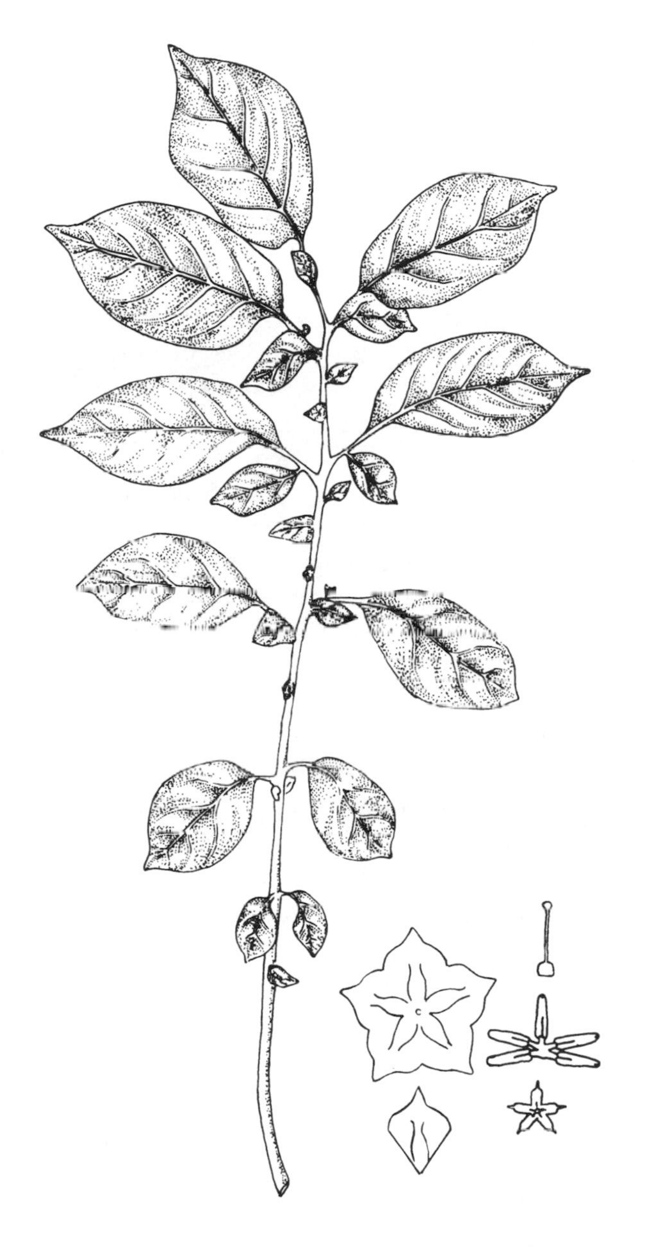

Figure 168. – Leaf and floral dissection of *Solanum tuberosum* subsp. *andigena* var. *longibaccatum* f. *cevallosii* (*Ochoa 3982*), × ½.

Caracollo, n.v. *Sakampaya* (EET-1587). Oruro, 3706 m alt., local market, 2n=4x=48, n.v. *Sackampaya* (OCH-10496).

Department Potosi
Province Chayanta: Ataccollo, 2n=48, n.v. *Juny Papa* (CIP-702530). Cajon Mayo, n.v. *Sacampaya* (HAW-6178). Maragua, 3700 m alt., 2n=48, n.v. *Sacampaya* (OCH-10602). Ravelo, 3100 m alt., 2n=48, n.v. *Yurac Sacampaya* (OCH-10643).

Province Frias: Frias, n.v. *Sackampaya* (EET-1082). Azangaro, 3800 m alt., n.v. *Sacampaya* (HAW-5689). Azangaro, 3800 m alt., n.v. *Sacampaya* (OCH-10505). Cerdas, n.v. *Sacampaya* (CIP-702540). Tinguipaya, n.v. *Juchuy Sacampaya* (CIP-702552). Tinguipaya, 3900 m alt., 2n=48, n.v. *Sacampaya* (OCH-10549). Tinguipaya, 3900 m alt., 2n=48, n.v. *Yurac Sacampaya* (OCH-10548). Near Totora, n.v. *Sacampaya* (HAW-6134). Santa Lucia, 3800 m alt. (OCH-10684A). Manquiri, 2n=48, n.v. *Sacampaya* (OCH-10668).

Province Linares: Alcatuyo, 3800 m alt., n.v. *Malcachu* (OCH-10586). Karachipampa, 4000 m alt., n.v. *Sacampaya* (OCH-10479). Palca, 3750 m alt., n.v. *Sacampaya* (OCH-10629) [all three collections have 2n=48].

Province Quijarro: Condoriri, 3900 m alt., 2n=48, n.v. *Sacampaya* (OCH-10678).

38C(b). Solanum tuberosum subsp. **andigena** var. **longibaccatum** f. **pallidum** (Juzepczuk and Bukasov) Ochoa.

S. andigenum Juz. et Buk. var. *longibaccatum* f. *pallidum* Juz. et Buk., in Bukasov, The Potatoes of South America and their Breeding Possibilities, p.74. 1933, *nom.nud.* Based on *Juzepczuk 1525*, Bolivia. La Paz (WIR, not seen).

Plants vegetatively similar to form *cevallosii*. Tubers with yellow flesh and light smoky-pink periderm and light pink-violet buds, otherwise as in form *cevallosii*. Upper part of pedicel 4 mm long. Calyx 7 mm long, the lobes attenuate or very shortly acuminate. Corolla very light pink-lilac. Anthers approximately 6 mm long. Style 9 mm long.

Specimens Examined

HERBARIUM COLLECTIONS

Department La Paz
Province Ingavi: Huanccollo, 3650 m alt., n.v. *Sackampaya*; *Ochoa 3987* (OCH).

Province Murillo: La Paz, 3650 m alt., Camacho Market, brought from Pucarani, n.v. *Sackampaya*; *Ochoa 3993* (OCH).

GERMPLASM COLLECTIONS

Department La Paz

Province Aroma: Patacamaya, n.v. *Sakampaya Rosada* (EET-1251). Quelkata, n.v. *Wila Sakampaya* (EET-1486).

Province Omasuyos: Huatajata, n.v. *Wila Sakampaya* (EET-1560).

Department Potosi

Province Linares: Puna, 3850 m alt., n.v. *Puca Sackampaya* (OCH-10619).

38C(c). Solanum tuberosum subsp. **andigena** var. **longibaccatum** f. **chojo-sajama** Ochoa f. nov., Phytologia 65(2): 103-113. 1988. FIG. 169.

Stems pigmented dark purple-violet. Tubers with yellow flesh, slender-subcylindric, straight to subfalcate or falcate, with dark purple-violet

Figure 169. – Tubers of *Solanum tuberosum* subsp. *andigena* var. *longibaccatum* f. *chojo-sajama*, ca. ×1.

periderm, deep and semideep eyes and purple buds, similar to form *cevallosii*. Leaves 9-12.5 cm long by 6.5-9 cm wide, less segmented than form *cevallosii*, with 4(-5) pairs of leaflets and 5-8 interjected leaflets. First upper pair of lateral leaflets 5.5-6 cm long by 3-3.5 cm wide, slightly larger than the terminal leaflet. Second upper pair of lateral leaflets 4.5-4.8 cm long by 2.6-3.6 cm wide. Calyx 7-8 mm long, the rectangular lobes abruptly narrowed at apex to short, 1.5 mm-long pointed acumens. Corolla occasionally hexagonal, light violet with white acumens and yellowish-green nectar guides.

Specimens Examined

Department La Paz

Province Omasuyos: Huatajata, 3850 m alt., n.v. *Chojo Sajampa*; *Ochoa 3970* (OCH). Huatajata, n.v. *Lloccalla*; *Ochoa 3971* (OCH, type of f. *chojo-sajama*).

38D. **Solanum tuberosum** subsp. **andigena** var. **aymaranum** (Juzepczuk and Bukasov) Ochoa.

S. andigenum Juz. et Buk. var. *aymaranum* Juz. et Buk., in Bukasov, The Potatoes of South America and their Breeding Possibilities, p. 68. 1933, *nom. nud.* Based on *Juzepczuk 1621*, Bolivia. La Paz (WIR).

Plants vigorous. Stems to 70 cm tall, decumbent, very branched, narrowly winged. Tubers ovate to ovate-compressed with generally pigmented periderm and pink or pink-violet buds. Leaves with 4-5(-6) pairs of leaflets and generally with many interjected leaflets. Leaflets elliptic-lanceolate or broadly elliptic to ovate, borne on short petiolules. Corolla violet, with broadly triangular acumens, the latter shorter and broader than in var. *longibaccatum*. Staminal column ovate-conical; anthers 6-7 mm long, filaments, 1.5-2.5 mm long. Style 10 mm long; stigma capitate, less than 1 mm in diameter, cleft.

Specimens Examined

HERBARIUM COLLECTION

Department La Paz

Province Murillo: La Paz; *Juzepczuk 1621* (WIR).

38D(a). Solanum tuberosum subsp. **andigena** var. **aymaranum** f. **huaca-lajra** (Juzepczuk and Bukasov) Ochoa. Fig. 170.

S. andigenum Juz. et Buk. var. *aymaranum* Juz. et Buk. f. *huaca-lajra* Juz. et Buk., in Bukasov, The Potatoes of South America and their Breeding Possibilities, p. 68, 1933, *nom. nud.* Based on *Juzepczuk 1809-a*, Bolivia. La Paz (WIR, not seen).

Stems strongly pigmented, with narrow, straight wings. Tubers with light beige or yellowish flesh, oval-compressed or long-compressed, the dark red-wine periderm with very dark purple, irregularly shaped areas, with superficial eyes and dark red-wine buds. Leaves 13-18 cm long by 5.5-8 cm wide, dark green. Leaflets 2.8-4.5 cm long by 1.9-3.4 cm wide, broadly elliptic, the apex pointed or shortly acuminate and the base rounded or broadly cordate, borne on 15-20 mm-long petiolules, the latter frequently sustaining 1-3 interjected leaflets. Peduncles 3-4 cm long. Pedicels articulated in the upper one-third of the stalk, the upper part 4-5 mm long and the lower 12 15 mm long. Calyx 8-9 mm long, the lobes abruptly acuminate and terminating in pointed 2-3 mm-long acumens. Corolla 3-3.5 cm in diameter, violet-lilac or lilac. Style 11-12 mm long, exerted 3 mm; stigma thick, capitate. Fruits ovoid, abundant.

Specimens Examined

HERBARIUM COLLECTIONS

Department La Paz

Province Ingavi: Tiahuanaco, 3850 m alt., n.v. *Pala*; *Ochoa 3144* (OCH). Tiahuanaco, n.v. *Pala*; *CPP-1767* (OCH), *CPP-1786* (OCH).

Province Los Andes: Pucarani, 3870 m alt., n.v. *Pala Chocke*; *Ochoa 2794* (OCH).

Province Murillo: La Paz; *Juzepczuk 1809a, 1809c* (WIR). La Paz, 3650 m alt., Camacho Market, n.v. *Huaca Lajra*; *Ochoa 3992* (OCH).

GERMPLASM COLLECTIONS

Department La Paz

Province Aroma: Calamarca, n.v. *Wila Huacalajra* (EET-1178). Lahuachaca, n.v. *Wila Huacalajra* (EET-1166). Patacamaya, n.v. *Wila Huacalajra* (EET-1299). Quelkata, n.v. *Wila Huacalajra* (EET-1400). Vilaque, 3800 m alt., n.v. *Wila Pala* (HUA-880).

Province Camacho: Puerto Acosta, 3880 m alt., 2n=48, n.v. *Huaca Lajra* (CO-2301).

Province Ingavi: Patarani, 3850 m alt., 2n=48, n.v. *Pala* (CO-2178). Sullca Taca, n.v. *Pala* (CIP-702692, CIP-702578).

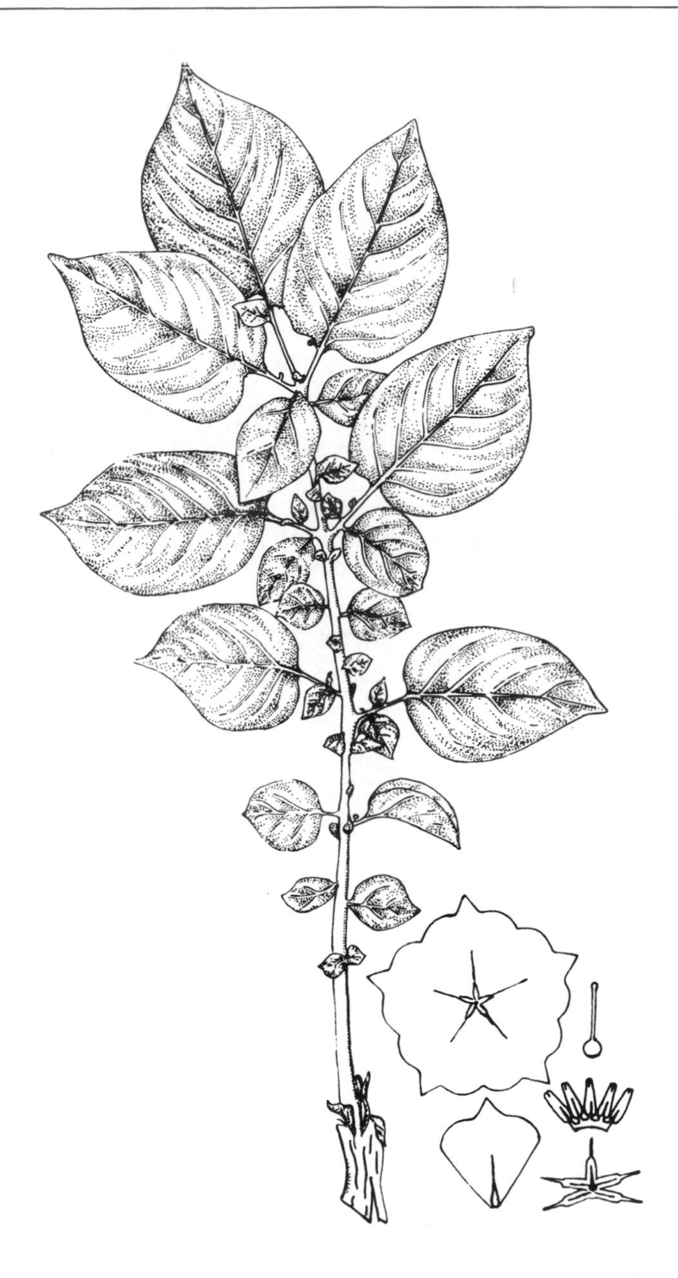

Figure 170. – Leaf and floral dissection of *Solanum tuberosum* subsp. *andigena* var. *aymaranum* f. *huaca-lajra* (*Ochoa 3992*), ca. ×1.

Province Los Andes: Cullacachi, 3850 m alt., 2n=48, n.v. *Huaca Lajra* (CO-2237). Cura Pucara, 3860 m alt., near Tambillo, n.v. *Huaca Lajra* (HAW 6194). Keallani, 3880 m alt., 2n=48, n.v. *Huaca Lajra* (CO-2142).

Province Manco Kapac: Copacabana, n.v. *Huacalajra* (EET-1413).

Province Murillo: La Paz, n.v. *Pala* (HAW-5579). Viacha, near La Paz, 3800 m alt., 2n=48, n.v. *Pala* (OCH-10523). La Paz, 3650 m alt., Buenos Aires Market, 2n=48, n.v. *Pala* (CO-2201).

Province Omasuyos: Huarisata, 3900 m alt., n.v. *Huanuri* or *Pala* (CIP-702318); Huarisata, n.v. *Huacalajra* (EET-1519).

Province Pacajes: Caquiaviri, 3960 m alt., 2n=48, n.v. *Huaca Lajra* (CO-2376). Nazacara, 3860 m alt., 2n=48, n.v. *Huaca Lajra* (CO-2405).

Department Oruro

Province Carangas: Totora, n.v. *Soliman* (CIP-702701).

Province Cercado: Belen, 2n=48, n.v. *Huaca Lajra* (HUA-807).

38D(b). Solanum tuberosum subsp. **andigena** var. **aymaranum** f. **amajaya** Ochoa f. nov., Phytologia 65(2): 103-113. 1988.

Stem more intensely pigmented than in the preceding form. Tubers long, slender, subcylindrical or compressed, densely mottled blue-violet except for the uniformly white medulla and the cortical ring that surrounds the external phloem, with deep or semideep eyes and an obtuse and enlarged apex, and narrow, pointed base.

This form is related to *Solanum tuberosum* subsp. *andigena* var. *aymaranum* f. *huaca-zapato*.

The vernacular name of this form, *Amajaya* or *Amajaa*, was included by Bertonio (1612) in his classic dictionary of the Aymara language. He describes this cultivar as 'one of the best known and most delicious potatoes of the altiplano.'

Specimens Examined

HERBARIUM COLLECTIONS

Department Potosi

Province Frias: Azangaro, 3800 m alt., n.v. *Yana Runa*; *Ochoa 10508* (OCH, type of f. *amajaya*).

GERMPLASM COLLECTIONS

Department La Paz

Province Aroma: Occopampa, 3850 m alt., n.v. *Anahuaya* (CO-2455).
Province Los Andes: Tambillo, 3880 m alt., n.v. *Amajaya* (CO-2159).

Department Potosi

Province Frias: Tinguipaya, 3900 m alt., n.v. *Amajaya* (OCH-10644). Cuchi Chiri, 4200 m alt., n.v. *Amajayo* (OCH-10652).

38D(c). Solanum tuberosum subsp. **andigena** var. **aymaranum** f. **huacazapato** Ochoa f. nov., Phytologia 65(2): 103-113. 1988. PLATE XXV-6.

Tubers with white flesh, long, thick or subcylindrical, with enlarged and obtuse apex and narrowed, pointed base, superficial eyes, light blue-violet to dark purple periderm and buds colored blue-violet at base and apex. Leaf dissection similar to *S. tuberosum* subsp. *andigena* var. *aymaranum*. Leaves 18-25 cm long by 10.5-14 cm wide. Leaflets 4.5-5.7 cm long by 2.8-4 cm wide, broadly elliptic, with a subacute apex.

Specimens Examined

HERBARIUM COLLECTIONS

Department La Paz

Province Los Andes: Huancané, 3900 m alt., n.v. *Amajaya*; *CPP-1817* (OCH). Pucarani, 3900 m alt., n.v. *Papa Runa*; *CPP-1843* (OCH, type of f. *huacazapato*).

Department Oruro

Province Poopó: Urmiri, 3750 m alt., n.v. *Huaca Zapato*; *Ochoa 3947* (OCH).

GERMPLASM COLLECTION

Department Potosi

Province Linares: Hornos, 3400 m alt., 2n=48, n.v. *Yana Runa* (OCH-10666).

38D(d). Solanum tuberosum subsp. **andigena** var. **aymaranum** f. **kunurana** Ochoa f. nov., Phytologia 65(2): 103-113. 1988. FIGS. 171, 172.

Plants robust. Stems 40-50 cm tall. Tubers with white-beige flesh and pinkish medula, isodiametric to slightly compressed, with dark pink-violet skin, deep eyes and dark pink buds. Leaves short and broad, with usually 4 pairs of leaflets and 10-12 interjected leaflets. Leaflets elliptic, the apex shortly acuminate and the base rounded or subcordate, borne on 15 mm-long petiolules, the latter frequently sustaining 1-2 interjected leaflets. Pedicels 15-20 mm long. Pedicels articulated near base of calyx, the upper part 3-4 mm long and the lower 12-16 mm long. Calyx 6-7 mm long, the lobes shortly acuminate. Corolla small, approximately 3 cm in diameter, violet-lilac with white acumens. Anthers 6-6.5 mm long. Style 10 mm long; stigma broadly capitate.

Figure 171. – Tuber of *Solanum tuberosum* subsp. *andigena* var. *aymaranum* f. *kunurana*, ca. ×1.

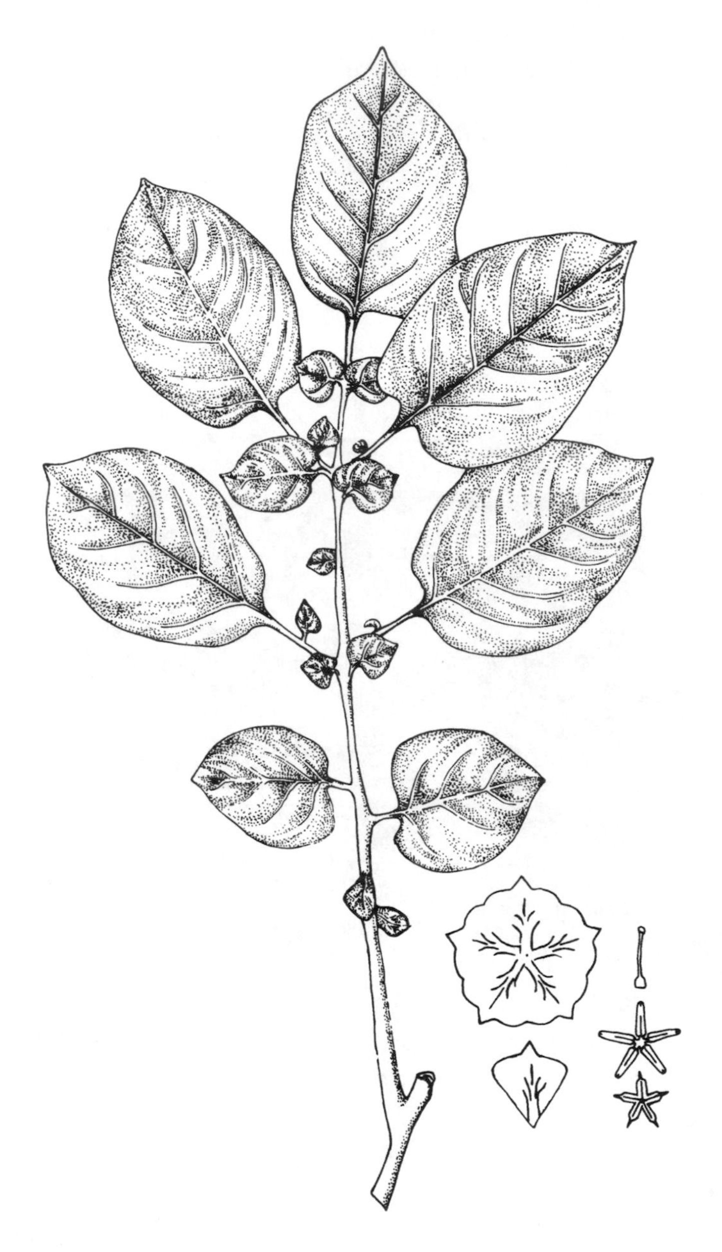

Figure 172. – Leaf and floral dissection of *Solanum tuberosum* subsp. *andigena* var. *aymaranum* f. *kunurana* (*Ochoa 12115*), ×1.

Specimens Examined

HERBARIUM COLLECTIONS

Department Potosi

Province Chayanta: Ravelo, 3500 m, n.v. *Sunsa*; *Ochoa 12115* (OCH, type of f. *kunurana*).

Province Saavedra: Betanzos, 3400 m alt., n.v. *Kunurana*; *Ochoa 12119* (OCH). Lequezana, 3300 m alt., n.v. *Kunurana*; *Ochoa 12112* (OCH).

GERMPLASM COLLECTIONS

Department Oruro

Province Carangas: Calacota, n.v. *Kunurana* (EET-1596).

Province Cercado: Caracollo, n.v. *Kunurana* (EET-1462).

Department Potosi

Province Frias: Lecherias, 3800 m alt., n.v. *Cunarana* (HAW-5687). Tinguipaya, 3900 m alt., 2n=48, n.v. *Kunurana* (OCH-10559).

Province Linares: Pacasi, 3900 m alt., 2n=48, n.v. *Kunurana* (OCH-10580).

Province Saavedra: Cerdas, 3400 m alt., 2n=48, n.v. *Kunurana* (OCH-10560).

38D(e). **Solanum tuberosum** subsp. **andigena** var. **aymaranum** f. **overita** Ochoa f. nov., Phytologia 65(2): 103-113. 1988..

Plants vigorous. Stems erect, narrowly winged. Tubers with light beige flesh, long, subcylindric to strongly compressed and with obtuse ends, periderm bicolored, yellowish-white or dark blue-violet to violet-purple, deep eyes and dark pink-violet buds. Leaves dark green, with 5-6 pairs of leaflets and numerous interjected leaflets of various sizes. Leaflets broadly elliptic-lanceolate, the apex shortly acuminate and the base obliquely rounded, borne on long petiolules, the latter sustaining one or more interjected leaflets. Peduncles up to 15 cm long. Pedicels 25-35 mm long, articulated in the upper one-third of the stalk, the upper part 8-10 mm long and the lower part 17-25 mm long. Corolla lilac with white acumens. Fruit globose.

This form is probably a natural hybrid of *S. tuberosum* subsp. *andigena* var. *aymaranum* and *S. tuberosum* subsp. *andigena* var. *bolivianum*.

Specimens Examined

HERBARIUM COLLECTION

Department La Paz

Province Murillo: La Paz, 3650 m alt., Camacho Market, brought from altiplano; *Ochoa 3991* (OCH, type of f. *overita*).

38D(f). **Solanum tuberosum** subsp. **andigena** var. **aymaranum** f. **surico** Ochoa f. nov., Phytologia 65(2): 103-113. 1988.

Plants robust, of medium stature. Stems 60 cm tall, slightly and irregularly pigmented. Tubers with yellowish-white flesh, densely mottled with dark blue-violet, oblong, the yellowish-white skin with blue-violet stripes about the eyes and apical end. Corolla dark purple.

Although this tetraploid cultivar of *S. tuberosum* subsp. *andigena* is frequently called by its vernacular name of *Surimana*, this form should not be confused with triploid cultivars of *S.* × *chaucha* of the same name.

Specimens Examined

HERBARIUM COLLECTION

Department La Paz

Province Camacho: Puerto Acosta, 3850 m alt., n.v. *Surimana Negra*; *Ochoa 2803* (OCH, type of f. *surico*).

GERMPLASM COLLECTIONS

Department La Paz

Province Larecaja: Tacacoma, 2n=48, n.v. *Surimana* (CIP-702688).

Province Manco Capacc: La Paz (?) [tubers not vermiform], 2n=48, n.v. *Khati* (HAW-5574).

Province Murillo: Tiahuanaco, 2n=48, n.v. *Ajanhuiri* (HAW-5557). Tiahuanaco, n.v. *Khati* or *Huaico* (HAW-5555).

Province Omasuyos: Huarisata, 3900 m alt., n.v. *Surico* or *Surinama* (CIP-702317). Huarisata, n.v. *Surimana* (CIP-702302, CIP-702308).

38D(g). **Solanum tuberosum** subsp. **andigena** var. **aymaranum** f. **tinguipaya** Ochoa f. nov., Phytologia 65(2): 103-113. 1988.

Tuber flesh blue-violet except for the white medula and ring between the external phloem and periderm, round or slightly compressed, with yellowish-white skin and semideep eyes. Leaves 12-15 cm long by 6-11 cm wide, with 4-5 pairs of leaflets and 6-10 interjected leaflets. Leaflets 4-5.5 cm long by 2-3.5 cm wide. Peduncles 6-10 cm long, 5-8-flowered.

This form is one of the many cultivars of hybrid origin that derive from crosses between *S. tuberosum* subsp. *andigena* var. *aymaranum* and *S. tuberosum* subsp. *andigena* var. *chiar-imilla*.

In the potato production zones of Potosi, this form is known as *Yurac Imilla*. In Tinguipaya, however, it has been traditionally known as *Luru*.

The two known collections of this form are identical electrophoretically.

Specimens Examined

HERBARIUM COLLECTIONS

Department Potosi

Province Frias: Tinguipaya, 3900 m alt., n.v. *Luru*; *Ochoa 10648* (OCH, type of f. *tinguipaya*).

Province Saavedra: Betanzos, 3400 m alt., n.v. *Yurac Imilla*; *Ochoa 10662* (OCH).

38D(h). Solanum tuberosum subsp. andigena var. aymaranum f. wila-huaycku Ochoa f. nov., Phytologia 65(2): 103-113. 1988.

Plants vigorous. Stems 60-70 cm tall. Tubers with white flesh, long and thick, subcylindrical or cylindric, somewhat irregular (bulging), with thick, obtuse apices and narrow, slightly pointed bases, light pink periderm, deep eyes and white buds, the latter with purple apices and bases. Leaves dark green, with 4-5 pairs of leaflets and several interjected leaflets. Leaflets elliptic-lanceolate to ovate. Pedicels 17-25 mm long, articulation high, the upper part 5-6 mm long and the lower part 12-15 mm long. Calyx symmetrical, the lobes shortly acuminate. Corolla light violet with white acumens, 2.5-3 cm in diameter.

Sold in La Paz in the Camacho, Buenos Aires, and Lanza de La Paz Markets under the name of *Wila Wuaycku*, or under the name *Wila Phiñu*, which is often used for another potato cultivar.

Specimens Examined

Department La Paz

Province Murillo: La Paz, Camacho Market, brought from Waywasi, Aroma, 2n=48; *Ochoa 3897* (OCH, type of f. *wila-huaycku*).

38D(i). Solanum tuberosum subsp. **andigena** var. **aymaranum** f. **janck'o-kkoyllu** Ochoa f. nov., Phytologia 65(2): 103-113. 1988.

Tubers with light beige flesh, long-compressed, with semideep eyes, the periderm bicolored, yellowish-white with a tinge of smoky violet-purple at the apical end, buds white above, light blue at base. Leaves 10.6-13.5 cm long by 8-9 cm wide. Terminal leaflets slightly larger than laterals, 4.2-4.8 cm long by 2.5-3 cm wide. First and second pairs of lateral leaflets 3.5-4.3 cm long by 1.8-2.5 cm wide.

The Aymara name of this cultigen, *Janck'o-kkoyllu*, signifies 'cloudy or white eye cataract.' This taxon is closely related to *S. tuberosum* subsp. *andigena* var. *aymaranum* f. *huaca-lajra*.

The electrophoretic analysis of the two germplasm samples cited below gave identical results.

Specimens Examined

HERBARIUM COLLECTIONS

Department La Paz

Province Murillo: Huancané, 3900 m alt., n.v. *Jank'o Imilla*; *CPP-1814*, (OCH, type of *jancck'o-kkoyllu*). Huatajata, 3840 m alt., n.v. *Jank'o Imilla*; *CPP-1799, 1800* (OCH). Huatajata Market, 3840 m alt., n.v. *Jank'o Imilla* (CPP-1802, OCH).

Province Omasuyos: Huarisata, 3800 m alt., n.v. *Jank'o Imilla*; *CPP-1866* (OCH).

Department Oruro

Province Carangas: Saucari-Toledo, 3700 m alt., n.v. *Hango Coillo*; *HUA-814* (CIP).

GERMPLASM COLLECTIONS

Department Oruro

Province Carangas: Toledo-Corque, 3700 m alt., n.v. *Hango Imilla* (HUA-817). Toledo, 3700 m alt., n.v. *Hanko Coillo* (HUA-852) [both 2n=48].

38D(j). Solanum tuberosum subsp. **andigena** var. **aymaranum** f. **milagro** Ochoa f. nov., Phytologia 65(2): 103-113. 1988.

Tubers with white flesh, oval-compressed to oblong-subcylindrical, with superficial and semideep eyes and pink buds, periderm yellowish-white with splashes of smoky pink. Corolla white, 3-3.5 cm in diameter.

This cultigen may have been derived from crosses between *S. tuberosum* subsp. *andigena* vars. *chiar-imilla* and *aymaranum*.

Specimens Examined

HERBARIUM COLLECTIONS

Department La Paz

Province Murillo: Huatajata, 3850 m alt., n.v. *Milagro*; *CPP-1807* (OCH, type of f. *milagro*).

GERMPLASM COLLECTIONS

Department La Paz

Province Ingavi: Cachuma, 4000 m alt., 2n=48 (CO-2342).
Province Los Andes: Yana Cachi, 3900 m alt., 2n=48 (CO-2229).
Province Omasuyos: Hualata, 3840 m alt. (CIP-702292).

Department Oruro

Province Cercado: Oruro, 3706 m alt., local market (OCH-10487). Oruro, local market, n.v. *Imilla* (CO-2509). Oruro, local market (CO-2472, CO-2499) [all four of these collections 2n=48].

38D(k). **Solanum tuberosum** subsp. **andigena** var. **aymaranum** f. **huichinkka** Ochoa f. nov., Phytologia 65(2): 103-113. 1988.

Tubers having blue-violet cortex except for the medulla and the ring between the external phloem and periderm, oval-compressed, slightly irregular (bulging), with very dark blue-violet periderm, semideep eyes and dark blue-violet buds.

This cultigen is possibly derived from natural crosses of *S. tuberosum* subsp. *andigena* vars. *aymaranum* and *bolivianum*.

Specimens Examined

HERBARIUM COLLECTION

Department Potosi

Province Frias: Gran Peña, 3800 m alt., n.v. *Huchinkka*; *Ochoa 10671* (OCH, type of f. *huichinkka*).

GERMPLASM COLLECTION

Department Potosi

Province Frias: Azangaro, 3800 m alt., 2n=48, n.v. *Yana Runa*; *Ochoa 10508A* (OCH).

38E. **Solanum tuberosum** subsp. **andigena** var. **lelekkoya** Ochoa var. nov., Phytologia 65(2): 103-113. 1988.

Plants robust. Stems thick, to 20 mm in diameter at base, light green, broadly-straight or sinuously winged. Stolons usually long, to 1.5 m or more in length. Tubers round to oblong or oval-compressed to elongated-compressed with superficial eyes, pure white periderm or with violet-purple stripes at apex, purple buds, white flesh. Leaves, 14-20.5 cm long by 8.5-10.5 cm wide, with 4-5 pairs of leaflets and 5-9 interjected leaflets; leaflets, 4.5-6.5 cm long by 2.5-4 cm wide. Peduncles 8-12 cm long, 6-10 or more flowered. Pedicels articulated in the upper one-third of stalk. Calyx 7-9 mm long, usually symmetrical, light green or slightly pigmented. Corolla rotate to rotate-pentagonal, blue-violet or light blue. Staminal column truncate-conical to subcylindrical-conical; anthers 5.5-6.5 mm long, borne on 1.5-2.5 mm-long filaments. Style 9-10 mm long; stigma 0.8-0.9 mm long, capitate to oval capitate. Fruits abundant, globose, 2-3 cm in diameter.

This variety occurs as a weed in maize fields. It is grown abundantly at altitudes between 2600 and 3200 m in the valleys of La Paz, Sorata, Timusi, Curupampa, and Palca. Although this variety has a number of vernacular names, it is most commonly referred to as *Lelekkoya* or *Semillu*.

All chromosome counts that have been made of this variety at CIP indicate a somatic number of 2n=48.

Specimens Examined

Department La Paz

Province Larecaja: Valle de Tipuani, n.v. *Lilicoya*, March [?], 1851; *Weddell s.n.* (P). Curupampa, 2700 m alt., near Sorata, n.v. *Lelekkoya*; *Ochoa and Salas 11792, 11793* (OCH, type of var. *lelekkoya*). Huallpapampa, 2700 m alt., near Sorata, n.v. *Semillu*; *Ochoa and Salas 11807* (OCH).

Province Murillo: Palca, 3200 m alt., n.v. *Lelekkoya*; *Ochoa 3194* (OCH).

Department Chuquisaca

Province Oropeza: Ckelkamayo, 3100 m alt., n.v. *Alcco Papa*; *Ochoa 11919, 11920* (OCH). Guadalupe, 3100 m alt.; *Ochoa 11921* (OCH).

38E(a). Solanum tuberosum subsp. **andigena** var. **lelekkoya** f. **laram-lelekkoya** Ochoa f. nov., Phytologia 65(2): 103-113. 1988.

Tubers with white flesh, oblong to oval-compressed, with light blue-violet periderm, semideep or superficial eyes and blue-violet buds. Leaves short and broad, with 4 pairs of leaflets and several interjected leaflets of different sizes. Terminal leaflets much larger than the lateral, broadly ovate-lanceolate to suborbicular, the apex pointed to shortly acuminate and the base subcordate. Lateral leaflets broadly elliptic to elliptic-lanceolate, borne on 10 mm-long petiolules, the latter frequently sustaining 2-4 interjected leaflets. Pseudostipular leaves large, broadly falcate, clasping, 12 mm long by 8 mm wide. Corolla dark violet, up to 4 cm in diameter.

This cultigen is related to *S. tuberosum* subsp. *andigena* var. *lelekkoya* f. *chiar-lelekkoya*.

Specimens Examined

Department Chuquisaca

Province Sud Cinti: Higueras, 2600 m alt.; *Ochoa 11976* (OCH, type of f. *laram-lelekkoya*).

Department La Paz

Province Murillo: Some 30 km from La Paz on road to Palca, 3200 m alt., between large rocks, n.v. *Laram Lelekkoya Chocke*; *Ochoa 3995* (OCH).

38E(b). Solanum tuberosum subsp. **andigena** var. **lelekkoya** f. **chiar-lelekkoya** Ochoa f. nov., Phytologia 65(2): 103-113. 1988.

Stems strongly pigmented. Tubers with yellow flesh and dark blue-violet vascular ring, long subcylindrical, with obtuse apex and narrow base, semi-deep eyes, and dark blue-violet (almost black) periderm, the latter dotted with lenticels. Leaves dark green.

The vernacular name of this cultigen is *Semillu*.

Specimens Examined

Department La Paz

Province Larecaja: Curupampa, near Sorata, 2700 m alt., among maize, n.v. *Chiar Lelekkoya* or *Semillu*; *Ochoa and Salas 11795* (OCH, type of f. *chiar-lelekkoya*).

38F. Solanum tuberosum subsp. **andigena** var. **malcachu** Ochoa var. nov., Phytologia 65(2): 103-113. 1988. FIGS. 173, 174.

Stems irregularly lightly pigmented. Tubers with white flesh, long-compressed, lenticulate, more narrowed at apex than base, the periderm light violet or violet-purple or occasionally with transverse yellowish-white stripes and with superficial eyes and white buds, the latter pale bluish-purple or pinkish-purple at apex and base. Leaves short and broad, with 4-5 pairs of leaflets and 5-9 or more interjected leaflets of two or three different sizes. Terminal leaflet larger and broader than the laterals. Lateral and terminal leaflets with pointed or sharply acuminate apices and symmetrical to obliquely rounded bases, borne on 5-8 mm-long petiolules. Corolla violet.

This potato differs from all other native Bolivian varieties of *S. tuberosum* subsp. *andigena* by its unusually colored tubers and distinctive leaf and leaflet shapes.

Specimens Examined

HERBARIUM COLLECTIONS

Department Potosi

Province Chayanta: Ravelo, 3500 m alt.; *Ochoa 12110, 12111* (OCH).

Province Linares: Esquire, n.v. *Tullumalcachu*; *Ochoa 12117* (OCH). Esquire, 2600 m alt12111 (OCH).

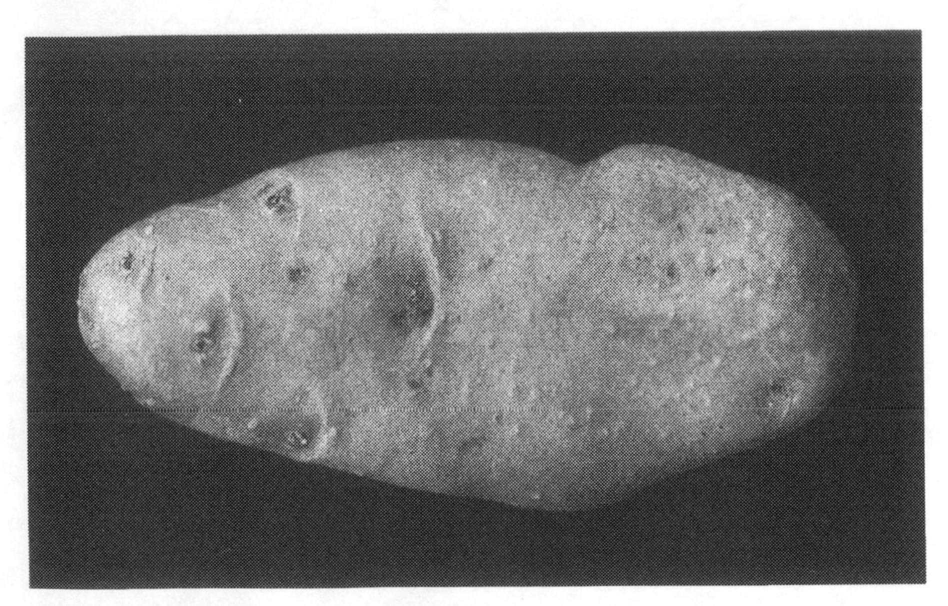

Figure 173. – Tuber of *Solanum tuberosum* subsp. *andigena* var. *malcachu*, ×1.

Province Linares: Esquire, n.v. *Tullumalcachu*; *Ochoa 12117* (OCH). Esquire, 2600 m alt., n.v. *Wiramalcachu*; *Ochoa 12118* (OCH).

Province Saavedra: Lequezana, 3300 m alt., n.v. *Malcacha*; *Ochoa 3959* (OCH, type of var. *malcachu*).

GERMPLASM COLLECTIONS

Department Potosi

Province Frias: Paljacancha, 3600 m alt., 2n=48, n.v. *Malcacho* (HAW-6123). Between Cotagaita and Potosi, 3945 mvince Frias: Paljacancha, 3600 m alt., 2n=48, n.v. *Malcacho* (HAW-6123). Between Cotagaita and Potosi, 3945 m alt., (HAW-6122). Santa Lucia, 3800 m alt., 2n=48, n.v. *Malcachu* (OCH-10684).

Province Linares: Alcatuyo, 3800 m alt., n.v. *Malcachu* (OCH-10587). Alcatuyo, n.v. *Sutu Malcachu* (OCH-10591). Puna, 3850 m alt., n.v. *Malcachu* or *Yurac Malcachu* (OCH-10617, 10625). Utacalla, 3750 m alt., n.v. *Malcachu* (OCH-10635) [the last five collections 2n=48].

Province Nor Chichas: Between Cotagaita and Potosi, 3500 m-alt., n.v. *Sacampaya* (HAW-6117). Ramada, 3250 m alt., between Cotagaita and Potosi, n.v. *Malcacha* (HAW-6110).

Province Saavedra: Betanzos, n.v. *Malcacho* (EET-1224, 1228). Tocsupaya, n.v. *Malcacho* (EET-1078).

Figure 174. – Tuber of *Solanum tuberosum* subsp. *andigena* var. *malcachu*, ×1.

38G. Solanum tuberosum subsp. **andigena** var. **runa** Ochoa var. nov., Phytologia 65(2): 103-113. 1988. FIGS. 175, 176; PLATE XXV-7.

S. andigenum Juz. et Buk. subsp. *runa* Buk. et Lechn., in Bukasov (ed.), Flora of Cult. Pl., Vol IX, Potato, Leningrad, pp. 224-225. 1971, *nom. nud.* Based on *Knappe 0004 and K-5590* (WIR [?], not seen).

Stems light green to lightly pigmented, 70-80 cm tall, branched, somewhat decumbent. Tubers with light beige flesh, oblong to long-compressed, narrowed and obtuse at apex and broader and subquadrate at base, with semideep to deep eyes, light brown or whitish periderm and very light pink eyes and brows, buds light pink at base and apex. Leaves 20-29 cm long by 11-15 cm wide, with 4-5 pairs of leaflets and 6-11 interjected leaflets. Terminal leaflet somewhat longer and wider than the laterals. Upper lateral leaflets 4-6.8 cm long by 2-5.3 cm wide, broadly elliptic, the apex pointed or shortly acuminate and the base rounded, borne on 7-10(-14) mm-long petiolules, the latter

Figure 175. – Tuber of *Solanum tuberosum* subsp. *andigena* var. *runa*, ca. ×1.

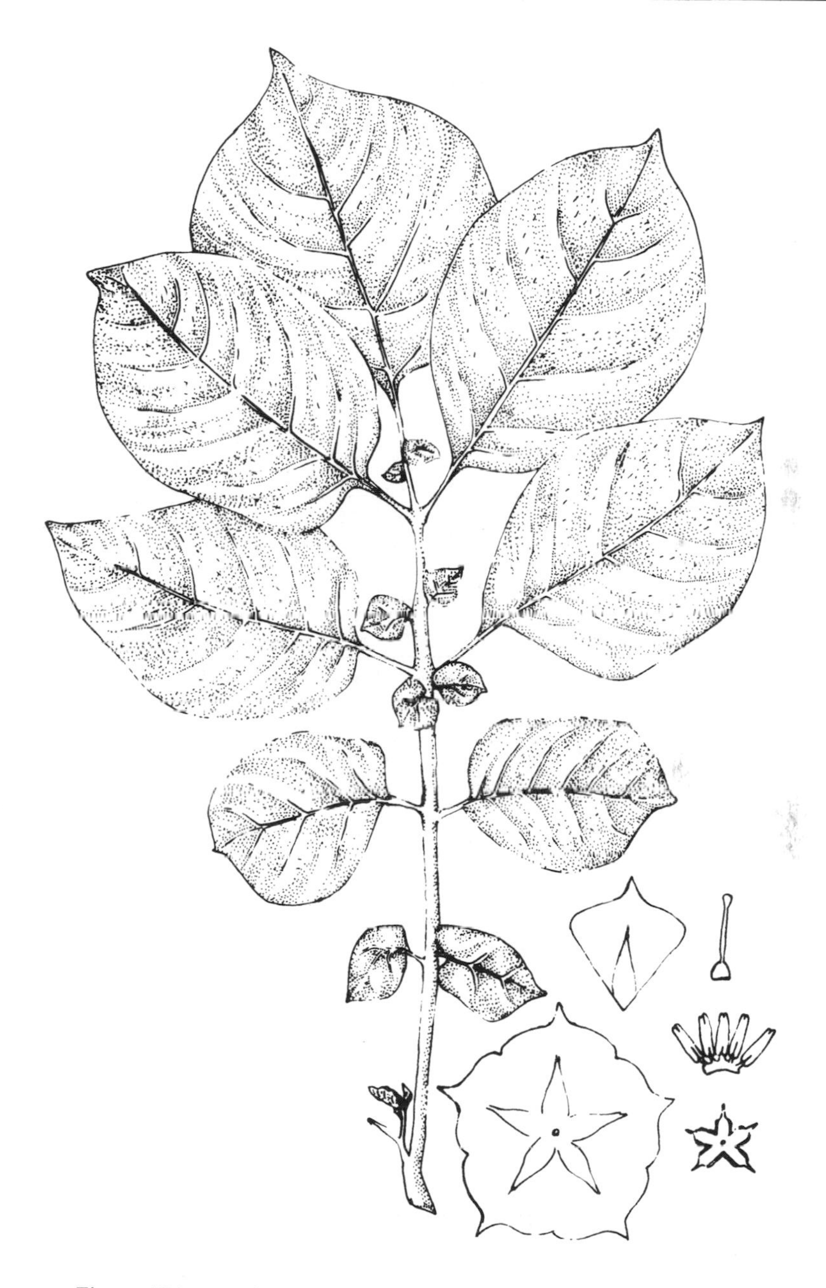

Figure 176. – Leaf and floral dissection of *Solanum tuberosum* subsp. *andigena* var. *runa* (*Ochoa 3954*), ca. ×1.

sustaining 1-2 interjected leaflets. Peduncles 10 cm long, 6-9 flowered. Pedicels 15-20 mm long, articulated in the upper one-third of the stalk, the upper part 4-5 mm long and the lower part 10-15 mm long. Calyx 7-8 mm long, the lobes shortly acuminate. Corolla lilac or violet-lilac with whitish, narrowly triangular-lanceolate acumens. Anthers 6.5-7 mm long. Style 11 mm long; stigma capitate, small.

The variety *runa* is cultivated chiefly between 2700 and 3200 m altitude in Cochabamba and the Province of Valle Grande, Department of Santa Cruz.

Specimens Examined

HERBARIUM COLLECTION

Department Cochabamba

Province Cercado: Cochabamba, La Cancha, local market, brought from Punata, n.v. *Runa Colorada*; *Ochoa 3999* (OCH, type of var. *runa*).

GERMPLASM COLLECTIONS

Department Cochabamba

Province Cercado: Cochabamba, local market, 2700 m alt., 2n=48, n.v. *Huaicku* (CO-2529).

Province Chapare: Colomi, n.v. *Runa* (HAW-5646).

Department Oruro

Province Cercado: Pasto Pampa, n.v. *Runa Blanca* (EET-1619).

Department Potosi

Province Chayanta: Cajón Mayu, Ravelo, n.v. *Runa* (HAW-6177). Cajón Mayu, n.v. *Yurac Runa* (EET-1601). Maragua, 3700 m alt., 2n=48, n.v. *Huaicu* (OCH-10600).

Province Frias: Tinguipaya, 2n=48, n.v. *Vilulaira* (OCH-10650) [this collection and OCH-10600 were also identical electrophoretically].

Province Linares: Puna, 3850 m alt., 2n=48, n.v. *Runa Cachuti* (OCH-10620).

Department Tarija

Province Mendez: Iscayachi, n.v. *Runa* (EET-1250).

38G(a). Solanum tuberosum subsp. **andigena** var. **runa** f. **azul-runa** Ochoa f. nov., Phytologia 65(2): 103-113. 1988. FIG. 176.

Stems strongly to lightly pigmented below center. Tubers with white to light beige flesh, and irregularly blue-violet to uniformly blue-violet periderm, semideep to deep eyes and purple buds.

Specimens Examined

HERBARIUM COLLECTIONS

Department La Paz

Province Larecaja: Achacachi, n.v. *Huaicu*; *Hawkes 5558* (CIP).

Department Oruro

Province Carangas: Saucari-Toledo, 3700 m alt., n.v. *Huanco Sullu*; *Huamán 846* (CIP).

Department Potosí

Province Chayanta: Ravelo, 3500 m alt., n.v. *Runa*; *Ochoa 3960* (OCH).

Province Frias: Challactiri, 3900 m alt., 2n=48, n.v. *Abajeña*; *Ochoa 10566* (OCH). Tinguipaya, n.v. *Yana Alka*; *CIP-702574* (CIP). Above Yocalla, 3450 m alt.; *Hawkes 6139* (CIP).

Province Saavedra: Lequezana, 3300 m alt.; *Ochoa 3955* (OCH). Lequezana, n.v. *Runa Papa*; *Ochoa 3954* (OCH, type of f. *azul-runa*) [both 2n=48].

GERMPLASM COLLECTIONS

Department Cochabamba

Province Chapare: Colomi, n.v. *Runa* (EET-1046, 1047). Colomi, n.v. *Azul Runa* (EET-1038). Melga, n.v. *Azul Runa* (HAW-5640).

Province Cercado: Cochabamba, local market, 2n=48 (CO-2526).

Province Jordán: Cliza, n.v. *Runa* (HAW-5647).

Department La Paz

Province Aroma: Occopampa, 3850 m alt., 2n=48, n.v. *Huancku Sullu* (CO-2468).

Province Ingavi: Tiahuanaco, n.v. *Papa Runa* (EET-1020).

Department Oruro

Province Abaroa: Challapata, 3770 m alt., 2n=48 (CO-2521).

Province Carangas: Saucari-Toledo, 3700 m alt., n.v. *Runa Papa* (HUA-848).

Department Potosi

Province Chayanta: Tocsupaya, n.v. *Collpa Runa* (EET-1075). Macha, n.v. *Runa* (EET-1239). Ravelo, n.v. *Runa* (EET-1060). Atacollo, n.v. *Yana Runa* (CIP-702518).

Province Frias: Lecherias, 3800 m alt., n.v. *Yana Runa* (HAW-5690). Frias, n.v. *Yana Runa* (EET-1083). Carma, 3800 m alt., 2n=48, n.v. *Runa* (OCH-10690). Challactiri, 3900 m alt., 2n=48, n.v. *Runa* (OCH-10570). Near Totora, n.v. *Runa Papa* (HAW-6131). Manquiri, 3800 m alt., 2n=48, n.v. *Yana Runa* (OCH-10669).

Province Linares: Utacalla, 3750 m alt., 2n=48, n.v. *Runa* (OCH-10638).

Province Saavedra: Betanzos, n.v. *Runa* (EET-1249). Cerda, 3900 m alt., n.v. *Runa* (OCH-10564). Lequezana, n.v. *Runa* (EET-1263).

38G(b). **Solanum tuberosum** subsp. **andigena** var. **runa** f. **abajeña** Ochoa.

Stems strongly to lightly pigmented. Tubers with light beige flesh, oval-compressed to oblong-compressed, with dark blue-violet periderm (occasionally colored yellowish-white at base), superficial or semideep eyes and blue-violet buds, the latter with white to creamy-white tips. Leaves short and broad, with 4 pairs of leaflets and 6–10 interjected leaflets.

Electrophoretic analyses of the following six collections gave essentially identical results: OCH-10577, OCH-10595, CIP-702542, OCH-10598, OCH-10653, and OCH-10546.

This cultigen is related to *S. tuberosum* subsp. *andigena* var. *runa* f. *azul-runa*.

Specimens Examined

HERBARIUM COLLECTIONS

Department Potosi

Province Frias: Cerdas, n.v. *Abajeña Papa*; *CIP-702542* (CIP). Tinguipaya, 3900 m alt., n.v. *Yana luki*; *Ochoa 10546* (OCH).

Province Saavedra: Lequezana, 3300 m alt., n.v. *Abajeña*; *Ochoa 3956* (OCH). Lequezana, 3300 m alt., n.v. *Anil Papa*; *Ochoa 3957* (OCH).

GERMPLASM COLLECTIONS

Department La Paz

Province Murillo: La Paz Market, 3650 m alt., brought from altiplano, 2n=48, n.v. *Abajeña* (OCH-10535).

Department Potosi

Province Chayanta: Maragua, 3700 m alt., 2n=48, n.v. *Puca Sallimani* (OCH-10595). Maragua, 2n=48, n.v. *Puca Lurum* (OCH-10598). Maragua, n.v. *Yana Luky* (OCH-10614).

Province Frias: Azangaro, 3800 m alt., n.v. *Abajeña* (OCH-10500). Azangaro, n.v. *Pumeka* (OCH-10504). Challactiri, 3900 m alt., n.v. *Huaiku* (OCH-10577). Gran Peña, 3800 m alt., n.v. *Abajeña* (OCH-10670). Lecherias, 3800 m alt., n.v. *Abajeña* (HAW-5705). Tinguipaya, n.v. *Abajeña Papa* (CIP-702575). Totora, n.v. *Abajeña* (HAW-6132). Tinguipaya, 3900 m alt., 2n=48, n.v. *Yana Abajeña* (OCH-10651). Cuchi Chiri, 4200 m alt., 2n=48, n.v. *Anil Papa* (OCH-10653).

Province Linares: Alcatuyo, 3800 m alt., n.v. *Abajeña* (OCH-10582). Karachipampa, 4000 m alt., n.v. *Abajeña* (OCH-10475). Pacani, 3900 m alt., n.v. *Abajeña* (OCH-10578). Utacalla, 3750 m alt., n.v. *Abajeña* (OCH-10634) [each of these last four collections 2n=48].

38H. **Solanum tuberosum** subsp. **andigena** var. **sicha** Ochoa var. nov., Phytologia 65(2): 103-113. 1988. FIG. 177; PLATE XXIV-6.

Plants vigorous. Stems 70 cm tall, very branched, irregularly dark to lightly pigmented, with broad, straight, light green wings. Tubers with bicolored flesh, dark blue-violet with bright yellow center, round, irregular (or bulging), with black skin, large deep eyes and dark blue-violet buds. Leaves 12-16 cm long by 7-9 cm wide, with 5 pairs of leaflets and 9-14 interjected leaflets. Leaflets 3.5-5.5 cm long by 2.3-3.3 cm wide, diminishing gradually in size toward base, the first upper pair occasionally decurrent on the rachis, sparsely pilose, ovate-lanceolate to elliptic-lanceolate, the apex obtuse and the base oblique, borne on 2-3 mm-long petiolules. Peduncles 10-15 cm long, few-flowered (3-5). Pedicels articulated in the upper one-third of stalk. Calyx 6.5-7 mm long, with very shortly acuminate, subquadrate lobes. Corolla 3 cm in diameter, rotate-pentagonal, dark violet. Staminal column cylindrical-conical, compact, slightly asymmetrical; anthers 6 mm long, yellowish-orange, borne on 1-2 mm-long filaments. Style short, 8.5-9 mm long; stigma capitate. Fruit globose.

Chromosome number 2n=4x=48.

This very rare variety, reputed to have excellent taste and culinary qualities, is high in dry matter and cooks very rapidly. A germplasm sample of this variety was collected for the first time in 1956 near Millipaya, above Sorata by the author (CO-2264). Today this cultigen is extinct from that locality.

Due to the tuber form and color of the periderm, this variety superficially resembles *S. tuberosum* subsp. *andigena* var. *chiar-imilla* f. *nigrum*.

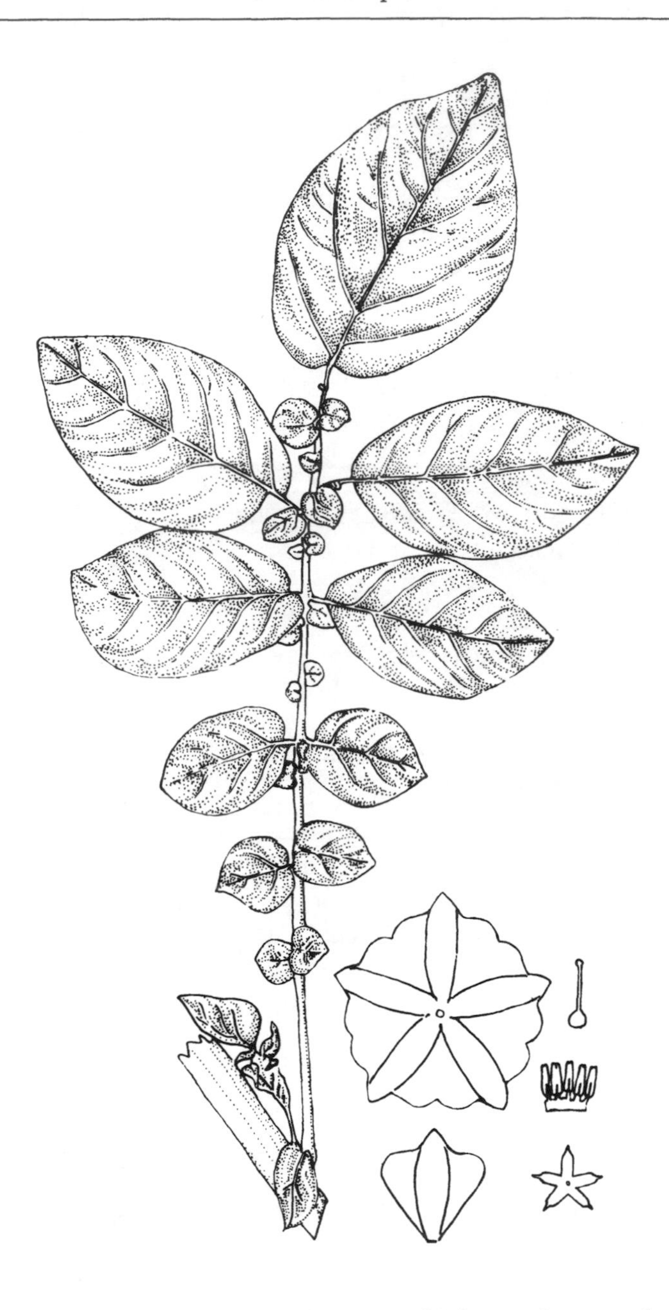

Figure 177. – Leaf and floral dissection of *Solanum tuberosum* subsp. *andigena* var. *sicha* (*Ochoa 3898*), ×1.

Specimens Examined

HERBARIUM COLLECTIONS

Department La Paz

Province Inquisivi: Near Quime, 2700 m alt., n.v. *Sicha*; *Ochoa and Salas 15517* (OCH, type of var. *sicha*).

Province Murillo: La Paz, 3650 m alt., local market, n.v. *Sicha*; *Ochoa 3898* (OCH).

38I. Solanum tuberosum subsp. **andigena** var. **sipancachi** Ochoa var. nov., Phytologia 65(2): 103-113. 1988.

Plants robust. Stems 50-60 cm tall, thick, 15 mm or more in diameter at base, light green, the axils slightly pigmented, with 1.5-2 mm-broad straight wings. Tubers with light beige flesh, oval-compressed, somewhat irregular (bulging), the yellowish-white periderm with violet stripes at base, deep eyes and creamy-white buds, the latter with light violet apices and bases. Leaves 20-25 cm long by 7-10 cm wide, usually with 4 pairs of leaflets and 7-11 interjected leaflets. Leaflets 3.5-5.5 cm long by 2-3 cm wide, diminishing gradually in size toward base, broadly elliptic-lanceolate, the apex shortly acuminate and the base rounded, borne on 3-5 mm-long petiolules. Corolla violet.

Electrophoretic analyses of the two collections, OCH-10589 and OCH-10659, gave identical results.

Specimens Examined

HERBARIUM COLLECTIONS

Department Potosi

Province Frias: Tinguipaya, 3900 m alt., n.v. *Sipancachi*; *Ochoa 10550* (OCH).

Province Linares: Alcatuyo, 3800 m alt., n.v. *Sipancachi*; *Ochoa 10589* (OCH).

Province Saavedra: Cerda, 3900 m alt., n.v. *Sipancachi*; *Ochoa 10561* (OCH). Lequezana, 3300 m alt., n.v. *Sipancachi*; *Ochoa 3958* (OCH, type of var. *sipancachi*).

GERMPLASM COLLECTIONS

Department Chuquisaca

Province Frias: Azangaro, 3800 m alt., 2n=48 (OCH-10506). Yoccalla (EET-1186).

Province Linares: Alcatuyo, 3800 m alt., 2n=48 (OCH-10590).

Province Saavedra: Betanzos (EET-1212, 1234). Betanzos, 3407 m alt., 2n=48 (OCH-10659).
Province Yamparaez: Carabuco (EET-1074).

38J. **Solanum tuberosum** subsp. **andigena** var. **stenophyllum** (Juzepczuk and Bukasov) Ochoa.

FIGS. 178, 179.

S. andigenum Juz. et Buk. var. *stenophyllum* Juz. et Buk., in Bukasov, The Potatoes of South America and their Breeding Possibilities, p. 74. 1933, *nom. nud.* Based on *Juzepczuk 1682-a*, Bolivia. La Paz (WIR).

Plants erect. Tubers with white flesh, oblong or occasionally slightly compressed with pure white periderm or occasionally white periderm with a tinge of smoky purple at apex, superficial to semideep eyes and pink buds. Leaves, 15.5-17.5 cm long by 8.5-10 cm wide with 5-6 pairs of leaflets and 8-12 interjected leaflets. Leaflets narrowly ovate-lanceolate to elliptic-lanceolate, subacuminate or very shortly acuminate at apex, 3-4(-5) cm long by 1.5-2.0 (-2.7) cm wide, terminal leaflets somewhat larger than laterals. Peduncles 8-12 cm long and up to 3.5 mm in diameter at base. Pedicels, 20-25 mm long, articulated in the upper one-quarter of the stalk, the upper part about 5 mm long. Calyx small, 6-7 mm long, with symmetrical or bilabiate lobes, the latter shortly acuminate and terminated by pointed 1.5-2 mm-long acumens. Corolla violet, usually 3 cm in diameter with short, broad lobes. Staminal column cylindrical conical, anthers 6.5-7 mm long. Fruit globose, frequently dark violet at base.

This cultigen shares some characteristics in common with *S. tuberosum* subsp. *andigena* var. *aymaranum*. It differs from that variety, however, in leaf and calyx shape and in having shorter corolla lobes and a cylindrical anther column.

The type collection of *S. tuberosum* subsp. *andigena* var. *stenophyllum*, collected by Juzepczuk (No. 1682-a) in the vicinity of La Paz, is housed in the Leningrad Herbarium (WIR). A specimen in the author's private herbarium (n.v. *K'oyu*; *Ochoa 3894*) compares favorably with Juzepczuk's collection. In Aymara, *K'oyu* means 'bruise caused by a blow.'

Specimens Examined

HERBARIUM COLLECTIONS

Department La Paz

Province Ingavi: Huancollo, 3850 m alt., 2n=48, n.v. *K'oyu*; *Ochoa 3894* (OCH). Viacha, 3840 m alt., n.v. *K'oyu*; *CPP-1954* (OCH).

Province Los Andes: Pucarani, 3900 m alt., 2n=48, n.v. *K'oyu*; *CPP-1822* (OCH).

Province Murillo: La Paz; *Juzepczuk 1682a* (WIR). La Paz, 3650 m alt., Buenos Aires Market, 2n=48, n.v. *K'oyu*; *Ochoa 3896* (OCH). La Paz, Camacho Market, 3650 m alt., 2n=48, n.v. *K'oyu*; *Ochoa 3895* (OCH).

Department Potosi

Province Saavedra: Lequezana, 3300 m alt., 2n=48, n.v. *Pucañahui*; *Ochoa 12114* (OCH).

GERMPLASM COLLECTIONS

Department La Paz

Province Aroma: Calamarca, 3600 m alt., n.v. *Ccoyo* (HUA-873). Lahuachaca, n.v. *Coyu* (EET-1163, 1177). Patacamaya, n.v. *Coyu* (EET-1441). Vilaque, 3800 m alt., n.v. *Ccoyo* (HUA-891). Wayhuasi, n.v. *Coyu* (EET-1357).

Province Ingavi: Azafranal, n.v. *Coyu* (EET-1349). Guaqui, n.v. *Janko Coyu* (EET-1155). Tiahuanaco, n.v. *Coyu* (EET 1326).

Figure 178. – Tuber of *Solanum tuberosum* subsp. *andigena* var. *stenophyllum*, ca. ×1.

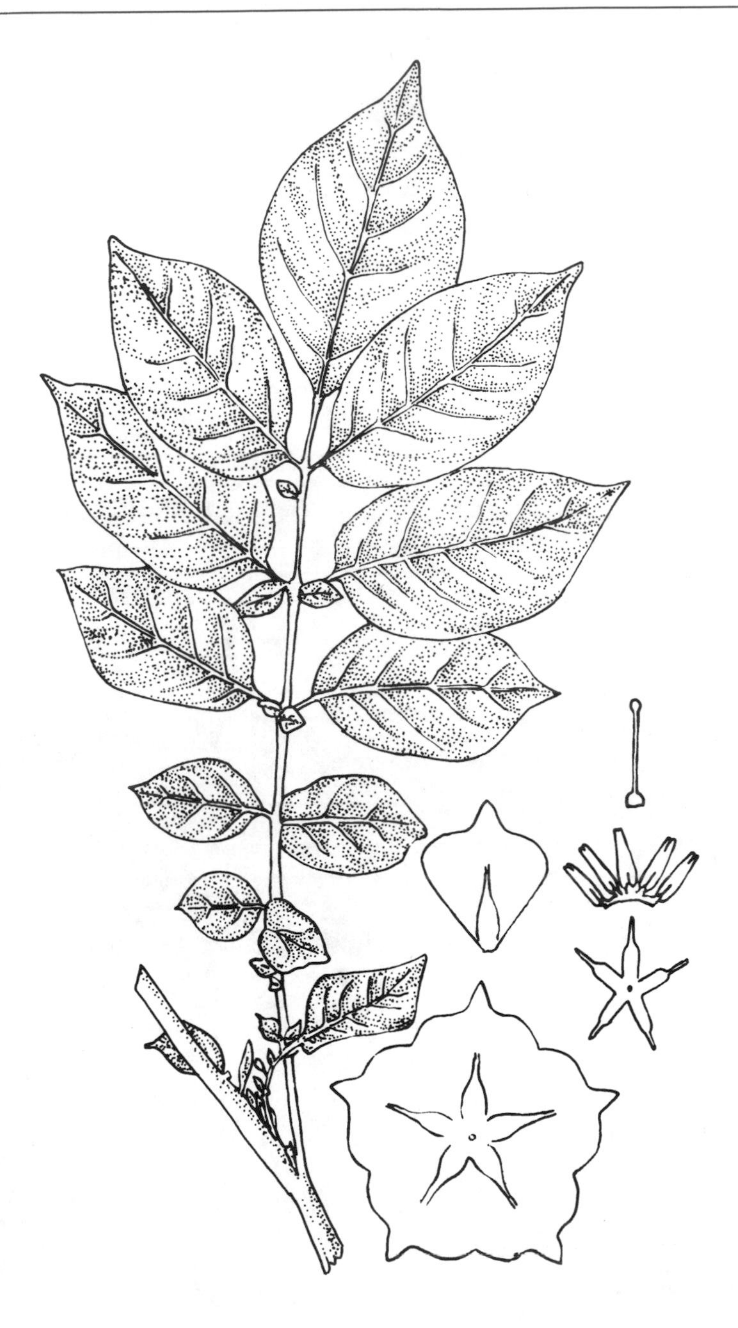

Figure 179. – Leaf and floral dissection of *Solanum tuberosum* subsp. *andigena* var. *stenophyllum* (*Ochoa 3894*), ca. ×1.

Province Los Andes: Cura Pucara, 3860 m alt., 2n=48, n.v. *Coyo Papa* (HAW-6190). Laja, n.v. *Coyu* (EET-1440).
Province Murillo: El Alto, n.v. *Coyu* (EET-1387).
Province Omasuyos: Near Estrecho de Tiquina, 3960 m alt., 2n=48, n.v. *Ccoyu* (CO-2252). Warisata, n.v. *Coyu* (EET-1471).
Province Pacajes: Caquiaviri, 3900 m alt., 2n=48, n.v. *Ccoyu* (HUA-784).

Department Oruro
Province Cercado: Lequepalca, n.v. *Coyu* (EET-1389).

38J(a). Solanum tuberosum subsp. **andigena** var. **stenophyllum** f. **wila-k'oyu** Ochoa f. nov., Phytologia 65(2): 103-113. 1988.

Plants vegetatively similar to the preceding form. Tubers with light beige flesh, oblong to round, with light pink periderm, red semideep to deep eyes, buds dark lilac at base.
The Aymara name of this plant, *Wila-K'oyu*, means 'red bruise.'

Specimens Examined

HERBARIUM COLLECTIONS

Department La Paz
Province Larecaja: Pucará, n.v. *Koyu*; *CIP-702691* (OCH, CIP, type of f. *wila-k'oyu*.

GERMPLASM COLLECTIONS

Department La Paz
Province Murillo: El Alto, n.v. *Wila Coyu* (EET-1573). La Paz, 3840 m alt., near Viacha, 2n=48, n.v. *K'oyu Papa* (OCH-10529).
Province Omasuyos: Warisata, n.v. *Wila Coyu* (EET-1541).

Department Potosi
Province Frias: Atacollo, n.v. *Koyu Papa* (CIP-702531).

38J(b). Solaum tuberosum subsp. **andigena** var. **stenophyllum** f. **pulo** Ochoa f. nov., Phytologia 65(2): 103-113. 1988.

Stems 70 cm or more tall, decumbent, light green. Tubers with white flesh and periderm, oblong to oval-compressed. Leaves very dissected. Leaflets narrow, resembling some forms of *S. stenotomum*. Fruit globose, light green.

Chromosome number: 2n=4x=48.

Electrophoretic analyses of the following four collections gave identical results: CIP-702293, 702310, 702316, and 702704.

Specimens Examined

HERBARIUM COLLECTIONS

Department La Paz

Province Omasuyos: Huarisata, 3900 m alt., n.v. *Pulu*; *CIP-702316* (CIP).

Department Potosi

Province Alonso Ibañes: Sacaca, n.v. *Pole*; CIP-702704 (OCH, type of f. *pulo*).

GERMPLASM COLLECTIONS

Department La Paz

Province Larecaja: Ccochapata, 3600 m alt., 2n=48, n.v. *Jarro Phuco* (O/S-11813). Paccolla, 4150 m alt., 2n=48, n.v. *Pulo* (CO-2256).

Province Los Andes: Cullacachi, 3850 m alt., 2n=48, n.v. *Pulo Monda* (CO-2234).

Province Murillo: Huaillara, 3800 m alt., 2n=48, n.v. *Pulo* (CO-2190).

Province Omasuyos: Hualata, 3850 m alt., n.v. *Pulu* (CIP-702293). Huarisata, 3900 m alt., 2n=48, s.n. (CIP-702310).

38K. **Solanum tuberosum** subsp. **andigena** var. **muru'kewillu** Ochoa var. nov., Phytologia 65(2): 103-113. 1988. PLATE XXII-1.

Plants vigorous. Stem to 70 cm tall, pigmented along the lower two-thirds of its length, the upper part sparsely mottled dark blue-violet. Tubers with white flesh, long, slender, subcylindrical, straight or slightly to strongly falcate, somewhat irregular with deep or semideep eyes and dark blue-violet periderm, white to light purple halos around the eyes and white buds, the latter with blue-violet apices and bases. Leaves with 5(-6) pairs of leaflets and several interjected leaflets. Leaflets elliptic-lanceolate to ovate, shortly acuminate, borne on 2-5 mm-long petiolules. Peduncles 5-8 cm long. Pedicels articulated in the upper one-third of stalk, the upper part 5-6 mm long. Calyx 10-11 mm long, the narrowly rectangular lobes abruptly narrowed at apex to 1.5 mm-long acumens. Corolla dark violet, 3.5 cm in diameter. Anthers 7.0-7.5 mm long, filaments 1.5-2.5 mm long. Style 10-11 mm long, rather thick, densely papillose near base; stigma capitate, about 1 mm in diameter.

Chromosome number: 2n=4x=48.

This variety, cultivated in the Bolivian altiplano, is known by the Aymara name of *Anu Ikiska*, which means 'sleeping dog,' alluding to the shape of the tuber. In southern Peru where this potato variety is more commonly grown, it is known by the native name of *Muru-'Kewillu*, which in Quechua means 'spotted-twisted,' alluding to the shape and color of the tuber.

Specimens Examined

HERBARIUM COLLECTIONS

Department La Paz

Province Murillo: La Paz, 3650 m alt., local market, brought from altiplano, n.v. *Anu-ikisca*; *Ochoa 3893* (OCH, paratype of var. *muru'kewillu*).

GERMPLASM COLLECTION

Department La Paz

Province Murillo: Huaillara, 3800 m alt., 2n=48, n.v. *Acero Chocke* (CO-2194).

38L. Solanum tuberosum subsp. andigena var. taraco Ochoa var, nov., Phytologia 65(2): 103-113. 1988.

Plants vigorous, of medium stature. Stems 40 to 50 cm tall. Stolons short, 5-10 cm long. Tubers with light beige or light yellow flesh, smooth, oval-compressed, with red, or dark red-violet or occasionally variegated periderm, dark pink-violet buds and contrasting narrow halo areas around the superficial or rarely semideep eyes, the latter uniformly dark violet. Leaves 15-20 cm long by 8-9.5 cm wide. Leaflets 5.5-6.5 cm long by 3.6-4.6 cm wide. Peduncle 5-8 cm long, 7-10-flowered. Corolla violet.

Chromosome number 2n=4x=48.

The variety *taraco* is related to *S. tuberosum* subsp. *andigena* var. *bolivianum*. Four forms are cultivated in Bolivia.

Specimens Examined

HERBARIUM COLLECTIONS

Department La Paz

Province Aroma: Lahuachaca, n.v. *Sotamari*; *Hawkes 5584* (CIP).

Province Murillo: Huatajata [not Prov. Aroma], 3900 m alt., n.v. *Wila-taraco*; *Ochoa 2791* (OCH, type of var. *taraco*).

GERMPLASM COLLECTIONS

Department La Paz

Province Aroma: Calamarca, 3800 m alt., 2n=48, n.v. *Yana Pala* (HUA-776).

Province Camacho: Tajani, 3950 m alt., 2n=48, n.v. *Wila Taraco* (CO-2320).

Province Ingavi: Azafranal, n.v. *Taracko* (EET-1344).

Province Lareeaja: Paccolla, 4100 m alt., 2n=48, n.v. *Wila Taraco* (CO-2255).

Province Murillo: Huaillara, 3800 m alt., 2n=48, n.v. *Wila Taraco* (CO-2189).

Province Omasuyos: Sisasani, 4160 m alt., 2n=48, n.v. *Sotamari* (CO-2282). Sisasani, n.v. *Sotamari* (EET-1180). Warisata, n.v. *Taracko* (EET=1481).

Province Pacajes: Caquiaviri, 3960 m alt., 2n=48, n.v. *Sotamari* (CO-2370). Nazacara, 3860 m alt., 2n=48, n.v. *Sotamari* (CO-2406).

Department Oruro

Province Cercado: Oruro, 3700 m alt., local market, 2n=48, n.v. *Yana Pali* (OCH-10494). Wancoyo, n.v. *Morocco Sutamari* (EET-1463).

Department Potosi

Province Frias: Indefinite, n.v. *Sotamari* (EET-1064). Tinguipaya, 3900 m alt., 2n=48, n.v. *Sutamari* (OCH-10552).

38L(a). Solanum tuberosum subsp. andigena var. taraco f. yurac-taraco Ochoa f. nov., Phytologia 65(2): 103-113. 1988.

Tubers with light yellow flesh, oval-compressed, with yellowish-white periderm, superficial eyes and whitish buds, the latter light pink at base.

Specimens Examined

HERBARIUM COLLECTION

Department Potosi

Province Frias: Santa Lucia, 3800 m alt., n.v. *Puca Nahui* or *Yurac Taraco*; *Ochoa 10685* (OCH, type of f. *yurac-taraco*).

GERMPLASM COLLECTIONS

Department Cochabamba

Province Mizque: Mizque-Colpani, n.v. *Puca Nahui* (CIP-702278).

Department Oruro
Province Cercado: Lequelpaca, n.v. *Yurac Sotamari* (EET-1453).

Department Potosi
Province Chayanta: Cajón Mayu, 2n=48, n.v. *Sotamari* (HAW-6182).
Province Linares: Palca, 3750 m alt., 2n=48, n.v. *Puca Nahui* (OCH-10632).

38L(b). Solanum tuberosum subsp. **andigena** var. **taraco** f. **alkka-silla** Ochoa, f. nov., Phytologia 65(2): 103-113. 1988. FIG. 180.

Tubers with light beige flesh, oval-compressed, with superficial to semideep eyes and dark blue-violet buds, the periderm with well-defined, very dark blue-violet areas alternating with yellowish-white, or with the distal one-third

Figure 180. – Tuber of *Solanum tuberosum* subsp. *andigena* var. *taraco* f. *alkka-silla*, ca. ×1.

of the tuber colored yellowish-white and the rest very dark blue-violet with yellowish-white halos.

Chromosome number 2n=4x=48.

Specimens Examined

Department Potosi

Province Chayanta: Ravelo, 3500 m alt., n.v. *Domingo*; *Ochoa 3990* (OCH).

Province Saavedra: Esquiri, 2600 m alt., n.v. *Jalkasillami*; *Ochoa 3989* (OCH, type of f. *alkka-silla*).

38M(a). Solanum tuberosum subsp. andigena var. qhuchi-aca f. qhuchi-chupa Ochoa.

Tubers with white flesh and an internal blue-violet vascular ring, long slender-cylindrical or subcylindrical, strongly irregular (bulging), enlarged and obtuse at apex and slender and more pointed at base, with black periderm, semideep and deep eyes and swollen or enlarged brows, the latter arched and ciliate as in the form of human eyebrows.

Chromosome number: 2n=4x=48.

The variety name of this potato in the Quechuan language means 'pig excrement.' This cultigen, at present, is known only from the Department of Potosi in Bolivia, but it is probably also present in the Bolivian altiplano. It is more widely grown in the Departments of Apurimac, Cusco, and Puno in southern Peru.

Specimens Examined

GERMPLASM COLLECTION

Department Potosi

Province Frias: Azangaro, 3800 m alt., n.v. *Qhuchi-chupa* (OCH-10512).

Natural-Occurring Forms of S. tuberosum subsp. andigena

The author has, on various occasions, collected plants of *S. tuberosum* subsp. *andigena* in Bolivia growing spontaneously in natural vegetation, most frequently in grasslands dominated by *Stipa ichu*. These wild forms appear to be established as a direct result of the dispersal of plant seeds by wind and birds. It is unlikely that these represent hold-overs from fields that had been cultivated the previous year.

The Aymara name for spontaneously occurring forms of *S. tuberosum* subsp. *andigena* is *apharu*, meaning 'wild.'

The following collections of naturally occurring forms of *S. tuberosum* subsp. *andigena* have 2n=4x=48 chromosomes.

Specimens Examined

Department La Paz

Province Omasuyos: Rio Ceckia, 3800 m alt.; *Ochoa and Salas 11819* (OCH). Sorucachi, 3900 m alt.; *Ochoa and Salas 11827* (OCH). Jahuirlaca, 3800 m alt.; *Ochoa and Salas 15477* (OCH). Parahua, 3800 m alt.; *Ochoa and Salas 15478* (OCH).

Exsiccatae

Numerical list of wild taxa

1 *S. acaule*
2 *S. berthaultii*
3 *S. chacoense*
4 *S. flavoviridens*
5 *S. litusinum*
6 *S. tarijense*
7 *S. yungasense*
8 *S. infundibuliforme*
9 *S. circaeifolium*
9a var. *capsicibaccatum*
10 *S. soestii*
11 *S. bombycinum*
12 *S. neovavilovii*
13 *S. violaceimarmoratum*
14 *S. boliviense*
15 *S. megistacrolobum*
15a var. *toralapanum*
16 *S. acharachense*
17 *S. alandiae*
18 *S. brevicaule*
19 *S. candelarianum*
20 *S. candolleanum*
20a *f. sihuanpampinum*
21 *S. doddsii*
22 *S. gandarillasii*
23 *S. leptophyes*
24 *S. microdontum*
24a var. *metriophyllum*
24b var. *montepuncoense*
25 *S. neocardenasii*
26 *S. okadae*
27 *S. oplocense*
28 *S. sparsipilum*
29 *S. sucrense*
30 *S. vidaurrei*
31 *S. vilgultorum*

Exsiccatae of wild taxa

The numbers in parentheses correspond to the numbers of the taxa in the present treatment. See numerical list of taxa.

Alandia, S. and R. Rimpau,
148–150(15); 331(22); 333, 335(15); 344(14)
Alandia, S., R. Rimpau and H. Ross,
180(15)
Anonymous,
WIS-1730(9a)

Asplund, E.,
1914(15); 1204(7); 2044(23); 2048, 2049, 2107(1); 2223, 2413(15); 2436, 2437(1); 2461, 2571(15); 3110(8); 3240, 3303(23); 5832, 5835(1)
Astley, D.,
12(28); 21, 58(27); 92(23)
Balls, E.K.,
6146(28); 6201(1); 6202, 6242(18); 6339(1); 6381(18); 6404(1); 6537(28)
Balls E.K., W.B. Gourlay and J.G. Hawkes,
5890, 5895, 5903(28); 6093(6); B6100(15); 6125(24a); 6126(8); 6127(27); 6147(14); 6149(28); 6169, 6170(29); 6177(23); 6188(14); 6189(28); 6190(14); 6222(28); 6241(15); 6275(13); 6297(2); B6100(15)
Bang, M.,
1100, 1101(18); 2509(9); 2519p.p.(13)
Bridarolli, A.,
4322(1)
Brooke, W.M.A.,
s.n.(15a); 3002(15a); 3003(18); 3004(1); 3005(2); 3006(28); 3008(18); 3009(23); 3010, 3011(18); 3011a(15a); 3013(9a); 3014p.p.(1); 3014p.p.(18); 3015(23); 3016(1); 3017(18); 3019(1); 3020p.p.(18); 3020p.p.(23); 3030(26); 3031, 3032(1); 3033(15a); 3034(28); 3035(1); 3037(9a); 3050(15a); 3051, 3052(9a); 3053(15a); 3063p.p.(1); 6158(9a); 6172(13); 6185(15a); 6380(1); 6428(9); 6716(13)
Brucher, H.,
s.n.(15a); 563(28); 571(15a)
Buchtien, O.,
617(7); 764(13); 771(28); 3982(23); 4693(7); 4696(23); 5547(7); 5858(1); s.n.(13)
Cárdenas, M.,
185(1); 358(15a); 397(15); 606(1); 689(13); 2208(2); 3262, 3502, 3503, 3504, 3505p.p. (15a); 3507, 3508(28); 3510(18); 3513(2); 3515p.p.(6); 3518, 3519, 3520(28); 3600(9a); 3637, 3638(13); 3639(9); 3640(28); 3641(23); 3680, 3681, 3682(15a); 3683(8); 3684, 3685(1); 3686p.p. (18); 3687(31); 3688(16); 3691(15a); 3693(18); 3694(9a); 4501(15); 4503(2); 4513(23); 4515, 4516(15a); 5067(24); 5068(22); 5069(6); 5070, 5071(2); 5072, 5073(3); 5075(30); 5076(3); 5077(6); 5078(27); 5079(17); 5080(19); 5087(24); 5474(1); 5514(9a); 5515(9); 5516(27); 5517(24); 5518(18); 5519, 5520(9a); 6112(15a); 6113(19); 6114 Spec. 1 (18); 6114 Spec. 2 (23); 6115 (9a); 6116(13); 6117(24); 6118(13); 6119(8); 6120(27); 6121(14); 6229(28); s.n-Colomi, s.n-Potosi(1); s.n-PI 243501 (17); s.n ca. La Recoleta-HNF 0080(2); s.n. Liriuni-s.n.-Mollepujru(9a); s.n. Cucho Ingenio(8); s.n. Sucre, s.n. Tarabuco(14); s.n. Mollepujro, s.n. Pte. San Miguel-HNF0011(18); s.n. near Sorata (20); s.n.- Miraflores, s.n-Potosi, s.n-Toralapa-HNF 0087 (15); s.n-Colomi, s.n-ca. Tiraque(15a); s.n-Sivingamayo(29)
Cárdenas, M., R. Rimpau and H. Ross,
500, 504a(9a)
Correll, D.S., E.H. Brücher, K.S. Dodds and G.J. Paxman,
B600, B601(1); B602, B603(15); B605A, B605B (15a); B606(2); B607, B608(18); B609(9a); B610(8); B611(18); B612(28); B613(9a); B614(8); B615(17); B616(21); B617(22); B618A(17); B619(6); B620, B620-1, B620-2(22); B621, B622, B623, B624(2); B625, B626, B627(28); B629(28); B630(2); B632(27); B633, B634(14); B635(28); B636(1); B637, B638(8); B639(15); B642(30); B645 (23); B649(27); B650, B651(8); B652 (15); B653(1); B654(15); B656(18); B657(27); B658(28); B659(27)

Correll, D.S., K.S. Dodds, E.H. Brücher, M. Cárdenas and G.J. Paxman,
B604(28); B605A, B605B(15a)
Corro, A.,
s.n.(7)
Cutler, H.C.,
7424(28); 7690(18)
Cutler, H.C. and M. Cárdenas,
9059(9a)
Cutler, H.C., M. Cárdenas and H. Gandarillas,
7600(8); 7602, 7623(18)
D'Orbigny, A.,
1170(14); 1494(1)
Fiebrig, K.,
2618(15); 3429(1)
Gandarillas, H.,
57(8); 60(9a)
Gandarillas, H., R. Rimpau and H. Ross,
614(9)
Hammarlund, C.,
126(23); 127(15); 128(1); 129(15a)
Hawkes, J.G., I. Aviles and Hoopes,
6727(26)
Hawkes, J.G., J.P. Hjerting and I. Aviles,
6438, 6451, 6453(6); 6468(18); 6496(25); 6510(9a); 6521(19); 6532(9a); 6562, 6572(15a); 6581(9a); 6585(26); 6588(9a); 6589, 6613(15a); 6619, 6621(18); 6622(15a); 6690, 6701(18)
Hawkes, J.G., J.P. Hjerting, P.J. Cribb and Z. Huamán,
4168(7); 4234, 4236(1); 4242(28); 4247(15a); 4251(18); 4270(15a); 4271(1); 4279(2); 4292(24a); 4300, 4304, 4307(6); 4315(27); 4321(8); 4324(1); 4337(27); 4339(23); 4344, 4346(28); 4349(2); 4359(24); 4396(9a); 4399(28); 4408(18); 4580(27)
Hawkes, J.G., A.M. van Harten and J. Landeo,
5703, 5730(18)
Hill, A.M.,
350(15)
Hitchcock, A.S.,
22832(18)
Hjerting, J.P., P.J. Cribb and Z. Huamán,
4317, 4323(15); 4411(2); 4412(28); 4416(1); 4418(28); 4422(2); 4424, 4428(28); 4436, 4474(13); 4486(9a); 4526(24); 4534(21); 4535(17); 4538(6); 4542, 4553(14); 4556(15a); 4563(2); 4567(29); 4578, 4579(27); 4590(15); 4594(29); 4596, 4615, 4636(27); 4667, 4673(6); 4695(15a); 4713, 4719(27); 4748(9a); 5012(9); 5040, 5042(13)
Hjerting, J.P., E. Petersen and J.O. Reche,
1035(1)
Kuntze, O.,
s.n.(1); s.n.(13)
Mandon, G.,
397(20); 398(15); 399(31); 400(9); 401(1)

Ochoa, C.M.,
637(9); 657(2); 659(18); 661, 662(28); 666, 668, 671(2); 672(9a); 674(2); 2074(28); 2076, 2078(15); 2274(20); 2414, 11899-A(1); 11908(13); 11914(15a); 11915(9a); 11916(15a); 11917(26); 11918(15); 11926(29); 11927, 11928(27); 11929, 11930, 11933, 11933-A(14); 11933-B(23); 11934(18); 11935, 11937, 11938(14); 11940(28); 11942(8); 11945(27); 11946(28); 11947, 11948(27); 11952(30); 11954(14); 11960(15a); 11961(1); 11962, 11963, 11964 (15a); 11965(1); 11966(8); 11968(8); 11969, 11970, 11971, 11972(27); 11973, 11974, 11977(8); 11979, 11980(1); 11981(15); 11982, 11983, 11984(1); 11985(15); 11986(1); 11987, 11988(15); 11989, 11990(1); 11991(15); 11992(23); 11993, 11994, 12000, 12001(6); 12003(30); 12004(21); 12005(6); 12006(2); 12007, 12009(6); 12010(22); 12010-A, 12010-B(27); 12011, 12012, 12013, 12014(17); 12014A(27); 12016, 12017, 12018(17), 12019(28); 12025(24b); 12026(3); 12027(5); 12027-A(9a); 12028(28); 12029(2); 12030, 12031(28); 12033(1); 15572, 15576(27); 15584(6); 15901(16); 16150(7)

Ochoa, C.M. and R. Cagigao,
15571, 15573, 15574, 15575, 15577(29); 15578, 15579, 15580(2); 15581, 15582(29); 15583(6)

Ochoa, C.M. and P. Jatala,
12098(15a)

Ochoa, C.M. and A. Salas,
11796(9); 11805(20); 11806(9); 11814(20); 11818(1); 11820(28); 11823, 11824, 11825, 11826, 11829, 11830, 11831, 11832, 11833(1); 11835(20a); 11895(1); 11896, 11897(20); 11900(4); 11901(13); 11909(9); 11910, 11910A(1); 11911(28); 11912(1); 11913(20); 14945, 14946, 14947, 14948, 14949, 14950, 14951, 14952, 14953, 14954(1); 14955. 14958. 14959(20); 14960(1); 14961, 14962(12); 14964(11); 14967, 14968, 14969(12); 14983, 14984, 14986, 14990(20); 14993, 14994(12); 15002, 15004, 15011, 15012, 15013(20); 15455(23); 15460(4); 15463(20); 15465, 15466, 15467, 15468(1); 15469, 15470(20); 15476(1); 15481, 15482, 15483, 15487, 15488(26); 15489(9a); 15497(18); 15498, 15499, 15500, 15501(26); 15502, 15503, 15504, 15505(10); 15506, 15507, 15508(26); 15509(24a); 15518(15a); 15519, 15520(1); 15521(15); 15522(28); 15526, 15527(17); 15528(2); 15529(6); 15534(24); 15539(19); 15540(24); 15542, 15543(19); 15545(24); 15546(31); 15547, 15552(24); 15553(19); 15554(24); 15555, 15556(25); 15557, 15558(22); 15559, 15560(25); 15561(6); 15562(17); 15565(2); 15566(17); 15567(9a); 15568(2); 15569, 15570(1); 15572(28); 15588(22); 15592, 15593, 15596, 15598(6); 15599(2)

Pentland, J.B.,
s.n.(15)

Petersen, E. and J.P. Hjerting.
1054(23)

Petersen, E., J.P. Hjerting and J.O. Reche,
995, 1028(15); 1033(23)

Pflanz, H.,
94(15a); 95(1)

Ross, H.,
201(15); EBS-1795, EBS-1847(14)

Ross, H., S. Alandia and R. Rimpau,
129(15)

Ross, H. and M. Cárdenas,
209(22)

Ross, H., M. Cárdenas, S. Alandia and R. Rimpau,
259, 268(22)
Ross, H. and A. Vidaurre,
195(15)
Shepard, R.S.
231(1); 244(15)
Smith, E.E. and H. Torrico,
1830-H-T(13)
Soest, L.J.M., K.A. Okada and C. Alarcón,
7A, 7E, 15(13); 19 (15a); 27(9); 31, 32(9a); 53, 54(1); 56(18); 65(20)
Steinback, J.,
9488(13)
Ugent, D.,
4557, 4558(15); 4581, 4582, 4582-1c, 4583-2, 4584-1b, 4584-2b, 4585, 4586-1a, 4586-2a(28); 4614, 4615-1, 4615-2, 4616-1, 4616-2, 4644-1, 4644-2, 4645, 4645-1, 4645-2, 4645-3(2); 4652-1, 4652-2, 4652-3, 4652-4, 4653, 4654, 4655(18); 4656(28); 4673, 4675, 4676, 4677, 4679, 4680, 4681, 4682, 4685, 4687, 4688, 4689, 4690, 4691, 4695, 4696, 4697, 4699, 4700(18); 4701, 4704-7, 4709-11, 4712, 4713, 4716, 4719, 4720, 4722, 4725, 4728-29, 4730, 4731-32, 4733, 4734, 4735(9a); 4770, 4771(18); 4772, 4773, 4774(1); 4792-1, 2, 3, 4, 5, 4793, 4795-4597, 4798, 4799, 4800(15a); 4801(15), 4805, 4806, 4808, 4809-1, 2, 3, 4810, 4813, 4817-18, 4819, 4820, 4822, 4826-28, 4829(15a); 4832(18); 4838-39(15); 4840-41, 4842-44, 4845, 4848-49, 4851-52, 4853, 4854-58, 4859-61(15a); 4930-39(28); 4994(2); 5014, 5016, 5017, 5018, 5019, 5022, 5024, 5026, 5028, 5030, 5031, 5032(13); 5070-72, 5077-82, 5086, 5087-89, 5090, 5091-93(15a)
Ugent, D. and M. Cárdenas,
4560, 4561, 4561-1, 4561-2, 4562, 4562-1, 4562-2(2); 4579-1, 4579-2, 4580(28); 4870, 4885, 4886, 4889(6); 4890(2); 4891, 4892, 4893(22); 4896, 4897-4901, 4902-4908(14); 4909(17); 4911(18); 4913, 4914(23); 4927(18); 4928, 4929(14); 4932(23); 4934, 4937(18); 4956-60, 4961-4963, 4964-4967, 4968-4972, 4973-4975(14)
Ugent, D. and A. Vidaurre,
5153, 5154, 5156, 5163-64, 5165, 5166, 5167-68, 5170(1)
Verne, C.,
s.n.(1)
Vidaurre, A.,
s.n.(15)
Weddell, H.A.,
3725(18); s.n.(15a)
West, J.,
8338(24a)
Wight, W.F.,
362(23)

Numerical list of cultivated Taxa

32 *S.* × *ajanhuiri*
32A var. *ajanhuiri*
32Aa f. *janck'o-ajanhuiri*
32B var. *yari*
32Ba f. *janck'o-yari*
33 *S. phureja*
33A var. *phureja*
33Aa f. *viuda*

33B var. *caeruleus*
33C var. *sanguineus*
33Ca f. *puina*
33D var. *erlansonii*
33E var. *flavum*
33Ea f. *sayhuanimayo*
33F var. *janck'o-phureja*
33Fa f. *timusi*
33G var. *macmillanii*
33H var. *quearanum*
33I var. *rubro-rosea*
33Ia f. *orbiculata*
34 *S. stenotomum*
34A var. *stenotomum*
34Aa f. *alkka-phiñu*
34Ab f. *añahuaya-culi*
34Ac f. *chiar-ckati*
34Ad f. *chiar-phiñu*
34Ae f. *pacajes*
34Af f. *pulu-wayk'u*
34B var. *ari-chuwa*
34C var. *ccami*
34D var. *chaquiña*
34E var. *chojllu*
34Ea f. *janck'o-chojllu*
34Eb f. *wila-chojllu*
34F var. *kkamara*
34G var. *luru*
34H var. *pitiquina*
34Ha f. *alkka-pitiquiña*
34Hb f. *azureo-ckati*
34Hc f. *ch'asca*
34Hd f. *laram*
34He f. *phiti-kalla*
34Hf f. *quime*
34I var. *zapallo*
34Ia f. *churipuya*
34Ib f. *kkarachi-pampa*
35 *S.* × *juzepczukii*
35A var. *ckaisalla*
35Aa f. *cko'yu-ckaisalla*
35Ab f. *jank'o-ckaisalla*
35Ac f. *wila-ckaisalla*
35B var. *lucki*
35Ba f. *lucki-pechuma*
35Bb f. *lucki-pinkula*
35E var. *sisu*
35Ea f. *janck'o-sisu*
36 *S.* × *curtilobum*
36A var. *curtilobum*
36Aa f. *china-malko*
37 S. × *chaucha*
37A var. *ckati*
37B var. *ccoe-sullu*
37C var. *kkoyllu*
37D var. *piña*
37Da f. *chulluco*
37E var. *puca-suitu*
37F var. *surimana*
38 *S. tuberosum* subsp. *andigena*
38A var. *chiar-imilla*
38Aa f. *nigrum*
38Ab f. *sani-imilla*
38Ac f. *isla-imilla*
38Ad f. *ccompis*
38Ae f. *janck'o-imilla*
38Af f. *janck'o-chockella*
38Ag f. *alkka-imilla*
38Ah f. *wila-imilla*
38Ai f. *wila-monda*
38B var. *bolivianum*
38Ba f. *chiar-pala*
38Bb f. *janck'o-pala*
38Bc f. *wila-pala*
38C var. *longibaccatum*
38Ca f. *cevallosii*
38Cb f. *pallidum*
38Cc f. *chojo-sajama*
38D var. *aymaranum*
38Da f. *huaca-lajra*
38Db f. *amajaya*
38Dc f. *huaca-zapato*
38Dd f. *kunurana*
38De f. *overita*
38Df f. *surico*
38Dg f. *tinguipaya*
38Dh f. *wila-huayku*
38Di f. *janck'o-kkoyllu*
38Dj f. *milagro*
38Dk f. *huichinkka*
38E var. *lelekkoya*
38Ea f. *laram-lelekkoya*
38Eb f. *chiar-lelekkoya*
38F var. *malcachu*
38G var. *runa*
38Ga f. *azul-runa*
38Gb f. *abajeña*
38H var. *sicha*

38I var. *sipancachi*
38J var. *stenophyllum*
38Ja f. *wila-k'oyo*
38Jb f. *pulo*
38K var. *muru-kewillu*
38L var. *taraco*
38La f. *yurac-taraco*
38Lb f. *alkka-silla*

Exsiccatae of cultivated taxa

The numbers in parentheses correspond to the numbers of the taxa in the present treatment. See numeral list of taxa.

Balls, E.K.,
6394(32Aa)
CIP,
700891(34H); 701283(35A); 702265(38Ad); 702267(38Ab); 702287, 702288(33Aa); 702313(38Ae); 702316(38Jb); 702436(36); 702522(37B); 702525(38Ag); 702538(38Ai); 702542(38Gb); 702574(38Ga); 702601(34); 702616(38Af); 702631, 702635(35Ba); 702691(38Ja); 702696(38Ah); 702704(38Jb); 702802(32A)
Cutler, H.C.,
7425(36Aa)
Cutler, H.C., M. Cárdenas and H. Gandarillas,
7630(33F)
Erlanson, C.O.,
K-451(33D)
Erlanson, C.O. and H.G. MacMillan,
454(32A)
Hawkes, J.G.
5558(38Ga); 5584(38L); 5593(34); 5600(38Ag); 5663(38Ai); 5667(38Ab); 5688, 5693(38Ad); 5696(38Ag); 5697(38Ad); 6139(38Ga); 6170(38Ad); 6180(38A)
Huamán, Z.,
781(34); 788, 789(35B); 809(35Ab); 814(38Di); 815(35Ac); 821(35Aa); 833(34Ad); 846(38Ga); 854(38Ab); 893(38Ad)
Juzepczuk, S.V.,
826(38Ad); 1518(32); 1526(38C); 1556(37); 1556-b(34); 1562, 1582(38A); 1621(38D); 1637(35A); 1640(38A); 1641(35); 1642(36); 1643(32); 1644(37); 1645(34C); 1647(35); 1654, 1655(33); 1656(38B); 1658(37F); 1660-B(38Aa); 1661(32); 1661-B(32Aa); 1663(38Ad); 1678(38C); 1678 B(35); 1682 A(38J); 1685(33); 1707(36); 1710(37); 1710-b(34); 1744(32); 1801(33Aa); 1807(38Ad); 1809 a, 1809 c(38Da); 1810, 1813(33Aa); 1815, 1815-B(33)
MacMillan, H.G.,
S.N.(33G)
MacMillan, H.G. and C.O. Erlanson,
429(35A)
Ochoa, C.M., U.N.A. Coleccion Peruana de Papa (=CPP-),
1720(38Ah); 1767, 1786(38Da); 1799, 1800(38Di); 1801(38Ab); 1802(38Di); 1807(38Dj); 1810(38Ah); 1814(38Di); 1815(34H); 1817(38Dc); 1819(38Ac); 1822(38J); 1834(38Ab); 1843(38Dc); 1866(38Di); 1877, 1888(38Ah); 1889(38Ba); 1898, 1901(38A); 1910(38Ab); 1932(38Ag); 1954(38J); 2009(38Ah); 2021, 2021-A(38Aa); 2075(38A); 2096(38Ab); 2117(38A)

Ochoa, C.M.,
2397(34Ad); 2790(34I); 2791(38L); 2793(34Ha); 2794(38Da); 2795(34I); 2796(38Ab); 2797(34H); 2798(38Ac); 2799(34); 2803 (38Df); 2806(35); 2807(34); 2808(35); 3112(32); 3140(34H); 3142(34); 3144(38Da); 3145(34I); 3146(33); 3148(34Ha); 3149(34Ad); 3150(34I); 3152(34Ia); 3154(34Aa); 3155(33); 3194(38E); 3529(34I); 3534(34); 3535(34H); 3536(33Aa); 3537(34); 3551(34Ae); 3552(34Eb); 3553(34B); 3564, 3755(34H); 3927(34Hd); 3881(32Aa); 3882, 3883, 3884(34); 3885(32); 3886(32Aa); 3887, 3888(32B); 3889, 3890(32); 3893(38K); 3894, 3895, 3896(38J); 3897(38Dh); 3898(38H); 3899(38Ac); 3900(38Ae); 3901, 3902(38Ag); 3903, 3904(38Aa); 3906(38Ac); 3908(38A); 3909(37A); 3910(33Fa); 3911(34Aa); 3912(34E); 3913(33E); 3914(34Ad); 3915(33E); 3918(34Ad); 3919(34H); 3920(34F); 3921, 3922(35E); 3923(35Ea); 3924(34Ad); 3925(34Ac); 3926(34Hd); 3928(34He); 3929(37A); 3930(34He); 3931(36); 3933(38A); 3934(36Aa); 3935(38A); 3936(36Aa); 3937(34B); 3938(34Ea); 3939, 3940(36); 3941(38Ae); 3942(34Eb); 3944(32Ba); 3945(35); 3946(37F); 3947(38Dc); 3948(37C); 3949(34G); 3950(37B); 3951(38Bb); 3952(38Bc); 3953(38Ca); 3954, 3955(38Ga); 3956, 3957(38Gb); 3958(38I); 3959(38F); 3960(38Ga); 3961(38Bc); 3962(38A); 3963(38Aa); 3964(38Ae); 3965, 3966(38Ab); 3967, 3968(38Ai); 3969(38Ac); 3970, 3971(38Cc); 3972, 3973(38Ab); 3974(38Ai); 3975, 3976(38Ac); 3977, 3978(38Ba); 3979, 3980(38Bc); 3981(38Ba); 3982(38Ca); 3983(38Bc); 3984, 3985(38Ca); 3986(38Bc); 3987(38Cb); 3989(38Lb); 3991(38De); 3992(38Da); 3993(38Cb); 3995(38Ea); 3996(38Bb); 3997(38Bc); 3999(38G); 10481(34Ib); 10483(34Ad); 10484(34H); 10493(34Af); 10507(38Aa); 10508(38Db); 10520(34Af); 10525(34); 10528(33Aa); 10537(34Ab); 10541(37B); 10546(38Gb); 10550(38I): 10558(38Ag); 10561(38I); 10566(38Ga); 10571(35Bb); 10589(38I); 10603(37E); 10608(34Hc); 10612(38Aa); 10622(37E); 10626(38Ag); 10628(32B); 10641(38Aa); 10646(37B); 10648(38Dg); 10657(38Aa); 10661(38Ag); 10662(38Dg); 10671(38Dk); 10675(36Aa); 10677(34B); 10680(32B); 10685(38La); 11919, 11920, 11921(38E); 11976(38Ea); 12110, 12111(38F); 12112(38Dd); 12114(38J); 12115(38Dd); 12117, 12118(38F); 12119(38Dd); 12130(37D); s.n.(38Bb)

Ochoa, C.M. and A. Salas,
S.N.(33Ca); 11787, 11788(33Aa); 11792, 11793(38E); 11795(38Eb); 11797, 11801(34D); 11807(38E); 11815, 11816(33F); 11838, 11840(33Aa); 14970(33E); 14971(33I); 14972(33E); 14973(33B); 14974(33I); 14975(33I); 14976(33H); 14999(33I); 15000(33E); 15001(33Ea); 15007(33I); 15009, 15010(33Ia); 15458(33F); 15459, 15471(33Aa); 15472(33F); 15473(33); 15513(34Hf); 15514(34Hb); 15515(33C); 15517(38H)

Schick, R.,
K-480(32)

Vavilov, N.I.,
K-100(33); K-97(32Aa); K-99(33)

Weddall, H.A.,
S.N. (38E);

Zhukowski, P.M.,
K-3372(32)

Appendix A

LIST OF ABBREVIATIONS

The following section lists the herbaria and germplasm stations cited in the text, as well as the collectors and cultivated potato species mentioned in Appendices B and C.

Herbaria

BAL	Buenos Aires, Argentina: Estación Experimental de Balcarce.
BM	London, Great Britain: British Museum (Natural History), Cromwell Road, London SW7 5BD.
C	Copenhagen, Denmark: Botanical Museum and Herbarium, Gothersgade 130, DK-1128 Copenhagen.
CA	Cochabamba, Bolivia: Herbarium Cardenasianum, in Herbario Facultad de Ciencias Agrícolas, Universidad de Cochabamba.
CIP	Lima, Peru: International Potato Center, P.O. Box 5969, Lima.
CPC	Edinburgh, Great Britain: Commonwealth Potato Collection, formerly at Potato Genetics Station (P.G.S.), Cambridge, now at Scottish Plant Breeding Station, Edinburgh.
CUZ	Cuzco, Peru (*see* VAR).
E	Edinburgh, Great Britain: Royal Botanic Garden, Edinburgh EH3 5LR.
F	Chicago, Illinois, U.S.A.: John G. Searle Herbarium, Field Museum of Natural History, Roosevelt Road at Lake Shore Drive, Chicago, IL 60605.
G	Geneva, Switzerland: Conservatoire et Jardin Botaniques, C.P., CH-1211, Genève 21.
GH	Cambridge, Massachusetts, U.S.A.: Gray Herbarium of Harvard University, 22 Divinity Avenue, Cambridge, MA 02138.

GM	Gottwaldov, Czechoslovakia: Oblastní Muzeum Jihovýchodní Moravy, Lesná u Gottwaldova P. Stípa, Gottwaldov.
GOET	Göttingen, Federal Republic of Germany: Systematisch-Geobotanisches Institut, Universität Göttingen, Unt. Karspüle 2, D 34 Göttingen.
HBG	Hamburg, Federal Republic of Germany: Staatsinstitut für allgemeine Botanik, Jungiusstrasse 6, Hamburg 36.
HNF	Cochabamba, Bolivia: Herbario Nacional Forestal, Jardín Botánico de Cochabamba.
JGH	Birmingham, Great Britain: Personal herbarium of J. G. Hawkes at Birmingham University.
K	Kew, Great Britain: The Herbarium and Library, Royal Botanic Gardens, Kew, Richmond, Surrey TW9 3AE.
LD	Lund, Sweden: Botanical Museum, Ö Vallgatan 19, S-223 61 Lund.
LE	Leningrad, U.S.S.R.: Herbarium of the Department of Higher Plants, V.L. Komarov Botanical Institute of the Academy of Sciences of the U.S.S.R., 197022, Prof. Popov Street 2, Leningrad P-22.
LL	Renner, Texas, U.S.A.: Lundell Herbarium, Texas Research Foundation, Renner, TX 75079, housed at University of Texas in Austin.
LP	La Plata, Buenos Aires, Argentina: Facultad de Ciencias Naturales y Museo, División Plantas Vasculares, Universidad Nacional de la Plata.
M	Munich, Federal Republic of Germany: Botanische Staatssammlung, Menzingerstrasse 67, D-8 München 19.
MAX	Cologne-Vogelsang, Federal Republic of Germany: Max-Planck-Institut.
MO	St. Louis, Missouri, U.S.A.: Missouri Botanical Garden, 2315 Tower Grove Avenue, St. Louis, MO 63110.
MPU	Montpellier, France: Institut de Botanique, Université de Montpellier, 5 Rue Auguste Broussonet, Montpellier.
NA	Washington, D.C., U.S.A.: Herbarium, United States National Arboretum, D.C. 20002.
NY	New York, New York, U.S.A.: Herbarium, New York Botanical Garden, Bronx Park, Bronx, NY 10458.
O	Oslo, Norway: Botanisk Museum, Trondheimsvn, 23 B, Oslo 5.
OCH	Lima, Peru: Private herbarium of C. Ochoa, Pasteur 1491, Zona-14, Lima.
P	Paris, France: Muséum National d'Histoire Naturelle, Laboratoire de Phanérogamie, 16 Rue Buffon, 75005 Paris.

PH	Philadelphia, Pennsylvania, U.S.A.: Department of Botany, Academy of Natural Sciences, 19th and Parkway, Philadelphia, PA 19103.
PI	Pisa, Italy: Istituto Botánico dell'Universita, Via Luca Ghini 5, 56100 Pisa.
S	Stockholm, Sweden: Section for Botany, Swedish Museum of Natural History (Naturhistoriska riksmuseet), S-104 05 Stockholm 50.
SI	San Isidro, Argentina: Instituto de Botánica Darwinion, San Isidro, Prov. Buenos Aires.
U	Utrecht, Netherlands: Institute for Systematic Botany, Tweede Transitorium, Heidelberglaan, 2, de Uithof, Utrecht.
UC	Berkeley, California, U.S.A.: University of California Herbarium, Department of Botany, University of California, Berkeley, CA 94720.
UPS	Uppsala, Sweden: Institute of Systematic Botany, University of Uppsala, P.O. Box 541, S-751 04 21 Uppsala.
US	Washington, D.C., U.S.A.: U.S. National Herbarium, Department of Botany, Smithsonian Institution, D.C. 20560.
VAR	Cuzco, Peru: Private Herbarium of C. Vargas, P.O. Box 79, Cuzco (*see* CUZ).
W	Vienna, Austria: Naturhistoriches Museum, Burgring 7, A-1014 Wien.
WIR	Leningrad, U.S.S.R.: All-Union Institute of Plant Industry Herbarium, Herzen Street 44, 190000 Leningrad.
WIS	Madison, Wisconsin, U.S.A.: University of Wisconsin Herbarium, 245 Birge Hall, 430 Lincoln Drive, Madison, WI 53706.
WU	Vienna, Austria: Botanisches Institut und Botanischer Garten der Universität Wien, Rennweg 14, A-1030 Wien.
Z	Zurich, Switzerland: Botanischer Garten und Institut für Systematische Botanik der Universität Zürich, 40 Pelikanstrasse, 8001 Zürich (after 1975: Zollikerstr. 105, 8008 Zürich).

Germplasm Stations

CIP	Lima, Peru: International Potato Center, P.O. Box 5969, Lima.
CPP	Lima, Peru: Peruvian Potato Collection, National Agrarian University (UNA)-La Molina, P.O. Box 456, Lima.
EET	Cochabamba, Bolivia: Toralapa Experimental Station, Toralapa.
SB-PI	Sturgeon Bay, Wisconsin, U.S.A.: Potato Introduction Station, Sturgeon Bay, WI 54235.

Collectors (Chapter 4)

BAL-1	E. K. Balls
CO	C. Ochoa (used only in reference to 1958 Lake Titicaca collections)
CUT	H. C. Cutler
ERL	C. O. Erlanson
E/M	Erlanson and MacMillan
HAW	J. G. Hawkes, A. M. van Harten, and J. Landeo
HUA	Z. Huamán
JUZ	S. V. Juzepczuk
MLL	H. G. MacMillan
M/E	MacMillan and Erlanson
OCH	C. M. Ochoa
O/S	C. M. Ochoa and A. Salas
SCH	R. Schick
VAV	N. I. Vavilov
WED	H. A. Weddell
ZHU	P. M. Zhukowski

Cultivated Species

ADG	*Solanum tuberosum* subsp. *andigena*
AJN	*Solanum* × *ajanhuiri*
CHA	*Solanum* × *chaucha*
CUR	*Solanum* × *curtilobum*
GON	*Solanum goniocalyx*
JUZ	*Solanum* × *juzepczukii*
PHU	*Solanum phureja*
STN	*Solanum stenotomum*
TUB	*Solanum tuberosum* subsp. *tuberosum*

Departments (Appendix C)

CH	Chuquisaca
CO	Cochabamba
LP	La Paz
OR	Oruro
PO	Potosi
TA	Tarija

Appendix B

SUMMARY OF GERMPLASM COLLECTIONS PRESENTED IN CHAPTER IV

Collector/ institution	Cultivated Species							
	ADG	AJN	CUR	CHA	JUZ	PHU	STN	Total
BAL-1	–	1	–	–	–	–	–	1
CIP	44	39	7	2	10	2	4	108
CO	70	1	–	9	4	–	61	145
CPP	30	–	–	–	–	–	1	31
CUT	–	–	1	–	–	1	–	2
EET	151	8	–	3	–	–	–	162
ERL	–	–	–	–	–	1	–	1
E/M	–	1	–	–	–	–	–	1
HAW	75	2	1	5	–	–	9	92
HUA	33	18	2	4	23	–	10	90
JUZ	15	6	2	3	4	8	4	42
MLL	–	–	–	–	–	1	–	1
M/E	–	–	–	–	1	–	–	1
OCH	198	15	6	19	7	7	60	312
O/S	19	–	–	–	–	26	4	49
SCH	–	1	–	–	–	–	–	1
VAV	–	1	–	–	–	2	–	3
WED	1	–	–	–	–	–	–	1
ZHU	–	–	–	–	1	–	–	1

Appendix C

ADDITIONAL GERMPLASM COLLECTIONS

The following list gives the taxonomic determinations by species as identified by the author; the collectors (*see* Appendix A for list of abbreviations) and accession numbers; chromosome numbers; native names; and the localities, altitudes, and departments for the collections maintained by the International Potato Center (CIP), Lima Peru; the Peruvian Potato Collection (CPP), National Agrarian University-La Molina, Lima, Peru; and the Toralapa Experimental Station, Cochabamba, Bolivia (EET or BOL). The departments referred to are Cochabamba (CO), Chuquisaca (CH), La Paz (LP), Oruro (OR), Potosi (PO), and Tarija (TA).

This is followed by a summary (Tables 1 and 2) of the more than 1100 collections of Bolivian cultivated potatoes included in this Appendix.

Solanum × *ajanhuiri*

Collection	2n	Native Name	Locality	Alt.	Dept.
CIP 702652			Condoriri		OR
CIP 702653					
CIP 702663			Condoriri		OR
CIP 702665					LP
CIP 702670					
CIP 702550		Acahuiri	Tinguipaya		PO
CO 2143	24	Ajahuiri	Kcallani	3880	LP
CO 2169	24	Ajahuiri	Tiahuanaco	3880	LP
CO 2182	24	Ajahuiri	Patarani	3850	LP
CO 2227	24	Ajahuiri	Pucarani	3900	LP

Solanum × ajanhuiri (cont.)

Collection	2n	Native Name	Locality	Alt.	Dept.
CO 2303	24	Ajahuiri	Unahuaya	3960	LP
CO 2340	24	Ajahuiri	Cachuma	4000	LP
CO 2346	24	Ajahuiri	Chama	4100	LP
CO 2427		Ajahuiri	S.A. Machaca	3980	LP
CO 2436	24	Ajahuiri	Maso Cruz	4020	LP
CO 2462		Ajahuiri	Occopampa	3850	LP
CO 2493	24	Ajahuiri	Mdo. Oruro	3700	OR
CPP 1749	24	Ajahuiri	Tiahuanaco	3880	LP
CPP 1794		Ajahuiri	Mdo. La Paz	3600	LP
CPP 1809	24	Ajahuiri	Huatajata	3600	LP
CPP 1982	24	Ajahuiri	Corocoro	3890	LP
CPP 1920	24	Ajahuiri Blanca	Pto. Acosta	3840	LP
CPP 323		Ajahuiri Blanco			LP
CPP 2080	24	Ajahuiri Blanco	J.D. Machaca	3890	LP
CPP 536		Ajahuiri Negro	Panduro		LP
CIP 702321		Ajanhuiri	Caquiaviri	4000	LP
CIP 702322	24	Ajanhuiri	Caquiaviri	4000	LP
CIP 702660		Ajanhuiri	Aypa Paruyo		LP
OCH 12135	24	Ajanhuiri	Huanccollo	3650	LP
CPP 2017	24	Ajanjuiri	Caquiaviri	3960	LP
CIP 702669		Ajanwiri	Kaata		LP
CIP 702675		Ajanwiri Blanca	Calacoto		LP
CPP 1930	24	Ajawiri	Escoma	3880	LP
CPP 1978	24	Ajawiri	Corocoro	3890	LP
CPP 2051	24	Ajawiri	S.A. Machaca	3980	LP
CPP 532		Ajhuiri	Panduro		LP
CPP 322	24	Ajhuiri Colorado			LP
OCH 10593	24	Amaya	Maragua	3700	PO
HUA 1039		Ancho Yari	Caquiaviri	3900	LP
CIP 702649		Chanu Yari	Escara		OR
EET 0003		Chiar Ajahuiri	Ocke		OR
EET 1325		Chiar Katy	El Alto		LP
CO 2196	24	Ckati	Huaillara	3800	LP
CPP 2026	24	Hanq'o Ajahuiri	Caquiaviri	3960	LP
CO 2390	24	Jancko Ajahuiri	Caquiaviri	3960	LP
CO 2207	24	Jancko Ckati	Mdo. La Paz	3600	LP
CO 2149	24	Janko Ajahuiri	Kcallani	3880	LP
CIP 702799		Laran Jari	Caracollo		OR
CIP 702801		Orko Ajanhuiri	Lacoyo		LP
HUA 824		Pinta Yari	V. Bonito	3800	OR
CO 2344	24	Sisu Yari	Chama	4100	LP
HUA 1040		Wila Yari	Caquiaviri	3900	LP
CO 2492	24	Yari	Mdo. Oruro	3700	OR
CPP 1960	24	Yari	Q. Qhella	3880	LP
CPP 1961	24	Yuro	Q. Qhella	3880	LP

Solanum phureja

Collection	2n	Native Name	Locality	Alt.	Dept.
CIP 702609	24		Mdo. Rodrig.	3600	LP
O/S 15516	24		Quime	2900	LP
CO 2267	24	Alcca Phureja	Millipaya	3550	LP
CIP 702268		Alkka Phureja	Chilcapampa	3600	LP
CPP 1933		Ccoyo Negra	Pallapi	3850	LP
CIP 702269		Phureja	Sorata-Acha.	3600	LP
CIP 702289		Phureja	Mdo. Rodrig.	3600	LP
CIP 702290		Phureja	Mdo. Rodrig.	3800	LP
CIP 702291		Phureja	Mdo. Rodrig.	3600	LP
CIP 702319		Phureja	La Paz	3800	LP
CO 2261	24	Phureja	Millipaya	3550	LP
CO 2272	24	Phureja	Ckotanani	3700	LP
CO 2273	24	Phureja	Tacacoma	3610	LP
CPP 1787		Phureja	Mdo. La Paz	3600	LP
CPP 1789		Phureja	Mdo. La Paz	3600	LP
CPP 1931		Phureja	Ullachapi	3880	LP
OCH 12138	24	Phureja	Mdo. Rodrig.	3800	LP
O/S 11803	24	Phureja	Ccochapata	3200	LP
O/S 11836	24	Phureja	Catalone	3000	LP
O/S 11902	24	Phureja	Charasani	3200	LP
O/S 11903	24	Phureja	Charasani	3200	LP
O/S 11904		Phureja	Charasani	3200	LP
O/S 11905		Phureja	Charasani	3200	LP
O/S 11906	24	Phureja	Charasani	3200	LP
O/S 11907		Phureja	Charasani	3200	LP
O/S 11810		Phureja negra	Ccochapata	3200	LP
OCH 10524	24	Pitiquilla	La Paz	3800	LP
OCH 10524-A		Pitiquilla	La Paz	3800	LP
OCH 10524-B		Pitiquilla	La Paz	3800	LP
O/S 14971-A	24	Puca Chaucha	Queara	3400	LP
OCH 10544		Phureja	Tinguipaya	3900	PO

Solanum stenotomum

Collection	2n	Native Name	Locality	Alt.	Dept.
CIP 702580			Condoriri		OR
CIP 702610			Patacamaya		LP
CIP 702797			Janco Cala	4000	CO
HAW 6111			Ramada	3250	PO
OCH 10654-A	24		Cuchi Chiri	4200	PO
CO 2502	24	Achac Chuima	Mdo. Oruro	3700	OR
CO 2487	24	Achacana	Mdo. Oruro	3700	OR
OCH 12136	24	Alcca Checcke	Huanccollo	3850	LP
CO 2369		Alcca Imilla	Caquiaviri	3960	LP
CO 2466	24	Alcca Imilla	Occopampa	3850	LP
CO 2490	24	Allca Imilla	Mdo. Oruro	3700	OR
CO 2393	24	Alcca Monda	Caquiaviri	3960	LP
CO 2481	24	Alcca Zapallo	Mdo. Oruro	3700	OR
HUA 884		Alcka Pinu	Vilaque	3800	LP
OCH 10498	24	Alka Imilla	Oruro	3706	OR
OCH 10681	24	Alka Koyllu	Santa Lucia	3800	PO
CIP 702533		Alka Luru	Tinguipaya		PO
EET 1570		Alka Pitikalla	El Alto		LP
CPP 2057	24	Alpaca Chuchuli	S.A. Machaca	3980	LP
CPP 2039	24	Alq'a Phyno	S.A. Machaca	3980	LP
CPP 1988	24	Alqa	Corocoro	3890	LP
CPP 1967	24	Alqa Kuli	Q. Quqlla	3880	LP
CPP 1991	24	Alqa Phino	Corocoro	3890	LP
CPP 2062	24	Alqa Phinu	Nasacara	3860	LP
EET 1160		Anahuaya	Conani		LP
CPP 2038	24	Anaway Phyno	S.A. Machaca	3980	LP
CPP 1926		Arcquero	Pto. Acosta	3840	LP
CO 2507	24	Ari Chua	Mdo. Oruro	3700	OR
OCH 10645		Ataculi	Tinguipaya	3900	PO
EET 1122		Canastilla			
OCH 10538	24	Canastilla	Tinguipaya	3900	PO
OCH 10539	24	Canastilla	Tinguipaya	3900	PO
CIP 702520		Canastilla Papa	Atacollo		PO
EET 1386		Chiar Pituhuay.	Quelkata		LP
EET 1465		Chiar Pituhuay.	Huatajata		LP
CPP 1827		Chiara Pinola	Pucarani	3830	LP
CPP 2123	24	Chije Pitiquina	Taraco	3840	LP
CPP 1705		Chiquina Roja	Pituta	3880	LP
OCH 10674	24	Ch. Puca Nahui	Condoriri	3900	PO
CPP 2015	24	Choclo	Caquiaviri	3960	LP
CPP 2024	24	Choclo	Caquiaviri	3960	LP
HUA 899		Chojllo	Vilaque	3800	LP
CO 2351	24	Chojllu	Chama	4100	LP
CO 2453	24	Chojllu	Occopampa	3850	LP

Solanum stenotomum (cont.)

Collection	2n	Native Name	Locality	Alt.	Dept.
CO 2454	24	Chojllu	Occopampa	3850	LP
CO 2477	24	Chojllu	Mdo. Oruro	3700	OR
EET 1243		Chojllu	Calamarca		LP
HAW 5585		Chojllu	Lahuachaca		LP
HAW 5598		Chojllu	Lahuachaca		LP
CPP 1860		Churipuya	Hualata Ch.	3850	LP
CO 2500	24	Chuso Nahui	Mdo. Oruro	3700	OR
CO 2321	24	Ckellu Puya	Tajani	3960	LP
CO 2329	24	Ckellu Puya	Italaque	3590	LP
CO 2484	24	Cole	Mdo. Oruro	3700	OR
CO 2527	24	Cuchi Aca	Mdo. Cochab.	2800	CO
CIP 702592		Cunurana	Caquiaviri		LP
CO 2432	24	Cunurana	Ballivian	4030	LP
CPP 1972	24	Duraznillo	Corocoro	3890	LP
CPP 2056	24	Duraznillo	S.A. Machaca	3980	LP
EET 1297		Duraznillo	Caracollo		OR
CPP 2018		Durazno	Caquiaviri	3960	LP
HUA 783		Durazno	Caquiaviri	3900	LP
CIP 702547		Espanol Papa	Tinguipaya		PO
OCH 10556		Espanola	Tinguipaya	3900	PO
HUA 862		Hanko Pitucaya	Sica Sica	3800	LP
CO 2483	24	Huaicku	Mdo. Oruro	3700	OR
CO 2504	24	Huaicku	Mdo. Oruro	3700	OR
CO 2382	24	Huaman Ppecke	Caquiaviri	3960	LP
OCH 10484-A		Huayki Papena	Oruro	3706	OR
CO 2306	24	Huila Ajahuiri	Unahuaya	3960	LP
CIP 702588		Huila Ajanhuiri	Llimpi		LP
HUA 822		I. Wila Layrani	Toledo-Corq.	3799	OR
CO 2334	24	Jancko Monda	Cachuma	4000	LP
CO 2355	24	Jancko Phinu	Chama	4100	LP
CO 2368	24	Jancko Phinu	Caquiaviri	3960	LP
CIP 702515		Janka Papa	Tinguipaya		PO
HUA 871		Janko Pinu	Calamarca	3600	LP
HUA 890		Janko Pinu	Vilaque	3800	LP
HAW 6179		Kachun Huakhac.	Cajon Mayu		PO
OCH 10654	24	Katari Papa	Cuchi Chiri	4200	PO
CO 2151	24	Kcati Chocke	Kcallani	3880	LP
CIP 702623		Keta o Katu	Pomani		LP
EET 1100		Khachum Huacac.			PO
CPP 1999		Khuchi Nasa	E. Yarihuaya	3880	LP
HAW 6164		Koillu	Ravelo		PO
HAW 6165		Koillu	Ravelo		PO
HAW 5644		Koyllu	Tiraque		CO
HAW 5698		Koyllu	Lecherias	3800	PO

Solanum stenotomum (cont.)

Collection	2n	Native Name	Locality	Alt.	Dept.
HAW 5642		Koyllu Imilla	Tiraque	5642	CO
CPP 1959	24	Kuli	Q. Qhella	3880	LP
CPP 1998	24	Kuli	E. Yarihuaya	3880	LP
HAW 6169		Kununara	Huchun. Pamp.		PO
CPP 527	24	Kunurana	Mayutambo		PO
EET 1062		Kunurana			
OCH 10557	24	Laphia	Tinguipaya	3900	PO
OCH 10609		Laphia	Maragua	3700	PO
EET 1584		Laram Pali	Obacuyo		
CPP 2046	24	Maman Peke	S.A. Machaca	3980	LP
CPP 2035	24	Maman Peque	S.A. Machaca	3980	LP
CO 2422	24	Maman Ppecke	S.A. Machaca	3980	LP
HUA 829		Maman Tegue	V. Bonito	3800	OR
OCH 10630	24	Manzana	Palca	3750	PO
CPP 519		Membrilla	Chinoli		PO
CIP 702624		Morock Luqui	Pomani		LP
OCH 10532	24	Muruka	La Paz	3800	LP
CPP 1963		Orureno	Q. Qhella	3880	LP
CO 2425	24	Pacco Ckahua	S.A. Machaca	3980	LP
CO 2495	24	Pacco Culi	Mdo. Oruro	3700	OR
EET 1024		Pala			LP
EET 1029		Pala			LP
CPP 2022	24	Paqo Khawa	Caquiaviri	3960	LP
CPP 1957		Paqoniway	Q. Qhella	3880	LP
CPP 1964		Paqoniway	Q. Qhella	3880	LP
CPP 1993	24	Phina Nasari	Corocoro	3890	LP
CPP 1990	24	Phino Blanca	Corocoro	3890	LP
CIP 702596		Phinu	Ayoayo		LP
CO 2209	24	Phinu	Mdo. La Paz	3600	LP
CPP 2000	24	Phinu	E. Yarihuaya	3880	LP
CPP 1775	24	Phyno	Tiahuanaco	3880	LP
CPP 2029	24	Phyno	Caquiaviri	3960	LP
CPP 2043	24	Phyno	S.A. Machaca	3980	LP
CPP 2030	24	Phyno Rojo	Caquiaviri	3960	LP
CPP 1985		Phynu	Corocoro	3890	LP
CPP 1906	24	Pichuya Negra	Pto. Acosta	3840	LP
CPP 2010	24	Pina	Caquiaviri	3960	LP
CPP 1739	24	Pino	Tiahuanaco	3880	LP
EET 1025		Pinu Negra	Viacha		LP
CIP 702541		Pinu Papa	La Paz		LP
CPP 1837	24	Piticalla	Pucarani	3830	LP
CPP 1845	24	Piticalla	Pucarani	3830	LP
CPP 2007	24	Pitikina	Caquiaviri	3960	LP
CPP 1780	24	Pitiquilla	Tiahuanaco	3880	LP

Solanum stenotomum (cont.)

Collection	2n	Native Name	Locality	Alt.	Dept.
CPP 1784	24	Pitiquilla	Tiahuanaco	3880	LP
CPP 1842	24	Pitiquilla	Pucarani	3830	LP
CPP 1857	24	Pitiquilla	Huarina	3830	LP
CPP 1947	24	Pitiq. Negra	Pallapi	3850	LP
CO 2373	24	Potockeri	Caquiaviri	3960	LP
CO 2361	24	Ppotockeri	Aipa	4000	LP
OCH 10597-A		Puca Luru	Maragua	3700	PO
OCH 10477	24	Puca Manzana	Karachip.	4000	PO
OCH 10478	24	Pulu Papa	Karachip.	4000	PO
CPP 2048	24	Qami	S.A. Machaca	3980	LP
CPP 2036		Quello Imilla	S.A. Machaca	3980	LP
CPP 1795	24	Quene	Mdo. La Paz	3800	LP
CO 2419	24	Rosca	S.A. Machaca	3980	LP
CPP 2053	24	Rosca	S.A. Machaca	3980	LP
CPP 1777	24	Runtuya	Tiahuanaco	3880	LP
CPP 1759		s/n	Tiahuanaco	3880	LP
OCH 10518-A		s/n	Mdo. La Paz	3800	LP
OCH 12128	24	s/n	Mdo. La Paz	3800	LP
OCH 12129	24	s/n	Mdo. La Paz	3800	LP
O/S 14977	24	s/n	Queara	3400	LP
CIP 702583		Sallama	Cebadillas		PO
OCH 10601	24	Sapallu	Maragua	3700	PO
CO 2339	24	Senorita	Cachuma	4000	LP
CIP 702556		Sofia Papa	Tinguipaya		PO
OCH 10480	24	Solama	Karachip.	4000	PO
EET 1320		Surimana	El Alto		LP
EET 1360		Surimana	Desaguadero		LP
HAW 5561		Surimana	Mdo. La Paz		LP
OCH 10521	24	Suso Koyllo	La Paz	3800	LP
CO 2451	24	Tami	Calamarca	3900	LP
HUA 892		Titihuichinca	Vilaque	3800	LP
HUA 847		Titikhoma	Toledo	3700	OR
CO 2185	24	Tolcca	Patarani	3850	LP
CO 2283	24	Tolcca Ckocka	Sisasani	4160	LP
CPP 2005		Wacharallallawa	E. Yarihuaya	3880	LP
CO 2333	24	Wancko Phinu	Cachuma	4000	LP
CPP 2102		Wawachara	Taraco	3840	LP
HUA 1038		Wila Ajanhuiri	Caquiaviri	3900	LP
HUA 868		Wila Kunurana	Calamarca	3600	LP
HUA 894		Wila Kunurana	Vilaque	3800	LP
EET 1405		Wila Pajra	Ca. Copacab.		
CPP 1977	24	Wila Phiño	Corocoro	3890	LP
CPP 1994	24	Wila Phiño	Corocoro	3890	LP
CPP 2034	24	Wila Phiño	S.A. Machaca	3960	LP

Solanum stenotomum (cont.)

Collection	2n	Native Name	Locality	Alt.	Dept.
EET 1518		Wila Phiñu	El Alto		LP
EET 1391		Wila Pitikalla	El Alto		LP
EET 1458		W. Pituhuayaca	Panduro		LP
CPP 1908	24	Wila Puya	Pto. Acosta	3840	LP
CPP 1918	24	Wila Puya	Pto. Acosta	3840	LP
CIP 702700		Wila Tonkoraya	Totora		OR
OCH 10679	24	Yalka Coillu	Condoriri	3900	PO
CIP 702586		Yana Koyllu	Chimpapata		PO
CPP 1992	24	Yuramana	Corocoro	3890	LP
CIP 702602		Zapallo			
CIP 702603		Zapallo			
CIP 702604		Zapallo	Mdo. La Paz	3800	LP
CIP 702605		Zapallo	Patacamaya		LP
CIP 702606		Zapallo	Patacamaya		LP
CO 2208	24	Zapallo	Mdo. La Paz	3600	LP
CO 2297	24	Zapallo	Pto. Acosta	3880	LP
CO 2325		Zapallo	Soracko	4150	LP
CO 2326	24	Zapallo	Soracko	4150	LP
CO 2354	24	Zapallo	Chama	4100	LP
CO 2417	24	Zapallo	S.A. Machaca	3980	LP
CO 2467	24	Zapallo	Occopampa	3850	LP
CO 2491	24	Zapallo	Mdo. Oruro	3700	OR
CPP 1762	24	Zapallo	Tiahuanaco	3880	LP
CPP 1899	24	Zapallo	Pto. Acosta	3840	LP
CPP 1969	24	Zapallo	Q. Quqlla	3880	LP
CPP 2052	24	Zapallo	S.A. Machaca	3980	LP
EET 1543		Zapallo	Waiwasi		LP
HAW 5553		Zapallo	Pacota		LP
HUA 867		Zapallo	Calamarca	3600	LP
OCH 1995	24	Zapallo	Patamanta	3900	LP

Solanum × *juzepczukii*

Collection	2n	Native Name	Locality	Alt.	Dept.
CIP 702618			Achacachi	4150	LP
CIP 702619					
CIP 702796					
CIP 702798			Belen		LP
CIP 702662		Ajanwiri	Lirko		LP
CPP 1880	36	Ajawiri Luki	Huarisata	3830	LP
CPP 1874	36	Ancco Luki	Huarisata	3830	LP
CPP 1873	36	Azul Luki	Huarisata	3830	LP
CPP 1897	36	Ccochama	Pto. Acosta	3840	LP
CIP 702305		Chimi Luque	Huarisata	3900	LP
CO 2241	36	Chirisaya Lucki	Huancane	3900	LP
CO 2311	36	Chocke Pito	Unahuaya	3960	LP
EET 1522		Choquepitu	Tiahuanaco		LP
CO 2413	36	Ckauna Lucki	Omaccollo	3870	LP
CPP 1711	36	Hancco Luqui	Pituta	3880	LP
HUA 827		Hango Luque	V. Bonito	3800	OR
CPP 2063	36	Hanqo Qaisa	Nasacara	3860	LP
CPP 1913	36	Huata Cacho	Pto. Acosta	3840	LP
CPP 2055	36	Huevo	S.A. Machaca	3980	LP
CO 2399	36	Huila Nasari	Nazacara	3860	LP
CO 2449	36	Jancko Kaisa	Calamarca	3900	LP
CO 2435	36	Jancko Lucki	Maso Cruz	4020	LP
CIP 702280		Jancko Luky	Jancko Nunu		OR
CO 2403	36	Jancko Nasari	Nasacara	3860	LP
CO 2446	36	Janckorai	Calamarca	3900	LP
EET 1398		Janko Pinca	Vila Vila		OR
CPP 2050		Juchi Jipilla	S.A. Machaca	3980	LP
CO 2360	36	Kaisa	Jihuackuta	3960	LP
CO 2367	36	Kaisa	Caquiaviri	3960	LP
CO 2407	36	Kaisa	Nasacara	3860	LP
CO 2416	36	Kaisa	S.A. Machaca	3980	LP
CPP 2040	36	Kaisa	S.A. Machaca	3980	LP
HUA 808		Kaisalla	Toledo	3700	OR
CPP 2049		Kama Luki	S.A. Machaca	3980	LP
CO 2394	36	Kauna Lucki	Caquiaviri	3960	LP
CO 2429	36	Kauna Lucki	S.A. Machaca	3980	LP
CIP 702627		Kauna Luqui	Kalla		LP
CIP 702621		Kaysalla	Coromata		LP
CIP 702638		Kaysalla	Patacamaya		LP
CIP 702639		Kaysalla Blanca	Patacamaya		LP
CPP 1912		Kaysalla Luki	Pto. Acosta	3840	LP
EET 1421		Keta Luky	Collana		LP
CPP 2054	36	Kjetla	S.A. Machaca	3980	LP
CIP 702646		Kola	Patacamaya		LP
EET 1374		Laram Choquep.	Huancaroma		OR

Solanum × juzepczukii (cont.)

Collection	2n	Native Name	Locality	Alt.	Dept.
CO 2254	36	Laram Lucki	Paccolla	4150	LP
CO 2438	36	Laram Lucki	Maso Cruz	4020	LP
CO 2168	36	Lucki	Tiahuanaco	3880	LP
CO 2225	36	Lucki	Pucarani	3900	LP
CO 2307	36	Lucki	Unahuaya	3960	LP
CO 2331	36	Lucki	Cachuma	4000	LP
CO 2353	36	Lucki	Chama	4100	LP
CO 2465	36	Lucki	Occopampa	3850	LP
CO 2523	36	Lucki	Challapata	3770	OR
OCH 12137		Lucki	Huanccollo	3850	LP
CO 2358	36	Lucki Pinkula	Jihuackuta	3960	LP
OCH 12133		Lucky	Mdo. La Paz	3800	LP
CPP 1782	36	Luki	Tiahuanaco	3880	LP
CPP 1851		Luki	Pucarani	3830	LP
CPP 1852	36	Luki	Pucarani	3830	LP
CPP 1975	36	Luki	Corocoro	3890	LP
CPP 1979		Luki	Corocoro	3890	LP
CPP 1981		Luki	Corocoro	3890	LP
CPP 2027	36	Luki	Caquiaviri	3960	LP
CPP 2058		Luki	Nasacara	3860	LP
OCH 10555	36	Luki	Tinguipaya	3900	PO
CPP 1868		Luki Azul	Huarisata	3830	LP
CPP 1989	36	Luki Blanca	Corocoro	3890	LP
CPP 1973	36	Luki Blanco	Corocoro	3890	LP
CPP 1981-B		Luki Negra	Corocoro	3890	LP
CPP 1955		Luki Pingo	Viacha	3960	LP
CPP 1955-A		L. Qaisa Violeta	Viacha	3960	L
CPP 2084	36	Luki Rosado	J.D. Machaca	3890	LP
CPP 2085	36	Luki Rosado	J.D. Machaca	3890	LP
CPP 1995	36	Luki Violeta	Corocoro	3890	LP
CPP 2086		Luki Violeta	J.D. Machaca	3890	LP
EET 1193		Luky	Huancane		OR
EET 1302		Luky	Patacamaya		LP
EET 1485		Luky	Irutambo		CO
OCH 10486	36	Luky	Oruro	3706	OR
OCH 10694	36	Luky	Cuchu Chiri	4200	PO
HUA 810		Luque	Toledo	3700	OR
HUA 828		Luque	V. Bonito	3800	OR
CIP 702629		Luqui	Vilaque		LP
CIP 702642		Luqui	Patacamaya		LP
CIP 702643		Luqui	Patacamaya		LP
CIP 702644		Luqui	Patacamaya		LP
CPP 1708		Luqui Lila	Pituta	3880	LP
CPP 1713		Luqui Rosada	Pituta	3880	LP

Solanum × juzepczukii (cont.)

Collection	2n	Native Name	Locality	Alt.	Dept.
CPP 1871		Morocco Luki	Huarisata	3830	LP
EET 1422		Morocko Luky	Pasto Gdc.		OR
CPP 1841	36	Moroco Luki	Pucarani	3830	LP
CPP 1853	36	Moroqo Luki	Pucarani	3830	LP
CO 2398	36	Mulluncko	Nazcara	3860	LP
CPP 2037		Muroko Luki	S.A. Machaca	3980	LP
CIP 702634		Nasare Luqui	Torno		LP
CPP 2081	36	Nasari	J.D. Machaca	3890	LP
EET 1288		Pacco Toro	Collana		L
CO 2457	36	Pingo	Occopampa	3850	LP
CO 2260	36	Pocco Tturu	Paccolla	4150	LP
CO 2439	36	Pocco Tturu	Jackara	3960	LP
CO 2177	36	Pocco Turu	Patarani	3850	LP
CPP 2042	36	Q'aisa Blanca	S.A. Machaca	3980	LP
CPP 2023	36	Q'aisa Overa	Caquiaviri	3960	LP
CPP 1981-A		Qalsa	Corocoro	3890	LP
CPP 1980	36	Qaysa	Corocoro	3890	LP
CO 2448	36	Rayi Lucki	Calamarca	3900	LP
OCH 10495	36	Runa	Oruro	3706	OR
CPP 1929	36	Samka	Pto. Acosta	3840	LP
HUA 842		Sautu Luke	Toledo	3700	OR
CIP 702648		Shiri Luqui	Vilaque		LP
CPP 2065	36	Sisu	Nasacara	3860	LP
CPP 2072	36	Sisu	Nasacara	3860	LP
CPP 1836	36	Suito Luki	Pucarani	3830	LP
CO 2428	36	Torillo Huajra	S.A. Machaca	3980	LP
CPP 2047		Torillo Wagra	S.A. Machaca	3980	LP
CO 2524	36	Tuni	Mdo. Cochab.	2800	CO
CO 2525	36	Tuni	Mdo. Cochab.	2800	CO
CPP 1923	36	Wila Kaysalla	Pto. Acosta	3840	LP
CIP 702632		Wila Luqui	Vilaque		LP
CPP 1875	36	W. Nairami Luki	Huarisata	3830	LP
CPP 1823	36	Wila Luqui	Pucarani	3830	LP
CPP 2066	36	Yari	Nasacara	3860	LP
CPP 1892	36	Yurak Kaisalla	Pto. Acosta	3840	LP

Solanum × curtilobum

Collection	2n	Native Name	Locality	Alt.	Dept.
CIP 702612					
CIP 702613					
CIP 702617					
CO 2437	60	Ccoloya	Maso Cruz	4020	LP
CO 2218	60	Chocke Pito	Pucarani	3900	LP
CO 2349	60	Chocke Pito	Chama	4100	LP
CO 2441	60	Chocke Pito	Jackara	3960	LP
CO 2167	60	Chocke Pitu	Tiahuanaco	3880	LP
CO 2240	60	Choke Pito	Huancane	3900	LP
CPP 1737	60	Choquepito	Tiahuanaco	3880	LP
CPP 1966		Choquepito	Q. Quqlla	3880	LP
CPP 2076		Choquepito	J.D. Machaca	3860	LP
CPP 2097	60	Choquepito	Taraco	3840	LP
CIP 702283		Choquep. Blanco	Belen		LP
HAW 6195		Choquepitu	Cura Pucara	3860	LP
EET 1158		Choquep. Blanca	Belen		LP
CO 2378	60	Ckcta	Caquiaviri	3960	LP
EET 1403		Janko Choquep.	Huancaroma		OR
CO 2392	60	L. Chocke Pito	Caquiaviri	3960	LP
CO 2357	60	L. Chocke Pito	Jihuackuta	3960	LP
CO 2442	60	L. Chocke Pito	Jackara	3960	LP
CO 2152	60	Lucki	Kcallani	3880	LP
CO 2170	60	Lucki	Tiahuanaco	3880	LP
CO 2187	60	Lucki	Patarani	3850	LP
CO 2259	60	Lucki	Paccolla	4150	LP
CPP 1783	60	Luki	Tiahuanaco	3880	LP
CPP 2087	60	Luki	Sta. Rosa	3850	LP
CPP 1838	60	Luki Ovalada	Pucarani	3850	LP
EET 1628-A		Luky	Sacaca		PO
CPP 1706	60	Luqui Blanca	Pituta	3850	LP
CPP 1712	60	Luqui Lila	Pituta	3880	LP
CPP 1716	60	Luqui Lila	Tiahuanaco	3880	LP
CPP 1709		Luqui Morado	Pituta	3880	LP
CPP 1928	60	Wana Laka	Pto. Acosta	3840	LP

Solanum × chaucha

Collection	2n	Native Name	Locality	Alt.	Dept.
CIP 702714					
CIP 702715					
HAW 5709			Lecherias	3800	PO
CO 2518	36	Achacana	Challapata	3770	OR
CO 2356	36	Alcca Monda	Jihuackuta	3960	LP
CO 2144	36	Ccoyo	Kcallani	3800	LP
OCH 10583		Challi	Alcatuyo	3800	PO
HAW 6181		Charansiqui	Cajon Mayu		PO
HUA 818		Chiar Phinu	Toledo-Corq.	3700	OR
CO 2275	36	Chocke Kaisall	Ullacsantia	3870	LP
CO 2411	36	Ckara Monda	Omaccollo	3870	LP
CO 2219	36	Ckeni	Pucarani	3900	LP
CO 2489	36	Copacabana	Mdo. Oruro	3700	OR
CO 2371	36	Culi	Caquiaviri	3960	LP
HAW 6171		Dominguito	Suruhana		PO
OCH 10584		Florentino	Alcatuyo	3800	PO
CO 2513	36	Huaca Nahui	Challapata	3770	OR
CO 2474	36	Huaicku Papa	Mdo. Oruro	3700	OR
HAW 5559		Huaico	Mdo. La Paz		LP
EET 1478		Huancusullu	Copacabana		LP
CO 2323	36	Huila Tamatta	Cariquina	3940	LP
CO 2444	36	Jancko Ckeni	Jackara	3960	LP
CO 2276	36	Jancko Imilla	Ullacsantia	3870	LP
CO 2365	36	Jancko Imilla	Caquiaviri	3960	LP
CO 2456	36	Jancko Pala	Occopampa	3850	LP
HAW 5573		Khati	Ingavi		LP
EET 1089		Koyllu			PO
HAW 6173		Kulurana	Suruhana		PO
CIP 702585		Kunurama	Chimpapata		PO
CPP 2045		L. Laqayo Nasari	S.A. Machaca	3980	LP
CO 2452	36	Macha Chocke	Occopampa	3850	LP
CPP 2011		Maman P'eque	Caquiaviri	3960	LP
HAW 6166		Michi Aca	Ravelo		PO
CO 2424	36	Monda	S.A. Machac	3980	LP
CPP 2019	36	Nasari	Caquiaviri	3960	LP
CPP 2060	36	Nasari	Nasacara	3860	LP
CPP 2067	36	Nasari	Nasacara	3860	LP
CO 2519	36	Negra	Challapta	3770	OR
OCH 10697	36	Oka Papa	Challa Kkollu	4000	PO
CIP 702693		Paca	Patarani		LP
CO 2281	36	Pacco Ckahua	Ullacsantia	3870	LP
CO 2210	36	Pala	Mdo. La Paz	3600	LP
CO 2338	36	Pala	Cachuma	4000	LP
HUA 775		Paluya	Calamarca	3800	LP
CO 2242	36	Pepino	Huancane	3900	LP

Solanum × chaucha (cont.)

Collection	2n	Native Name	Locality	Alt.	Dept.
CO 2389	36	Pepino	Caquiavir	3960	LP
CO 2443	36	Pichuya	Jackara	3960	LP
OCH 3943	36		Oruro	3700	OR
CIP 701076		Puca Misquilla			
CIP 702590		Runa Toclla	Culpina		CH
CO 2235	36	Sali Ckeni	Cullacachi	3850	LP
CIP 702711		Sano			
CIP 701052		Santus Cachu	Mdo. La Paz	3800	LP
HUA 887		Sicha	Vilaque	3800	LP
CO 2161	36	Sorimana	Tambillo	3880	LP
HAW 6172		Sulimani	Suruhana		PO
CIP 702301		Surimana	Huarisata	3900	LP
CO 2337	36	Surimana	Cachuma	4000	LP
EET 1445		Surimana	El Alto		LP
EET 1614		Surimana	Sacaca		PO
HAW 5566		Surimana			LP
HAW 6185		Surimana	Cura Pucara	3860	LP
EET 1595		Sutamari	Chayanta		PO
CPP 2061		Wila Nasari	Nasacara	3860	LP
EET 1367		Wila Surimana	El Alto		LP
CIP 702551		Yana Achacana	Tinquipaya		PO
EET 1050		Yana Coyllu	Tiraque		CO
EET 1145		Yana Sutamari			CO
CO 2299	36	Yurima	Puerto Acosta	3880	LP

Solanum tuberosum subsp. *andigena*

Collection	2n	Native Name	Locality	Alt.	Dept.
CIP 701084					
CIP 702270			Chilcapampa	3600	LP
CIP 702271			Chilcapampa	3600	LP
CIP 702272			Koari	3550	CO
CIP 702273			Quilacollo	4000	CO
CIP 702284			Moza		OR
CIP 702309			Huarisata	3900	LP
CIP 702595					
CIP 702599					
CIP 702607					
CIP 702702			La Paz		LP
CIP 702716					
CIP 702719					

Solanum tuberosum subsp. *andigena (cont.)*

Collection	2n	Native Name	Locality	Alt.	Dept.
HAW 5719			Camargo	3800	PO
HAW 5722			Padcaya	3450	CO
HAW 6108			Rio Blanco	3200	PO
HAW 6121			Potosi	3945	PO
HAW 6142			Yacalla	3450	PO
HAW 6143			Yacalla	3450	PO
HAW 6144			Yacalla	3450	PO
HAW 6146			Yacalla	3450	PO
HUA 752			Caquiaviri	3900	LP
HUA 792			Guaqui	3800	LP
HUA 793			Guaqui	3800	LP
HUA 797			Tiahuanaco	3800	LP
OCH 10489-A			Oruro	3706	OR
OCH 10514-A			La Paz	3800	LP
OCH 10519-A			La Paz	3800	LP
OCH 10529-A			La Paz	3800	LP
OCH 10569-A			Challactiri	3900	PO
OCH 10576-A			Challactiri	3900	PO
OCH 10583-A			Alcatuyo	3800	PO
OCH 10610-A			Maragua	3700	PO
OCH 10616-A			Puna	3850	PO
OCH 10631-A			Palca	3750	PO
OCH 10636-A			Utacalla	3750	PO
OCH 10642-A			Ravelo	3100	PO
OCH 10649-A			Tinquipaya	3900	PO
OCH 10685-A			Santa Lucia	3800	PO
OCH 10693-A			Turuchupi	3200	PO
EET 1065		Abajena			
HAW 6168		Abajena	Hunchuna Pampa		PO
CPP 1752		Ajahuiri	Tiahuanaco	3880	LP
HUA 795		Ajahuiri	Guaqui	3800	LP
CO 2157	48	Ajanhuiri	Tambillo	3880	LP
CO 2238	48	Ajanhuiri	Cullacanchi	3850	LP
OCH 10482	48	Ajanhuiri	Oruro	3706	OR
CO 2459	48	Ajipa	Occopampa	3850	LP
CIP 701075		Alca Jairmi			
CO 2230	48	Alcca Chocke	Yana Cachi	3900	LP
CPP 1883	48	Alcca Imilla	Achacachi	3820	LP
HAW 5639		Alcca Imilla	Melga		CO
HAW 6148		Alcca Imilla	Yacalla	3450	PO
HAW 6133		Alcca Papa	Totora		PO
OCH 12116	48	Alcca Papa	Lequezana	3300	PO
CO 2266	48	Alcca Taraco	Millipaya	3550	LP
OCH 10568		Alka Coillu	Challactiri	3900	PO
OCH 10567	48	Alka Imilla	Challactiri	3900	PO
OCH 10682	48	Alka Imilla	Santa Lucia	3800	PO

Solanum tuberosum subsp. *andigena (cont.)*

Collection	2n	Native Name	Locality	Alt.	Dept.
EET 1477		Alka Pala	Copacabana		LP
EET 1429		Alka Pali	Quelkata		LP
EET 1488		Alka Taracko	Warisata		LP
EET 1316		Alka Tarma	Desaguadero		LP
OCH 10594	48	Alkamari	Maragua	3700	PO
CIP 702295		Allcca Pala	Hualata	3850	LP
CPP 1750	48	Alqa Imilla	Tiahuanaco	3880	LP
CPP 1796	48	Alto Semilla	Mdo. Huatajut	3840	LP
HUA 812		Amajalla	Toledo	3700	OR
CPP 537	48	Amajana	Mayutambo		PO
CPP 1938	48	Amajaya	Pillapi	3850	LP
CPP 1907	48	Amajaya	Puerto Acosta	3840	LP
CPP 2121	48	Amajayo	Taraco	3840	LP
CIP 702539		Amajayu Papa	Cerda		PO
CPP 1748	48	Amaqhaya	Tiahuanaco	3880	LP
CPP 2013	48	Anahuayculi	Caquiaviri	3960	LP
CPP 2016	48	Anahuayculi	Caquiaviri	3960	LP
HAW 6125		Ancañahui	Paljacancha	3600	PO
CPP 1896		Ancco Imilla	Puerto Acosta	3840	LP
CPP 1879		Anccosani	Huarisata	3800	LP
OCH 10653-A	48	Anil Papa	Cuchi Chiri	4200	PO
CIP 702312		Aozanco	Huarisata	3900	LP
O/S 11828	48	Apharu	Sorucachi	3900	LP
CPP 1909		Arekero	Puerto Acosta	3840	LP
OCH 10696		Asamaya	Cruce Oruro	4000	PO
CIP 702697		Atacama	Colcha		PO
CIP 702698		Atacama	Pelcoya		PO
CO 2310	48	Azarona	Unahuaya	3960	LP
EET 1424		Azul Sutamari	Lequepalca		OR
CPP 1953	48	Blanca	Viacha	3940	LP
CPP 1870	48	Blanca Jaspeada	Huarisata	3830	LP
CIP 701050		Bole			
CPP 1872	48	Buti Imilla	Huarisata	3800	LP
CPP 1821	48	Calla Imilla	Pucarani	3850	LP
CPP 2106		Camarona	Taraco	3840	LP
EET 1455		Carlos	Lequepalca		OR
CO 2469	48	Carlosa	Mdo. Oruro	3700	OR
CPP 2089		Casa Blanca	Santa Rosa	3850	LP
EET 1559		Casablanca	Huata Pampa		
EET 1565		Casablanca	Jocopampa		
OCH 10691	48	Casablanca	Vilacaya	3800	PO
OCH 12134	48	Ccami	Mdo. La Paz	3800	LP
CIP 701065		Ccillo Acoto			
CPP 1738	48	Ccopo	Tiahuanaco	3880	LP
CPP 1949	48	Ccoyllo	Pillapi	3850	LP
CPP 1717	48	Ccoyo	Tiahuanaco	3880	LP

Solanum tuberosum subsp. *andigena (cont.)*

Collection	2n	Native Name	Locality	Alt.	Dept.
CPP 1891	48	Ccoyo	Achacachi	3820	LP
CPP 2132	48	Ccoyo	Patamanta	3850	LP
HAW 5692		Ch'cca Imilla	Lecherias	3800	PO
HAW 6137		Chajli	Totora		PO
CPP 1971	48	Chapcha	Quella Quolla	3880	LP
CPP 1863	48	Chaquena	Hualata Chico	3850	LP
CPP 1996	48	Chaucha	E. Yarihuaya	3880	LP
EET 1671		Chaucha	Julpe		CO
CIP 702306		Chequeno	Huarisata	3900	LP
EET 1175		Chiar Alka I.	Calamarca		LP
CO 2280	48	Chiar Amajaya	Ullacsantia	3870	LP
CO 2204	48	Chiar Imilla	Mdo. La Paz	3600	LP
CPP 1727		Chiar Imilla	Tiahuanaco	3800	LP
HUA 843		Chiar Imilla	Toledo	3700	OR
HAW 5591		Chiar Sinka	Lahuachaca		LP
CO 2173	48	Chiar Surimana	Patarani	3850	LP
CPP 1887		Chiara Alcca I.	Achacachi	3820	LP
CPP 1816		Chiara Imilla	Huancane	3830	LP
CPP 1825	48	Chiara Imilla	Pucarani	3830	LP
CPP 1833		Chiara Imilla	Pucarani	3830	LP
CPP 1854		Chiara Imilla	Huarina	3830	LP
CPP 1854 A		Chiara Imilla	Huarina	3830	LP
CPP 1890	48	Chiara Imilla	Achacachi	3820	LP
CPP 1901 A		Chiara Imilla	Puerto Acosta	3840	LP
CPP 2088		Chiara Imilla	Santa Rosa	3850	LP
CPP 1881		Chiara Pichuya	Huarisata	3830	LP
CPP 1862		Chiara Polo	Hualata Chico	3900	LP
OCH 2792	48	Chiara Sali	Huancane	3900	LP
OCH 2800		Chiara Surimana	Huarina	3850	LP
CO 2319	48	Chiara Taraco	Tajani	3960	LP
CPP 1743		Chiari Imilla	Tiahuanaco	3880	LP
CPP 1744		Chiari Imilla	Tiahuanaco	3880	LP
CIP 702589		Chilena Blanca	Chairumani		LP
OCH 10663	48	Chirca	Miraflores	3800	PO
OCH 12132	48	Chojllu	Mdo. La Paz	3800	LP
CPP 1793		Chojo	Mdo. La Paz	3600	LP
CPP 2115		Choqllo	Taraco	3840	LP
CPP 1905	48	Choque	Puerto Acosta	3840	LP
CPP 1936		Choque Incho	Pillapi	3850	LP
HUA 850		Choquella	Toledo	3700	OR
CPP 1719	48	Choquepito	Tiahuanaco	3880	LP
CPP 1904		Chuji	Puerto Acosta	3840	LP
CPP 1734		Chuki Imilla	Tiahuanaco	3880	LP
CO 2250	48	Chumpi Imilla	Near Tiquina	3960	LP
CPP 1861		Churi Imilla	Hualata Chico	3850	LP
CO 2289	48	Cintan Pollera	Sisasani	4160	LP

Solanum tuberosum subsp. *andigena (cont.)*

Collection	2n	Native Name	Locality	Alt.	Dept.
OCH 2802		Ckaisa	Coroc.-Viacha	3850	LP
O/S 15496		Ckaya Chocke	Canayapu	2700	LP
CPP 1921		Ckena	Puerto Acosta	3840	LP
CO 2171	48	Ckeni	Patarani	3850	LP
OCH 10689		Coillu	Carma	3800	PO
OCH 10689-A		Coillu	Carma	3800	PO
CIP 702694		Coipa	Ticoniri		LP
CIP 702703		Cole	Sacaca		PO
OCH 12122	48	Collareja	Tinguipaya	3900	PO
CPP 1746		Colloj	Tiahuanaco	3880	LP
OCH 10569		Condor Anka	Challactiri	3900	PO
OCH 3988	48	Condor Chuchuli	Urmiri		OR
CPP 2098		Condor Kauna	Taraco	3840	LP
CPP 1830	48	Condor Naira	Pucarani	3830	LP
CPP 1745		Conejo Peque	Tiahuanaco	3880	LP
CPP 1740	48	Qoquene	Tiahuanaco	3880	LP
OCH 10540		Cuchiaca	Tinguipaya	3900	PO
CIP 701054		Cuchipa Accan			
CO 2181	48	Cuyo	Patarani	3850	LP
CPP 1902		Damiana	Puerto Acosta	3840	LP
CPP 530		Dominguillo	Chinoli		PO
CPP 1997	48	Duraznillo	E. Yarihuaya	3850	LP
CIP 701081		Florida			
CIP 701080		Garmendia			
CIP 701048		Goiwa			
CPP 1903		H. Imilla	Puerto Acosta	3840	LP
CPP 1858		Hanco Surimana	Huarina	3900	LP
CIP 702279	48	Hancu Sullu	Moza		OR
HUA 872		Hanko Munta	Calamarca	3600	LP
CPP 1726		Hanq'o Imilla	Tiahuanaco	3880	LP
CPP 1728	48	Hanq'o Imilla	Tiahuanaco	3880	LP
OCH 3147		Hanq'o Imilla	Huancane	3900	LP
CPP 2114		Hanqu Surimana	Taraco	3840	LP
OCH 3556	48	Hanqo Yako	Taraco	3840	LP
OCH 10631		Holandesa	Palca	3750	PO
OCH 10599		Huaca	Maragua	3700	PO
CPP 1940		Huaca Nuno	Pillapi	3860	LP
OCH 10585	48	Huaca Zapatu	Alcatuyo	3800	PO
HUA 801		Huacalagra	Belena	3700	OR
CPP 2122	48	Huacalajra	Taraco	3840	LP
EET 1282		Huachua Chara	Laja		LP
CO 2216		Huaicka	Pucarani	3900	LP
CO 2265	48	Huaili	Millipaya	3550	LP
CO 2236	48	Huailla Chia	Cullacachi	3850	LP
CPP 1894		Huantaiso	Puerto Acosta	3840	LP
CIP 702690	48	Huanuri			LP

Solanum tuberosum subsp. *andigena (cont.)*

Collection	2n	Native Name	Locality	Alt.	Dept.
CO 2200	48	Huaraquipa	Huaillara	3800	LP
CPP 2125		Huaraquipa	Taraco	3840	LP
CO 2150		Huarisaya	Kcallani	3880	LP
CO 2180	48	Huarisaya	Patarani	3850	LP
CO 2348	48	Huarisaya	Chama	4100	LP
CO 2434	48	Huarisaya	Viacha	3960	LP
CPP 1729		Huarisaya	Tiahuanaco	3880	LP
CPP 1741		Huarisaya	Tiahuanaco	3880	LP
CO 2239	48	Huarisaya	Huancane	3900	LP
CPP 1730		Huayllachiusa	Tiahuanaco	3880	LP
OCH 10640		Huayku	Ravelo	3100	PO
CO 2195	48	Huila Chocke	Huaillara	3800	LP
CO 2148	48	Huila Imilla	Kcallani	3880	LP
CO 2163	48	Huila Imilla	Tambillo	3880	LP
CO 2179	48	Huila Imilla	Patarani	3850	LP
CO 2197		Huila Imilla	Huaillara	3800	LP
CO 2203		Huila Imilla	Mdo. La Paz	3600	LP
CO 2223	48	Huila Imilla	Pucarani	3900	LP
CO 2312	48	Huila Imilla	Unahuaya	3960	LP
CO 2366	48	Huila Imilla	Caquiaviri	3960	LP
HAW 6189		Huila Imilla	Cura Pucara	3860	LP
HAW 5707		Huila Laira	Lecherias	3800	PO
CO 2341	48	Huila Monda	Cachuma	4000	LP
CPP 1820		Huila Pala	Huancane	3830	LP
CO 2244	48	Huila Pata	Huancane	3900	LP
CO 2153	48	Huila Sacko	Tambillo	3880	LP
CO 2188	48	Huila Sicha	Huaillara	3800	LP
CPP 1828		Huilla Calla	Pucarani	3850	LP
CO 2315		Huilla Imilla	Tajani	3960	LP
CO 2160	48	Huislla Packi	Tambillo	3880	LP
CO 2359	48	Huislla Ppacki	Jihuackuta	3960	LP
CO 2386	48	Huislla Ppacki	Caquiaviri	3960	LP
CIP 702276		Imilla	Punata		CO
CIP 702709		Imilla	Punata		CO
CPP 2129	48	Imilla Blanca	Taraco	3840	LP
HAW 5572		Imilla Blanca	Ingavi		LP
OCH 10513	48	Imilla Blanca	La Paz	3800	LP
OCH 10513-A		Imilla Blanca	La Paz	3800	LP
OCH 10477-A		Imilla Manzana	Karachipampa	4000	PO
CIP 702685		Imilla Roja	Huanuri		OR
CPP 1722	48	Imilla Roja	Tiahuanaco	3880	LP
EET 1059		Imilla Roja	Ravelo		PO
EET 1381		Imilla Roja	Quelkata		LP
OCH 10517	48	Imilla Roja	La Paz	3800	LP
EET 1197		Imilla Rosada	Tambillo		LP
HAW 5731		Jalca	Culpina	3000	CO

Solanum tuberosum subsp. *andigena (cont.)*

Collection	2n	Native Name	Locality	Alt.	Dept.
CPP 526	48	Jalka Imilla	Chinoli		PO
CIP 702517		Jalka Papa	Atacollo		PO
CIP 702519		Jalka Papa	Atacollo		PO
OCH 12123	48	Jalka Papa	Betanzos	3400	PO
CPP 535		Jalka Sillama	Chinoli		PO
CO 2247	48	Jancko	Near Tiquina	3960	LP
CO 2164	48	Janc. Huarisaya	Tiahuanaco	3880	LP
CO 2316	48	Jancko Llocco	Tajani	3960	LP
CO 2193	48	Jancko Surimana	Huaillara	3800	LP
CIP 702564		Juchuy Paly	Tinguipaya		PO
OCH 10526		Kallaguayitas	La Paz	3800	LP
CIP 702546		Kallpa Runa	Ravelo		PO
HAW 5665		Kallpa Runa	Tocsupaya		CO
OCH 10553	48	Kantipapa	Tinguipaya	3900	PO
HUA 813		Karasapa Coillo	Toledo	3700	OR
CIP 702633		Kaysalla Luqui	Patacamaya		LP
CPP 2059		Kcene	Nasacara	3860	LP
HAW 5576		Khati	La Paz		LP
CIP 702523		K. Sipancachi	Tinguipaya		PO
CPP 1970		Khumacana	Quella Qoqlla	3880	LP
CPP 2006	48	Kjoyo	Caquiaviri	3960	LP
CPP 1968	48	Kondor Kauna	Quella Quqlla	3880	LP
HAW 5590		Koyllo	Lahuachaca		LP
CO 2290	48	Kuchama	Sisasani	4160	LP
OCH 10616		Kunurana	Puna	3850	PO
CIP 702572		Lapia Papa	Atacolla		PO
CPP 1798		Larwincho	Huatajut	3840	LP
CO 2284	48	Laurayo	Sisasani	4160	LP
CO 2268	48	Leleckoya	Millipaya	3550	LP
CO 2271	48	Leleckoya	Iminapi-Sorata	2800	LP
CPP 2116		Leque Cayu	Taraco	3840	LP
O/S 11794	48	Lleccoya	Curupampa	2700	LP
OCH 2801		Llocalla	Huarina	3900	LP
CO 2294	48	Llocco	Puerto Acosta	3880	LP
CPP 1785		Lloqalla	Tiahuanaco	3850	LP
CPP 1958	48	Lloqallito	Qhella Qhella	3880	LP
CPP 2113	48	Llucta Pura	Taraco	3840	LP
CPP 2008		Loq'a	Caquiaviri	3960	LP
HAW 6135		Maconko	Totora		PO
OCH 10606	48	Makasa	Maragua	3700	PO
HAW 5662		Malcacho	Tarabuco		CO
OCH 10531	48	Malcachu Puca	La Paz	3800	LP
CPP 524		Maluchi	Chinoli		PO
OCH 10605	48	Malula	Maragua	3700	PO
HUA 837		Maman Pekhe	Toledo	3700	OR
HUA 806		Maman Teque	Belena	3700	OR

Solanum tuberosum subsp. *andigena (cont.)*

Collection	2n	Native Name	Locality	Alt.	Dept.
CPP 531		Manzana	Chinoli		PO
CPP 1942		Manzana	Pillapi	3860	LP
OCH 12124	48	Manzana	Tinguipaya	3900	PO
HAW 5668		Marcacho	Tocsopaya		CO
CPP 2001	48	Monda	E. Yarihuaya	3880	LP
CPP 2025		Monda	Caquiaviri	3960	LP
CPP 2028		Monda	Caquiaviri	3960	LP
CPP 2064	48	Monda	Nasacara	3860	LP
CPP 2012	48	Monda Blanca	Caquiaviri	3960	LP
CPP 2031		Monda Colorada	Caquiaviri	3960	LP
CIP 702636		Morocho	Vilaque		LP
CIP 702548		Mula Papa	Tinguipaya		PO
CIP 702524		Nagui Caballo	Atacollo		PO
CO 2217	48	Nasa Huiri	Pucarani	3900	LP
CPP 2074		Ojani Lloqalla	Nasacara	3860	LP
CPP 1956	48	Oqonaira	Qhella Qhella	3880	LP
CPP 1944	48	Pacco Imilla	Pillapi	3850	LP
CPP 1943		Pacco Imilla B.	Pillapi	3850	LP
CPP 1951		Pacco Qawa	Pillapi	3850	LP
CO 2508	48	Pacena	Mdo. Oruro	3700	OR
EET 1379		Pacena	Pasto Grande		OR
CIP 701078	48	Paciencia			
CIP 701070	48	Paco			
CPP 2093		Paco Qhawa	Taraco	3840	LP
CPP 1779	48	Pacocahua	Tiahuanaco	3850	LP
CIP 702695		Pako Uaira	Llosa		LP
CPP 1755	48	Pala	Tiahuanaco	3880	LP
CPP 1778		Pala	Tiahuanaco	3880	LP
CPP 2068		Pala	Nasakara	3860	LP
HAW 5569		Pala	Mdo. La Paz		LP
CO 2440	48	Pala Monda	Jackara	3960	LP
CIP 702300		Pala Negra	Hualata	3850	LP
HUA 879		Pala Negra	Vilaque	3800	LP
CPP 2070	48	Pala o Lloqo	Nasacara	3860	LP
CPP 1761	48	Pala Pala	Tiahuanaco	3880	LP
CPP 1710		Pala Roja	Pituta	3880	LP
OCH 12120	48	Pali	Puna	3850	PO
OCH 10676		Palka Imilla	Condoriri	3900	PO
CPP 1893	48	Palma	Puerto Acosta	3840	LP
CPP 1917	48	Palma	Puerto Acosta	3840	LP
OCH 10692		Palqui	Chaqui	3900	PO
CO 2285	48	Paltalla	Sisasani	4160	LP
CPP 522		Panti Imilla	Chinoli		PO
CIP 702689	48	Papa Blanca			LP
EET 1490		Papa Blanca	Wacuyo		LP
EET 1521		Papa Blanca	Huatajata		LP

Solanum tuberosum subsp. *andigena (cont.)*

Collection	2n	Native Name	Locality	Alt.	Dept.
CIP 702297	48	Papa Colorada	Hualata	3850	LP
CIP 702298		Papa Dalia	Hualata	3850	LP
CIP 702275		Papa Epizana	Punata		CO
HUA 857		Papa Gendarme	Toledo	3700	OR
CO 2458	48	Papa Isla	Occopampa	3850	LP
CPP 1803		Papa Isla	Huatajata	3600	LP
CPP 1856		Papa Isla	Huarina	3830	LP
HUA 855		Papa Kunarana	Toledo	3700	OR
CIP 702678		Papa Negra	Santa Rosa		LP
CIP 702708		Papa Negra	Tacacoma		LP
CPP 1707		Papa Negra	Pituta	3880	LP
CIP 701060	48	Papa Roja			
HAW 5556		Papa Runa	Tiahuanaco		LP
CPP 1927		P.S. Choquepito	Puerto Acosta	3840	LP
CPP 525	48	Papa Tuni	Mojororillo		PO
CPP 1885		Parara	Achacachi	3820	LP
EET 1589		Pasenaaloma	Calchani		CO
CIP 701053	48	Paspa Suncho			
CPP 2041		Pepino	S.A. Machaca	3980	LP
CPP 1939		Phala	Pillapi	3850	LP
CPP 2083	48	Phala	J.D. Machaca	3890	LP
CPP 1771		Phimila	Tiahuanaco	3880	LP
CPP 2079		Phino	J.D. Machaca	3890	LP
CO 2228	48	Phinu	Yana-Cachi	3900	LP
CIP 702435		Phureja	Palcoco		LP
CPP 1865		Phureja	Huarisata	3800	LP
OCH 2746	48	Phureja	Mdo. La Paz	3800	LP
CIP 702707		Phureja Rosada	Tacacoma		LP
CPP 2044		Phyño	S.A. Machaca	3980	LP
CPP 2032		Phyño Negro	Caquiaviri	3960	LP
OCH 10563		Pichca	Cerda	3900	PO
CIP 702568		Pichea Papa	Cerda		PO
CPP 1804		Pichicanahua	Huatajata	3600	LP
CIP 702591		Pichuya	Santa Ana		LP
CO 2309	48	Pichuya	Unahuaya	3960	LP
CPP 1747		Pichuya	Tiahuanaco	3880	LP
CPP 1764	48	Pichuya	Tiahuanaco	3880	LP
CPP 1791		Pichuya	Tiahuanaco	3880	LP
CPP 1813	48	Pichuya	Huancane	3830	LP
CPP 1849		Pichuya	Pucarani	3830	LP
CPP 1886		Pichuya	Achacachi	3800	LP
CPP 1886-A		Pichuya	Achacachi	3800	LP
CPP 1886-B		Pichuya	Achacachi	3800	LP
CPP 1946	48	Pichuya	Pillapi	3860	LP
CPP 2092		Pichuya	Taraco	3840	LP
CPP 1844		Pinola	Pucarani	3830	LP

Solanum tuberosum subsp. *andigena (cont.)*

Collection	2n	Native Name	Locality	Alt.	Dept.
CPP 2109	48	Pintanuro	Taraco	3840	LP
HAW 5594		Pituhuayaca	Lahuachaca		LP
HAW 6174		Pitu Papa	Suruhana		PO
CPP 1846		Pituhuayaca	Pucarani	3900	LP
CPP 1941		Polo	Pillapi	3850	LP
CIP 702314	48	Polo	Huarisata	3900	LP
CPP 1847		Polochoque	Pucarani	3830	LP
CO 2155	48	Polonia	Tambillo	3880	LP
EET 1427		Polonia	Quelkata		LP
HAW 6124		Puca	Paljacancha	3600	PO
HAW 5700		P. Chasca Nahui	Lecherias	3800	PO
CO 2496	48	Puca Huaicku	Mdo. Oruro	3700	OR
CO 2516	48	Puca Imilla	Challapata	3770	OR
EET 1039		Puca Imilla	Chapare		CO
EET 1088		Puca Imilla			PO
EET 1204		Puca Imilla	Otavi		PO
EET 1244		Puca Imilla	Ravelo		PO
CIP 701047		Puca Lomo			
OCH 10597	48	Puca Luru	Maragua	3700	PO
CIP 702534		Puca Muscara	Tinguipaya		PO
CPP 529	48	Puca Nahui	Chinoli		PO
OCH 12121		Puca Nahui	Puna	3850	PO
EET 1006		Puca Paceña	Independencia		CO
CPP 538	48	Puca Palama	Mayutambo		PO
EET 1621		Puca Papa	Tawa Reja		OR
CIP 701051	48	Puca Pitiquiña			
CIP 701046		Puca Tacclla			
CIP 701044		Puca Yana			
CPP 521		Puca Imilla	Chinoli		PO
OCH 10542		Puku Kunurana	Tinguipaya	3900	PO
CO 2387	48	Pulum Puta	Caquiaviri	3960	LP
CPP 1788		Puma	Mdo. La Paz	3600	LP
OCH 10545	48	Pureka Nagui	Tinguipaya	3900	PO
CPP 1806		Puro Imilla	Huatajata	3840	LP
CPP 1974		Q'aysa Blanca	Corocoro	3890	LP
CPP 2071		Q'oyo	Nasacara	3860	LP
CPP 2082	48	Q'oyo	J.D. Machaca	3890	LP
CPP 2107		Q'oyo	Taraco	3840	LP
CPP 2119	48	Q'oyo	Taraco	3840	LP
CPP 1952		Qhali	Viacha	3960	LP
CPP 1742		Qoyo	Tiahuanaco	3880	LP
CPP 1756	48	Qoyoqoyo	Tiahuanaco	3880	LP
OCH 10581		Quchulli	Pasasi	3900	PO
O/S 15474	48	Quipa Chocke	Parahua	3880	LP
EET 1546		Runa	Munani		
HAW 5732		Runa	Culpina	3000	CO

Solanum tuberosum subsp. *andigena (cont.)*

Collection	2n	Native Name	Locality	Alt.	Dept.
OCH 3998		Runa	Mdo. Cochab.	2800	CO
OCH 12126	48	Runa	Mdo. Bs. Aires	3700	LP
CPP 1850		Runa Jaspeada	Pucarani	3830	LP
CIP 702264	48		Mdo. Palcoco		LP
CIP 702266			Palcoco		LP
CPP 1714			Pituta	3880	LP
CPP 1758	48		Tiahuanaco	3880	LP
CPP 1831	48		Pucarani	3830	LP
CPP 1911	48		Puerto Acosta	3840	LP
CPP 1935			Pillapi	3860	LP
CPP 1945			Pillapi	3850	LP
CPP 1962			Qhella Qhella	3880	LP
OCH 2805			Q. Yuriwaya	3880	LP
OCH 2809			E. Yariwaya	3850	LP
OCH 11922	48		Sucre-Guadal.	3100	CH
OCH 11931			Sucre-Tarab.	3000	CH
OCH 11939	48		Cyto. Recoleta		CH
OCH 11953	48		Vcd. Camargo		LP
OCH 11956	48		Vcd. Camargo	2900	CH
OCH 11975	48		Tupiza-Tarija		TA
OCH 12113			Lequezana	3300	PO
OCH 12131			Mdo. La Paz	3800	LP
CO 2270	48		Sorata	2800	LP
CPP 520		Sacampaya	Mayatambo		PO
CPP 1715		Sackampaya	Pituta	3880	LP
CO 2154	48	Sacko	Tambillo	3880	LP
CO 2221	48	Sacko	Pucarani	3900	LP
CO 2460	48	Sacko Imilla	Occopampa	3850	LP
HAW 5691		Sairi	Lecherias	3800	PO
CIP 702594	48	Sako	Choromata		LP
CPP 1768		Sakoqhaisa	Tiahuanaco	3880	LP
CIP 701067	48	Sale			
CIP 701069		Sale			
CIP 701068	48	Sale			
CPP 1808		Sali	Huatajata	3600	LP
CPP 1867		Sali	Huarisata	3830	LP
CPP 1812	48	Sali Huila I.	Huancane	3830	LP
CPP 1900		Sali Imilla	Puerto Acosta	3840	LP
CIP 702587		Sallama	Corque		OR
CO 2515	48	Sallama	Challapata	3770	OR
CPP 2078		Sani Imilla	J.D. Machaca	3890	LP
CIP 702686		Sanko Pulu	Payrumani		LP
OCH 10519	48	Sapanas	La Paz	3800	LP
CPP 1736		Saq'ampaya	Tiahuanaco	3800	LP
CPP 2091		Saq'ampaya	Taraco	3840	LP
CPP 1984		Saq'o	Corocoro	3890	LP

Solanum tuberosum subsp. *andigena (cont.)*

Collection	2n	Native Name	Locality	Alt.	Dept.
CPP 1986	48	Saq'o Blanca	Corocoro	3890	LP
CPP 1805	48	Saqampaya	Huatajata	3600	LP
CPP 2101		Saqampaya	Taraco	3840	LP
CPP 1769		Saqo	Tiahuanaco	3850	LP
CPP 1770		Saqo	Tiahuanaco	3850	LP
CPP 1773		Saqo	Tiahuanaco	3850	LP
OCH 10642	48	Sata Imilla	Ravelo	3100	PO
OCH 10623	48	Sata	Puna	3850	PO
CO 2264		Sicha	Millipaya	3550	LP
OCH 10621		Silverio	Puna	3850	PO
EET 1179		Sipa	Puerto Acosta		LP
EET 1554		Sipa Blanca	Janko Amaya		LP
HAW 5664		Sipancachi	Tarabuco		CO
CIP 701066		Soli			
CPP 1987		Soq's Qate Ph.	Corocoro	3850	LP
CPP 2033		Sotomari	Caquiaviri	3960	LP
OCH 10592	48	Suchi	Alcatuyo	3800	PO
CPP 1895		Sullu Caua	Puerto Acosta	3840	LP
CPP 1835		Sumi Imilla	Pucarani	3830	LP
CPP 1757		Surimana	Tiahuanaco	3880	LP
CPP 1765	48	Surimana	Tiahuanaco	3880	LP
CPP 1772	48	Surimana	Tiahuanaco	3880	LP
CPP 1774		Surimana	Tiahuanaco	3880	LP
CPP 2099		Surimana	Taraco	3840	LP
CPP 1948		Surimana Wila	Pillapi	3850	LP
OCH 10655	48	Suru	Cuchi Chiri	4200	PO
OCH 10519-B	48	Suso Koillo	La Paz	3800	LP
HUA 858		Sutamara	Sica Sica	3800	LP
CPP 1965		Satamariy	Qhella Qhella	3880	LP
CPP 2124		Sutamary	Taraco	3840	LP
EET 1425		Tantta	Patacamaya		LP
CIP 702315		Teke Teke	Huarisata	3900	LP
CPP 2100		Titi Wichinca	Taraco	3840	LP
CPP 1916		Toma Kahua	Puerto Acosta	3840	LP
CO 2450	48	Tomata	Calamarca	3900	LP
OCH 10588	48	Tonco Turu	Alcatuyo	3800	PO
CIP 702549		Troluke	Tinguipaya		PO
CIP 701083		Trumbus			
CPP 528		Trune	Chinoli		PO
CO 2293	48	Tuni	Puerto Acosta	3880	LP
HUA 825		Turacari	Valle Bonito	3800	OR
CPP 1760	48	Tutamari	Tiahuanaco	3880	LP
CPP 1797	48	Tutamari	Mdo. Huatajut	3840	LP
CPP 2090		Tutamari	Santa Rosa	3850	LP
CPP 2095		Tutamari	Taraco	3840	LP
CIP 702557	48	Vilahaira	Tinguipaya		PO

Solanum tuberosum subsp. *andigena (cont.)*

Collection	2n	Native Name	Locality	Alt.	Dept.
CPP 1937		Waca Lagra	Pillapi	3850	LP
CPP 1915		Waca Lajra	Puerto Acosta	3840	LP
CPP 2111	48	Waka Lajra	Taraco	3840	LP
CPP 1934	48	Wala Chia	Pillapi	3850	LP
CPP 2103	48	Wallachia	Taraco	3840	LP
CPP 1826	48	Warisa	Pucarani	3830	LP
CIP 702304		Warisaya	Huarisata	3900	LP
CIP 702311	48	Warisaya	Huarisata	3900	LP
CPP 1766		Wariscka	Tiahuanaco	3880	LP
CPP 1731	48	Warisaya	Tiahuanaco	3880	LP
CPP 1950		W. Toq'oruro	Pillapi	3860	LP
HUA 830		Wila Coyllo	Valle Bonito	3800	OR
CIP 702307		Wila Imilla	Huarisata	3900	LP
CPP 1732	48	Wila Imilla	Tiahuanaco	3880	LP
CPP 1733		Wila Imilla	Tiahuanaco	3880	LP
CPP 1922		Wila Imilla	Puerto Acosta	3840	LP
CPP 1925		Wila Imilla	Puerto Acosta	3840	LP
CPP 1976	48	Wila Imilla	Corocoro	3890	LP
CPP 2110	48	Wila Imilla	Taraco	3840	LP
CPP 2127		Wila Imilla	Taraco	3840	LP
EET 1220		Wila Imilla	Lahuachaca		LP
EET 1290		Wila Imilla	Laja		LP
EET 1338		Wila Imilla	El Alto		LP
EET 1346		Wila Imilla	Tiahuanaco		LP
EET 1406		Wila Imilla	Quelkata		LP
EET 1539		Wila Imilla	Huatajata		LP
HUA 799		Wila Imilla	Tiahuanaco	3800	LP
HUA 844		Wila Imilla	Toledo	3700	OR
CPP 1983	48	Wila Monda	Corocoro	3890	LP
EET 1371		Wila Pacena	Quelkata		LP
EET 1384		Wila Pacena	Quelkata		LP
CPP 1721		Wila Pala	Tiahuanaco	3880	LP
CPP 1876	48	Wila Pala	Huarisata	3830	LP
CPP 1884		Wila Pala	Achacachi	3820	LP
HUA 860		Wila Pala	Sica Sica	3800	LP
CPP 1878		Wila Pichuya	Huarisata	3800	LP
EET 1333		Wila Pinu	Waywasi		LP
CPP 1924		Wila Ruya	Puerto Acosta	3840	LP
EET 1474		Wila Sipa	Janco Amaya		LP
CPP 2105	48	Wila Surimana	Taraco	3840	LP
CPP 1864	48	Wila Taraco	Hualata Chico	3850	LP
CIP 702303		Wila Taralco	Huarisata	3900	LP
CPP 1718		Wila Yaku	Tiahuanaco	3880	LP
CPP 2112		Wila Yaku	Taraco	3840	LP
CPP 1724	48	Wilapalapala	Tiahuanaco	3880	LP
CPP 1725	48	Wilapitikilla	Tiahuanaco	3880	LP

Solanum tuberosum subsp. *andigena (cont.)*

Collection	2n	Native Name	Locality	Alt.	Dept.
CPP 1723		Wilasoqo	Tiahuanaco	3880	LP
CPP 2069		Willca Paro	Nasacara	3860	LP
CPP 2073	48	Wislla Paki	Nasacara	3860	LP
HUA 874		Wislla Paqui	Calamarca	3600	LP
HUA 878		Wislla Paqui	Vilaque	3800	LP
CPP 1882		Wuiricaro	Achacachi	3820	LP
CPP 1751		Yaca	Tiahuanaco	3880	LP
CPP 1776	48	Yaco	Tiahuanaco	3880	LP
CPP 1763	48	Yako	Tiahuanaco	3880	LP
OCH 3143		Yako	Tiahuanaco	3850	LP
OCH 10660		Yalka Sillani	Betanzos	3407	PO
OCH 10651-A		Yana Abajena	Tinguipaya	3900	PO
OCH 10651-B		Yana Abajena	Tinguipaya	3900	PO
OCH 10656	48	Yana Aku	Cuchi Chiri	4200	PO
OCH 10575		Yana Chiricuti	Challactiri	3900	PO
CO 2506	48	Yana Huaicku	Mdo. Oruro	3700	OR
CIP 702571		Yana Imilla	Atacolla		PO
CO 2512		Yana Imilla	Challapata	3770	OR
OCH 10613	48	Yana Imilla	Maragua	3700	PO
HAW 5650		Yana Koyllu	Tiraque		CO
CIP 701062	48	Yana Lomo			
CIP 702537		Yana Moscara	Tinguipaya		PO
OCH 10667	48	Yana Paloma	Hornos	3400	PO
OCH 12125		Yana Paloma	Tinguipaya	3900	PO
CIP 701063	48	Yana Palta			
CPP 534		Yana Papa	Chinoli		PO
CPP 523	48	Yana Runa	Chinoli		PO
HAW 5661		Yana Runa	Potolo		CO
OCH 10574	48	Yari	Challactiri	3900	PO
OCH 2810	48	Yari Overa	Corocoro	3850	LP
CPP 2003		Yari Papa	E. Yarihuaya	3880	LP
CIP 702569		Yocalla Papa	Cerda		PO
OCH 10576	48	Yurac Chiricuti	Challactiri	3900	PO
OCH 10509	48	Yurac Imilla	Azangaro	3800	PO
OCH 10615		Yurac Luky	Maragua	3700	PO
CO 2471	48	Yurac Pali	Mdo. Oruro	3700	OR
OCH 10510		Yurac Paloma	Azangaro	3800	PO
CIP 701074		Yurac Rumpus			
CIP 702516		Yuracc Imilla	Atacillo		PO
OCH 10499	48	Yuracc Pali	Oruro	3706	OR
CPP 533		Yuracc Runa	Chinoli		PO
CO 2277	48	Yurima	Ullacsantia	3870	LP
CPP 2118		Yurima	Taraco	3840	LP
CPP 1790	48	Zapallo	Mdo. La Paz	3600	LP

Table 1. Summary of Bolivian potatoes accessioned by the International Potato Center (CIP), the Peruvian Potato Collection (CPP), and the Toralapa Experimental Station (EET).

Germplasm station	Cultivated Species							
	ADG	AJN	CUR	CHA	JUZ	PHU	STN	Total
CIP	90	14	4	10	19	7	22	166
CPP	242	17	12	6	52	4	65	398
EET	40	2	3	8	9	–	20	82
Total	372	33	19	24	80	11	107	646

Table 2. Summary of living potato collections contributed by the following collectors: J. G. Hawkes, A. M. van Harten, and J. Landeo (HAW); Z. Huamán (HUA); C. M. Ochoa (CO, OCH); and C. M. Ochoa and A. Salas (O/S).

Collector	Cultivated Species							
	ADG	AJN	CUR	CHA	JUZ	PHU	STN	Total
HAW	37	–	1	10	–	–	12	60
HUA	24	3	–	3	5	–	14	49
CO	73	17	14	28	33	4	46	215
OCH	103	2	–	4	7	5	26	147
O/S	4	–	–	–	–	12	1	17
Total	241	22	15	45	45	21	99	488

Literature cited

Adams, J. B. 1946. Aphid resistance in potatoes. Amer. Potato J. 23:1-22.

Alandia, B. S. 1951. Ensayos de resistencia a *Phytophtora infestans* con especies silvestres de papa. Folia Universitaria, La Paz, Cochabamba 5(5):145-150.

Anonymous. 1915. Inventory of seeds and plants imported. U.S. Department of Agriculture Bureau of Plant Industry 36:14-18 (Wash., D.C.).

Anonymous. 1934. U.S.D.A. Inventory No. 112. Plant material introduced by the Division of Foreign Plant Introduction, Bureau of Plant Industry, July 1 to September 30, 1932. (Nos. 100468-101157), Washington, D.C.

Argandoña, F. M. 1971. Descubridores y exploradores de Bolivia. Edit. Los Amigos del Libro. La Paz, Cochabamba. 249 pp.

Asplund, E. 1926. Contributions to the flora of the Bolivian Andes. Arkiv f. Botanik, K. Svenska Vetenskapsakademien. Band 20 A. No. 7. 38 pp.

Astley, D. and J. G. Hawkes. 1979. The nature of the Bolivian weed potato species *Solanum sucrense* Hawkes. Euphytica 28(3):685-696.

Baker, J. G. 1884. A review of the tuber-bearing species of *Solanum*. Bot. J. Linn. Soc. 20:489-507. Illus. 6.

Bavyko, N. F. 1982. Cultivated potato species of South America, their areal and value for breeding. Bull. Appl. Genet. Breed. 73(2):109-114.

Bazan de Segura, C. and C. Ochoa. 1959. Búsqueda de fuentes de resistencia en papa al hongo *Verticillium albo-atrum*. Informe Mensual Estac. Exp. Agric. La Molina (Lima) 33:378, 381-387.

Berthault, P. 1911. Recherches botaniques sur les variétés cultivées du *Solanum tuberosum* et les espèces sauvages de *Solanum* tubérifères voisins. Ann. Sci. Agron. et Etrang. (Paris), Ser. 3 6(2):1-59, 87-143, 173-216, 248-291. Illus.

Bertonio, L. 1612. Vocabulario de la lengua Aymara. Impr. Fco. del Canto, Juli, Prov. Chucuito, Puno, Peru.

Bitter, G. 1912. Solana nova vel minus cognita. Reprium nov. Spec. Regni veg. 10:533, 535, 536; 11:18, 381, 385-394.

Bitter, G. 1913. Solana nova vel minus cognita. Reprium nov. Spec. Regni veg. 12:5-6, 152-153, 448-450, 453-454, 533-535.

Blomquist, A. W. and Lauer, F. I. 1962. Evaluation of frost resistance using detached leaves in population of *Solanum acaule* × *S. tuberosum* hybrids. Amer. Potato J. 38:389.

Bravo, E. A. 1939. La papa en Sud Yungas Departamento de La Paz. Geo 3(19):38.
Brown, C. R., L. Salazar, C. Ochoa, R. Chávez, L. Schilde-Rentschler, and C. Lizárraga. 1984a. Ploidy manipulation of a new source of resistance to PLRV from *Solanum acaule*. pp. 288-289. *In* Triennial Conference of the European Association for Potato Research, 9a, Interlaken, Switzerland.
Brown, C. R., L. Salazar, C. Ochoa, and C. Chuquillanqui. 1984b. Strain-specific immunity to PVX_{HB} is controlled by a single dominant gene. pp. 249-250. *In* Triennial Conference of the European Association for Potato Research, 9a, Interlaken, Switzerland.
Brücher, E. H. 1956. Kritische Betrachtungen zur Nomenklatur argentinischer Wildkartoffeln. I. Die Serie Commersoniana. Univ. Nac. Cuyo. An. Dept. Invest. Cien. Sec. Biol. (Mendoza, Argentina) 26:97-106.
Brücher, E. H. 1957a. Kritische Betrachtungen zur Nomenklatur argentinischer Wildkartoffeln. III. Die Serie Cuneoalata. Der Züchter 27:77-80.
Brücher, E. H. 1957b. Kritische Betrachtungen zur Nomenklatur argentinischer Wildkartoffeln. IV. Die Serie Tuberosa. Der Züchter 27:353-357.
Brücher, E. H. 1959a. Kritische Betrachtungen zur Nomenklatur argentinischer Wildkartoffeln. V. Die Serie Acaulia. Der Züchter 29:149-156.
Brücher, E. H. 1959b. Kritische Betrachtungen zur Nomenklatur argentinischer Wildkartoffeln. VI. Die Serie Alticola (=Megistacroloba). Der Züchter 29:257-264.
Brücher, E. H. 1964. Kritische Betrachtungen zur Nomenklatur argentinischer Wildkartoffeln. VII. *Solanum setulosistylum* Bitter, eine seit 50 Jahren falsch interpretierte Spezies der Serie Commersoniana. Der Züchter 34:27-32.
Brücher, E. H. 1966. Kritische Betrachtungen zur Nomenklatur argentinischer Wildkartoffeln. VIII. *Solanum bijugum* und dessen synonyma. Der Züchter 36:103-106.
Brücher, E. H. 1974. The section *Tuberarium* of the genus *Solanum* in Paraguay. Ber. Deutsch. Bot. Ges. Bol. 87:405-420.
Brücher, E. H. 1985. *Solanum leptophyes* Bitter, eine häufig verwechselte südamerikanische Wildkartoffel mit bemerkenswerter Nematoden-Resistenz. Vereinigung f. Angewandte Botanik, Göttingen 59:113-123.
Brücher, E. H. and H. Ross. 1953. La importancia de las especies tuberiferas de *Solanum* del noroeste argentino como fuente de resistencia a las enfermedades. Lilloa 26:453-488. 3 illus.
Buchtien, O. 1910. Contribuciones a la flora de Bolivia. Dirección general de estadística y estudios geográficos, sección Museo Nacional. Tall. Tip. Lit. de J. M. Gamarra, La Paz, Bolivia. 197 pp.
Bukasov, S. M. 1932. Frost resistance in the potato. Trudy prikl. Bot. Genet. Selek Ser. 2,3:287-297.
Bukasov, S. M. 1933. The potatoes of South America and their breeding possibilities. Lenin Acad. Agric. Sci., U.S.S.R Inst. Plant Industry (Suppl. 58 to Bull. Appl. Bot. Genet. Plant Breed., Leningrad), 192 pp. Illus. (In Russian, with English summary.)
Bukasov, S. M. 1934. The great crisis in potato breeding. Bull. Appl. Bot., Leningrad, Ser. A 10:51-60.
Bukasov, S. M. 1937. Theoretical bases of plant breeding (in N. I. Vavilov) 3:1-76.
Bukasov, S. M. 1938. Interspecific hybridization in the potato. Bull. Acad. Sci. U.S.S.R., pp. 711-732.
Bukasov, S. M. 1939. The origin of potato species. Physis (Buenos Aires) 18:41-44.

Bukasov, S. M. 1940a. Classification of potato species on dry matter content. Soviet Plant Industry Record (Vest. sots. Rasteniev.) 4:144–145.
Bukasov, S. M. 1940b. New wild potato species from Argentina and Uruguay. Soviet Plant Industry Record (Vest. sots. Rasteniev.) 4:3–12.
Bukasov, S. M. 1941. The origin of species of potatoes. Soviet Plant Industry Record (Vest. sots. Rasteniev.) 1:157–164.
Bukasov, S. M. 1955a. Systematics of potato species. *In* Problemy Botaniki II:317–326.
Bukasov, S. M. 1955b. Production of wart-resistant potato varieties. Fruit and Vegetable Gardens 10:90–93.
Bukasov, S. M. 1960. Methods of wide hybridization in potato breeding. *In* Wide hybridization in plants. Coll. Repts. Akademiya NAUK U.S.S.R., Moscow.
Bukasov, S. M. 1971a. Systematics of potato species of the section Tuberarium (Dun) Buk. of the genus *Solanum* L. Bull. Bot. Genet. Plant Breed., Leningrad 46(1):3–44.
Bukasov, S. M. 1971b. Classification of wild species. pp. 5–40. *In* S. M. Bukasov, ed., Flora of Cultivated Plants, Vol. IX, The Potato. Leningrad: Publ. House Kolos.
Bukasov, S. M. 1978. Systematics of the potato. Bull Appl. Bot. Genet. Breed. 62(1):1–42.
Bukasov, S. M. and A. Y. Kameraz. 1959. Bases of potato breeding. Gosudar. Izdatel. Selsko. Liter. Moskva-Leningrad. 528 pp.
Bukasov, S. M. and V. S. Lechnovitch. 1935. Importancia en la fitotecnia de las papas indígenas de la América del Sur. Rev. Argentina. Agron. 2(7):173–183.
Cabrera, A. L. and A. Willink. 1973. Biogeografía de América Latina. OEA. Programa Regional de Desarrollo Científico y Tecnológico. Serie de Biología. Monografía No. 13. 117 pp.
Cadman, G. H. 1942. Autotetraploid inheritance in the potato: some new evidence. J. Genet. 44:33–52.
Calderoni, A. V. and C. J. Induni. 1952. Resistencia a Sarna (*Streptomyces scabies* (Thaxter, Waksman et Henrici)) en 127 variedades y clones de papa. pp. 67–70. *In* Memoria de la Primera Reunión de Papa Realizada en la Estación Experimental de Balcarce, Argentina.
Camadro, E. L. and S. J. Peloquin. 1981. Cross incompatibility between two sympatric polyploid *Solanum* species. Theor. Appl. Genet. 60(2):65–70.
Cárdenas, M. 1944. Enumeración de las papas silvestres de Bolivia. Rev. Agric. (Univ. Cochabamba, Bolivia) 2(2):27–37. Illus.
Cárdenas, M. 1956. New species of *Solanum* (*Tuberbarium-Hyperbasarthrum*) from Bolivia. Bol. Soc. Peruana Bot. 5:9–42. Illus.
Cárdenas, M. 1958. Importancia de las especies silvestres en el mejoramiento de la papa. pp. 325–332. *In* Reunión Latinoamericana de Fitotecnia, 5a, Santiago, Chile.
Cárdenas, M. 1963. Germoplasma de papas cultivadas acumulado en Bolivia durante los últimos seis años. Folia Universitaria (Cochabamba) 9:58–87.
Cárdenas, M. 1966. The South American potential potato germplasm. Amer. Potato J. 43(10):367–370.
Cárdenas, M. 1968. A new species of wild potato from Cochabamba, *Solanum ruiz-cevallosii*. Rev. Agric. (Univ. Cochabamba, Bolivia) 11:13–14. Illus.
Cárdenas, M. and J. G. Hawkes. 1945. New and little-known wild potato species from Bolivia and Peru. Bot. J. Linn. Soc. 53:91–108. Figs. 1–10.
Cevallos Tovar, W. 1914. Clasificación de la papa de Bolivia. La Paz. Dir. Gen. de Estadística y Estudios Geográficos 10(88):124–137. Color illus.

Choudhury, H. C. 1944. Cytological and genetical studies in the genus *Solanum*. II. Wild and cultivated 'diploid' potatoes. Trans. Roy Soc. Edinburgh 61:199-219.

Ciampi, L. and L. Sequeira. 1980. Multiplication of *Pseudomonas solanacearum* in resistant potato plants and the establishment of latent infections. Amer. Potato J. 57:319-329.

Cockerham, G. 1943. Potato breeding for virus resistance. Ann. Appl. Biol. 30:105-108.

Correll, D. S. 1961. New species and some nomenclatural changes in section Tuberarium of *Solanum*. Wrightia 2:169-197.

Correll, D. S. 1962. The Potato and Its Wild Relatives: Section Tuberarium of the Genus *Solanum*. Renner, Texas: Texas Research Foundation (A Series of Botanical Studies, Vol. 4). 606 pp.

Cribb, P. J. 1972. Studies on the origin of *Solanum tuberosum* L. subsp. *andigena* (Juz. et Buk.) Hawkes. The cultivated tetraploid potato of South America. Ph.D. Thesis, University of Birmingham, U.K.

D'Arcy, W. G. 1972. *Solanaceae* studies. II. Typification of subdivisions of *Solanum*. Ann. Missouri Bot. Garden 59(2):262-278.

Dearborn, C. H. 1969. Alaska Frostless: an inherently frost resistant potato variety. Amer. Potato J. 46:1-4.

De Candolle, A. P. 1886. Nouvelles recherches sur le type sauvage de la pomme de terre. Bibliothèque Universelle. Archive des Sciences Physiques et Naturelles. Ser. 3(15):425-438.

D'Orbigny, A. 1945. Viaje a la América Meridional. Editorial Futuro. Buenos Aires, Argentina. Volumes I-IV. 1614 pp.

Dodds, K. 1962. Classification of cultivated potatoes. *In* D. S. Correll, The Potato and Its Wild Relatives. pp. 517-539. Renner, Texas: Texas Research Foundation (A Series of Botanical Studies, Vol. 4). 606 pp.

Dodds, K. 1965. The history and relationships of cultivated potatoes. pp. 123-141. *In* J. Hutchinson, Essays on Crop Plant Evolution. London: Cambridge University Press. 204 pp.

Dodds, K. and G. J. Paxman. 1962. The genetic system of cultivated diploid potatoes. Evolution 16(2):154-167.

Don, G. 1838. A General System of Gardening and Botany [often cited as Gen. Syst.] 4:400 (London).

Dremliug, L. A. 1937. A frost resistant triple potato hybrid *Solanum acaule* × *S. tuberosum* (Fürstenkrone) × *S. tuberosum* (Centifolia). Dokl. Akad. Nauk U.S.S.R. 16:423-426.

Dumortier, B. C. 1827. Florula belgique [often cited as Fl. Belg.], p. 39. Tournay, Belgium.

Dunal, M. F. 1852. *In* A. De Candolle, Prodromus (*Solanaceae*). Paris 13(1):43.

Dunnet, J. M. 1959. Variation in pathogenicity of the potato root eelworm: technique and results of testing wild potatoes for resistance. pp. 106-120. Tagunsberichte No. 20, Deutsch. Akad. Landw. Wiss. Berlin.

Ehlenfeldt, M. K. and R. E. Hanneman, Jr. 1984. The use of endosperm balance number and 2n gametes to transfer exotic germplasm in potato. Theor. Appl. Genet. 68:155-161.

Ellenby, C. 1952. Reistance to the potato-root eelworm, *Heterodera rostochiensis* Wollenweber. Nature, London 170:1016.

Ellenby, C. 1954. Tuber forming species and varieties of the genus *Solanum* tested for resistance to the potato-root eelworm, *Heterodera rostochiensis* Wollenweber. Euphytica 3:195-202.

Estrada, R. N. 1977. Breeding frost-resistant potatoes for the tropical highlands. pp. 333-341. *In* P. H. Li and A. Sakai, eds., Plant Cold Hardiness and Freezing Stress. New York: Academic Press.

Estrada, R. N. 1980. Frost resistant potato hybrids via *Solanum acaule* Bitt. diploid-tetraploid crosses. Amer. Potato J. 57:609-619.

Estrada, R. N. 1984. Acaphu: a tetraploid, fertile breeding line, selected from an *S. acaule* × *S. phureja* cross. Amer. Potato J. 61:1-7.

Fernandez-Northcote, E. N. 1983. Prospects for stability of resistance to potato virus Y. *In* W. J. Hooker, ed., Research for the Potato in the Year 2000. Lima, Peru: International Potato Center. 197 pp.

Fiebrig, K. 1910. Ein Beitrag zur Pflanzengeographie Boliviens. Bot. Jahrb 45:1-68.

Firbas, H. and H. Ross. 1961. Züchtung auf Frostresistenz bei der Kartoffel. II. Über die Frostresistenz des Laubes von Wildarten und Primitivformen der Kartoffel und ihre Beziehung zur Höhenlage des Artareals. Z. Pflanzenz. 45:259-299.

Fisel, U. and W. Hanagarth. 1983. Estudio ecológico en una comunidad del altiplano boliviano. Ecología en Bolivia. Rev. Inst. Ecología 4:1-17.

French, E. R. and L. de Lindo. 1980. Resistance to *Erwinia chrysanthemi* in tubers of *S. tuberosum* ssp. *andigena*. Fitopatologia 15(1):34.

French, E. R. and L. de Lindo. 1982. Resistance to *Pseudomonas solanacearum* in potato: specificity and temperature sensitivity. Phytopatology 72:1408-1412.

Gandarillas, H. 1961. Nueva fuente de genes para el mejoramiento de la papa en Bolivia. pp. 410-411. *In* Reunión Latinoamericana de Fitotecnia, 5a, Buenos Aires, Argentina. Actas. Buenos Aires. 2 v.

Gautney, T. L. and F. L. Haynes. 1983. Recurrent selection for heat tolerance in diploid potatoes (*Solanum tuberosum* ssp. *phureja* and *stenotomum*). Amer. Potato J. 60:537-542.

Geransenkova, E. D. 1974. Characterization of some potato species by their resistance to leafroll virus (L). Bull. Appl. Bot. Genet. Plant Breed., Leningrad 53(2):166-175. (In Russian with English summary.)

Gibson, R. W. 1971. Glandular hair providing resistance to aphids in certain wild potato species. Ann. Appl. Biol. 68:113-119.

Gibson, R. W. 1974. Aphid-trapping glandular hairs or hybrids of *Solanum tuberosum* and *S. berthaultii*. Potato Res. 17:152-154.

Gibson, R. W. 1976. Glandular hairs of *Solanum polyadenium* lessen damage by the Colorado beetle. Ann. Appl. Biol. 82(1):147-150.

Gómez, P. L., R. L. Plaisted, and B. B. Brodie. 1983a. Inheritance of the resistance to *Meloidogyne incognita*, *M. javanica*, *M. arenaria* in potatoes. Amer. Potato J. 60:339-351.

Gómez, P. L., R. L. Plaisted, and H. D. Thurston. 1983b. Combining resistance to *Meloidogyne incognita*, *M. javanica*, *M. arenaria* and *Pseudomonas solanacearum* in potatoes. Amer. Potato J. 60:353-360.

Hammarlund, C. 1943. Sydamerikanska Strövtag. Stockholm Lindfors Bokforlags A.B.

Hanneman, R. E., Jr. 1983. Assignment of endosperm balance numbers (EBN) to the tuber-bearing *Solanum* species. Amer. Potato J. 60:809-810.

Hawkes, J. G. 1944. Potato collecting expeditions in Mexico and South America. II. Systematic classification of the collections. Bull. Imp. Bur. Plant Breed. Genet., Cambridge. 142 pp. Illus.
Hawkes, J. G. 1945. The indigenous American potatoes and their value in plant breeding. Empire J. Exp. Agric. 13:11-40.
Hawkes, J. G. 1954. New *Solanum* species in subsection *Hyperbasarthrum* Bitt. Ann. Mag. Nat. Hist., Ser. 12, 7:689-710. Illus.
Hawkes, J. G. 1956a. A revision of the tuber-bearing *Solanums*. pp. 37-109. *In* Scottish Plant Breeding Station Annual Report.
Hawkes, J. G. 1956b. Taxonomic studies on the tuber-bearing *Solanums*. I. *Solanum tuberosum* and the tetraploid species complex. Proc. Linn. Soc. London 166:97-144.
Hawkes, J. G. 1962. The origin of *Solanum juzepczukii* Buk. and *S. curtilobum* Juz. et Buk. Z. Pflanzenz. 47:1-14.
Hawkes, J. G. 1963. A revision of the tuber-bearing *Solanums*. 2nd edit. pp. 76-181. Reprint. *In* Scottish Plant Breeding Station Record.
Hawkes, J. G. 1979. Recent concepts in the evolution of the tuber-bearing *Solanums*. pp. 126-140. *In* Report of the Planning Conference on the Exploration, Taxonomy, and Maintenance of Potato Germ Plasm III. Lima, Peru: International Potato Center. 193 pp.
Hawkes, J. G. and J. P. Hjerting. 1960. Some wild potato species from Argentina. Phyton. (Graz) 9:140-146.
Hawkes, J. G. and J. P. Hjerting. 1969. The potatoes of Argentina, Brazil, Paraguay and Uruguay. London: Oxford University Press. 525 pp. Illus., maps.
Hawkes, J. G. and J. P. Hjerting. 1983. New tuber-bearing *Solanum* taxa from Bolivia and Northern Argentina. Bot. J. Linn. Soc. 86:405-417. Fig. 7.
Hawkes, J. G. and J. P. Hjerting. 1985a. Two new wild potato species from Bolivia. Bot. J. Linn. Soc. 90:105-112. Figs. 1-4.
Hawkes, J. G. and J. P. Hjerting. 1985b. New *Solanum* taxa from Bolivia: A brief note. Bot. J. Linn. Soc. 91:445-446.
Haynes, F. L. 1972. The use of cultivated diploid *Solanum* species in potato breeding. pp. 100-110. *In* E. R. French, ed., Prospects for the Potato in the Developing World. Lima, Peru: International Potato Center.
Hermsen, J. G. Th. and M. S. Ramanna. 1969. Meiosis in different F_1-hybrids of *Solanum acaule* Bitt. × *S. bulbocastanum* Dun. and its bearing on genome relationships, fertility and breeding behaviour. Euphytica 18:27-35.
Herzog, Th. 1923. Die Pflanzenwelt der bolivischen Anden und ihres östlichen Vorlandes. Leipzig. 258 pp.
Howard, H. W. 1955. Breeding potatoes for resistance to root eelworm (*Heterodera rostochiensis*). Heredity 9:150.
Howard, H. W., C. S. Cole, and J. M. Fuller. 1970. Further sources of resistance to *Heterodera rostochiensis* Woll. in the *andigena* potatoes. Euphytica 19:210-216.
Huamán, Z. 1975. The origin and nature of *Solanum ajanhuiri* Juz. et Buk., a South American cultivated potato. Ph.D. Thesis, University of Birmingham, U.K. 193 pp.
Huamán, Z. 1979a. Review of planning conference on the utilization of genetic resources: status of germ plasm maintenence and computerization. pp. 36-39. *In* Report of the Planning Conference on the Exploration, Taxonomy, and Maintenance of Potato Germ Plasm III. Lima, Peru: International Potato Center.
Huamán, Z. 1979b. Computerized data management on the primitive cultivated

collection maintained at CIP. pp. 153-161. *In* Report of the Planning Conference on the Exploration, Taxonomy, and Maintenance of Potato Germ Plasm III. Lima, Peru: International Potato Center.

Huamán, Z. 1986. Conservation of potato genetic resources at CIP. CIP Circular 14(2):1-7, Lima, Peru.

Huamán, Z., J. G. Hawkes, and P. R. Rowe. 1980. *Solanum ajanhuiri*: an important diploid potato cultivated in the Andean altiplano. Econ. Bot. 34(4):335-343.

Huamán, Z., J. G. Hawkes, and P. R. Rowe. 1982. A biosystematic study of the origin of the cultivated diploid potato, *Solanum* × *ajanhuiri* Juz. and Buk. Euphytica 31:665-676.

Huamán, Z., J. G. Hawkes, and P. R. Rowe. 1983. Chromatographic studies on the origin of the cultivated diploid potato *Solanum* × *ajanhuiri* Juz. and Buk. Amer. Potato J. 60:361-368.

Hueck, K. 1978. Los bosques de Sudamerica. Ecología, composición e importancia económica. Soc. Alemana de Cooperación Técnica (GTZ). Rep. Fed. Alemana.

International Potato Center. 1973. Workshop on Germplasm Exploration and Taxonomy of Potatoes. Lima, Peru. 35 pp.

International Potato Center. 1975. Annual Report 1974. p. 46. Lima, Peru.

International Potato Center. 1976. Report of the Planning Conference on the Exploration and Maintenance of Germplasm Resources. Lima, Peru. 130 pp.

International Potato Center. 1977. Annual Report 1976. pp. 10, 42-43. Lima, Peru.

International Potato Center. 1978. Annual Report 1977. pp. 67-69, 74, 76-81. Lima, Peru.

International Potato Center. 1979a. Annual Report 1978. pp. 33, 38-40. Lima, Peru.

International Potato Center. 1979b. Report of the Planning Conference on the Exploration, Taxonomy, and Maintenance of Potato Germ Plasm III. Lima, Peru. 193 pp.

International Potato Center. 1980. Annual Report 1979. pp. 35, 42, 46. Lima, Peru.

International Potato Center. 1981. Annual Report 1980. p. 40. Lima, Peru.

International Potato Center. 1982. Annual Report 1981. pp. 31, 35-36, 40-41. Lima, Peru.

International Potato Center. 1983. Annual Report 1982. pp. 37, 48-49, 60. Lima, Peru.

International Potato Center. 1984a. Annual Report 1983. p. 28. Lima, Peru.

International Potato Center. 1984b. Potatoes for the Developing World. Lima, Peru. 148 pp.

International Potato Center. 1985. Annual Report 1984. pp. 36-37. Lima, Peru.

Jackson, M. T., J. G. Hawkes, and P. R. Rowe. 1977. The nature of *Solanum* × *chaucha* Juz. and Buk., a triploid cultivated potato of the South American Andes. Euphytica 26:775-783.

Jackson, M. T., P. R. Rowe, and J. G. Hawkes. 1978. Crossability relationships of Andean potato varieties of three ploidy levels. Euphytica 27:541-551.

Jatala, P. and C. Martin. 1978. Interactions of *Meloidogyne incognita acrita* and *Pseudomonas solanacearum* on *Solanum chacoense* and *Solanum sparsipilum*. p. 178. *In* Proceedings of the American Phytopathological Society, Vol. 4, 1977, St. Paul, Minnesota. 254 pp.

Jaworski, C. A., S. C. Phatak, S. R. Ghate, and R. D. Gitaitis. 1984. *Solanum sucrense* and *Solanum tuberosum*, bacterial wilt-tolerant potato germplasm. HortScience 19(2):312-313.

Johns, T. A. 1985. Chemical ecology of the Aymara of western Bolivia: selection for glycoalkaloids in the *Solanum* × *ajanhuiri* domestication complex. Ph.D. Thesis, University of Michigan, U.S.A. 312 pp.

Johnston, S. A. and R. E. Hanneman, Jr. 1980. Support of the endosperm balance number hypothesis utilizing some tuber-bearing *Solanum* species. Amer. Potato J. 57:7-14.

Johnston, S. A. and R. E. Hanneman, Jr. 1982. Manipulations of endosperm balance number overcome crossing barriers between diploid *Solanum* species. Science 217:446-448.

Johnston, S. A., T. P. M. den Nijs, S. J. Peloquin, and R. E. Hanneman, Jr. 1980. The significance of genic balance to endosperm development in interspecific crosses. Theor. Appl. Genet. 57:5-9.

Juzepczuk, S. W. and S. M. Bukasov. 1929. A contribution to the question of the origin of the potato. Proc. U.S.S.R. Congress Genet. Plant and Animal Breed. 3:593-611. (In Russian with English summary.)

Kameraz, A. J. 1957. Wart resistance of interspecific potato hybrids. Vest. sel. Nauk 6:35-42.

Knoch, K. 1930. Klimakunde von Südamerika Band II, Teil G. pp. 136-144. *In* W. Köppen, Graz, und R. Geiger, Handbuch der Klimatologie, München. Berlin: Kraus Reprint, Neudeln/Liechtenstein, 1972.

Köppen, W. 1931. Grundriss der Klimakunde. Berlin und Leipzig.

Lebedeva, N. A., V. A. Lebedeva, and T. T. Dabletbekova. 1978. *Solanum chacoense* in breeding virus-X resistant potato forms. Biol. Bull. U.S.S.R. Acad. Sci. 5:229-231.

Lechnovitch, V. S. 1971. Classification of cultivated species. pp. 41-304. *In* S. M. Bukasov, ed., Flora of Cultivated Plants, Vol. IX, The Potato. Leningrad: Publ. House Kolos.

Linnaeus, C. 1753. Species Plantarum. pp. 184-188. Stockholm.

Lopez, L. E. 1979. Collection of primitive Colombian cultivars. pp. 60-68. *In* Report of the Planning Conference on the Exploration, Taxonomy, and Maintenance of Potato Germ Plasm III. Lima, Peru: International Potato Center. 193 pp.

Lorini, J. and Liberman, M. 1983. El clima de la Provincia de Aroma del Departamento de La Paz, Bolivia. Ecologia en Bolivia. Rev. del Inst. Ecología 4:19-29.

Martin, C. and R. E. French. 1980. Desarrollo de cultivares de papa con resistencia a *Pseudomonas solanacearum* and *Phytophthora infestans*. Fitopatologia 15(1):33.

Martin, C., E. R. French, and H. Mendoza. 1980. Additional sources of resistance in potatoes to *Pseudomonas solanacearum*. Fitopatologia 15:33-34.

Mastenbroek, C. 1956. Some experiences in breeding frost-tolerant potatoes. Euphytica 5:289-297.

Mehlenbacher, S. A. and R. L. Plaisted. 1983. Heritability of glandular trichome characteristics in a *Solanum tuberosum* × *S. berthaultii* hybrid population. pp. 128-130. *In* Proceedings of the International Congress Research for the Potato in the Year 2000. Lima, Peru: International Potato Center. 199 pp.

Mehlenbacher, S. A., R. L. Plaisted, and W. M. Tingey. 1983. Inheritance of glandular trichomes in crosses with *Solanum berthaultii*. Amer. Potato J. 60:699-708.

Mendiburu, A. O. and S. J. Peloquin. 1977. Bilateral sexual polyploidization in potatoes. Euphytica 26:573-583.

Mendoza, H. A. and P. Jatala. 1978. Breeding for resistance to the root-knot nematode. pp. 41-116. *In* Report of the 2nd Nematode Planning Conference. Lima, Peru: International Potato Center. 193 pp.

Mock, D. W. S. and S. J. Peloquin. 1975. Three mechanisms of 2n pollen formation in diploid potatoes. Canadian J. Genet. Cytology 17:217-225.
Muñoz, F. J., R. L. Plaisted, and H. D. Thurston. 1975. Resistance to potato virus Y in *Solanum tuberosum* ssp. *andigena*. Amer. Potato J. 52:107-115.
Muñoz Reyes, J. 1977. Geografía de Bolivia. La Paz. Acad. Nac. de Ciencias de Bolivia. 478 pp. Illus. and maps.
Ochoa, C. M. 1951. Algunos estudios sobre papas peruanas como base para un programa de mejoramiento en el país. Agronomía 65:31-38.
Ochoa, C. M. 1956. *Solanum candolleanum* Berth., una papa silvestre poco conocida. Lima, Dir. Grl. de Agric. Bol. 19-22:24-26. Illus.
Ochoa, C. M. 1958. Expedición colectora de papas cultivadas a la Cuenca del Lago Titicaca; I. Determinación sistemática y número cromosómico del material colectado. Prog. Coop. de Exper. Agrop. Minist. Agric. Lima Investigaciones en Papa, No. 1. 18 pp. Map.
Ochoa, C. M. 1962. Los *Solanum* tuberiferos silvestres del Perú (Sect. *Tuberarium*, sub-sect. *Hyperbasarthrum*). Lima: Editorial Villanueva. 296 pp.
Ochoa, C. M. 1964. Una nueva especie de papa silvestre de la serie politopica Cuneoalata. Anales Científicos 2(4):391-395. Illus.
Ochoa, C. M. 1965. Determinación sistemática y recuentos cromosómicos de las papas indígenas cultivadas en el centro del Perú. Anales Científicos 3(2):136 and 138; 3(2):103-163.
Ochoa, C. M. 1975a. Las papas cultivadas triploides de *Solanum* × *chaucha* y su distribución geográfica en el Perú. Anales Científicos 8(1-2):31-44.
Ochoa, C. M. 1975b. Potato collecting expeditions in Chile, Bolivia and Peru, and the genetic erosion of indigenous cultivars. pp. 167-173. *In* O. H. Frankel and J. G. Hawkes, eds., Crop Genetic Resources for Today and Tomorrow. International Biological Programme 2. Cambridge: Cambridge University Press. 492 pp.
Ochoa, C. M. 1979a. Review of recommendations of March 1976 conference: status of *Solanum* expeditions, collections and classifications. pp. 23-25. *In* Report of the Planning Conference on the Exploration, Taxonomy, and Maintenance of Potato Germ Plasm III. Lima, Peru: International Potato Center. 193 pp.
Ochoa, C. M. 1979b. Collection of primitive potato cultivars. pp. 40-59. Exploration, taxonomy and maintenance of potato germplasm. *In* Report of the Planning Conference on the Exploration, Taxonomy, and Maintenance of Potato Germ Plasm III. Lima, Peru: International Potato Center. 193 pp.
Ochoa, C. M. 1979c. Collection and taxonomy of Andean wild potato (Peru and Bolivia). pp. 114-125. *In* Report of the Planning Conference on the Exploration, Taxonomy, and Maintenance of Potato Germ Plasm III. Lima, Peru: International Potato Center. 193 pp.
Ochoa, C. M. 1979d. Exploración colectora de papas silvestres en Bolivia. Biota 11(91):324-330. Map.
Ochoa, C. M. 1980a. A new tuber-bearing *Solanum* potentially useful for breeding for aphid resistance. Amer. Potato J. 57:387-390.
Ochoa, C. M. 1980b. New taxa of *Solanum* from Peru and Bolivia. Phytologia 46:223-225.
Ochoa, C. M. 1981. Two new tuber-bearing *Solanum* from South America. Phytologia 48:229-232.
Ochoa, C. M. 1982. A new variety of the Bolivian tuber-bearing *Solanum capsicibaccatum*. Phytologia 50:181-182.

Ochoa, C. M. 1983a. A new taxon and name changes in *Solanum* (Sect. *Petota*). Phytologia 54(5):391-392.
Ochoa, C. M. 1983b. *Solanum bombycinum*, a new tuber-bearing tetraploid species from Bolivia. Amer. Potato J. 60(11):849-852.
Ochoa, C. M. 1983c. *Solanum neovavilovii*: a new wild potato species from Bolivia. Amer. Potato J. 60:919-923.
Ochoa, C. M. 1984a. *Solanum venatoris* (Sect. *Petota*), a new species from Bolivia. Phytologia 55(5):297-298.
Ochoa, C. M. 1984b. Karyotaxonomic studies on wild Bolivian tuber-bearing *Solanum* (Sect. *Petota*) I. Phytologia 55:17-40.
Ochoa, C. M. 1985. Karyotaxonomic studies on wild Bolivian tuber-bearing *Solanum* (Sect. *Petota*) II. Phytologia 57:315-324.
Ochoa, C. M. 1988. New Bolivian taxa of *Solanum* (Sect. Petota). Phytologia 65:103-113.
Okada, K. A. 1973. Colección de papas silvestres, variedades nativas cultivadas e híbridos interespecíficos artificiales. Lista de semillas No. 2. Instituto Nacional de Tecnología Agropecuaria, Balcarce, Argentina. 47 pp.
Okada, K. A. 1979. Collection and taxonomy of the Argentine tuber-bearing *Solanums*. pp. 98-113. *In* Report of the Planning Conference on the Exploration, Taxonomy, and Maintenance of Potato Germ Plasm III. Lima, Peru: International Potato Center. 193 pp.
Pandey, K. K. 1960. Self-incompatibility in *Solanum megistacrolobum* Bitt. Phyton 14:13-19.
Quinn, A. A., D. W. S. Mok, and S. J. Peloquin. 1974. Distribution and significance of diplandroids among the diploid *Solanum*. Amer. Potato J. 51:16-21.
Ramanna, M. S. and M. M. F. Abdalla. 1970. Fertility, late blight resistance and genome relationship in an interspecific hybrid, *Solanum polytrichon* Rydb. × *S. phureja* Juz. and Buk. Euphytica 19:317-326.
Ross, H. 1954. Über die extreme resistenz von *Solanum acaule* gegen das X-Virus. Mitt. aus der Biolog. Bundesanstalt Berlin-Dahlem 80:144-145.
Ross, H. 1958. Kartoffel. II Ausgangsmaterial für die Züchtung. pp. 43-59. *In* H. Kappert and W. Rudorf, eds., Handbuch der Pflanzenzüchtung. Berlin: Paul Parey. 351 pp.
Ross, H. 1960a. German botanical and agricultural expedition to the Andes 1954. FAO Plant Introduction Newsletter (Rome) 8.
Ross, H. 1960b. Über die Zugehörigkeit der knollentrageden *Solanum*-arten zu den pflanzengeographischen Formationen Südamerikas und damit verbundene Resistenzfragen. Z. Pflanzenz. 43(3):217-240. (In German with English summary.)
Ross, H. 1966. The use of wild *Solanum* species in German potato breeding of the past and today. Amer. Potato J. 43:63-79.
Ross, H. and M. L. Baerecke. 1950. III. Selection for resistance to mosaic virus (diseases) in wild species and in hybrids of wild species of potatoes. Amer. Potato J. 27:275-284.
Ross, H. and C. A. Huijsman. 1969. Über die Resistenz von *Solanum Tuberarium*-Arten gegen europäische Rassen den Kartoffelnematoden (*Heterodera rostochiensis* Woll.). Theor. Appl. Genet. (Züchter) 39:113-122.
Ross, H. and R. Rimpau. 1959. Über die Deutsche Botanisch-Landwirtschaftliche Anden-Expedition. (II. Genzentren-Expedition nach Südamerika des Kaiser-Wilhelm-bzw. Max-Planck-Institut für Züchtungsforschung,

Köln-Vogelsang, und Karyogeographische Expedition des Botanischen Instituts der Universität Köln).
Rothacker, D. 1961. Die wilden und kultivierten mittel-und südamerikanischen Kartoffel-species. pp. 353-558. *In* R. Schick and M. Klinkowski, eds., Die Kartoffel 1.
Rothacker, D. and W. A. Müller. 1960. Arbeiten zur Züchtung krebsresistenter Kartoffeln. II. Untersuchung kultivierter südamerikanischer Kartoffel-species auf ihr Verhalten gegenüber dem Krebsbiotyp G_1. Züchter 30:340-343.
Rothacker, D. and H. Stelter. 1961. *Solanum tarijense* Hawk.– eine weitere gegen den Kartoffelnematoden (*Heterodera rostochiensis* (Wollenweber) resistente Species. Naturwissenschaften 48:742-743.
Rothacker, D., H. Stelter, and, W. J. Junges. 1966. Investigations on the collection of wild and cultured potato species in the Gross Lüsewitz Plant Breeding Institute. V. On the occurrence of resistance to potato nematode, *Heterodera rostochiensis* Woll. Z. Pflanzenz. 2:101-131.
Rusby, H. H. 1893. An enumeration of the plants collected in Bolivia by Miguel Bang. Mem. Torrey Bot. Club 3(3):1-67.
Rybin, V. A. 1929. The karyological investigations on some wild as well as native cultivated potatoes of America. Proc. U.S.S.R. Congress Genetics 3:467-468. (In Russian, English summary.)
Rybin, V. A. 1933. Cytological investigations of the South American cultivated and wild potatoes and its significance for plant breeding. Bull. Appl. Bot., Genet. Plant Breed., Leningrad, Ser. 2(2):3-100.
Rydberg, P. A. 1924. The section Tuberarium of the genus *Solanum* in Mexico and Central America. Bull. Torrey Bot. Club 51:145-154, 167-176.
Schaper, P. 1953. Beitrag zur Resistenz des *Solanum chacoense* (Bitt.) gegen den Kartoffelkäfer (*Leptinotarsa decemlineata* Say.). Züchter 23:115-121.
Schick, R. 1931. Kort verslag van enn reis door de Andesgebieden van Zuid Amerika en de in deze gebieden gekweekte aardappelssorten. Landbouwk. Tijdschr. Groningen 43:1133-1136.
Schmiediche, P. 1977a. Biosystematic studies on the cultivated frost-resistant potato species *Solanum* × *juzepczukii* Buk. and *Solanum curtilobum* Juz. and Buk. Ph.D. Thesis, University of Birmingham, U.K. 180 pp.
Schmiediche, P. 1977b. Evaluation of clones of *Solanum acaule* for resistance to *Heterodera pallida* and *H. rostochiensis*. Nematropica 7(1):6-7.
Schmiediche, P., J. G. Hawkes, and C. Ochoa. 1980. Breeding of the cultivated potato species *Solanum* × *juzepczukii* Buk. and *Solanum* × *curtilobum* Juz. and Buk. I. A study of the natural variation of *S.* × *curtilobum* and their wild progenitor *S. acaule* Bitt. Euphytica 29(3):685-704.
Schmiediche, P., J. G. Hawkes, and C. Ochoa. 1982. The breeding of the cultivated potato species *Solanum* × *juzepczukii* and *S.* × *curtilobum*. II. The resynthesis of *S.* × *juzepczukii* and *S.* × *curtilobum*. Euphytica 31(3):695-707.
Schmiediche, P. and C. Martin. 1983. Widening of the genetic base of the potato for resistance to bacterial wilt (*Pseudomonas solanacearum*). pp. 172-173. *In* Proceedings of the International Congress Research for the Potato in the Year 2000. Lima, Peru: International Potato Center.
Scurrah, M. de and J. Franco. 1985. Breeding for resistance to *Globodera pallida* at CIP. Bull. European and Mediterranean Plant Protection Organization 15:167-173.
Sequeira, L. 1980. Development of resistance to bacterial wilt derived from *S. phureja*.

pp. 55-62. *In* Report of the Planning Conference on the Developments in the Control of Bacterial Diseases of Potatoes. Lima, Peru: International Potato Center. 137 pp.

Sleesman, J. P. 1940. Resistance in wild potatoes to attack by the potato leaf-hopper and the potato flea beetle. Amer. Potato J. 17:9-12.

Soest, L. J. M. van, J. G. Hawkes, and W. Hondelmann. 1983. Potato collecting expeditions to Bolivia and the importance of Bolivian germplasm for plant breeding. Z. Pflanzenz. 91:154-168.

Soest, L. J. M. van, W. Hondelmann, and J. G. Hawkes. 1980. Potato collecting in Bolivia. Plant Genet. Resources Newsletter 43:32-35.

Soest L. J. M. van, H. J. Rumpenhorst, and C. A. Huijsman. 1983. Resistance to potato cyst-nematodes in tuber-bearing *Solanum* species and its geographical distribution. Euphytica 32(1):65-74.

Stelter, H. and D. Rothacker. 1965. Eine Bemerkungen zu der Nematoden-resistenz der Arten *S. multidissectum* Hawk., *S. kurtzianum* Bitt. und Wittm. und *S. juzepczukii* Buk. Züchter 35:180-186.

Stelzner, G. 1943. Über die Fertilitätsverhältnisse bei Bastardierungen zwischen frostfesten wildkartoffeln *S. acaule* und der Kulturkartoffel *S. tuberosum*. Züchter 15:143-144.

Swaminathan, M. S. 1951. Notes on induced polyploids in the tuber-bearing *Solanum* species and their crossability with *S. tuberosum*. Amer. Potato J. 28:472-489.

Swaminathan, M. S. 1954. Nature of polyploidy in some 48 chromosome species of the genus *Solanum* section *Tuberarium*. Genetics 39:59-76.

Tay, C. S. 1979. Evolutionary studies on the cultivated potatoes *Solanum stenotomum* Juz. and Buk., *S. goniocalyx* Juz. and Buk. and *S. phureja* Juz. and Buk. Ph.D. Thesis, University of Birmingham, U.K.

Thung, T. H. 1947. Potato diseases and hybridization. Phytopathology 37:373-381.

Tingey, W. M., R. L. Plaisted, J. M. Laubengayer, and S. A. Mehlenbacher. 1982. Green peach aphid resistance by glandular trichomes in *Solanum tuberosum* × *S. berthaultii* hybrids. Amer. Potato J. 59:241-251.

Tingey, W. M., A. Shawn, S. A. Mehlenbacher, and J. E. Laubengayer. 1981. Occurrence of glandular trichomes in wild *Solanum* species. Amer. Potato J. 58:81-83.

Torka, M. 1948. Die Resistenz von *S. chacoense* Bitt. gegen *Leptinotarsa decemlineata* Say und ihre Bedeutung für die Kartoffelzüchtung. Z. Pflanzenz. 28:63-78.

Torka, M. 1950. Breeding potatoes with resistance to the Colorado beetle. Amer. Potato J. 27:263-271.

Toxopeus, H. J. and C. A. Huijsman. 1952. Genotypical background of resistance to *Heterodera rostochiensis* in *Solanum tuberosum* var. *andigenum*. Nature (London) 170:1016.

Toxopeus, H. J. and C. A. Huijsman. 1953. Breeding for resistance to potato-root eelworm. I. Preliminary data concerning the inheritance and nature of resistance. Euphytica 2:180-186.

Turkensteen, L. J. 1979. Tizón foliar de la papa en el Perú: Resistencia e interacción con Tizón tardío en *Solanum phureja*, *S. tuberosum* ssp. *andigena*, e híbridos de *S. demissum*. Fitopatologia 14(1):29-32.

Ugent, D. 1970. *Solanum raphanifolium*, a Peruvian wild potato species of hybrid origin. Bot. Gazette 131(3):225-233.

Ugent, D. 1981. Biogeography and origin of *Solanum acaule* Bitter. Phytologia 48(1):85-95.

Unzueta, O., J. A. Tosi, Jr., and L. R. Holdridge. 1975. Mapa Ecológico de Bolivia (Tomo I, 187 pp.; Tomo II, 312 pp.). La Paz, Bolivia: Ministerio de Asuntos Campesinos y Agropecuarios.
Vargas, C. C. 1949. Las Papas Sudperuanas, Parte I. Peru: Publ. Univ. Nac. Cusco. 144 pp. Illus.
Vargas, C. C. 1952. Algunos resultados de investigación con papas nativas peruanas. Primera Asamblea Latino Americana de Fitologistas. Foll. Misc. Ofic. Stud. Esp. Sec. Agric. y Ganad. (México) 3:187-200.
Vargas, C. C. 1956. Las Papas Sudperuanas, Parte II. p. 20. Figs. 1-22. Peru: Publ. Univ. Nac. Cusco. 66 pp.
Virsoo, E. V. 1956. El valor amilaceo de las papas autóctonas de Bolivia y Perú. Rev. Agron. Noroeste Argent. 2:197-224.
Vlasova, E. A. 1974. Rhizoctonia disease of potato and the evaluation of resistance in wild *Solanum* species. Bull. Appl. Bot. Genet. Plant Breed., Leningrad 53(2):150-165. (In Russian with English summary.)
Wangenheim, K. H. von. 1954. Zur Ursache der Kreuzungsschwierigkeiten zwischen *Solanum tuberosum* L. und *S. acaule* Bitt. bzw. *S. stoloniferum* Schlechtd. et Bouché. Z. Pflanzenz. 34:7-48.
Weddell, H. A. 1867. Notice sur M. G. Mandon. Bull. Soc. Bot. France 24:10-12.
Wight, W. F. 1916. Origin, introduction and primitive culture of the potato. Proceedings 3rd Annual Meeting Potato Assoc. Amer. 3:35-52.
Zadina, J. 1958. Resistance of wild potatoes to *Spongospora* powdery scab (*S. subterranea* Johnson). Sb. csl. Akad. zemed. Ved., Rada Rostlinna vyroba 31:1115-1126.
Zavaleta, M. 1968. Colección propia de papa altiplanica boliviana. Fitotecnia Latinoamer. (Venezuela) 5(1):111-117.

Index of scientific names

Valid names are in roman type and *synonyms are in italics.*
Page numbers in **bold** type refer to main entries.